Lecture Notes in Computer Science   9713

Commenced Publication in 1973
Founding and Former Series Editors:
Gerhard Goos, Juris Hartmanis, and Jan van Leeuwen

More information about this series at http://www.springer.com/series/7407

Ying Tan · Yuhui Shi
Li Li (Eds.)

Advances in Swarm Intelligence

7th International Conference, ICSI 2016
Bali, Indonesia, June 25–30, 2016
Proceedings, Part II

 Springer

Editors
Ying Tan
Peking University
Beijing
China

Li Li
Shenzhen University
Shenzhen
China

Yuhui Shi
Xi'an Jiaotong-Liverpool University
Suzhou
China

ISSN 0302-9743   ISSN 1611-3349   (electronic)
Lecture Notes in Computer Science
ISBN 978-3-319-41008-1   ISBN 978-3-319-41009-8   (eBook)
DOI 10.1007/978-3-319-41009-8

Library of Congress Control Number: 2016942017

LNCS Sublibrary: SL1 – Theoretical Computer Science and General Issues

Printed on acid-free paper

This Springer imprint is published by Springer Nature
The registered company is Springer International Publishing AG Switzerland

Preface

This book and its companion volumes, LNCS vols. 9712 and 9713, constitute the proceedings of the 7th International Conference on Swarm Intelligence (ICSI 2016) held during June 25–30, 2016, in Bali, Indonesia.

The theme of ICSI 2016 was "Serving Life with Intelligence and Data Science." ICSI 2016 provided an excellent opportunity and/or an academic forum for academics and practitioners to present and discuss the latest scientific results and methods, innovative ideas, and advantages in theories, technologies, and applications in swarm intelligence. The technical program covered all aspects of swarm intelligence and related areas.

ICSI 2016 was the seventh international gathering in the world for researchers working on all aspects of swarm intelligence, following successful events in Beijing (ICSI-CCI 2015), Hefei (ICSI 2014), Harbin (ICSI 2013), Shenzhen (ICSI 2012), Chongqing (ICSI 2011), and Beijing (ICSI 2010), which provided a high-level academic forum for participants to disseminate their new research findings and discuss emerging areas of research. It also created a stimulating environment for participants to interact and exchange information on future challenges and opportunities in the field of swarm intelligence research. ICSI 2016 was held in conjunction with the International Conference on Data Mining and Big Data (DMBD 2016) at Bali, Indonesia, for sharing common mutual ideas, promoting transverse fusion, and stimulating innovation.

Bali is a famous Indonesian island with the provincial capital at Denpasar. Lying between Java to the west and Lombok to the east, this island is renowned for its volcanic lakes, spectacular rice terraces, stunning tropical beaches, ancient temples and palaces, as well as dance and elaborate religious festivals. Bali is also the largest tourist destination in the country and is renowned for its highly developed arts, including traditional and modern dance, sculpture, painting, leather, metalworking, and music. Since the late 20th century, the province has had a big rise in tourism. Bali received the Best Island Award from *Travel and Leisure* in 2010. The island of Bali won because of its attractive surroundings (both mountain and coastal areas), diverse tourist attractions, excellent international and local restaurants, and the friendliness of the local people. According to BBC Travel released in 2011, Bali is one of the world's best islands, rank second after Greece!

ICSI 2016 received 231 submissions from about 693 authors in 42 countries and regions (Algeria, Australia, Austria, Bangladesh, Bolivia, Brazil, Brunei Darussalam, Canada, Chile, China, Colombia, Fiji, France, Germany, Hong Kong, India, Indonesia, Italy, Japan, Kazakhstan, Republic of Korea, Macao, Malaysia, Mexico, The Netherlands, New Zealand, Nigeria, Oman, Pakistan, Peru, Portugal, Russian Federation, Serbia, Singapore, South Africa, Spain, Chinese Taiwan, Thailand, Tunisia, Turkey, United Arab Emirates, USA) across six continents (Asia, Europe, North America, South America, Africa, and Oceania). Each submission was reviewed by at least two

Plenary Session Co-chairs

Nikola Kasabov   Auckland University of Technology, New Zealand
Rachid Chelouah   EISTI, France

Tutorial Co-chairs

Milan Tuba   University of Belgrade, Serbia
Dunwei Gong   China University of Mining and Technology, China
Li Li   Shenzhen University, China

Symposia Co-chairs

Maoguo Gong   Northwest Polytechnical University, China
Yan Pei   University of Aziz, Japan

Publicity Co-chairs

Yew-Soon Ong   Nanyang Technological University, Singapore
Carlos A. Coello Coello CINVESTAV-IPN, Mexico
Pramod Kumar Singh   Indian Institute of Information Technology
                           and Management, India
Yaochu Jin   University of Surrey, UK
Fernando Buarque   Universidade of Pernambuco, Brazil
Eugene Semenkin   Siberian Aerospace University, Russia
Somnuk Phon-Amnuaisuk   Institut Teknologi Brunei, Brunei

Finance and Registration Co-chairs

Andreas Janecek   University of Vienna, Austria
Chao Deng   Peking University, China
Suicheng Gu   Google Corporation, USA

Program Committee

Mohd Helmy Abd Wahab   Universiti Tun Hussein Onn, Malaysia
Lounis Adouane   LASMEA, France
Ramakrishna Akella   University of California, USA
Miltiadis Alamaniotis   Purdue University, USA
Rafael Alcala   University of Granada, Spain
Peter Andras   Keele University, UK
Esther Andrés   INTA, Spain
Helio Barbosa   Laboratório Nacional de Computação Científica, Brazil
Anasse Bari   New York University, USA
Carmelo J.A. Bastos Filho University of Pernambuco, Brazil
Christian Blum   IKERBASQUE, Basque Foundation for Science, Spain

Vladimir Bukhtoyarov	Siberian State Aerospace University, Russia
David Camacho	Universidad Autonoma de Madrid, Spain
Bin Cao	Tsinghua University, China
Jinde Cao	Southeast University
Kit Yan Chan	Curtin University, Australia
Chien-Hsing Chen	Ling Tung University, Taiwan
Liang Chen	University of Northern British Columbia, Canada
Walter Chen	National Taipei University of Technology, Taiwan
Shi Cheng	The University of Nottingham Ningbo, China
Manuel Chica	European Centre for Soft Computing, Spain
Carlos Coello Coello	CINVESTAV-IPN, Mexico
Jose Alfredo Ferreira Costa	UFRN – Universidade Federal do Rio Grande do Norte, Brazil
Micael S. Couceiro	Polytechnic Institute of Coimbra, Portugal
Prithviraj Dasgupta	University of Nebraska, Omaha, USA
Kusum Deep	Indian Institute of Technology Roorkee, India
Mingcong Deng	Tokyo University of Agriculture and Technology, Japan
Ke Ding	Baidu Corporation, China
Yongsheng Dong	Henan University of Science and Technology, China
Haibin Duan	Beijing University of Aeronautics and Astronautics, China
Mark Embrechts	Rensselaer Polytechnic Institute, USA
Andries Engelbrecht	University of Pretoria, South Africa
Jianwu Fang	Xi'an Institute of Optics and Precision Mechanics of CAS, China
Shangce Gao	University of Toyama, Japan
Ying Gao	Guangzhou University, China
Beatriz Aurora Garro Licon	IIMAS-UNAM, Mexico
Maoguo Gong	Xidian University, China
Amel Grissa	Ecole Nationale d'Ingénieurs de Tunis, Tunisia
Shenshen Gu	Shanghai University, China
Yinan Guo	Chinese University of Mining and Technology, China
Fei Han	Jiangsu University, China
Haibo He	University of Rhode Island, USA
Shan He	University of Birmingham, UK
Lu Hongtao	Shanghai Jiao Tong University, China
Mo Hongwei	Harbin Engineering University, China
Reza Hoseinnezhad	RMIT University, Australia
Jun Hu	Chinese Academy of Sciences, China
Teturo Itami	Hirokoku University, Japan
Andreas Janecek	University of Vienna, Austria
Yunyi Jia	Clemson University, USA
Changan Jiang	Ritsumeikan University, Japan
Mingyan Jiang	Shandong University, China
Licheng Jiao	Xidian University, China

Colin Johnson	University of Kent, UK
Matthew Joordens	Deakin University, Australia
Ahmed Kattan	Umm Al-qura University, Saudi Arabia
Liangjun Ke	Xi'an Jiaotong University, China
Arun Khosla	National Institute of Technology, Jalandhar, Punjab, India
Slawomir Koziel	Reykjavik University, Iceland
Thanatchai Kulworawanichpong	Suranaree University of Technology, Thailand
Rajesh Kumar	MNIT, India
Hung La	University of Nevada, USA
Germano Lambert-Torres	PS Solutions, Brazil
Xiujuan Lei	Shaanxi Normal University, China
Bin Li	University of Science and Technology of China, China
Xiaodong Li	RMIT University, Australia
Xuelong Li	Chinese Academy of Sciences, China
Andrei Lihu	Politehnica University of Timisoara, Romania
Fernando B. De Lima Neto	University of Pernambuco, Brazil
Bin Liu	Nanjing University of Post and Telecommunications, China
Ju Liu	Shandong University, China
Qun Liu	Chongqing University of Posts and Communications, China
Wenlian Lu	Fudan University, China
Wenjian Luo	University of Science and Technology of China, China
Chengying Mao	Jiangxi University of Finance and Economics, China
Michalis Mavrovouniotis	De Montfort University, UK
Mohamed Arezki Mellal	M'Hamed Bougara University, Algeria
Bernd Meyer	Monash University, Australia
Martin Middendorf	University of Leipzig, Germany
Sanaz Mostaghim	Institute IWS, Germany
Krishnendu Mukhopadhyaya	Indian Statistical Institute, India
Ben Niu	Shenzhen University, China
Yew-Soon Ong	Nanyang Technological University, Singapore
Feng Pan	Beijing Institute of Technology, China
Jeng-Shyang Pan Pan	National Kaohsiung University of Applied Sciences, Taiwan
Quan-Ke Pan	Nanyang Technological University, Singapore
Shahram Payandeh	Simon Fraser University, Canada
Yan Pei	The University of Aizu, Japan
Somnuk Phon-Amnuaisuk	Institut Teknologi, Brunei
Ghazaleh Pour Sadrollah	Monash University, Australia
Radu-Emil Precup	Politehnica University of Timisoara, Romania
Kai Qin	RMIT University, Australia
Quande Qin	Shenzhen University, China

Boyang Qu	Zhongyuan University of Technology, China
Robert Reynolds	Wayne State University, USA
Guangchen Ruan	Indiana University, USA
Eugene Santos	Dartmouth College, USA
Kevin Seppi	Brigham Young University, USA
Luneque Silva Junior	Federal University of Rio de Janeiro, Brazil
Pramod Kumar Singh	ABV-IIITM Gwalior, India
Ponnuthurai Suganthan	Nanyang Technological University, Singapore
Hideyuki Takagi	Kyushu University, Japan
Ying Tan	Peking University, China
Qian Tang	Xidian University, China
Qirong Tang	University of Stuttgart, Germany
Liang Tao	University of Science and Technology of China, China
Christos Tjortjis	International Hellenic University, Greece
Milan Tuba	University of Belgrade, Serbia
Mario Ventresca	Purdue University, USA
Bing Wang	BSC IEE, UK
Cong Wang	Notheastern University, China
Gai-Ge Wang	Jiangsu Normal University, China
Guoyin Wang	Chongqing University of Posts and Telecommunications, China
Jiahai Wang	Sun Yat-sen University, China
Lei Wang	Tongji University, China
Qi Wang	Northwestern Polytechnical University, China
Xiaoying Wang	Changshu Institute of Technology, China
Yong Wang	Central South University
Zhanshan Wang	Northeastern University, China
Zhenzhen Wang	Jinling Institute of Technology, China
Ka-Chun Wong	City University of Hong Kong, SAR China
Shunren Xia	Zhejiang University, China
Bo Xing	University of Johannesburg, South Africa
Benlian Xu	Changshu Institute of Technology, China
Jin Xu	Peking University, China
Bing Xue	Victoria University of Wellington, New Zealand
Xiao Yan	Chinese Academy of Sciences, China
Yingjie Yang	De Montfort University, UK
Wei-Chang Yeh	National Tsing Hua University, Taiwan
Kiwon Yeom	NASA Ames Research Center, USA
Peng-Yeng Yin	National Chi Nan University, Taiwan
Zhuhong You	Shenzhen University, China
Yang Yu	Nanjing University, China
Yuan Yuan	Chinese Academy of Sciences, China
Zhigang Zeng	Huazhong University of Science and Technology, China
Zhi-Hui Zhan	Sun Yat-sen University, China
Jie Zhang	Newcastle University, UK

Jun Zhang   Waseda University, Japan
Junqi Zhang   Tongji University, China
Lifeng Zhang   Renmin University of China, China
Mengjie Zhang   Victoria University of Wellington, New Zealand
Qieshi Zhang   Waseda University, Japan
Qiangfu Zhao   The University of Aizu, Japan
Shaoqiu Zheng   Peking University, China
Yujun Zheng   Zhejiang University of Technology, China
Zhongyang Zheng   Peking University, China
Cui Zhihua   Complex System and Computational Intelligence
                             Laboratory, China
Guokang Zhu   Shanghai University of Electric Power, China
Xingquan Zuo   Beijing University of Posts and Telecommunications,
                             China

Additional Reviewers

Bari, Anasse   Jia, Guanbo   Portugal, David
Chen, Zonggan   Jing, Sun   Shan, Qihe
Cheng, Shi   Lee, Jie   Shang, Ke
Dai, Hongwei   Li, Junzhi   Tao, Fazhan
Dehzangi, Abdollah   Li, Mengyang   Wan, Ying
Ding, Sanbo   Li, Yaoyi   Wang, Junyi
Feng, Jinwang   Lian, Cheng   Weibo, Yang
Ghafari, Seyed Mohssen   Lin, Jianzhe   Yan, Shankai
Guo, Xing   Liu, Xiaofang   Yao, Wei
Guyue, Mi   Liu, Zhenbao   Yu, Chao
Hu, Jianqiang   Lv, Gang   Zhang, Yong
Hu, Weiwei   Lyu, Yueming   Zi-Jia, Wang
Hu, Zihao   Paiva, Fábio

Contents – Part II

Clustering Algorithm

Classification

Image Classification and Encryption

Data Mining

Sensor Networks and Social Networks

Neural Networks

Swarm intelligence in Management Decision Making and Operations Research

Robot Control

Swarm Robotics

Intelligent Energy and Communications Systems

Intelligent Interactive and Tutoring Systems

Contents – Part I

Hybrid Search Optimization

Particle Swarm Optimization

PSO Applications

Ant Colony Optimization

Brain Storm Optimization

Fireworks Algorithms

Multi-Objective Optimization

Large-Scale Global Optimization

Biometrics

Scheduling and Planning

Hyper-heuristics for the Flexible Job Shop Scheduling Problem with Additional Constraints

Jacomine Grobler[1]([✉]) and Andries P. Engelbrecht[2]

[1] Department of Industrial and Systems Engineering,
University of Pretoria, Pretoria, South Africa
jacomine.grobler@gmail.com
[2] Department of Computer Science, University of Pretoria, Pretoria, South Africa
engel@cs.up.ac.za

Abstract. This paper investigates a highly relevant real world scheduling problem, namely the multi-objective flexible job shop scheduling problem (FJSP) with sequence-dependent set-up times, auxiliary resources and machine down time. A hyper-heuristic algorithm is presented which makes use of a set of meta-heuristic algorithms which are self-adaptively selected at different stages of the optimization process to optimize a set of candidate solutions. This meta-hyper-heuristic algorithm was tested on a number of real world production scheduling data sets and was also benchmarked against the previous state-of-the-art job shop scheduling algorithms applied to this specific problem. In addition to the competitive results obtained, the self-adaptive nature of the algorithm avoids the resource intensive process of developing a meta-heuristic algorithm for one specific problem instance.

1 Introduction

Production scheduling is one of the most important issues in the planning and operation of manufacturing systems. A large number of meta-heuristic algorithms have already been developed to address this issue, however, it is difficult to predict which of the many algorithms will be best to address a specific scheduling problem. The development of a hyper-heuristic scheduling algorithm, which adaptively selects between a number of subalgorithms, is thus a valuable contribution resulting in reduced algorithm development time and better utilization of development resources.

In this paper such a hyper-heuristic algorithm is developed for the multi-objective flexible job shop scheduling problem with sequence-dependent set-up times, auxiliary resources and machine down time. With regards to previous work, a comprehensive review of the state-of-the-art in flexible job shop scheduling problems was recently published by Chaudhry and Khan [1]. Although a number of basic flexible job shop scheduling algorithms have already been successfully developed, more complex production scheduling problems such as the FJSP with additional constraints are still relatively sparse. The most popular additional complexities currently addressed include sequence-dependent setup

© Springer International Publishing Switzerland 2016
Y. Tan et al. (Eds.): ICSI 2016, Part II, LNCS 9713, pp. 3–10, 2016.
DOI: 10.1007/978-3-319-41009-8_1

times [2,3], break down intervals [4,5], fuzzy or stochastic input data [6,7], and the requirement for real-time rescheduling [4,8]. However, to the best of the authors' knowledge, the multi-objective flexible job shop scheduling problem with sequence-dependent set-up times, auxiliary resources and machine down time has only been addressed in [9]. Significant opportunities for additional research into this complex problem thus exists.

The resulting hyper-heuristic algorithm which was developed is based on the heterogeneous meta-hyper-heuristic (HMHH) algorithm of Grobler et al. [10]. Improvements in algorithm structure, with specific reference to the application of the production calendar, were made and newer more powerful meta-heuristic algorithms were used to improve on the existing state-of-the-art algorithms developed for this specific flexible job shop problem. The HMHH scheduling algorithm was also benchmarked against its constituent meta-heuristic algorithms using different sized datasets derived from real customer data. The HMHH algorithm showed great promise due to the good results obtained as well as the advantages associated with the self-adaptive nature of the algorithm. The paper is considered significant because, to the best of the authors' knowledge, it describes the first hyper-heuristic algorithm developed for this specific variation of the flexible job shop scheduling problem.

The rest of the paper is organized as follows: Sect. 2 describes the scheduling problem in more detail. Section 3 provides an overview of the HMHH production scheduling algorithm before Sect. 4 describes the experimental setup and results obtained. Finally, the paper is concluded in Sect. 5.

2   The Multi-objective Flexible Job Shop Scheduling Problem with Additional Constraints

The flexible job shop scheduling problem with sequence-dependent set-up times, auxiliary resources and machine down time can be formulated as follows: There is a set of J jobs that needs to be processed on d machines and e auxiliary resources. The set machines is denoted by $\boldsymbol{D} = \{\boldsymbol{D}_1, \boldsymbol{D}_2, \ldots, \boldsymbol{D}_d\}$ and the set of auxiliary resources are denoted by $\boldsymbol{E} = \{\boldsymbol{E}_1, \boldsymbol{E}_2, \ldots, \boldsymbol{E}_e\}$. Each job j consists of a sequence of operations N_j, where $\sum N_{j\,j=1}^{J} = n$. All operations in each job needs to be completed in the correct sequence to complete a job. The execution of each operation i requires one machine out of a set of primary resources denoted by $\boldsymbol{D}_i$ and one resource from a set of auxiliary resources denoted by $\boldsymbol{E}_i$. The problem thus focuses on determining both an assignment and a sequence of the operations on all machines that minimize some criteria. In addition, production calendars and resource down time intervals were addressed as described in [9].

Three objectives need to be minimized, namely makespan ($\max f_i \forall i \in \boldsymbol{L}$), earliness/tardiness ($\sum_{i=1}^{n} \max\{0, \nu_i - - f_i\} + \max\{0, f_i - \nu_i\} \forall i \in \boldsymbol{L}$) and queue time ($\sum_{j=1}^{J} \max\{0, f_j - \tau_j - p_j\}$). Here ν_i denotes the due date of operation i, $\boldsymbol{L}$ is the set containing the last operations of all jobs and $\boldsymbol{\Upsilon}$ is the set containing the first operations of all jobs.

3 The Heterogeneous Meta-hyper-heuristic Scheduling Algorithm

Due to its excellent performance against other popular multi-method algorithms, the tabu-search based HMHH algorithm of [10] was used as a basis for the development of the HMHH scheduling algorithm. The HMHH algorithm divides a population of entities (decision variables) into a number of subpopulations which are evolved in parallel by a set of constituent algorithms. Each entity is able to access the solution components of other subpopulations, as if part of a common population of entities. The allocation of entities to constituent algorithms is updated on a dynamic basis throughout the optimization run. The idea is that an intelligent algorithm can be evolved which selects the appropriate constituent algorithm at each kth iteration to be applied to each entity within the context of the common parent population, to ensure that the population of entities converges to a high quality solution. The constituent algorithm allocation is maintained for k iterations, while the common parent population is continuously updated with new information and better solutions. Please refer to [10] for a detailed description of the HMHH algorithm.

The HMHH algorithm was, however, developed for a continuous search space, while the complicated production scheduling problem considered in this paper is combinatorial in nature. A previous analysis of mapping strategies and alternative particle representations [9] showed that a priority-based mapping mechanism is the most appropriate method for decoding continuous decision variables into a feasible schedule.

Each of the HMHH entities or decision variables, consists of a $2n$-dimensional vector. Dimensions 1 to n are the sequencing variables and is interpreted as the priority values of each of the operations. Dimensions $n + 1$ to $2n$ are used to represent the allocation of operations to resources. This is done by discretizing the search space as follows: For each operation i, the ith dimension of the entity is divided into D_i intervals, where D_i denotes the number of primary resources on which operation i can be processed. Since each interval is associated with a unique integer number or resource index, dimensions $n + 1$ to $2n$ of the position vector can easily be interpreted as resource allocation variables.

The operation priorities and resource allocations are then used as input to a schedule-building heuristic which attempts to schedule each operation at the earliest available time on its selected resource. Giffler and Thompson's heuristic [11] was extended to include the unique characteristics of the FJSP with additional constraints. This priority-based mechanism has a significant impact on the size of the problem search space since the decision variables can only be decoded into feasible schedules which comply with all the problem constraints, thereby eliminating the existence of infeasible solutions. This mechanism to decode the continuous variables optimized by the HMHH algorithm into schedules is described in more detail in [9]. After the schedule associated with each entity has been generated, the fitness function of the schedule can be calculated.

4   Empirical Evaluation of the Improved Priority-Based HMHH Scheduling Algorithm

During the development of the HMHH scheduling algorithm it became evident that there were a number of opportunities to improve on the priority-based PSO (P-PSO) and DE (P-DE) algorithms developed for the FJSP with additional constraints in 2010 [9]. At that stage the production calendar was applied to the best schedule after the optimization process was completed to save on computational time. This production calendar consisted of the downtime intervals applicable to all resources, for example evenings and weekends where the job shop would be closed. However, due to the increased availability of computational resources, it was hypothesized that it would now be possible to apply the production calendar to each candidate solution as part of the fitness function evaluation. This could potentially lead to improvement in the quality of the final solutions obtained.

This hypothesis was tested by adapting the decoding mechanism of decision variables into feasible schedules to include the production calendar and comparing this improved priority-based PSO (IP-PSO) algorithm to the P-PSO and P-DE algorithms. These three algorithms, the IP-PSO, P-PSO and P-DE algorithms, makes use of the same problem representation and problem mapping mechanism. The only difference lies in the application of the production calendar. The P-PSO and P-DE algorithms were considered suitable for comparison purposes in this paper since they significantly outperformed the existing rule-based algorithms currently used the South African manufacturing industry to solve the problem, as well as the state-of-the-art random keys genetic algorithm of Norman and Bean [12]. The evaluation was conducted on the three datasets described in [9] which are available for comparison purposes from the corresponding author. Since an extensive analysis of the most suitable control parameters for priority-based FJSP algorithms were already performed in [9], these parameters were used as is. Similarly, the constituent algorithm control parameters specified in [13] were used.

For all the experiments conducted in this paper, results for each algorithm were recorded over 30 independent simulation runs. The notation, μ and σ, denote the mean and standard deviation of the final solution obtained at I_{max}, F_1 denotes makespan, F_2 the earliness/tardiness criteria, and F_3 the queue time. Due to the complexity associated with the development of a dominance-based multi-objective hyper-heuristic and for comparison against the results obtained in [9], goal programming was used to address the multiple objectives in this initial investigation.

The results in Tables 1 and 2 show that the difference in application of the production calendar does not make a difference for the first smallest dataset. However, for the larger two datasets a difference in performance can be seen in Table 1. The production calendar will thus be incorporated into every fitness function evaluation when the performance of the improved priority-based HMHH (IP-HMHH) algorithm is evaluated.

Table 1. Comparison results of the P-PSO and P-DE algorithms with the IP-PSO algorithm.

Dataset	Objective	P-PSO		P-DE		IP-PSO	
		μ	σ	μ	σ	μ	σ
Dataset 56	$F1$	1653.71	44.2	1594.19	24.71	2037.7	17.247
	$F2$	3579.17	231.81	3861.5	231.19	3280.1	17.825
	$F3$	90.64	129.21	497.38	122.14	84.095	8.5436
	$F4$	4828.84	202.77	5459.02	328.29	4907.3	30.403
Dataset 100	$F1$	1858.57	11.29	2276.37	66.72	2083.6	72.773
	$F2$	7151.72	235.12	9754.2	227.96	5940.5	190.31
	$F3$	675.7	214.73	1408.47	214.59	88.972	85.166
	$F4$	9195.67	367.53	12952.48	418.3	7621.1	249.76
Dataset 256	$F1$	5010.93	45.57	5343.74	143.92	5472.3	186.87
	$F2$	38169.64	3065.49	30931.67	1018.09	26807	1849.8
	$F3$	12893.7	1788.01	5143.53	811.32	3445.9	729.69
	$F4$	55630.34	4088.77	40960.35	1392.98	35261	2351.2

The IP-HMHH scheduling algorithm as well as a popular variation, the exponentially increasing HMHH algorithm with *a priori* knowledge (UP-EIHH1) [13], was implemented and compared with the IP-HMHH's constituent meta-heuristic algorithms. IP-EIHH1 only differs from IP-HMHH in that the set of available meta-heuristics is manipulated over time. The best performing constituent algorithm is first allowed to work on the problem until minimal improvement is obtained and then other constituent algorithms are added at exponential time intervals. The results are provided in Table 2.

Statistical tests were also used to evaluate the significance of the results. The results in Table 3 were obtained by comparing the performance of the IP-HMHH algorithm and the IP-EIHH algorithm on each dataset to the performance of the constituent algorithms on the same dataset. For every comparison, a Mann-Whitney U test at 95 % significance was performed (using the two sets of 30 data points of the two algorithems under comparison) and if the first algorithm statistically significantly outperformed the second strategy, a win was recorded. If no statistical difference could be observed a draw was recorded. If the second algorithm outperformed the first algorithm, a loss was recorded for the first algorithm. The total number of wins, draws and losses were then recorded. As an example, (2-2-1) in row 2 column 1, indicates that the EIHH algorithm significantly outperformed PSO with regard to two datasets. Furthermore, two draws and one loss were recorded. From the results it can be seen that the EIHH algorithm statistically significantly outperforms the constituent meta-heuristics five times out of all the comparisons considered and is also outperformed five times. The HMHH algorithm, on the other hand, statistically significantly outperforms the constituent meta-heuristics nine times and is only outperformed three times.

Table 2. Comparison results of the different algorithms on customer datasets of five different sizes. Dataset n denotes a dataset consisting of n operations.

Dataset	Objective	IP-HMHH		IP-EIHH1		IP-CMAES		IP-GCPSO		IP-SaNSDE		IP-GA	
		μ	σ	μ	σ	μ	σ	μ	σ	μ	σ	μ	σ
Dataset 56	$F1$	2024.8	13.445	2031.2	6.9663	2036.7	6.5606	2037.7	17.247	2030.4	2.2581	2029.4	25.228
	$F2$	3256.3	26.676	3256.4	8.6729	3276.1	19.404	3280.1	17.825	3254.2	9.2906	3270.3	47.841
	$F3$	88.48	24.011	79.763	3.8656	81.286	3.7329	84.095	8.5436	79.876	1.4126	93.486	35.354
	$F4$	4874.9	17.272	4872.8	10.983	4899.5	20.184	4907.3	30.403	4869.9	10.341	4898.5	37.895
Dataset 100	$F1$	2058.4	41.569	2058.3	44.8	2060.4	66.39	2083.6	72.773	2046.6	9.8712	2165.1	128.78
	$F2$	5711.3	219.3	5671.1	120.71	5671.8	161.21	5940.5	190.31	5536	103.11	6106.7	364.06
	$F3$	76.175	88.951	40.006	62.035	130.65	137.03	88.972	85.166	75.639	85.866	162.26	138.98
	$F4$	7353.8	261.97	7277.2	151.73	7370.8	260.7	7621.1	249.76	7165.9	108.05	7942.5	493.9
Dataset 146	$F1$	5945.5	94.809	5932.2	45.743	5926.7	21.731	5970.8	97.199	5908.6	31.198	6067.3	150.93
	$F2$	11568	315.94	11425	274.34	11552	295.31	11851	413.74	11393	130.97	12028	416.64
	$F3$	104.64	95.868	58.803	53.143	127.53	62.753	118.79	122.82	37.655	34.185	284.3	172.19
	$F4$	17136	418.72	16934	297.14	17124	303.84	17458	595.69	16856	143.95	17898	598.5
Dataset 200	$F1$	6600.5	115.76	6597.2	100.84	6616.6	71.092	6675.5	114.52	6646.5	68.811	6774.7	232.5
	$F2$	16186	479.95	15786	446.49	15994	469.45	16981	754.77	16374	338.12	18168	1374.1
	$F3$	653.24	187.93	578.95	122.73	670.99	132.58	734.86	186.38	713.37	126.62	1070.3	442.95
	$F4$	22963	600.33	22485	553.03	22804	603.6	23916	856.12	23258	332.9	25539	1812
Dataset 256	$F1$	5422.6	171.49	5541.4	200.28	5706.7	219.88	5472.3	186.87	5614.4	233.61	5750.8	344.5
	$F2$	25605	1314.3	24102	806.69	24239	698.21	26807	1849.8	25702	639.57	28850	1633.7
	$F3$	2954.4	735.92	2343.4	457.78	2453.6	382.85	3445.9	729.69	3239.2	662.72	3891	1035.2
	$F4$	33515	1719.5	31519	982.75	31932	556.25	35261	2351.2	34090	900.32	38030	2135.3

Table 3. Hypotheses analysis of the two HMHH algorithms compared to the constituent meta-heuristics.

	IP-EIHH	IP-HMHH	TOTAL
IP-CMAES	1-3-1	1-3-1	2-6-2
IP-PSO	2-2-1	2-3-0	4-5-1
IP-SaNSDE	0-3-2	1-2-2	1-5-4
IP-GA	2-2-1	5-0-0	7-2-1
TOTAL	5-10-5	9-8-3	14-18-8

Only SaNSDE obtains significantly better results than the HMHH algorithm. This is probably due to the fact that EIHH is biased towards CMAES, as discussed in [13]. SaNSDE, however, outperforms CMAES on these datasets thus the HMHH algorithm, which provides each constituent algorithm with equal opportunity to be selected, performs bettter than the EIHH algorithm. It is not easy to predict ahead of time which meta-heuristic will be best for solving each specific problem instance. The fact that the self-adaptive HMHH algorithms produces competitive results when compared to the constituent meta-heuristics is thus very encouraging. In this case, SaNSDE is the best algorithm for the problem, however, this is definitely not always the case and the benefits of having a self-adaptive algorithm capable of selecting from a set of meta-heuristics, should not be underestimated.

5   Conclusion

This paper has presented a HMHH scheduling algorithm for the FJSP with sequence-dependent set-up times, auxiliary resources and machine down time. The impact of applying the production calendar post optimization versus at each function evaluation was investigated and a significant performance improvement was obtained when the calendar was applied at each function evaluation. Two variations of the HMHH scheduling algorithm was also benchmarked against four constituent meta-heuristic algorithms. The HMHH scheduling algorithm was shown to produce similar results to its constituent algorithms, but with much greater generality and a resulting reduction in algorithm development resources.

Future research opportunities exist in expanding the experimental work to a larger number of datasets of different sizes, investigating the performance of meta-heuristics on significantly larger scheduling problems, and investigating the use of more complex, better performing dominance based multi-objective optimization techniques to address the multiple objectives of the problem.

References

1. Chaudhry, I.A., Khan, A.A.: A research survey: review of flexible job shop scheduling techniques. Int. Trans. Oper. Res. **23**(3), 551–591 (2016)
2. Sadrzadeh, A.: Development of both the AIS and PSO for solving the flexible job shop scheduling problem. Arab. J. Sci. Eng. **38**, 3593–3604 (2013)
3. Defersha, F.M., Chen, M.: A parallel genetic algorithm for a flexible job-shop scheduling problem with sequence dependent setups. Int. J. Adv. Manuf. Technol. **49**, 263–279 (2010)
4. Jensen, M.T.: Generating robust and flexible job shop schedules using genetic algorithms. IEEE Trans. Evol. Comput. **7**, 275–288 (2003)
5. Li, L., Huo, J.: Multi-objective flexible job-shop scheduling problem in steel tubes production. Syst. Eng.-Theory Pract. **29**(8), 117–126 (2009)
6. Davarzani, Z., Akbarzadeh-T, M.R., Khairdoost, N.: Multiobjective artificial immune algorithm for flexible job shop scheduling problem. Int. J. Hybrid Inf. Technol. **5**, 75–88 (2012)
7. Lei, D.: A genetic algorithm for flexible job shop scheduling with fuzzy processing time. Int. J. Prod. Res. **48**, 2995–3013 (2010)
8. Fattahi, P., Fallahi, A.: Dynamic scheduling in flexible job shop systems by considering simultaneously efficiency and stability. CIRP J. Manuf. Sci. Technol. **2**, 114–123 (2010)
9. Grobler, J., Engelbrecht, A.P., Kendall, G., Yadavalli, V.S.S.: Metaheuristics for the multi-objective FJSP with sequence-dependent set-up times, auxiliary resources and machine down time. Ann. Oper. Res. **180**, 165–196 (2010)
10. Grobler, J., Engelbrecht, A.P., Kendall, G., Yadavalli, V.S.S.: Investigating the impact of alternative evolutionary selection strategies on multi-method global optimization. In: Proceedings of the 2011 IEEE Congress on Evolutionary Computation, pp. 2337–2344 (2011)
11. Giffler, J., Thompson, G.L.: Algorithms for solving production scheduling problems. Oper. Res. **8**, 487–503 (1960)
12. Norman, B.A., Bean, J.C.: A genetic algorithm methodology for complex scheduling problems. Naval Res. Logistics **46**, 199–211 (1999)
13. Grobler, J., Engelbrecht, A.P., Kendall, G., Yadavalli, V.S.S.: Heuristic space diversity control for improved meta-hyper-heuristic performance. Inf. Sci. **300**, 49–62 (2015)

On-Orbit Servicing Mission Planning for Multi-spacecraft Using CDPSO

Jianxin Zhang$^{(\boxtimes)}$, Ying Zhang, and Qiang Zhang$^{(\boxtimes)}$

Key Lab of Advanced Design and Intelligent Computing, Dalian University,
Ministry of Education, Dalian 116622, China
zjx99326@163.com, zhangq26@126.com

Abstract. Spacecraft mission planning can improve the collaborative work efficiency of the on-orbit servicing (OOS) spacecrafts. A chaos discrete particle swarm optimization (CDPSO) algorithm is applied according to the characteristics of multi-spacecraft collaborative mission planning problem. We design the new update formulae of position and velocity of the particles for the OOS optimization mission. By analyzing the critical index factors which contain the value of the target spacecrafts, the attrition of servicing spacecraft and consumption of time and fuel during the process of service, a mathematical model is formulated. Simulation results show that the algorithm can solve the multi-spacecraft mission planning problem under multiple constraints efficiently. It is expressive, flexible, extensible and feasible easily.

Keywords: On-orbit servicing · Spacecraft mission planning · Discrete particle swarm optimization · Chaos

1 Introduction

The on-orbit servicing (OOS) technology is the use of automated technology and robot technology for the on-orbit inspection, repair, refueling, upgrade, maintenance, assembling and release by the use of automation and robotics [1, 2]. The OOS technology can effectively prolong the life of spacecraft and reduce the costs. However, the carrying capacity of space vehicle is strictly limited. Spacecraft mission planning algorithms can improve the collaborative work efficiency of OOS, which makes them widely studied in many military developed countries.

With the maturity of OOS, it will be developed to a systematic way, and set up an OOS system which cooperative service by multiple on-orbit spacecraft [3]. In Ref. [4], the mathematical model of long-range maneuver transfer with impulse thrust is analyzed. It considers the time and energy cost and uses EDA to optimize. However the capability of single spacecraft is limited and multi-spacecraft cooperation is drawing more and more attention [5]. This paper presents a chaotic discrete particle swarm (CDPSO) algorithm for the multi-constrained problem about value of the target spacecrafts, attrition of servicing spacecraft and consumption of transfer time and fuel using Lambert two-impulse orbital transfer with unfixed transfer time. The algorithm combines the chaos algorithm with the discrete particle swarm optimization (DPSO) algorithm.

© Springer International Publishing Switzerland 2016
Y. Tan et al. (Eds.): ICSI 2016, Part II, LNCS 9713, pp. 11–18, 2016.
DOI: 10.1007/978-3-319-41009-8_2

2 Problem Description

The scenario and transfer process of OOS planning studied in this paper is shown in Fig. 1. The target spacecraft and the servicing spacecraft run in the predefined orbits respectively. If an OOS request is issued, a suitable Lambert double pulses orbit transfer to the target track would implemented from the servicing spacecraft maneuvers to the target spacecraft immediately to conduct the required OOS operation. Thus the servicing satellite is reusable [6].

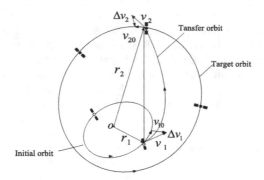

Fig. 1. Sketch of Lambert transfer

3 Mathematical Models

3.1 Decision Variables

Assuming N servicing spacecrafts on-orbit are deployed, and M different task priority target spacecrafts waiting for service at a given time, according to the characteristics of the planning problem, the decision variables can be defined as follows:

$$x_{nm} = \begin{cases} 1, & \text{assigned } n \text{ to } m \\ 0, & \text{otherwise} \end{cases} \tag{1}$$

where $n = 1, 2, \cdots, N; m = 1, 2, \cdots, M$. Thus, important performance indicators for measuring the pros and cons of the multi-spacecraft collaborative mission planning program are mainly included the following three parts in this paper [7–9].

3.2 Service Spacecraft's Consume Indicators

The value of attrition for servicing spacecraft N to serve target spacecraft M is a_{nm}, and $1 - a_{nm}$ represents its remaining service capacity. The objective is to find a feasible target assignment to minimize servicing spacecraft attrition. The problem can be formulated as follows:

$$\min \sum_{n \in N} \sum_{m \in M} a_{nm} x_{nm} \tag{2}$$

3.3 Target Spacecraft's Value Indicators

We use E to represent the value of target spacecraft. We use P_c as the probability of the above movement. The value can be expressed as $E_{nm} = E \bullet P_c \bullet (1 - a_{nm})$ when we provide services for target spacecraft m with servicing spacecraft n. The specific formula can be defined as follows:

$$\max \sum_{n \in N} \sum_{m \in M} E_{nm} x_{nm} \tag{3}$$

3.4 The Optimal Time and Fuel Consuming Indicators

Reasonable phasing maneuver can reduce energy and time consumption. Considering the fuel consumption and the transfer time, we establish the performance indicator which makes the weighted sum of energy and time minimized, i.e.

$$\min \sum_{n \in N} \sum_{m \in M} [k(|\Delta v_1| + |\Delta v_2|) + (1 - k)t] x_{nm} \tag{4}$$

where, Δv_1 is the velocity increment required for the initial rendezvous moment, Δv_2 is velocity increment required for terminal rendezvous moment, $0 \leq k \leq 1$ is the proportionality coefficient. The energy consumption required for Lambert double pulse orbital maneuvering can be calculated according to rendezvous orbit time. The calculation method of Δv_1 and Δv_2 is described in Ref. [10].

3.5 Objective Function and Constraints

Through the above analysis, multi-objective function for multi-spacecraft collaborative mission planning is established. Therefore, the method of weighted summation can be used to convert multi-objective decision problems to single-objective optimization problem. Due to the different dimensions of each objective function, dimension conversion is required. We transform each dimension into the value between 0 and 1, set the value of target spacecraft E between 0 and 1, and let $\Delta V_{nm} = \Delta v_{nm} / \Delta v_{max}$, $T_{nm} = t_{nm} / t_{max}$. In summary, servicing spacecraft task allocation model is as follows:

$$\max \sum_{n \in N} \sum_{m \in M} (\omega_1 \bullet E_{nm} - \omega_2 \bullet a_{nm} + \omega_3 \bullet T_{nm}) x_{nm} \tag{5}$$

where, $\omega_1, \omega_2, \omega_3$ are weight coefficients which represent the importance degree of each sub-objective functions. Multi-spacecraft mission planning should satisfy the following constraints:

(1) $\sum_{n=1}^{N} x_{nm} = 1$, target spacecraft can only accept one servicing spacecraft's service.
(2) For each target spacecraft, no matter what way the service is, income of the target value is not greater than the value of the target spacecraft $\sum_{n=1}^{N} P_c \bullet$ $(1 - a_{nm}) \bullet E \bullet x_{nm} \leq E_m$.

4 Hybrid Discrete Particle Swarm Optimization Algorithm

4.1 Discrete Particle Swarm Optimization

In this paper, the natural number coding mode is used. The length of each particle is equal to the total number of target spacecrafts. For example, the number of servicing spacecraft n is 5, the number of the target spacecraft m is 10, so a particle is represented as

Particle 2 3 4 1 3 5 4 2 3 1   Serving spacecraft number

1 2 3 4 5 6 7 8 9 10 Target spacecraft number

According to the characteristics of multi-spacecraft collaborative task allocation problem, the position and velocity update formula of PSO can be defined as [11].

$$X_i(t+1) = c_2 F_3(c_1 F_2(w \bullet F_1(X_i(t)), p_{best}(t), g_{best}(t)) \qquad (6)$$

where X_i is the position of the particle i; p_{best} is the best previous position, g_{best} is the position of the best particle in current swarm, and w represents inertia weight of the current iteration. Cognitive parameter c_1 is defined to adjust the flight step length of p_{best}, and social parameter c_2 is adopted to adjust the flight step length of g_{best}. $F_1(X_i(t))$ is the influence function for particle $X_i(t)$, $F_2(X_i(t), p_{best}(t))$ is the learning operation of $X_i(t)$ to $p_{best}(t)$ and $F_3(X_i(t), g_{best}(t))$ is the learning operation of $X_i(t)$ to $g_{best}(t)$. For inertia weight, we use linearly decreasing strategy as in reference [12], i.e.

$$w = (\omega_i - \omega_e)(period - t)/period + \omega_e \qquad (7)$$

where parameter t is the current number of iteration, $period$ is the largest number of iteration, ω_i is the initial inertia weight value, and ω_e is the inertia weight value when particles evolve to the largest number of iterations. The position updating formula can be divided into three parts as in reference [16].

4.2 Improved Discrete Particle Swarm Algorithm Combined with Chaos

The nature of chaos is random and unpredictable apparently, and it also possesses an element of regularity [13]. Due to the ergodicity of chaos, chaos optimization algorithm is easy to jump out of local optima, which is precisely to overcome the shortcoming of discrete particle swarm optimization (DPSO). One chaos discrete particle swarm optimization (CDPSO) algorithm based on Logistic map is proposed in this paper, which combines DPSO and chaos theory [14]. Here, chaos variables are obtained by the Logistic mapping method [15]. Its equation is as follows.

$$z_{n+1} = \mu z_n (1 - z_n) \tag{8}$$

where $\mu > 3.57$ and $z_n \notin \{0.0, 0.25, 0.5, 0.75, 1.0\}$. When $\mu = 4$, Eq. (8) produces a status of the pseudo-random distribution which is a completely chaotic state.

Here, chaos variables are obtained by the Logistic mapping method [15]. Its equation is as follows.

$$Z' = (1 - \beta)\Psi^* + \beta Z \tag{9}$$

where β distributed in the interval [0, 1] is an adjustment parameter. Z' is a chaotic variable that obtained by adding a small disturbance to the current optimal solution variable X^*. Z is a chaotic variable obtained by Logistic mapping. Ψ^* is the optimal chaotic variable obtained by the current optimal solution variable X^* mapping to [0,1],

$$\Psi^* = (X^* - X_{\min})/(X_{\max} - X_{\min}) \tag{10}$$

4.3 Implementation Process of On-Orbit Servicing Spacecraft Mission Planning Based on CDPSO

Flowchart of the CDPSO algorithm is shown in Fig. 2.

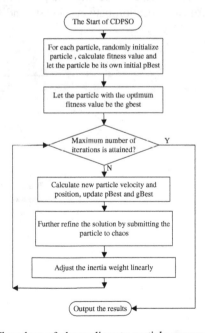

Fig. 2. Flowchart of chaos discrete particle swarm algorithm

The following two points are the additional supplements to the flowchart:

(1) Calculate the position of the new particles according to Eq. (6).
(2) Utilize the chaos to downsize the solution and optimize the optimum position g_{best} by chaos according to Eqs. (8), (9) and (10). First, we map g_{best} to Z_i which is located in interval [0,1]. Then, we produce the chaotic sequence using Logistic equation and convert the chaotic variables to the original solution region. Finally, we evaluate the new solution and choose the best solution.

5   Simulation

Several experiments were carried out to demonstrate the improved algorithm. In order to compare with the models and algorithms in Ref. [16], partial data of Ref. [16] are adopted. The following assumptions are made in this allocation problem. (1) Remaining service capacity by assigning the nth servicing spacecraft to serve the mth target spacecraft is given. (2) The value of target spacecraft m is E. (3) Service probability certainty P_c is 1. (4) The initial orbital parameters of each spacecraft are known. (5) Time constraint is 3000 s. (6) The time of orbital maneuver is set in $1000 \sim 3000$. Then, parameter values of CPSO are set as initial population size = 50, inertia weight $\omega_i = 0.9$, $\omega_e = 0.2$; (3) $c_1 = 0.5$, $c_2 = 0.5$, and $\beta = 0.39$.

For this scenario, we used DPSO and CDPSO to simulate the above question respectively, and compared their simulation results. Figure 3 shows the fitness change of the best assignment, the average assignment and the worst assignment in 50 iterations based on improved CDPSO. The x axis is the iteration number and the y axis is the fitness of particles. Table 1 represents the ultimate performance comparison results of the servicing spacecraft task allocation between DPSO and CDPSO.

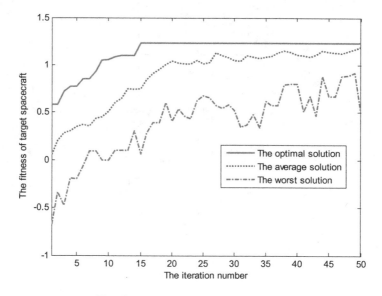

Fig. 3. Convergence curve of CDPSO

Table 1. Two algorithms' performance comparison

	DPSO	CDPSO
Optimal solution	1.2365	1.2365
Average solution	1.2190	1.2294
Worst solution	1.1691	1.1999
Convergence generation	22	15
Optimal allocation result	[2,4,4,2,4,1,2,3,2,4]	[2,4,4,2,4,1,2,3,2,4]

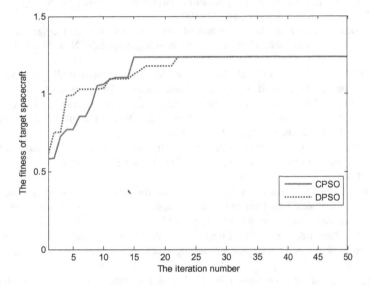

Fig. 4. Algorithm performance comparison

The convergence rate of the two algorithms and the performance of solution in Fig. 4, we know that the convergence behavior of CDPSO is better than DPSO, and it can quickly find the optimal solution. From the solution set of CDPSO changed with iterations number, it has converged to the optimal solution direction of less evolutionary generation, and this indicates that it has a higher optimize efficiency. From the curve of CDPSO with average solution, we can see that there is no phenomenon of convergence of all particles even in the late iteration stage. It enables the algorithm to continue to maintain strong optimization ability.

6 Conclusion

In spacecraft cooperative control, proper allocation of tasks is of great importance for the efficient utilization of the resources. The algorithm makes full use of local search ability of DPSO and global search ability of chaos, which makes the two algorithms get effective complementary and fast converges to the optimal solution, and significantly improves the performance of the algorithm. The simulations show that CDPSO is capable to solve on-orbit servicing spacecraft mission planning. The algorithm is simple, flexible and easy to implement and extend. It can quickly find the optimum allocation scheme. However, on-orbit service mission planning is a complex problem, and we will consider more factors in the future work.

References

1. Zhu, Y.W., Yang, L.P.: Study on approaching strategies for on-orbit servicing of satellites. Chin. Space Sci. Technol. **27**(1), 14–20 (2007)
2. Krolikowski, A., David, E.: Commercial on-orbit satellite servicing: national and international policy considerations raised by industry proposals. New Space **1**(1), 29–41 (2013)
3. Wu, Y.F., Liao, Y.R., Zhang, X.B.: Research on MAS-based mission allocating for on-orbit servicing spacecrafts. J. Acad. Equip. Command Technol. **20**(4), 48–53 (2009)
4. Hu, L., Sun, F., Xu, H., et al.: On-orbit long-range maneuver transfer via EDAs. In: 2008 IEEE Congress on Evolutionary Computation (IEEE World Congress on Computational Intelligence), CEC 2008, pp. 2343–2347 (2008)
5. Li, S., Mehra, R., Smith, R., et al.: Multi-spacecraft trajectory optimization and control using genetic algorithm techniques. In: Aerospace Conference Proceedings, vol. 7, pp. 99–108. IEEE (2000)
6. Yao, W., Chen, X., Huang, Y.Y., et al.: On-orbit servicing system assessment and optimization methods based on lifecycle simulation under mixed aleatory and epistemic uncertainties. Acta Astronaut. **87**, 107–126 (2013)
7. Pan, Q.K., FatihTasgetiren, M., Liang, Y.C.: A discrete particle swarm optimization algorithm for the no-wait flowshop scheduling problem. Comput. Oper. Res. **35**(9), 2807–2839 (2008)
8. Bethke, B., Valenti, M., Jonathan, P.H.: UAV task assignment. Robot. Autom. Mag. **15**(1), 39–44 (2008). IEEE
9. Dai, J., Cheng, J., Song, M.: Cooperative task assignment for heterogeneous multi-UAVs based on differential evolution algorithm. In: Intelligent Computing and Intelligent Systems, pp. 163–167. IEEE (2009)
10. Curtis, H.: Orbital Mechanics for Engineering Students. Elsevier, Amsterdam (2009)
11. Pan, Q.-K., Tasgetiren, M., Liang, Y.-C.: Minimizing total earliness and tardiness penalties with a common due date on a single-machine using a discrete particle swarm optimization algorithm. In: Dorigo, M., Gambardella, L.M., Birattari, M., Martinoli, A., Poli, R., Stützle, T. (eds.) ANTS 2006. LNCS, vol. 4150, pp. 460–467. Springer, Heidelberg (2006)
12. Goodman, J., Chandrakasan, P.A.: An energy-efficient reconfigurable public-key cryptography processor. IEEE J. Solid-State Circ. **36**(11), 1808–1820 (2001)
13. Alatas, B., Akin, E., Oze, A.B.: Chaos embedded particle swarm optimization algorithms. Chaos, Solitons Fractals **40**(4), 1715–1734 (2009)
14. Jiang, C.W., Bompard, E.: A hybrid method of chaotic particle swarm optimization and linear interior for reactive power optimisation. Math. Comput. Simul. **68**(1), 57–65 (2005)
15. Silva, C.P.: A survey of chaos and its applications. In: IEEE MTT-S International Microwave Symposium Digest, vol. 3, pp. 1871–1874 (1996)
16. Zhang, Q.X., Sun, F.C., Xu, B., et al.: Multiple spacecrafts on-orbit service task allocation based on DPSO. Chin. Space Sci. Technol. **32**(2), 68–76 (2012)

Solving the Test Task Scheduling Problem with a Genetic Algorithm Based on the Scheme Choice Rule

Jinhua Shi, Hui Lu[✉], and Kefei Mao

School of Electronic and Information Engineering, Beihang University,
Beijing 100191, People's Republic of China
mluhui@vip.163.com

Abstract. The test task scheduling problem (TTSP) is an essential issue in automatic test system. In this paper, a new non-integrated algorithm called GASCR which combines a genetic algorithm with a new rule for scheme selection is adopted to find optimal solutions. GASCR is a hierarchal approach based on the characteristics of TTSP because the given problem can be decomposed into task sequence and scheme choice. GA with the non-Abelian (Nabel) crossover and stochastic tournament (ST) selector is used to find a proper task sequence. The problem-specific scheme choice rule addresses the scheme choice. To evaluate the proposed method, we apply it on several benchmarks and the results are compared with some well-known algorithms. The experimental results show the competitiveness of the GASCR for solving TTSP.

Keywords: Test task scheduling problem · Genetic algorithm · Scheme choice rule · Hierarchical approach

1 Introduction

Since the mid-1950s, automatic test system has been applied in a variety of industries. With the development of it, time and equipment conflicts become a bottleneck because of the demands of cost, test efficiency and portability. The test task scheduling problem becomes a core issue for improving throughput, reducing time and optimizing resource allocation.

TTSP means a finite set of tasks with various schemes to be processed on several unrelated resources. The aim of TTSP is to minimize the test completion time for all available resources by sorting all tasks to an optimal sequence and obtaining the optimal scheme for each task. It is similar to the characteristics of other scheduling problems in industrial field, like flexible job shop problem (FJSP) and the resource scheduling problem.

It is natural to think of a hierarchical search algorithm to reduce the complexity of the given problem. Kir and Yazgan proposed a hierarchical approach consisting of meta-heuristic algorithms [1]. Yuan and Xu use a hierarchical strategy to minimize the makespan of multiobjective flexible job scheduling problem [2]. Another hierarchical

Y. Tan et al. (Eds.): ICSI 2016, Part II, LNCS 9713, pp. 19–27, 2016.
DOI: 10.1007/978-3-319-41009-8_3

method was proposed by Rohaninejad et al., which combine tabu search and firefly algorithms to solve capacitated job shop scheduling problem [3]. Bonyadi split the problem into two parts. The first part contains a population of processor numbers while the second one includes a population of task sequences [4]. In summary, the hierarchical structure divides the problem into several independent problems, all of which can be modified or improved without affecting another one. The hierarchical approach provides a framework for solving TTSP. However, the detailed method for TTSP is still undecided. Many literatures concerning heuristic methods can be used when dealing with NP-hard problems [5, 6]. In this paper, the genetic algorithm and dispatching rule are applied to TTSP to solve the task sequence and scheme choice respectively.

To solve the task sequence problem, a GA is applied based on following advantages. First, it has more resistance to convergence to a local optimum [7]. Second, the objective function can simply be defined over the problem domain and requires no specific assumptions on the objective function. Third, GA encompasses a range of different applications, such as the combinatorial optimization problems suggested by Albayrak and Allahverdi [8] and Fotakis and Sidiropoulos [9]. GA has been widely studied and applied in various fields. For example, Paulo et al. address the use of genetic algorithm applied to Markov Chains to estimate the best maintenance plan [10]. Pareek and Patidar proposed an encryption method for gray scale medical images based on the feature of genetic algorithm [11]. Li and Demeulemeester present a genetic algorithm for the robust resource leveling problem [12]. Kurdi proposed an effective new island model genetic algorithm for job shop scheduling problem [13].

Since the solution for the proper task sequence requires a certain amount of computation, it is advisable to limit scheme choices to reasonable ones. The dispatching rule is such a convenient method to make the scheme choice in concurrently with the task problem. Some dispatching rules are newly developed [14] by hybridization of common rules such as earliest due date and shortest processing time. They are applied at each moment to the jobs that can be scheduled according to the precedence relationships in the JSSP [15]. Here, the new dispatching rule is to solve the resource conflicts in TTSP. It is called the Scheme Choice Rule (SCR).

In addition, we propose the combination of GA and SCR rule (GASCR). The GA obtains a task order, which is provided to the dispatching rule to match a scheme choice. However, the exact time for a task cannot be completely determined in advance [16].

This paper is organized as follows. In Sect. 2, the model of TTSP is presented. In Sect. 3 a new genetic algorithm is proposed to solve the task sequence and the scheme choice. In Sect. 4, computational results are presented with a comparison to other approaches. Conclusion and future research are discussed in Sect. 5.

2 Problem Statement

The main characteristic of TTSP is that the scheme choice interacts with the task sequence and determines the resources in TTSP. The aim of TTSP is to arrange a sequence of tasks scheduled with certain schemes to minimize the total time by decreasing the idle time of equipment.

Here, we use $N \times M$ to denote TTSP. It means the execution of N tasks on M instruments. There are a set of tasks $T = \{t_j\}_{j=1}^{N}$ and a set of instruments $R = \{r_i\}_{i=1}^{M}$. The notifications P_j^i, S_j^i and C_j^i present the test time, the test start time, and the test completion time of task t_j on r_i, respectively. Therefore, we have $C_j^i \geq S_j^i + P_j^i$. For TTSP some instruments collaborate for one test task. A variable O_j^i is defined to express whether the task t_j occupies the instrument r_i and O_j^i is defined as:

$$O_j^i = \begin{cases} 1 & \text{if } t_j \text{ occupies } r_i \\ 0 & \text{others} \end{cases} \tag{1}$$

$W_j = \{w_j^k\}_{k=1}^{k_j}$ is used to denote the alternative schemes of task t_j, where k_j is the number of schemes of t_j. $K = \{k_j\}_{j=1}^{N}$ is the set containing the numbers of schemes that correspond to every task. Each w_j^k is a subset of R and can be represented as $w_j^k = \{r_{jk}^u\}_{u=1}^{u_{jk}}$, and u_{jk} is the number of instruments for w_j^k. Obviously, $\bigcup_{\substack{1 \leq k \leq k_j \\ 1 \leq j \leq n}} w_j^k = R$. The notification $P_j^k = \max_{r_i \in w_j^k} P_j^i$ express the test time of t_j for w_j^k.

The specific task sorting of T is TP and the set of their corresponding scheme choice is SC. The restriction of resources can be expressed as follows:

$$X_{jj*}^{kk*} = \begin{cases} 1 & \text{if } w_j^k \cap w_{j*}^{k*} \neq \emptyset \\ 0 & \text{others} \end{cases} \tag{2}$$

If $X_{jj*}^{kk*} = 1$, task t_j and task t_{j*} occupy the same resource and they cannot be executed in parallel. In addition, hypotheses considered in this paper are as following.

1. At a given time, an instrument can execute only for one task.
2. There is no preemption. It means that each task must be completed without interruption once it starts.
3. All tasks and instruments are available at time 0.
4. The setup time of instruments and the move time between the tasks can be negligible. In other words, $C_j^i = S_j^i + P_j^i$.
5. Assume $P_j^i = P_j^k$ to simplify the problem.

The optimization objective is the maximal test completion time and it is as follows.

$$f = \max_{\substack{1 \leq k \leq k_j \\ 1 \leq j \leq N}} C_j^k \tag{3}$$

Here, the notification $C_j^k = \max_{r_i \in w_j^k} C_j^i$ is the test completion time of t_j for w_j^k.

3 Proposed Method

The hierarchical method GASCR is based on the decomposition of TTSP into task sequence and scheme choice. GA focuses on the sequence of tasks using its advantage in the global search to investigate a diverse range of optimum task sequences. SCR makes a choice based on a fixed task sequence focusing on local research to find a better choice of schemes to obtain a minimal objective. A flowchart of the GASCR algorithm is given below.

Step 0: Initialize population $G(0) = TP^0, TP^1, TP^2, \ldots, TP^X$ where X is the number of individuals. Here, we set $generation = 0$.

Step 1: Choose the scheme SC using SCR rule.

Step 2: Compute the fitness f for each individual TP^i, $(i = 1, \ldots, X)$.

Step 3: Select X individuals for the next generation using stochastic tournament.

Step 4: Generate a new set of offspring with X individuals with crossover probability p_c, using the Nabel crossover operator.

Step 5: Mutate individuals with probability p_m.

Step 6: Set $generation = generation + 1$ and return to step 2 until a specified completion criterion is reached.

3.1 The Genetic Algorithm

The sequential task numbers are taken to represent a chromosome, where each number represents a gene. The number of initial individual is set to $X = N \sim 3N$. In the iteration process, ST selector, Nabel crossover, and inverse mutation are used to explore the best objective function value.

ST selector needs X pairs of chromosomes randomly chosen from a population. The element of each pair with lower fitness is selected for the mating pool, while the other is discarded. It has good ability for maintaining good chromosomes as well as certain part of the poor ones with high randomness. In comparison, the Deterministic Sampling (DS) is a deterministic selector which retains all individuals with better fitness level but totally ignores the worse individuals. The inferior chromosomes win the possibility to be the parents of the next offspring in the remainder stochastic sampling with replacement (RSSR). The roulette wheel selection performs badly if the differences among the fitness values are not noticeable [11].

The Nabel operator was created using non-Abelian group theory. It is the first choice to diverse of populations, compared with the C1 crossover, the Partially Mapped Crossover (PMX) [17]. The ability of the C1 to diversify the genes is weak. The Nabel operator rearranges the children of one parent according to the order of the elements of another parent. The offspring chromosomes of parent chromosomes *parent*1 and *parent*2 are

$$offspring2[i] = parent1[parent2[i]] \tag{4}$$

Here, $offspring1[i] = parent2[parent1[i]]$.

Inverse mutation is used to inverse two genes selected randomly with probability p_m [18]. The mutation can obtain diversity to search directions, and avoid convergence to local optima.

3.2   The Scheme Choice Rule

SCR is designed to find the proper scheme for the current scheduling task with the smallest completion time of all scheduled tasks. Before the rule is applied, each scheme combination should be considered to obtain a proper scheme choice for a fixed task sequence, and the number is as much as the multiplication of all scheme numbers. The amount of computation is 2^{20} if the total number of tasks is 20 and each task has two schemes. If the SCR rule is adopted, the amount decrease to 2×20. In addition, it also helps to solve the encoding problem of TTSP. We can ignore scheme choice when finding a task sequence. The problem-specific rule to determine the *SC* for a *TP* is shown as following.

- The first task in *TP* is to choose the scheme which has the minimum time.

- The following task t_j in *TP* has scheme k which ensures that $z = \max\left\{ C_j^k, 1 \leq k \leq k_j \right\}$ has a minimum value. The first scheme satisfying the general limitation is chosen if there exist alternatives which also have minimum z value.

To demonstrate the principle of SCR, an example is given in Table 1. A proper scheme sequence is obtained for a fixed task ordering $TP = \{1 \quad 4 \quad 2 \quad 3\}$ by the following steps. The first task is t_1. Since the second scheme has a shorter processing time than the first scheme, the scheme of task t_1 is 2. t_4 selects one scheme from the three possibilities. If the first scheme is chosen, t_4 can start at zero and its completion time is 4 units of time. If we select the second scheme, t_4 must wait until t_1 release the resources r_2, r_4. It finishes at 14 units of time. The completion of t_4 becomes 7 units of time if the last scheme is adopted. The first scheme is the best. The next task t_2 operates in an identical manner to t_4 and selects the third scheme. t_3 has only one choice of scheme. The scheme number of t_3 is 1. The final result is $SC=\{2 \quad 1 \quad 3 \quad 1\}$.

Table 1. Test schemes for test tasks for instance 4×4

Task	Scheme	Resource	Time	Task	Scheme	Source	Time
t_1	w_1^1	r_1, r_2	5	t_3	w_3^1	r_4	8
	w_1^2	r_2, r_4	3		w_4^1	r_1, r_3	4
t_2	w_2^1	r_1	6	t_4	w_4^2	r_2, r_4	11
	w_2^2	r_3	2		w_4^3	r_2, r_3	4

3.3   Performance Evaluation

A common evaluation metric for the performance of algorithms is the best value (BV). However, it is not comprehensive for the TTSP. The following metrics are used to assess the results.

1. The success rate (SR) measures the proportion of runs in which the algorithm is able to find the problem-specific optimum.
2. The average number of evaluations to solution (AES) represents the average number of evaluations that the algorithm requires to obtain success in those runs [19].
3. The mean best fitness measure (MBF) is the average of the best fitness values over all runs of the algorithm.

4   Computational Results

The size of the test task scheduling problem can not too huge based on characteristics of the test equipment and the test system. The test system is specific to some type of test equipment. As a result, the size of TTSP is limited and 40 tasks are rational. In this paper, the benchmarks are instance 20×8 [20] and instance 40×12 [21]. The first benchmark instance 20×18 is abstracted from a missile system. The second one instance 40×12 is randomly generated based on instance 20×8.

We adopt the hierarchical approaches applied to the problems similarly to the TTSP with good performances. The GASCR are compared with other hierarchical methods such as the Parallel Genetic Algorithm (PGA) [22] and the Genetic Algorithm with Simulated Annealing (GASA) [23].

In PGA sub-populations were allowed to evolve on several processors with the periodic exchange of individuals. One branch of GA is for the sequence of task and another is for resource dispatching. GASA combines GA and SA. After the crossover and mutation of the initial solutions, a simulated annealing-like procedure was applied to find the well scheme choice for the task sequence in the population.

The population size is the number of tasks. The generation is equal to 150. The crossover probability and mutation probability are set at $p_c=0.8$ and $p_m=0.05$.

The performances of GASCR, PGA and GASA in SR, MBF and AES are shown in Table 2. The optimal curves are shown in Fig. 1 for instance 20×8 and 40×12.

Table 2. Comparison of performance of different algorithms.

	Instance 20×8				Instance 40×12			
	BV	SR	MBF	AES	BV	SR	MBF	AES
GASCR	28	0.5	29	38	40	0.4	40.6	30.25
PGA	31	0.2	33.8	117.5	42	0.1	44.8	86
GASA	32	0.2	33.9	112.5	47	0.1	49.2	68

From the results in Fig. 1, we see that the differences in quality of solutions for GASCR, PGA and GASA are very clear. GASCR is highly competitive in all three metrics SR, MBF and AES. GASCR can obtain the better initial population, whose fitness is far smaller than PGA and GASA. Second, GASCR has a faster convergence speed. The generation when the algorithm converges to is smaller than those of PGA and GASA. Third, the optimal value found by GASCR yields the best solution. In the instance 20×8, the fitness value of the initial population of GASCR is almost equal to

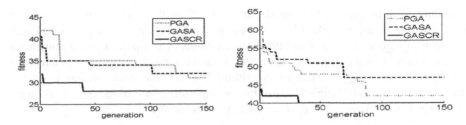

Fig. 1. The convergence curve of instance 20 × 8 and 40 × 12

the best optimal value of PGA and GASA. The effectiveness of GASCR can be attributed to the problem-specific genetic operators and the guidance of the SCR.

In addition, we can analyze the similarity and difference of the three algorithms. They are all based on the framework of GA and obtain the sequence of tasks using standard GA. The difference of them is the strategy of scheme choice. Our algorithm proposes the scheme choice rule, and PGA uses GA while GASA uses SA. Scheme choice rule reduce the solution space of test resource compared with GA and SA. The three algorithms have the same time complexity for obtaining sequence of tasks, and GASCR has a better time complexity for obtaining scheme. Therefore, GASCR has a higher performance than other two algorithms in time complexity.

The Gantt chart in Fig. 2 contains the scheduling processes for instance 20 × 8 and 40 × 12. The number on each rectangle is the number of task. This solution includes the task sequence *TP* and the scheme choice *SC* given below for instance 20 × 8.

$$TP=[4 \quad 15 \quad 11 \quad 6 \quad 5 \quad 20 \quad 17 \quad 13 \quad 16 \quad 2 \quad 9 \quad 3 \quad 18 \quad 10 \quad 1 \quad 8 \quad 19 \quad 12 \quad 14 \quad 7],$$
$$SC=[2 \quad 1 \quad 1 \quad 1 \quad 1 \quad 2 \quad 1 \quad 2 \quad 2 \quad 2 \quad 2 \quad 2 \quad 1 \quad 1 \quad 3 \quad 3 \quad 3 \quad 1 \quad 2 \quad 2].$$

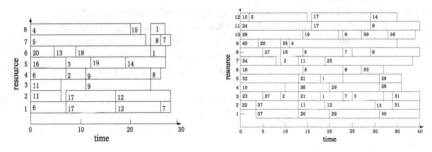

Fig. 2. The Gantt chart for instance 20 × 8 and 40 × 12.

5 Conclusions

GASCR uses GA to search the task sequence space, and SCR to select corresponding schemes under a fixed task sequence. This hierarchical method is based on the characteristics of TTSP, and it proves to be competitive when compared with the PGA and GASA.

Future work focuses on benchmarks of the TTSP, with more instances required to prove the effectiveness of the GASCR method for TTSP. In general, the good performance of algorithms tested on numerous benchmarks is a persuasive argument for their adoption. Another important aspect for future work would be to provide a method to test the theoretical efficiency of algorithm. The GA has been proved using the Markov chain, but most heuristic algorithms suffer from a lack of mathematical proof. Providing a method would lead to completeness for the TTSP.

Acknowledgments. This research is supported by the National Natural Science Foundation of China under Grant No. 61101153.

References

1. Kir, S., Yazgan, H.R.: A sequence dependent single machine scheduling problem with fuzzy axiomatic design for the penalty costs. Comput. Ind. Eng. **92**, 95–104 (2016)
2. Yuan, Y., Xu, H.: Multiobjective flexible job shop scheduling using memetic algorithms. IEEE Trans. Autom. Sci. Eng. **12**(1), 336–353 (2015)
3. Rohaninejad, M., Kheirkhah, A.S., Nouri, B.V., Fattahi, P.: Two hybrid tabu search–firefly algorithms for the capacitated job shop scheduling problem with sequence-dependent setup cost. Int. J. Comput. Integr. Manuf. **28**(5), 470–487 (2015)
4. Bonyadi, M.R., Moghaddam, M.E.: A bipartite genetic algorithm for multi-processor task scheduling. Int. J. Parallel Prog. **37**(5), 462–487 (2009)
5. Gao, J., Chen, R., Deng, W.: An efficient tabu search algorithm for the distributed permutation flowshop scheduling problem. Int. J. Prod. Res. **51**(3), 641–651 (2013)
6. Keskinturk, T., Yildirim, M.B., Barut, M.: An ant colony optimization algorithm for load balancing in parallel machines with sequence-dependent setup times. Comput. Oper. Res. **39**(6), 1225–1235 (2012)
7. Al-Hinai, N., ElMekkawy, T.Y.: An efficient hybridized genetic algorithm architecture for the flexible job shop scheduling problem. Flex. Serv. Manuf. J. **23**(1), 64–85 (2011)
8. Albayrak, M., Allahverdi, N.: Development a new mutation operator to solve the traveling salesman problem by aid of genetic algorithms. Expert Syst. Appl. **38**(3), 1313–1320 (2011)
9. Fotakis, D., Sidiropoulos, E.: Combined land-use and water allocation planning. Ann. Oper. Res. **219**(1), 169–185 (2012)
10. Paulo, P., Branco, F., Brito, J., de Silva, A.: BuildingsLife – the use of genetic algorithms for maintenance plan optimization. J. Clean. Prod. **121**, 94–98 (2016)
11. Pareek, N.K., Patidar, V.: Medical image protection using genetic algorithm operations. Soft Comput. **20**(2), 763–772 (2016)
12. Li, H., Demeulemeester, E.: A genetic algorithm for the robust resource leveling problem. J. Sched. **19**(1), 43–60 (2016)
13. Kurdi, M.: An effective new island model genetic algorithm for job shop scheduling problem. Comput. Oper. Res. **67**, 132–142 (2016)
14. Sels, V., Gheysen, N., Vanhoucke, M.: A comparison of priority rules for the job shop scheduling problem under different flow time- and tardiness-related objective functions. Int. J. Prod. Res., 1–16 (2011)
15. Ruiz, R., Serifoğlu, F.S., Urlings, T.: Modeling realistic hybrid flexible flowshop scheduling problems. Comput. Oper. Res. **35**, 1151–1175 (2008)

16. Xu, J., Zhang, Z.: A fuzzy random resource-constrained scheduling model with multiple projects and its application to a working procedure in a large-scale water conservancy and hydropower construction project. J. Sched. **15**(2), 253–272 (2012)
17. Zhang, L., Wang, L., Zheng, D.-Z.: An adaptive genetic algorithm with multiple operators for flowshop scheduling. Int. J. Adv. Manuf. Technol. **27**(5), 580–587 (2006)
18. Guo, Z.X., Wong, W.K., Leung, S.Y.S., Fan, J.T., Chan, S.F.: Genetic optimization of order scheduling with multiple uncertainties. Expert Syst. Appl. **35**(4), 1788–1801 (2007)
19. Laredo, J.L.J., Eiben, A.E., Steen, M., Merelo, J.J.: EvAg: a scalable peer-to-peer evolutionary algorithm. Genet. Program Evolvable Mach. **11**, 227–246 (2010)
20. Lu, H., Niu, R., et al.: A chaotic non-dominated sorting genetic algorithm for the multi-objective automatic test task scheduling problem. Appl. Soft Comput. **13**(5), 2790–2802 (2013)
21. Lu, H., Zhu, Z., et al.: A variable neighborhood MOEA/D for multiobjective test task scheduling problem. Math. Probl. Eng. **2014** (2014). 423621
22. Defersha, F.M., Chen, M.: A parallel genetic algorithm for a flexible job-shop scheduling problem with sequence dependent setups. Int. J. Adv. Manuf. Technol. **49**(1), 263–279 (2010)
23. Xia, R., Xiao, M.Q., Cheng, J.J., Fu, X.H.: Optimizing the multi-UUT parallel test task scheduling based on multi-objective GASA. In: The 8th International Conference on Electronic Measurement and Instruments, Xi'an, vol. 4, pp. 839–844 (2007)

Robust Dynamic Vehicle Routing Optimization with Time Windows

Yinan Guo, Jian Cheng[✉], and Junhua Ji

School of Information and Electronic Engineering,
China University of Mining and Technology, Xuzhou 221116, Jiangsu, China
chengjian@cumt.edu.cn

Abstract. Hard time window and randomly appeared dynamic clients are two key issues in dynamic vehicle routing problems. In existing planning methods, when the new service demand came up, global vehicle routing optimization method was triggered to find the optimal routes for non-served clients, which was time-consuming. Therefore, robust dynamic vehicle routing method based on local optimization strategies is proposed. Three highlights of the novel method are: (i) After constructing optimal robust virtual routes considering all clients, static vehicle routes for fixed clients are formed by removing all dynamic clients from robust virtual routes. (ii) The dynamically appeared clients append to be served according to their service demands and the vehicles' locations. Global vehicle routing optimization is triggered only when no suitable locations can be found for dynamic clients. (iii) A metric measuring the algorithms' robustness is given. The statistical results indicated that the routes obtained by the proposed method have better stability and robustness, but may be sub-optimum. Moreover, time-consuming global vehicle routing optimization is avoided as dynamic clients appear.

Keywords: Robust · Dynamic · Vehicle routing problem · Time windows · Ant colony algorithm

1 Introduction

The dynamic vehicle routing problem with time windows is the most practical and complex problem among the vehicle routing problems. Its goal is to find the optimal vehicle routes having the shortest driving distance or the least transport cost [1]. During the delivery cycle, as dynamic clients request new service demands, the distribution center needs to design the vehicles' routes in terms of their real-time location and the specific time windows of dynamic clients [2]. The time window limits the start and ending time of client's delivery tasks. It normally is divided into hard time windows and soft time windows [3]. We focus on dynamic vehicle routing problem with hard time windows. For hard time windows, the vehicle must arrive at the client earlier than the ending time. Otherwise, the delivery task is failed. If the vehicle reaches the client earlier than the start time, it shall wait for a while, which increases the time cost.

Many researches had been done to introduce intelligent optimization algorithms, such as ant colony algorithm, genetic algorithm, particle swarm optimization, to solve

© Springer International Publishing Switzerland 2016
Y. Tan et al. (Eds.): ICSI 2016, Part II, LNCS 9713, pp. 28–36, 2016.
DOI: 10.1007/978-3-319-41009-8_4

dynamic vehicle routing problem. Taillard and Badeau [4] proposed double ant colony system. Two colonies relying on independent pheromone respectively optimized two objective functions and shared the optimal routes during the evolution process. By adopting tabu search algorithm [5], a client was removed from one route and inserted into another vehicle's route with minimum cost in each iteration. Taniguchi et al. [6] adjusted and optimized the existing routes according to real-time road traffic information. An improved genetic algorithm using the novel crossover operator [7] was constructed so as to avoid the premature convergence. Liu [8] focused on the time dynamics of the vehicle. The time-depended vehicle routing problem was modeled and two methods to solve dynamic vehicle routing problems with dynamic demands were presented. The delivery cycle was divided into time slices and dynamic vehicle routing problem was transformed into continuous static planning problem.

In conclusion, two difficult issues for dynamic vehicle routing problems with time windows are the time constraint from hard time windows and dynamically appeared service demands. In existing planning methods, when the new service demand comes up, global vehicle routing optimization methods will be triggered to find the optimal routes for non-served deterministic clients and appeared dynamic client, which is time-consuming and complex. Therefore, robust dynamic vehicle routing method with time windows is proposed. Its main idea is to construct robust virtual routes for all clients, and then form static vehicle routes by remove all dynamic clients. As the new service requirement is requested, the static routes are directly adjusted based on robust virtual routes, which avoids global vehicle routing optimization.

2 Description of Dynamic Vehicle Routing Problems with Time Windows

In the paper, we focus on dynamic vehicles routing problem with single distribution center and clients with hard time windows. Especially, the delivered goods are only one type. Suppose there are one distribution center, M fixed clients and N dynamic clients. The distribution center locates in (x_1, y_1) and its delivery cycle is T_0. The deadline of its service time is l_0. Fixed clients have the service demands in each delivery cycle. But dynamic clients may be served in some delivery cycles according to certain probability. The coordinate of fixed and dynamic clients are respectively expressed by $(x_2, y_2), (x_3, y_3), \ldots, (x_i, y_i), \ldots, (x_{M+1}, y_{M+1})$ and $(x_{M+2}, y_{M+2}), (x_{M+3}, y_{M+3}), \ldots, (x_i, y_i), \ldots, (x_{M+N+1}, y_{M+N+1})$. The service demand of ith client is q_i and its service time windows is $ts_i \in [\underline{ts_i}, \overline{ts_i}]$. There are K vehicles with the same capacity. Their maximum capacity is Q and average driving speed is v. The start and end of each vehicle routing both are the distribution center. Assume that the probability of each dynamic client applying for the service during T_0 obeys the following distribution. $\overline{P_i}$ is maximum probability and $\lambda > 1$.

$$P_i(t) = \overline{P_i} e^{-\left(t - \frac{T_0}{\lambda}\right)^2}, i = M + 2, M + 3, \ldots M + N + 1. \tag{1}$$

Taken minimum the sum driving distance of all vehicles as the objective function, the driving distance between any two clients is defined as $D_{ij} = \left\| (x_i, y_i) - (x_j, y_j) \right\|_2$.

$x_{ijk} = 1$ means kth vehicle completes the service demand of ith and jth client in turn and goes through the route $(i \to j)$.

$$\min Z_R = \sum_{i=1}^{M} \sum_{j=1}^{M} \sum_{k=1}^{K} D_{ij} x_{ijk} + \sum_{i=M+1}^{M+N+1} \sum_{j=M+1}^{M+N+1} \sum_{k=1}^{K} P_i D_{ij} x_{ijk}. \qquad (2)$$

For dynamic vehicle routing problems, many constraints must be satisfied.

Constraints I: The load of each vehicle is not more than its maximum capacity, denoted by $\sum_{i=2}^{M+N+1} q_i y_{ki} \leq Q$. $y_{ki} = 1$ means ith client is served by kth vehicle.

Constraints II: Each client is only served by one vehicle. That is, $\sum_{k=1}^{K} y_{ki} \leq 1$.

Constraints III: Each vehicle leaves the served client and drives to next served client, expressed by $\sum_i x_{ijk} = y_{ki}$ and $\sum_j x_{ijk} = y_{kj}, i,j = 2,\ldots, M+N+1$.

Constraints IV: The vehicle must finish ith client's service demand during its time windows. That is, $t_j \in [\underline{ts_j}, \overline{ts_j}]$ and $\sum_{k=1}^{K} \sum_{i=1}^{M+N+1} x_{ijk}(t_i + t_{ij} + ts_i) = t_j, j = 2,3,\ldots M+N+1$.

Constraints IV: Maximum driving time of any vehicle cannot be later than the latest working time of distribution center, denoted by $\sum_{i=1}^{M+N+1} \sum_{j=1}^{M+N+1} x_{ijk}(t_{ij} + ts_i) \leq l_0$.

3 Robust Dynamic Vehicle Routing Methods Based on Ant Colony Algorithm and Local Optimization Strategy

At first, robust virtual routes and static vehicle routes are formed be ant colony algorithm. Secondly, as dynamic client appears, it is appended to the executing static route by robust local optimization strategy according to its service requirement and the vehicles' location.

Global vehicle routing optimization is done on all clients including dynamic clients and fixed clients by ant colony algorithm to find optimal robust virtual routes, denoted by $LB = \{Lb_1, Lb_2, \ldots, Lb_k\}$. Lb_k is the sub-route completed by a vehicle. As we known, all fixed clients must be accessed by the vehicle in a delivery cycle. Dynamic clients, consequently, are served after all fixed clients in robust virtual routes or formed a new route so as to affect the fixed clients as less as possible. Suppose there are c ants in the distribution center. Each ant is regarded as a vehicle. Whether kth ant c_k can transfer from ith to jth client depends on four factors including the sub-route's length, the pheromone concentration, the vehicle's capacity and time windows of the next client. Let τ_{ij} be the pheromone between ith and jth client. The transition probability of an ant from ith to jth client is defined as follows.

$$p_{ij}^k(t) = \begin{cases} \dfrac{[\tau_{ij}(t)]^{\alpha}[\eta_{ij}(t)]^{\beta}[\frac{T_Q}{ts_j(t)}]^{\gamma}}{\sum\limits_{g \in J_k(i)} \{[\tau_{ig}(t)]^{\alpha}[\eta_{ig}(t)]^{\beta}\}}, & j \in J_k(i) \\ 0, & others \end{cases} \qquad (3)$$

Here, $\eta_{ij}=1/D_{ij}$ reflects the expectation of ant transferring from ith to jth client. α, β and γ respectively reflect the importance. All feasible clients for ant c_k consist of the candidate set, expressed by $J_k(i) = \{1, 2, 3, \ldots, M+N+1\} - tabu_k$. The client passed by ant c_k is added to its tabu set $tabu_k$ so as to ensure each client served by only one vehicle. The route is feasible if its sum capacity does not exceed maximum load. Let $\rho \in (0, 1)$ be the evaporation coefficient. $\Delta\tau_{ij}^k = Q_P/L_k$ is the pheromone left by kth ant. $\Delta\tau_{ij}^k = 0$ if the ant does not go through the route $(i \rightarrow j)$. The pheromone is updated as follows.

$$\tau_{ij}(t+1) = \left[(1-\rho)\tau_{ij}(t) + \sum_{c_k=1}^{c}\Delta\tau_{ij}^k\right]\frac{T_0}{ts_j(t)}. \qquad (4)$$

If no dynamic clients request for the service in the delivery cycle, optimal robust virtual route is redundant. Hence, we define static vehicle route as $LS = \{Ls_1, Ls_2, \ldots, Ls_k\}$ by removing all dynamic clients from LB. That is, $Ls_k=\{Lb_k\backslash(x_i, y_i)\}, i = M+2, \ldots, M+N+1$.

Let $\delta \in (0, 1)$ be the service threshold. Once $P_i(t) > \delta$, dynamic client applies for the service. Suppose Ls_k^i is ith fixed client served by kth vehicle. Only when jth dynamic client sends the service request at $t_j < \sum_{l=1}^{i} ts_{(Ls_k^l)} + \sum_{l=1}^{i-1} D_{l,l+1}/v$, kth vehicle can complete its delivery task. $\sum_{l=1}^{i} ts_{(Ls_k^l)}$ and $\sum_{l=1}^{i-1} D_{l,l+1}/v$ respectively are the sum of delivery time to complete current delivery task and driving time to next client, which satisfy $\underline{ts_j} < \sum_{l=1}^{j-1} ts_{(Ls_k^l)} + \sum_{l=1}^{j-1} D_{l,l+1}/v < \overline{ts_j}$ and $\sum_{l=1}^{j} ts_{(Ls_k^l)} + \sum_{l=1}^{j-1} D_{l,l+1}/v < \overline{ts_j}$. If $j = Lb_k^l$ and $Ls_k^l \rightarrow Lb_k^l \subset Lb_k$, the new route expressed by $LN = \{Ls_1, Ls_2, \ldots, Ls_k\}$ is formed and $Ls_k = (Ls_k^1 \rightarrow Ls_k^2 \rightarrow \cdots \rightarrow Ls_k^i \rightarrow Lb_k^l \rightarrow Ls_k^{i+1})$.

If dynamic client applies for the delivery task at $\sum_{l=1}^{i-1} ts_{(Ls_k^l)} + \sum_{l=1}^{i-2} D_{l,l+} 1/v < td_j < \sum_{l=1}^{i} ts_{(Ls_k^l)} + \sum_{l=1}^{i} D_{l,l+1}/v$, the actual route cannot form directly in terms of optimal robust virtual route and local optimization strategy based on 2-opt algorithm is triggered to find the sub-route satisfying hard time windows. If we cannot find a new sub-route satisfying above constraint by mutation operation, the existing shortest sub-route meeting the need of time windows is preserved. That is, if the sub-route $(Ls_k^i \rightarrow Ls_k^{i+1} \rightarrow Ls_k^{i+2})$ is longer than the mutated sub-route $(Ls_k^i \rightarrow Ls_k^{i+2} \rightarrow Ls_k^{i+1})$, the latter will be the new feasible solution.

In summary, the core of proposed method is to construct optimal robust virtual vehicle route for all clients and form corresponding static vehicle routes. If no service request is given by dynamic client in the delivery cycle, static vehicle routes carry out. Otherwise, dynamic clients are directly appended to certain static sub-route in terms of robust virtual vehicle route as the service time of corresponding vehicle is enough to complete the delivery task of all non-served client. Obviously, it avoids time-consuming global vehicle routing optimization by using local optimization to find satisfied sub-route as soon as possible and fits for solving the practical real-time dynamic vehicle routing problems.

4 Comparison and Analysis of Simulation Results

In order to verify the rationality of the proposed method, two groups of experiments are done by Matlab 7.0. One is to analyze the adaptability of RDVR under different amount of dynamic clients. The other is to compare the performance of RDVR with dynamic vehicle routing method with time windows (DVR) [5]. In experiments, there are one distribution center expressed by O, fifteen fixed clients and fifteen dynamic clients. Their locations, demands, service time and time windows are shown in Table 1. $\overline{P_i} = 0.6$. The delivery cycle of distribution center $T_0 = 30$ s. Twenty vehicles with

Table 1. The information of clients

No.	Property	Coordinate	Demand	Service time	Time windows
1	Distribution centre	(18.7, 15.29)	0	0	[0,30]
2	Fixed node	(16.47,8.45)	3	1.8	[5.0,19.5]
3	Fixed node	(20.07, 10.14)	2.5	1.0	[4.1,14.9]
4	Fixed node	(19.39, 13.37)	5.5	2.3	[1.5,15.6]
5	Fixed node	(25.27, 14.24)	3	1.8	[6.0,19.6]
6	Fixed node	(22, 10.04)	1.5	1.2	[5.5,8.2]
7	Fixed node	(25.47, 17.02)	4	2.4	[3.6,19.8]
8	Fixed node	(15.79, 15.1)	2.5	1.5	[2.1,24.2]
9	Fixed node	(16.6, 12.38)	3	1.8	[2.5,27.1]
10	Fixed node	(14.05, 18.12)	2	1.2	[2.0,19.9]
11	Fixed node	(17.53, 17.38)	2.5	1.5	[2.1,15.6]
12	Fixed node	(23.52, 13.45)	3.5	2.1	[2.0,20.3]
13	Fixed node	(19.41, 18.13)	3	1.8	[2.3,15.5]
14	Fixed node	(22.11, 12.51)	5	2.0	[3.5,24.5]
15	Fixed node	(11.25, 11.04)	4.5	2.7	[3.5,28.8]
16	Fixed node	(14.17, 9.76)	2	1.3	[4.5,24.0]
17	Dynamic node	(22, 18.23)	3.5	1.2	[4.0,15]
18	Dynamic node	(12.21, 14.6)	4	1.5	[6.4,16]
19	Dynamic node	(13.82, 16.22)	1.8	1.5	[3.2,16]
20	Dynamic node	(16.12, 10.12)	3.2	1.0	[4.3,15]
21	Dynamic node	(22, 16.42)	4.6	1.5	[3.4,20]
22	Dynamic node	(23.55, 14.89)	2.2	1.0	[2.5,20]
23	Dynamic node	(14.12, 12.13)	2.4	1.0	[2.5,18]
24	Dynamic node	(16.23, 14.03)	3.1	1.0	[2.5,18]
25	Dynamic node	(12.33, 22.23)	2.5	1.0	[2.5,18]
26	Dynamic node	(16.57, 22.32)	2.4	1.0	[2.5,19]
27	Dynamic node	(18.17, 8.21)	1.8	1.0	[1.8,18]
28	Dynamic node	(20.14, 14.48)	1.5	1.0	[2.7,20]
29	Dynamic node	(21.42, 20.72)	1.2	1.0	[1.5,21]
30	Dynamic node	(10.22, 8.14)	2.1	1.0	[1.5,22]
31	Dynamic node	(24.07,19.88)	2.2	1.0	[2.1,23]

same capacity are in service and $Q = 15$. The parameters of ant colony algorithm are: $c = 60, \alpha = \beta = \gamma = 2$. Each experiment runs ten times and the termination iteration is 50.

Different from static vehicle routing problems, dynamic vehicle routing problems not only depends on the number of static clients and their spatial distribution, but also dynamic clients with random service demand. In order to rationally evaluate its complexity, the dynamic [9, 10] is defined to measure the possibility of dynamic clients can be served. Moreover, we propose the robustness to measure the ability of the optimal vehicle routes adapting to dynamically appeared clients. As the new service demand comes up, we hope to find at least one robust route to make the length of new vehicle route as small as possible. Assume that Ct_i and Bt_i denote the cost of vehicle route after and before ith dynamic client appears. Especially, the cost can be the route's total length, the driving time, the fuel consumption or the number of serviced clients. The robustness is defined as follows. Clearly, $RT \in (0, 1]$ and optimal vehicle routes more fit for dynamic environment as RT closes to 1.

$$RT = \frac{1}{N} \sum_{i=1}^{N} \frac{Bt_i}{Ct_i}. \tag{5}$$

Experiment I: The experiment is done to analyze the sensitivity of the proposed methods to the number of dynamic clients. We set $N = 5$, 10, 15. The time when dynamic clients ask for service are listed in Table 2. Especially, 18th dynamic client has no service demand as $N = 15$. The statistical results shown in Table 3 indicate that with the increasing of the clients' quantity, the route's length and the number of the vehicles are more while the robustness become worse. Under different amount of dynamic clients, the optimal static vehicle route excluding dynamic clients and the actual routes after local optimization as dynamic service demands appear are shown in

Table 2. The time when dynamic clients ask for service

$N = 5$	Node	17	18	19	20	21									
	Time	5	4	5	6	4									
$N = 10$	Node	17	18	19	20	21	22	23	24	25	26				
	Time	5	4	3	5	3	3	4	5	3	5				
$N = 15$	Node	17	18	19	20	21	22	23	24	25	26	27	28	29	30	31
	Time	3	–	6	5	6	4	4	3	3	5	3	3	4	5	5

Table 3. The performance of RDVR under different N

N	Length of static route	Length of dynamic route	The number of vehicles	Robustness
5	83.17	90.83	5	0.9157
10	108.9	131.47	5.5	0.8283
15	126.37	159.49	6	0.7923

Figs. 1, 2 and 3. Taken Fig. 3(c) as an example, three dynamic clients have the service requirements at $t = 4$. For static route($1 \rightarrow 6 \rightarrow 14 \rightarrow 17 \rightarrow 1$), the vehicle just completes the delivery task of all static clients and has enough time to arrive at 22th and 29th clients according to their time windows. So the subsequent route varies in terms of the robust virtual routes.

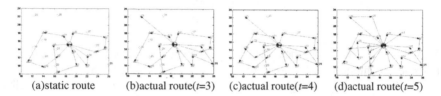

(a)static route   (b)actual route($t=4$)   (c)actual route($t=5$)   (d)actual route($t=6$)

Fig. 1. Robust vehicle routing planning methods with time windows ($N = 5$)

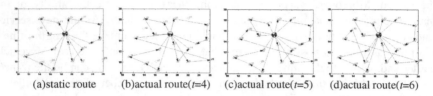

(a)static route   (b)actual route($t=3$)   (c)actual route($t=4$)   (d)actual route($t=5$)

Fig. 2. Robust vehicle routing planning methods with time windows ($N = 10$)

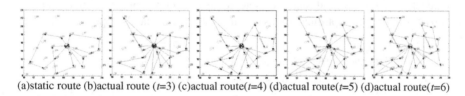

(a)static route (b)actual route ($t=3$) (c)actual route($t=4$) (d)actual route($t=5$) (d)actual route($t=6$)

Fig. 3. Robust vehicle routing planning methods with time windows ($N = 15$)

Experiment II: In order to fully compare RDVR with DVR, the experiments under different amount of dynamic clients are done. The statistical experimental results shown in Table 4 and Fig. 4 indicate that no matter how many dynamic clients, the length of vehicle routes obtained by RDVR is less than DVR and RDVR spends less time finding the optimal routes. The reason for that is robust local optimization strategy in RDVR considers dynamic clients in advance and avoids the complex global optimization for all non-served clients by changing fewer routes. Moreover, hard time windows have the obvious effect on the DVR, while the influence of dynamic service demand to route planning is less in RDVR. So the proposed method has better stability and delivery efficiency for all vehicles. It is suitable for the dynamic vehicle routing problems, in which dynamic clients appear in terms of certain probability (Fig. 4).

Table 4. Comparison of two algorithms' performances

N	Method	Length of route	The number of vehicles	Time consuming	Robustness
5	RDVR	765.31	3	0.5244	0.8065
	DVR	787.31	3	0.6564	0.5604
10	RDVR	927.66	3.5	0.6852	0.7843
	DVR	1049.2	4	0.9654	0.4204
15	RDVR	988.42	4	1.098	0.7895
	DVR	1248.3	4	1.312	0.3533

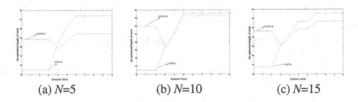

(a) $N=5$ (b) $N=10$ (c) $N=15$

Fig. 4. The routing length of two methods

5 Conclusions

Aiming at dynamic vehicle routing problems with hard time windows and randomly appeared dynamic clients, a novel robust dynamic vehicle routing optimization method is proposed. Firstly, optimal robust virtual routes considering all clients are constructed by ant colony algorithm, and then static vehicle routes are formed by removing all dynamic clients from robust virtual routes. As dynamic clients apply for the service, static vehicle routes are adjusted in terms of time windows and the vehicle's status. The time-consuming global vehicle routing optimization is triggered only when no suitable locations can be found in static vehicle routes for dynamic clients. The statistical results show that the proposed method can find more stable and robust routes with less computation cost, but may be sub-optimal solution as dynamic clients appear. Only one objective that route's length is considered. Robust dynamic vehicle routing method considering more than one objective is our further work.

Acknowledgement. This work was supported in part by National Natural Science Foundation of China under Grant 61573361, Research Program of Frontier Discipline of China University of Mining and Technology under Grant 2015XKQY19, the Innovation Team of China University of Mining and Technology under Grant 2015QN003 and National Basic Research Program of China under Grant 2014CB046300.

References

1. Zhang, L.-P., Chai, Y.-T.: Improved genetic algorithm for vehicle routing problems with time windows. Comput. Integr. Manuf. Syst. **8**(6), 451–454 (2002)
2. Wang, X.-B., Lu, H.-J., Chen, W.-T.: Improved ant colony algorithm for dynamic vehicle routing problems with time windows. Ind. Control Comput. **1**, 41–43 (2009)
3. Ma, W.-H.: Vehicle Routing Problem with Time Windows and Research on Its Heuristic Algorithm. HeFei University of Technology (2008)
4. Taillard, E., Badeau, P.: A tabu search heuristic for the vehicle routing problem with time windows. Transp. Sci. **31**, 170–186 (1999)
5. Cordeau, J.F., Laporte, G.: A unified tabu search heuristic for vehicle routing problems with time window. J. Oper. Res. Soc. **52**(8), 928–936 (2001)
6. Taniguchi, E., Ando, N., Okamoto, M.: Dynamic vehicle routing and scheduling with real time travel times on road network. In: Proceedings of the Eastern Asia Society for Transportation Studies, pp. 151–151 (2007)
7. Zhang, L., Chai, Y.: Improved genetic algorithm for vehicle routing problem. Syst. Eng. Theory Pract. **22**(8), 79–84 (2002)
8. Liu, X.: Research on the Vehicle Routing Problem. Huazhong University of Science and Technology (2007)
9. Guo, Y.-H., Zhong, X.-P.: Analysis of the dynamic vehicle routing problem's queuing model. J. Manag. Sci. **1**, 33–37 (2006)
10. Hong, L.-X.: Programming model and solution algorithm for the dynamic vehicle routing problem with time windows. Comput. Eng. Appl. **4**, 244–248 (2012)

Task Oriented Load Balancing Strategy for Service Resource Allocation in Cloud Environment

He Luo$^{(\boxtimes)}$, Zhengzheng Liang, Yanqiu Niu, and Xiang Fang

School of Management, Hefei University of Technology, Hefei, China
luohe@hfut.edu.cn

Abstract. Load balancing strategy is one of the most important issues for service resource allocation in order to balance tasks for different service resources. However, cloud computing has brought about many great changes to the traditional information service process when adjusting tasks among different service resources. In this paper, service alliance is introduced into the new information service model, which can provide a new communication mechanism for service providers and service users. Then, a task-oriented load balancing strategy, named double weighted least connection, is proposed. This strategy not only considers the usage of the service resources, but also takes account of the size of different tasks. Furthermore, a set of simulation experiments is discussed in order to evaluate the performance of different load balancing strategies in different situations.

Keywords: Cloud computing · Resource allocation · Load balancing · Service alliance · Double weighted least connection

1 Introduction

Cloud computing, a new type of information service environment, has been established on the Internet in the recent years [1] and developed by the core technologies including data center, Internet, mobile terminal and so on. Cloud computing also enables outside users to use service resources in a convenient and flexible way that individuals can be allowed to provide resources on-demand on a pay-as-you-go basis [2], like Amazon's Elastic Compute Cloud (EC2) [3], Google APP Engine [4] and Salesforce CRM [5]. The users in cloud computing can then make use of service resources like using water, power, gas, and other social public service resources. It is thus clear that cloud computing has greatly changed the traditional information service model and information service process. However, information service in cloud computing is a very complicated process [6] which involves the design of service model, the matching of service demands, the allocation of service resources, the evaluation of service quality, as well as the analysis, design and implementation of service system and so on. According to the continuous change of user's needs and the unceasing increase of service resources, the service resource allocation in cloud computing has become an important issue.

© Springer International Publishing Switzerland 2016
Y. Tan et al. (Eds.): ICSI 2016, Part II, LNCS 9713, pp. 37–46, 2016.
DOI: 10.1007/978-3-319-41009-8_5

In order to improve the utilization of service resources, cloud computing puts the dispersed service resources together into the resource pool, making a single service application providing services for different users in the multi-tenant way [7]. But when faced with the magnitude of service demands, the unreasonable situation of service resource allocation often arises. Some service resources may be idle and others over-loaded. This situation will cause the decreasing of service capacity, even system robustness. Therefore, the load balancing strategy is necessary for service resource allocation to make dynamic adjustments according to different needs of users and various types of service resources.

In this paper, we suggest a service alliance based information service model in cloud computing, and propose a double weighted least connection load balancing strategy for service resource allocation in cloud computing. The remainder of the paper is organized as follows. In Sect. 2, service alliance is introduced into the information service model, and a Double Weighted Least Connection load balancing strategy is proposed later in the Sect. 3. In Sect. 4, we conduct the set of simulation experiments and compare the results. The conclusion and future work are discussed in Sect. 5.

2 Related Work

The load balancing strategies for resource allocation have ready been applied in the P2P network [8], web network [9], parallel computation, grid computing [10], and so on. For instance, the P2P resource discovery mechanism in supporting dynamic load balancing [11], the statistical method of passive node load [12], load balancing cluster strategy for support proportion delay guarantee [13], the quantification model of load balancing performance [14], the mixed load balancing strategy for continuous task in the grid computing environment [15], and so on. In order to solve the load balancing problem, there are also several algorithms. For example, the load balancing algorithm in nonparametric regression and no-regret minimum delay [16], hierarchical iterative dynamic load balancing algorithm [17], the session adaptive load balancing algorithm [18], and so on. The uncertainty of service process in the existing distributed environment mainly reflects in the distribution of service and the uncertainty of service participants. Generally, service resource allocation is running under the information service model, but traditional information service models have been greatly changed by cloud computing, where service resources are obtained from distributed data centers of different regions. At the same time, the process of service resource allocation is related to the virtual resource pool where service resources have their centralized characteristics [19]. Therefore, a new kind of information service model is needed for cloud computing, and the coordinated load balancing strategy is also necessary for service resource allocation.

3 Information Service Model

Cloud computing has its own theoretical and technical basis. The development of cloud computing is on the basis of not only information technologies such as the virtualization technology, information terminal technology, but also multi-disciplines such as decision science, service science, management science, and so on. But cloud computing is not the simple combination of the above theories and technologies; it also provides a brand new environment for computing service and commercial service with a new service support platform. Thus, cloud computing is very different from the previous service environment and decision environment, there are many new changes brought about by cloud computing, especially in the information service model.

Cloud computing is a completely open service environment including the open system structure, dynamic service resources, open service platform and so on. Service providers, service users and service resources are free to enter or exit from cloud computing environment, resulting in the brand-new uncertainty and complexity in the information service. In the cloud computing environment, the information service model plays of 4 roles, service providers (SP), service users (SU), service resources (SR), service management platform (SMP) and their mutual relationship. In order to coordinate different roles with each other, we introduce the fifth role, service alliance (SA), into this model as shown in Fig. 1.

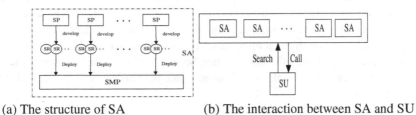

(a) The structure of SA   (b) The interaction between SA and SU

Fig. 1. Service alliance

Service alliance can integrate service providers with service resources and service management platform. Furthermore, service alliance also builds a communication mechanism for service users to interact.

In this model, service management platform does not have any resources but it can provide corresponding service resources by different service providers. Service alliance can then provide a collection of service resources for various service users. One service user can send resource request information to different service alliances. Service alliances need to match service resources for the resource request information, and assign the chosen service resources to certain service users. Service users do not need to care about the origin of service resources.

There are three kinds of basic operations in this model, deploy, search and call. Deploy means that service resources, developed by service providers, can be deployed to service management platform. Search indicates that service users can submit resource request information according to their needs via the service platform. Call shows that the service management platform can choose the right service resources for service users to meet users' needs under service strategy.

According to the above description, we can find that each service alliance has its service management platform and more than one service provider develops and deploys a variety of service resources on the service management platform. Service users can select different service providers in different service alliances to meet their needs. Although service resources are unlimited in service alliances from the aspect of cloud computing environment, service resources of one service provider are limited in one service alliance. So, when a large amount of service users are assigned to the limited service providers, service resources in some service alliances may be overloaded when other service alliances are still idle. Therefore, we propose a load balancing strategy for service resource in cloud environment.

4   Load Balancing Strategy

In the above model, service resource allocation involves several elements including service demands, service resources, service strategies and their interaction. However the dynamic and uncertainty of users' needs will increase the complexity of service resources allocation and decrease the capacity of service, leading to load unbalancing. Therefore, it becomes a dynamic scheduling problem which is NP-hard.

In this paper, the needs of users are processed in the batch mode, namely accepting a certain number of users' needs in a certain period of time. Furthermore, we take the Map/Reduce model to divide a larger task into several logic independent subtasks. Meanwhile, the hardware physical resources have also been divided into several virtual service resources so that the subtasks can then be assigned to virtual service resources. The followings are the definition of users' needs, service resources and their characteristics.

Definition 1. In cloud computing environment, the needs of users, namely cloudlet, will be transformed to a collection of n independent tasks $cloudlet_i$ in a batch, and can be expressed as: $cloudlet = \{cloudlet_1, cloudlet_2, \cdots cloudlet_n\}$.

Definition 2. In the $cloudlet_i$, $cloudletLength_i$ indicates the specification of task i, and can be expressed as $cloudLength_i = \{startTime, endTime, duration\}$.

Where $startTime$ means the earliest start time, $endTime$ means the latest completion time, and duration means the execution time.

Definition 3. In the cloud computing environment, the resource pool vm is composed of m distributed virtual service resources vm_i, and can be expressed as: $vm = \{vm_1, vm_2, \cdots vm_m\}$.

In one vm_j, $vmMips_j$ indicates the processing capacity of virtual service resource vm_j.

At this time, the load balancing strategy is to balance m virtual service resources when n independent tasks are successfully assigned during a batch process. When the certain period of time is set to the minimum, the model is equivalent to dynamic scheduling model. When the certain period of time is set to the maximum, the model can be converted to static scheduling one.

Some load balancing strategies can be useful for this model, such as Least Connection (LC) load balancing strategy, Weighted Least Connection (WLC) load balancing strategy. LC only estimates the load of service resources according to the number of current active resources connection instead of taking account of the state of resources themselves. While WLC overcomes the disadvantage of LC via corresponding weights and ensures the number of resources connection and weight of these resources in proportion. However, task size has been ignored by WLC. When a new task arises, WLC thinks that the number of the current task connection is 1, no matter the task size is long or short. Since different tasks have different specification, we suggest considering not only the weight of resource capacity but also the weight of task size to the process of load balancing, called Double Weighted Least Connection (DWLC) load balancing strategy. The detailed process is defined as follows:

Step 1: Initializing the parameters: Setting the number and attribute of tasks and virtual service resources.

Step 2: Calculating the weight: Calculating the weight k_i of the current virtual service resource vm_i.

$$k_i = \frac{vmMips_i}{\sum\limits_{j=1}^{m} vmMips_j}, i = 1, 2, \cdots, m \tag{1}$$

Step 3: Calculating single connection: Calculating the $cloudletTasks_j$ of the current $cloudlet_j$. $cloudletTasks_j$ indicates the number of $cloudlet_j$ connection, and it means the ratio of task $cloudlet_j$ size and unit task size.

$$cloudletTasks_j = \frac{cloudletLength_j}{averageLength} = \frac{cloudletLength_j}{\frac{\sum\limits_{i=1}^{n} cloudletLength_i}{n}} = \frac{cloudletLength_j}{\sum\limits_{i=1}^{n} cloudletLength_i} \times n$$

$$= w_j \times n \tag{2}$$

Where $averageLength$, namely standard size, indicates the unit task size. At the same time, the number of unit task connection is considered to be 1; w_j indicates the weight of $cloudlet_j$. n represents the number of tasks (j = 1, 2, ... , n).

Step 4: Calculating accepted connections: Calculating the $vmTasks_i$ of the current virtual service resource vm_i. $vmTasks_i$ is the number of task connection accepted by vm_i.

$$vmTasks_i = \sum_{j=1}^{n} cloudletTasks_j \bullet S_{ij} \tag{3}$$

Where S_{ij} is the element of the j-th column and the i-th row of the $m \times n$ 0-1 matrix, indicating whether the task $cloudlet_j$ is successfully assigned to the virtual service resources vm_i (i = 1, 2, ... , m).

Step 5: Calculating the utility value: Calculating the $Value_i$ of current vm_i, comparing the $Value_i$ of each vm_i, and choosing the best virtual service resources. $Value_i$ indicates the utility value of virtual service resources vm_i, the smaller the $Value_i$, the better the results. (i = 1, 2, ... , m)

$$Value_i = f(k_i, vmTasks_i) = \frac{vmTasks_i}{k_i} \tag{4}$$

Step 6: Repeating and Decision Making: Repeating Step 4 and Step 5, assigning these tasks to the best virtual service resources until the allocation of all tasks is finished. The following is the pseudo code of DWLC load balancing strategy.

for (int i =0; i < m; i ++)	//start from the first task
{	
idx = 0;	//the corresponding optimal re-
for (int j = 0; j < n; j ++)	sources
{	// start from the virtual service re-
if (Value_j >= Value_idx)	sources
idx = j;	
}	//determine whether it's better
distri[i] = idx;	// update better resources
vmTask_i += cloudletTasks_i;	
}	//record the corresponding resources
	// update the utility value of vmTaski

There also exist several parameters to evaluate the performance of load balancing in the service resource allocation process. They are defined as follows:

Definition 4. AET indicates the average execution time of the unit task.

$$AET = \frac{max\{tce_i\} - min\{tcs_j\}}{n} \tag{5}$$

Where $max\{tce_i\}$ indicates the end time of the last task, $min\{tcs_j\}$ represents the start time of the first task, and n represents the number of tasks.

Definition 5. LBV indicates the load balancing value of the unit task. The bigger the LBV, the higher the volatility, and the worse the ability of load balancing, and vice versa.

$$LBV = \frac{\sqrt{\frac{\sum_{i=1}^{m}(tvm_i - avmc)^2}{m}}}{n} \tag{6}$$

Where tvm_i indicates the load value of virtual service resources vm_i; $avmc$ indicates the average load value of virtual service resources. m represents the number of virtual service resources; n represents the number of tasks.

Definition 6. RV represents the ratio value of the number of tasks and the number of resources.

$$RV = \frac{n}{m} \tag{7}$$

Where n represents the number of tasks, and m represents the number of the virtual service resources.

5 Experiments

We conduct the set of simulation experiments to evaluate the three load balancing strategies discussed in the previous section. In these experiments, we choose CloudSim as the simulation platform. This choice was made because CloudSim not only supports the simulation framework of cloud computing resource management, but also supports modeling of objects in cloud computing.

We discussed the overall performance of the three strategies in the first experiment, and compared DWLC with WLC facing different number of tasks in the second experiment. The phenomena in these two experiments were analyzed in the last experiment. All the experiments are conducted in the simulation environment where the memory of the host is 3G; the hard is disk 650G; the CPU is 3.20 GHz; the OS is Windows XP; and the development tool is My Eclipse8.5 and jdk1.6.0 _10.

5.1 Experiment 1: Comparing DWLC with WLC

The overall performance of LC is inferior to that of DWLC and WLC, so LC will not be considered in the later experiments. Furthermore, during the first experiment, load balancing performance under DWLC and WLC is approximates when the number of virtual service resources is fixed. Therefore, in experiment 2, the number of virtual service resources is rising from 15 to 70, and the number of task is also changed as 200, 300, 400 and 500. Figure 2 shows the results of the average execution time of unit task AET under the above configuration.

In this experiment, the average execution time of unit task under DWLC is not longer than that of WLC. Furthermore, when the number of task exceeds 300, DWLC performs better than WLC with the increase of virtual service resources. The main

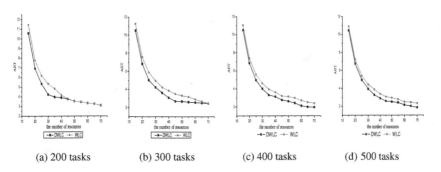

(a) 200 tasks   (b) 300 tasks   (c) 400 tasks   (d) 500 tasks

Fig. 2. The AET of DWLC and WLC

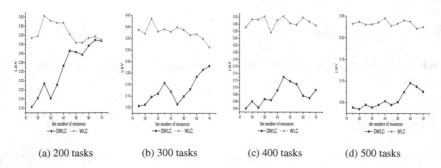

(a) 200 tasks   (b) 300 tasks   (c) 400 tasks   (d) 500 tasks

Fig. 3. The LBV of DWLC and WLC

reason is that DWLC considers the capacity of service resources and the task size as well. This helps DWLC to balance between large size task and small size one. The similar results of the load balancing value LBV are shown in Fig. 3, where the DWLC is at a definite advantage.

5.2   Experiment 2: Phenomena Analysis

In the Experiment 1, there is an interesting phenomenon. When the number of task rises from 50 to 200, DWLC and WLC both show an initial increase and then falling trend. So, we will analyze the reason of this phenomenon in experiments 3 where the number of virtual service resources is 50 and the number of task is rising from 50 to 200. The results are shown in Fig. 4.

At the beginning of this experiment, the number of tasks and the number of service resources are very close. When the number of tasks increases from 50 to 100, AET and LBV also become larger. Some service resources undertaking more tasks may account for load unbalancing. This situation is similar to the case of assigning 3 tasks to 2 service resources. The worst performance is that one resource takes all the 3 tasks and the other one takes none, while the best performance in this situation is that one resource takes 1 task and the other one takes 2. But all possible resource allocation results will cause a certain load unbalancing due to the ratio between the number of

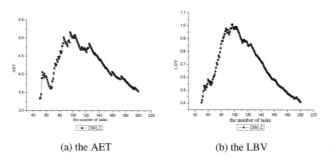

(a) the AET   (b) the LBV

Fig. 4. The AET and LBV of DWLC

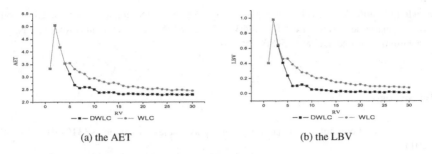

(a) the AET   (b) the LBV

Fig. 5. The AET and LBV of DWLC and WLC

tasks and the number of service resources (RV). To further analyze this situation, we investigate the process of RV ranging from 1 to 30 by using both DWLC and WLC load balancing strategies, and the results are shown in Fig. 5.

When RV is changing from 1 to 4, the AET and LBV of DWLC and WLC will fluctuate, and reach the maximum when RV is 2. Then the performance becomes better with the increase in RV, and DWLC performs much better than WLC. The similar phenomena can also be found in the experiment 2. For example, in the Fig. 3(a), the process of increasing service resources from 15 to 70 when facing 200 tasks can be viewed as drops RV from 13.33 to 2.85. The results in Fig. 3(a) are consistent with those in Fig. 5(b). This might be a disadvantage of DWLC, but taking into account the characteristics of the cloud computing, the RV is far beyond the interval [1, 4], and the fluctuations can be ignored in the real application. Therefore, the DWLC can meet the needs of users in an efficient way in case of limited service resources.

6 Conclusion

In this paper, we discussed the load unbalancing during the service resource allocation process in the cloud computing environment, and introduced service alliance into the information services model, where service alliance can help both service providers and service users to communicate with each other. This model provides a service management platform where service providers can deploy their service resources, and service users send their requests to service alliance in ignorance of the origin of the service resources. Then, a task-oriented load balancing strategy, called Double Weighted Least Connection, was proposed to improve the performance of service resource allocation, where the ability of service resources and the size of the task itself are considered. This helps service alliance to provide the service resources better. A set of experiment is discussed to compare DWLC with other load balancing strategies when the number of tasks and the number of resources are changing. The results also show that DWLC can play a better role in the case of large demands with limited resources.

Additionally, we are exploring other load balancing strategies, which can also save the energy during the process of service resource allocation. More and more factors need to be considered, such as the consumption of the resources, the utility of the task, the service quality and so on. Our long-term goal is to develop an efficient framework for load balancing during the service resource allocation process in cloud computing.

Acknowledgement. This work was supported in part by the National Natural Science Foundation of China under Grant 71401048, 71131002, 71472058, and by Anhui Provincial Natural Science Foundation under Grant 1508085MG140.

References

1. Fox, A.: Cloud computing – what's in it for me as a scientist? Science **331**(6016), 406–407 (2011)
2. Marston, S., Li, Z., Bandyopadhyay, S., et al.: Cloud computing - the business perspective. Decis. Support Syst. **51**(1), 176–189 (2011)
3. Amazon Web Service. http://aws.amazon.com
4. Google App. http://code.google.com
5. Salesforce. http://www.salesforce.com
6. Goscinski, A., Brock, M.: Toward dynamic and attribute based publication, discovery and selection for Cloud computing. Future Gener. Comput. Syst. **26**(7), 947–970 (2010)
7. Domingo, E.J., Niño, J.T., Lemos, A.L., et al.: CLOUDIO: a cloud computing-oriented multi-tenant architecture for business information system. In: The Third International Conference on Cloud Computing, pp. 532–533. IEEE (2010)
8. Wang, Y., Fu, T.Z., Chiu, D.M.: Design and evaluation of load balancing algorithms in P2P streaming protocols. Comput. Netw. **55**(18), 4043–4054 (2011)
9. Sousa, A.D., Santos, D., Matos, P., et al.: Load balancing optimization of capacitated networks with path protection. Electron. Notes Discrete Math. **36**(1), 1249–1256 (2010)
10. Foster, I., Zhao, Y., Raicu, I., Lu, S.: Cloud computing and grid computing 360-degree compared. In: The 2008 Grid Computing Environments Workshop, pp. 1–10. IEEE (2008)
11. Bao-yan, S., Nan, G., et al.: DLRD: a P2P grid resource discovery mechanism for dynamic load-balance. J. Commun. **29**(8), 94–99 (2008)
12. Wei, X., Xie, D.Q., Jiao, B.W., Liu, J.: Self-adaptive load balancing method in structured P2P protocol. J. Softw. **20**(3), 660–670 (2009)
13. Gao, A., Mu, D.J., Hu, Y.S.: Differentiated service and load balancing in web cluster. J. Electron. Inf. Technol. **33**(3), 555–562 (2011)
14. Lilun, Z., Hong, Y., Jianping, W., Junqiang, S.: Parallel load-balancing performance analysis based on maximal ratio of load offset. J. Comput. Res. Dev. **47**(6), 1125–1131 (2010)
15. Li, Y., Yang, Y., Ma, M., Zhou, L.: A hybrid load balancing strategy of sequential tasks for grid computing environments. Future Gener. Comput. Syst. **25**(8), 819–828 (2009)
16. Larroca, F., Rougier, J.L.: Minimum delay load-balancing via nonparametric regression and no-regret algorithms. J. Comput. Netw. **56**(4), 152–1166 (2012)
17. Wang, S.F., Zhou, Z., Wu, W.: A layered iterative load balancing algorithm for distributed virtual environment. J. Softw. **19**(9), 2471–2482 (2008)
18. Chen, Y.J., Lu, X.C.S., Zhi-Gang, X.Q.S.: A session-oriented adaptive load balancing algorithm. J. Softw. **19**(7), 1828–1836 (2008)
19. Liao, W.H., Shih, K.P., Wu, W.C.: A grid-based dynamic load balancing approach for data-centric storage in wireless sensor networks. J. Comput. Electr. Eng. **36**(1), 19–30 (2010)

Solving Flexible Job-Shop Scheduling Problem with Transfer Batches, Setup Times and Multiple Resources in Apparel Industry

Miguel Ortiz[1]([✉]), Dionicio Neira[1], Genett Jiménez[2],
and Hugo Hernández[3]

[1] Department of Industrial Engineering, Universidad de la Costa CUC,
Barranquilla, Colombia
{mortizl, dneiral}@cuc.edu.co
[2] Department of Industrial Engineering, Institución Universitaria ITSA,
Soledad, Colombia
gjimenez@itsa.edu.co
[3] Department of Business Management, Universidad Del Atlántico,
Puerto Colombia, Colombia
hugohernandezp@mail.uniatlantico.edu.co

Abstract. Apparel industry is characterized by the presence of flexible job-shop systems that have been structured to manufacture a wide range of customized products. However, Flexible Job-shop Scheduling is really challenging and even more complex when setup times, transfer batches and multiple resources are added. In this paper, we present an application of dispatching algorithm for the Flexible Job-shop Scheduling Problem (FJSP) presented in this industry. Days of delay, throughput, earlier date and monthly demand are used as rules of operation selection. A case study in apparel industry is shown to prove the validity of the proposed framework. Results evidence that this approach outperforms the company solution and other algorithms (PGDHS and HHS/LNS) upon reducing average tardiness by 61.1 %, 2.63 % and 1.77 % respectively. The inclusion of throughput in the model resulted in low tardiness for orders with high speed to make money. Promising directions for future research are also proposed.

Keywords: Flexible job-shop scheduling problem · Dispatching algorithm · Throughput · Apparel industry · Transfer batches · Setup times

1 Introduction

Empirical evidence indicates that customized goods offer bigger value than standardized goods [1]. In reply to this, some manufacturing enterprises offer a wide range of customized products to satisfy different customer requirements. However, this involves creating flexible production systems whose scheduling is highly complex. One of these systems is known as flexible job shop, defined as an extension of the classical job shop system which lets an operation to be processed by any machine from a specified set. This system has gained importance because companies need to produce more

© Springer International Publishing Switzerland 2016
Y. Tan et al. (Eds.): ICSI 2016, Part II, LNCS 9713, pp. 47–58, 2016.
DOI: 10.1007/978-3-319-41009-8_6

customized goods, which requires smaller batches, and machines capable to perform different operations.

Given the above, this arrangement can be found more often in apparel industry [1]. Nevertheless, the difficulty of addressing flexible job shop scheduling (FJSS) is a well-known and complex Non-Polynomial (NP) hard combinational optimization problem [3, 4]. According to Rakesh [2], the decisions to be made in a FJSS include the selection among optional machines on which to perform an operation or the selection among flexible process plan of a part-type. This is complemented with the objective of minimizing the makespan, average tardiness or mean flow time of parts [2].

FJSS is a prominent topic of study for researchers because of its theoretical, computational, and empirical significance since it was introduced. Owing to the complexity of FJSS, different solution techniques such as various meta-heuristics approaches, (Like Nature based heuristics, genetic algorithms, simulated annealing, among others) and heuristic approaches have been developed. The fact that a large number of small to medium companies operate at flexible job shops environment gives importance to the search for an efficient method to solve FJSS. Optimizing FJSS helps companies to increase its production efficiency; reduce cost and improve product quality [2]. It is also of particular interest to reduce average tardiness in high-throughput orders to ensure meaningful returns on investment (ROI) for companies with flexible job-shop systems since bad schedules could make clients return these kinds of orders, which generates superior affectations than low-throughput orders.

Particularly, this paper is organized as follows: Sect. 2 presents a literature review referred to some techniques that have been explored to solve FJSP. Section 3 describes dispatching algorithm with rules of operation selection, machine selection and robustness. Section 4 presents an application of this approach in apparel industry. As final point, in Sect. 5 conclusions are given.

2 Flexible Job Shop Problem (FJSP): A Literature Review

The main difference between the classical Job Shop Problem (JSP) and the Flexible Job Shop Problem (FJSP) lies in the fact that in JSP, each operation is processed on a predefined machine; on the other hand, each operation in the FJSP can be processed on one out of several machines [5, 6]. According to Demir and İşleyen [6] and Zhang et al. [5], the FJSP can be split into: "the routing sub-problem that assigns each operation to a machine selected out of a set of capable machines, the scheduling sub-problem that consists of sequencing the assigned operations on all machines in order to obtain a feasible schedule to minimize the predefined objective function".

Based on the survey of Sobeyko and Mönch [8], Brucker and Schlie [7] addressed the FJSP for the first time. They considered a job shop consisting of different multi-purpose machines. They proposed a polynomial algorithm for the special case of two jobs. Furthermore, Hurink et al. [9] used a tabu search approach to optimize the makespan (Cmax). A hierarchical algorithm for the flexible job shop scheduling problem was presented by Brandimarte [10] in which the operations were initially assigned to specific machines and a job shop scheduling problem was solved after this decision. In this research, the minimization of makespan and total weighted tardiness were considered as primary objectives. In addition to the formulation problem, another difficulty emerged with the problem solution when increased its size by including more machines, alternative machines, more jobs, and other considerations that turn out to be complex when have to be represented in accordance with the real production system. The need to solve this kind of formulations in a fast way, forces the use of meta-heuristics. Nevertheless, according to Calleja and Pastor [11] the advantage of having very capable machines might disappear.

It also important to indicate that flexibility has a positive effect on the manufacturing system performance. The processes should be flexible enough to overcome adversities related with operations, machines, and raw materials that potentially would alter the process performance and divert it from its goal (minimizing makespan, for example). According to this principle, Kim et al. [12] pointed that there are three types of flexibility to generate flexible process plans: operation flexibility, sequencing flexibility and processing flexibility. The inclusion of these flexibilities would make a process plan be better prepared to come through disturbances occurring at the shop floor. The main advantage of a flexible process plan relies on the fact that it helps to maximize flexibility preservation because the scheduler can choose an alternative way at any time during the production process. This allows the decision maker to react fast to any problem on the shop floor. The incorporation of flexibility in job-shop scheduling, increases dramatically the complexity of the problem that is already considered very hard and costly to solve. It is also important to highlight that these production schedules could be made many times in a week, reason by which it is important to find tools with quick solutions.

According to Wu [13], several researchers have studied the effects and relationships of flexibility on system performance. Different heuristic procedures such as dispatching rules, local search and meta-heuristics procedures have been applied to solve FJSS and find near optimal schedules. Some works are mentioned here. For example, Tanev et al. [14] proposed a hybrid evolutionary algorithm involving priority dispatching rules with a GA to generate a schedule for a flexible job-shop system (FJS) in a plastic injection process. One of the most utilized tools for solving the FJSS is genetic algorithms. Many authors have implemented genetic algorithms with different variations for this purpose. One of these approaches is presented by Pezzella et al. [15] where different strategies are integrated to select an initial population and individuals for reproduction are selected. The results evidenced that the integration of more strategies in a genetic approach leads to a better performance. On the other hand, Gao et al. [16] proposed a hybrid variable neighborhood descent genetic algorithm with the following objectives: makespan, maximal machine workload, and total workload. A combined adaptive randomized decomposition large neighborhood search scheme was also proposed by Pacino and Van Hentenryck [17] with the makespan minimization as the objective.

Many other authors addressed the FJSP with the use of dispatching rules, and heuristics like shifting bottleneck. Finally, it is important to point that Calleja and Pastor [11] developed a dispatching rule-based algorithm to minimize average tardiness for FJSP having transfer batches. This heuristic is used in this case study with the inclusion of throughput as second rule of operation selection [18, 19] since this measure has not been considered inside these algorithms. In conclusion, literature review reveals that the effect of the flexible process plans on average tardiness and other performance measures in job shop scheduling, has not been studied deeply, reason by which this paper represents a relevant contribution to the development of this research field.

3 Dispatching Algorithm for FJSP

This section describes the FJSP problem and the dispatching algorithm proposed by Calleja and Pastor [11] with the inclusion of throughput as second rule of operation selection to solve this problem in apparel industry. Throughput is included to ensure a minimum tardiness for orders with high speed to make money.

3.1 Problem Description

The Flexible Job-shop Problem may be defined as follows: Let $O = \{O_1, O_2, \ldots, O_n\}$ be a set of n orders, which can be performed by m machines $M = \{M_1, M_2, \ldots, M_m\}$. Each order $O_i(1 \leq i \leq n)$ consists of a predefined sequence of n_i operations $P_{ij}(j = 1, 2, \ldots, n_i)$. Each operation P_{ij} has to be manufactured on a machine chosen from a given set of available machines. The allocation of the operation P_{ij} to the machine $M_k(1 \leq k \leq m)$ involves the occupation of the last one at a time defined as t_{ijk} and R_{ki} has been defined as availability time of operation (j, i). On the other hand, fp_k has been denoted as the earlier time to start a new operation on the machine k (if no queue, infinite value is assigned).

In this paper, the objective is to minimize average tardiness of orders and get the lower tardiness scores in those with high throughput. The preliminary conditions considered in this system have been summarized as follows:

- Each machine can manufacture one operation at a time.
- Setup times and transfer batches are considered.
- Each operation can be performed without interruption on one of the available machines of the set.
- Priorities are assigned to the orders according to the next criteria and order of priority: days of delay, throughput, earlier date and monthly demand.
- All orders are released at time 0 and all machines are available at this time.
- Breakdowns are not taken into account.
- The sequence of operations for each order is predefined.

3.2 Steps of Dispatching Algorithm

3.2.1 Start

- Put the first operations of the parts with their respective $r_{1,i}$ values in the subset of candidate operations (E_i)
- For each machine k, estimate fp_k value.
- Calculate fp_{min} and its respective machine.

3.2.2 Machine Selection

- If $fp_{min} = \infty$, all operations have been scheduled.
- Otherwise, select a machine according to fp_{min} (Rule 1: Order with more days of delay). In case of a tie, select one according to Rules 2 (Order with the highest throughput), 3 (Order with earlier delivery date) and 4 (Order with the highest monthly demand) of operation selection.
- When selecting the machine, create the subset of candidate operations that is composed by the eligible operations that the machine is able to perform (E_j).

3.2.3 Operation Selection

- If there is only one candidate operation, this must be scheduled.
- Otherwise, apply the priority rules for operation selection to choose the operation to program.

3.2.4 Update

- Schedule the selected operation (j, i) setting its initial $(t_{start}(j, i))$ and final $(t_{final}(j, i))$ time (See Eqs. (1) and (2)).

$$t_{start}(j, i) = rp_{j,j,k}. \tag{1}$$

$$t_{final}(j, i) = t_{start}(j, i) + Dp_{j,i,k}. \tag{2}$$

Where D represents the number of ordered pieces and $p_{j,i,k}$ is the unit processing time of operation j of the order i in the machine k.
- Place the eligible operation in the subset of schedule operations with its start and final times.
- If it is not the final operation of order i, move its next operation from N_i (Unscheduled and unavailable operations) to E_j subset.

Transfer batches. Taking into account that k' is the machine associated to the next operation of order i, Q is the transfer batch size and t_q is the necessary time to move a transfer lot size Q, release date $r_{j+1,i}$ can be calculated as described in Table 1:

- Update f'_k values of the machine k'. If the machine k' has not already used, $f_k = 0$. If any operation in machine k has been scheduled, then, $f_k = t_{final}(j, i)$, i.e., the machine j will have an availability time f_k that is equal to final time of the last scheduled operation in the machine k.
- Calculate $rp_{j,i,k}$ according to Eq. (3):

$$rp_{j,i,k} = máx(r_{j,i}, f_k). \tag{3}$$

Table 1. Formulas to calculate release dates with transfer batches

Relation between k and k'	Relation between $p_{(j, i, k)}$ and $p_{(j+1, i, k)}$	$r_{(j+1, i, k)}$
$k = k'$	All possible relations between $p_{j,i,k}$ and $p_{j+1,i,k}$	$t_{final(j,i)}$
$k \neq k'$	$p_{(j+1,i,k')} \geq p_{(j,i,k)}$	$tstart(j, i) + (tq + Qp(j, i, k))/60$
	$p_{(j+1,i,k')} < p_{(j,i,k)}$	$tfinal(j, i) - Dp_{(j,i,k)} + (tq + Qp(j, i, k))/60$

- Calculate fp_{min} according to Eq. (4)

$$fp_{min} = \min\left(rp_{j+1,i}\right). \tag{4}$$

- Return to step 1.

3.2.5　Objective Function

Calculate average tardiness according to Eq. (5):

$$\text{Average tardiness} = \sum_{i=1}^{n} máx(0, Ci - di)/n. \tag{5}$$

C_i, represents the completion time of the order i and d_i is the delivery date.

4　A Case Study in Apparel Industry

A case study in a company from apparel industry is shown in this paper to prove the validity of the proposed approach. This company presents a production system with 6 processes as shown in Fig. 1.

In this application, 35 orders were considered with $Q = 25$ units and $t_q = 0.16$ min. PRINTING process has 1 machine (H), LONGITUDINAL SECTION is composed by 2 worker-machines (CO1, CO2), SIDE SEAM is integrated by 4 worker-machines (CL1, CL2, CL3, CL4), CROSS CUT has 2 worker-machines (CT1, CT2), HEAD SEWING is composed by 7 worker-machines (CC1, CC2, CC3, CC4, CC5, CC6, CC7) and CLEANING has 4 worker-machines (L1, L2, L3, L4). Each product has its own production sequence as shown in Table 2. Black cells indicate that the product does not need to be processed in that stage. Table 3 shows the orders received by the company in each product reference. Each cell indicates the order size and the delivery date of the order where "A" means August, "S" September and "O" October. In this way, "1A" indicates August 1st. If the cell contains a number in brackets, this number denotes the days of the delay corresponding to the order.

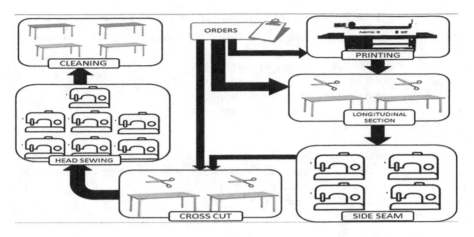

Fig. 1. Layout of flexible job-shop studied in apparel industry – Case study

Table 2. Processing times in min for each product reference in each process

PRODUCT REFERENCE	PRINTING	LONGITUDINAL SECTION	SIDE SEAM	CROSS CUT	HEAD SEWING	CLEANING
30 X 50		0.02	0.43	0.06	0.75	0.67
35 X 60		0.02	0.43	0.06	0.75	0.14
50 X 90		0.02	0.75	0.07	1	0.42
35 X 60 R		0.02	0.43	0.06	0.75	0.14
33 X 56 P	0.08	0.02	0.43	0.06	0.75	0.14
70 X 140				0.1	1.2	0.67
60 X 120				0.083	1.2	0.67
30 X 30		0.01	0.38	0.04	0.9	0.13
30 X 20 P	0.04	0.01	0.38	0.04	0.9	0.13
60 X 120 P	0.19			0.083	1.2	0.67
60 X 120 G				0.083	1.2	0.67
35 X 60 G		0.02	0.43	0.06	0.75	0.14
30 X 30 G		0.01	0.38	0.04	0.9	0.13
60 X 120 H		0.04	1	0.083	1.2	0.67
70 X 140 H		0.045	1.2	0.1	1.2	0.67
70 X 140 HG		0.045	1.2	0.1	1.2	0.67

On the other hand, Table 4 defines the throughput and monthly demand of each product reference to be considered at the moment of scheduling. It is noticed that product references 60 × 120 and 70 × 140 present the highest throughput rates which has to be taken into account by the algorithm to ensure low tardiness scores. Products with high throughputs represent better returns on investment (ROI) for the company.

Table 3. Orders per product reference and delivery dates

Product reference	Order 1	Order 2	Order 3	Order 4	Order 5	Order 6
30 × 50	12000 – 24S	12000 – 4O	4797 – 14S			
35 × 60	7800 – 1A (9)	2000 – 1A (2)	6000 – 9A	7800 – 26S		
50 × 90	24000 – 1A (5)	8000 – 28S	6000 – 28S			
35 × 60 R	24000 – 1A (8)	13405 – 16S	8500 – 16S			
33 × 56 P	20400 – 28S					
70 × 140	2400 – 1A (2)					
60 × 120	2400 – 1A (4)					
30 × 30	17000 – 4O	5000 – 26S	2000 – 22S	5000 – 23S	2000 – 12S	10000 - 2O
30 × 20 P	12600 – 4O	4400 – 4O				
60 × 120 P	5351 – 30S	15000 – 2O				
60 × 120 G	406 – 12S	1600 – 15S				
35 × 60 G	1600 – 15S					
30 × 30 G	1600 – 15S					
60 × 120 H	200 – 5S	150 – 9S				
70 × 140 H	100 – 5S					
70 × 140 HG	100 – 12S	200 – 12O				

Furthermore, Table 5 shows the first seven operations scheduled by dispatching algorithm. It can be noted that LONGITUDINAL SECTION operation for the product reference 35 × 60 (Order 1) has been selected as the first scheduled operation since this order presents 9 days of delay. At this time, both first operations and machines are available. In total, 167 operations were programmed under this algorithm and priority rules.

Figure 2 shows a comparative graph among proposed algorithm; company solution, Pareto-based grouping discrete harmony search algorithm (PGDHS) [20] and the integrated HHS/LNS approach [21]. It is shown that the proposed algorithm obtains the best results for tardiness performance measures since they are equal to or less than provided by company solution and the other algorithms. Average tardiness was improved by proposed algorithm with 3.17 days; while company solution, HHS/LNS approach and Pareto-based grouping discrete harmony search algorithm offered a schedule with 8.14, 3.26 and 3.23 days respectively. This represents an improvement percentage of 61.1 % which is meaningful for the company. Deviation standard of tardiness was reduced from 13.88 days to 11.2 days (Improvement percentage: −19.2 %). Compared to PGDHS and HHS/LNS, average tardiness was reduced by 2.63 % and 1.77 % correspondingly.

Upon evaluating the number of late orders (although it is not the primary aim of this algorithm), dispatching algorithm presented 6 late orders (17.1 %); while results from company solution showed 25 (71.4 %). Finally, tardiness of orders with high

Table 4. Throughput and monthly demand of each product reference

Product reference	Throughput ($/min)	Monthly demand (units/month)
30 × 50	13530	1360
35 × 60	18942	10991
50 × 90	21250	26145
35 × 60 R	20951	2229
33 × 56 P	16669	16669
70 × 140	71400	1998
60 × 120	39827	5044
30 × 30	20062	1407
30 × 20 P	5986	2436
60 × 120 P	39827	1387
60 × 120 G	88560	274
35 × 60 G	25830	1302
30 × 30 G	12177	37
60 × 120 H	110208	140
70 × 140 H	120540	179
70 × 140 HG	150005	86

Table 5. First 7 operations scheduled by dispatching algorithm with the inclusion of throughput

Product reference	Operation	Order number	Possible resources	Selected resource	tstart (h)	tfinal (h)
35 × 60	LONGITUDINAL SECTION	1	CO1, CO2	CO1	0	2,68
35 × 60 R	LONGITUDINAL SECTION	1	CO2	CO2	0	8,08
60 × 120 P	PRINTING	1	H	H	0	17,94
35 × 60	SIDE SEAM	1	CL1, CL2, CL3, CL4	CL1	0,01	55,94
35 × 60 R	SIDE SEAM	1	CL2, CL3, CL4	CL2	0,01	172,04
50 × 90	LONGITUDINAL SECTION	1	CO1	CO1	2,68	10,76
50 × 90	SIDE SEAM	1	CL3, CL4	CL3	2,69	302,72

throughput was also assessed in the study through a test for differences between means. With T = −2.07 and P-value of 0.03 (α = 0.05), it can be said that average tardiness of high-throughput orders of proposed algorithm (0.36 days) is statistically less than provided by company solution (1.82 days). A comparative graph (Fig. 3) for this evaluation between proposed method (PM) and company method (CM) is shown to confirm the statistical results.

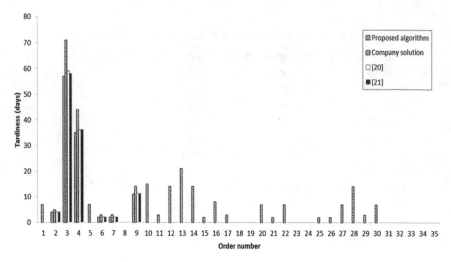

Fig. 2. Comparison of proposed algorithm, company solution and other algorithms

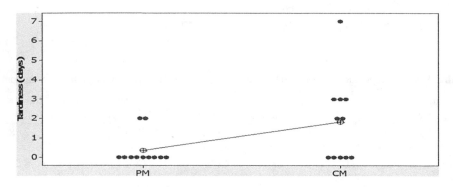

Fig. 3. Comparison between proposed algorithm and company solution for high-throughput orders

5 Conclusions

In this study, a dispatching algorithm [11] with the inclusion of throughput as second rule of operation selection has been proposed to solve FJSP in apparel industry. Application results prove the validity of the proposed approach with 61.1 % improvement in average tardiness, 19.2 % reduction in standard deviation of tardiness, 54.3 % reduction in late orders and a minor average tardiness for high-throughput orders in a company from apparel industry. It also offers better performance of tardiness measures than provided by PGDHS and HHS/LNS with 2.63 % and 1.77 % reduction correspondingly. This signifies a cost reduction for the company since some penalties for non-compliance are avoided. In addition, customer satisfaction is increased because of the reduction in average tardiness of orders. Lower average tardiness in high-throughput orders are less likely to be returned by clients; so that high return on investments can be ensured. For future research, it is recommended to explore scheduling changes upon varying transfer lot size; moreover, breakdowns can be included in order to evaluate strategies that allow reducing non-compliance, possible complaints and cost overruns.

References

1. Neufeld, J.S., Gupta, J.N., Buscher, U.: A comprehensive review of flowshop group scheduling literature. Comput. Oper. Res. **70**, 56–74 (2015)
2. Phanden, R.K., Jain, A.: Assessment of makespan performance for flexible process plans in job shop scheduling. In: IFA, pp. 1948–1953 (2015)
3. Conway, R.W., et al.: Theory of Scheduling. Addison Wesley, Reading (1967)
4. Mitchell, M.: An Introduction to Genetic Algorithms. PHI, New Delhi (2002)
5. Zhang, G., Gao, L., Shi, Y.: An effective genetic algorithm for the flexible job-shop scheduling problem. Expert Syst. Appl. **38**(4), 3563–3573 (2011)
6. Demir, Y., İşleyen, S.K.: Evaluation of mathematical models for flexible jobs-shop scheduling problems. Appl. Math. Model. **37**(3), 977–988 (2013)
7. Brucker, P., Schlie, R.: Job-shop scheduling with multi-purpose machines. Computing **45**(4), 369–375 (1990)
8. Sobeyko, O., Mönch, L.: Heuristic approaches for scheduling jobs in large-scale flexible job shops. Comput. Oper. Res. **68**, 97–109 (2015)
9. Hurink, J., Jurisch, B., Thole, M.: Tabu search for the job-shop scheduling problem with multiple-purpose machines. OR Spektrum **15**, 205–215 (1994)
10. Brandimarte, P.: Routing and scheduling in a flexible job shop by tabu search. Ann. Oper. Res. **41**, 157–183 (1993)
11. Calleja, G., Pastor, R.: A dispatching algorithm for flexible job-shop scheduling with transfer batches: an industrial application. Prod. Plann. Control: Manag. Oper. **25**(2), 93–109 (2014)
12. Kim, Y., Park, K., Ko, J.: A symbiotic evolutionary algorithm for the integration of process planning and job shop scheduling. Comput. Oper. Res. **30**, 1151–1171 (2003)
13. Wu, Z.: Multi-agent workload control and flexible job shop scheduling. Ph.D. thesis, University of South Florida (2005). http://scholarcommons.usf.edu/etd/921
14. Tanev, I.T., Uozumi, T., Morotome, Y.: Hybrid evolutionary algorithm-based real-world flexible job shop scheduling problem: application service provider approach. Appl. Soft Comput. **5**(1), 87–100 (2004)
15. Pezzella, F., Morganti, G., Ciachetti, A.: Genetic algorithm for the flexible job-shop scheduling problem. Comput. Oper. Res. **35**(10), 3202–3212 (2008)
16. Gao, J., Sun, L., Gen, M.: A hybrid genetic and variable neighborhood descent algorithm for flexible job shop scheduling problems. Comput. Oper. Res. **35**(9), 2892–2907 (2008)
17. Pacino, D., Van Hentenryck, P.: Large neighborhood search and adaptive randomized decompositions for flexible jobshop scheduling. In: Proceedings of the Twenty-Second International Joint Conference Artificial Intelligence, pp. 1997–2002 (2011)
18. Barrios, M.: Teoría de restricciones y modelación PL como herramientas de decisión estratégica para el incremento de la productividad en la línea de toallas de una compañía del sector textil y de confecciones. Prospectiva **11**(1), 21–30 (2013)

19. Herazo-Padilla, N., Montoya-Torres, J.R., Isaza, S.N., Alvarado-Valencia, J.: Simulation-optimization approach for the stochastic location-routing problem. J. Simul. 9(4), 296–311 (2015)
20. Gao, K.Z., Suganthan, P.N., Pan, Q.K., Chua, T.J., Cai, T.X., Chong, C.S.: Pareto-based grouping discrete harmony search algorithm for multi-objective flexible job shop scheduling. Inf. Sci. 289, 76–90 (2014)
21. Yuan, Y., Xu, H.: An integrated search heuristic for large-scale flexible job shop scheduling problems. Comput. Oper. Res. 40(12), 2864–2877 (2013)

A Comparative Analysis of Genetic Algorithms and QAP Formulation for Facility Layout Problem: An Application in a Real Context

Fabricio Niebles[1(✉)], Ivan Escobar[2], Luis Agudelo[2], and Genett Jimenez[3]

[1] Department of Business Management, Institucion Universitaria ITSA, Soledad, Colombia
fabniebles@itsa.edu.co
[2] Department of Industrial Engineering, Universidad de la Sabana, Bogotá, Colombia
{ivan.escobar,mauricio.agudelo}@unisabana.edu.co
[3] Department of Industrial Process Engineering, Institucion Universitaria ITSA, Soledad, Colombia
gjimenez@itsa.edu.co

Abstract. This paper considers the problem of locating facilities in manufacturing of electrical, telecommunications and building products. This is known as the Facility Layout Problem (FLP). This NP-hard problem has been largely studied in the scientific literature, and exact and approximate (heuristic and meta-heuristic) approaches have been used mainly to optimize one or more objectives. However, most of these studies do not consider real applications. Hence, in this work, we propose the use of Sule's Method and genetic algorithms, for facility layout in a real industry application in Colombia so that the total cost to move the required material between the facilities is minimized. As far as we know, this is the first work in which Sule's Method and genetic algorithms are used simultaneously for this combinatorial optimization problem. Computational experiments are carried out comparing the proposed approach versus QAP formulation. Additionally, the proposed approach was tested using well-known datasets from the literature in order to assure its efficiency.

Keywords: Genetic algorithms · Quadratic assignment problem · Facility layout problem

1 Introduction

In order to improve competitiveness, companies have to constantly improve their processes. Regarding to manufacturing operations (production and logistics), formal procedures play an important role in an effective resource management [1]. Therefore, Facility Layout Problem (FLP) is considered one of the most critical problems for a facility layout design [2]. Companies must tackle this issue when starting operations in a new facility, or when the process or facility has an important alteration. The problem, basically consists in the allocation of several resources in different locations [3]. In this

© Springer International Publishing Switzerland 2016
Y. Tan et al. (Eds.): ICSI 2016, Part II, LNCS 9713, pp. 59–75, 2016.
DOI: 10.1007/978-3-319-41009-8_7

problem numerous factors have to be considered, but distance between the resources and the flow of materials are considered the most relevant issues when solving FLP [4]. Bearing this in mind Dileep Sule proposed a method to determine a facility layout [5]. The objective of his proposal is material and people movements' reduction.

This paper considers analyzing the problem of locating facilities in a new manufacture facility that is going to make diverse products for electrical, telecommunications and building infrastructures. The company of this case study is planning to build a new facility and has estimated the allocation of the different assets and areas using basic tools as the spaghetti diagram. In this work, the use of Sule's Method and Genetic Algorithms for solving FLP is proposed in order to minimize the total cost to move the required material between the facilities. To the best of our knowledge, this is the first work in that is used Sule's Method and genetic algorithms simultaneously for this combinatorial optimization problem.

2   Literature Review

It has been proven that the FLP belongs to the class NP-hard [6], which means that this problem cannot be solved to optimality within polynomial bounded computation times. Even for smaller size problems, i.e. number of facilities less than or equal to 20 [7], the exact and conventional methods of resolution such as linear programming, integer and mixed programming, among others, are not efficient in terms of computing time to reach the optimal solution.

There are many academic works regarding to the utilization of exact and approximate (heuristic and meta-heuristic) approaches for solving the Facility Layout Problem considering one or more objectives. Among them: [8–10] in which it became clear that in order to deal with big problems (real life facility problems) the best way is to address this optimization problem with metaheuristics such as genetic algorithm, ant colony optimization among others. They present different surveys that expose various examples of methods to solve the Facility Layout Problem. In addition, [11] reviewed the state-of-the-art papers on facility layout problem with quadratic assignment model and mixed integer-programming models. More exhaustive surveys of the heuristic algorithms for the QAP can be found in [12–14]. However, due to the intrinsic complexity in Facility Layout Problem, which are of the NP-hard type -like we previously said- the attention of the researchers is focused on the development of heuristics and meta-heuristics for solving this problem with the less computational effort. According to [15] some of the most successful applications of metaheuristics for solving QAP heuristics do not demonstrate the same performances when the domain and/or the structure of the problem is changed and came to this conclusion after comparing the research from 1990 to 2014. Other applications of metaheuristics for solving QAP are described in other researches. In particular: ant colony optimization in [16–19]; evolution strategies [20], genetic algorithms [21–26]; greedy randomized adaptive search procedures [27]; hybrid heuristics [28–30]; iterated local search [31]; simulated annealing [32, 33]; tabu search and very large-scale neighborhood search [34, 35]. On the other hand, the CRAFT (Computerized Relative Allocation of Facilities Technique), used for the layout of facilities was first introduced by [36].

Due to the hardness of the QAP for heuristic methods [37], this problem is considered suitable as a testing platform for innovative intelligent optimization techniques or improvement methods like metaheuristics [38]. Hence, the design of the enhanced heuristic approaches for the QAP, remains as an active research field.

3 Problem Formulation

In this research, the FLP was tackled with the QAP formulation, which is useful to model the problem of allocating n workstations to n locations. The objective considered in this study was to minimize the associated cost with the distance and material flow between the facilities. Each of these assignments implied a certain cost. The determining factors for cost assignment are distance and the flow of materials between facilities. As these two are determining factors, it is essential to generate the mathematical formulation of the problem based on them, in the following way:

Parameters:
n: Amount of workstations and locations
f_{ij}: Flow of materials from workstation i to workstation j
d_{kl}: Distance between location k and location l
r_{ij}: Proximity relation between workstation i and workstation j

$$Zmin = \sum_i \sum_j \sum_k \sum_l f_{ij} d_{kl} W_{ik} W_{jl} + \sum_i \sum_j \sum_k \sum_l C_{kl} W_{ik} W_{jl} \tag{1}$$

Considering:

$$if \; r_{ij} < 0 \rightarrow C_{kl} = |r_{ij}| \div d_{kl} \tag{2}$$

$$if \; r_{ij} \geq 0 \rightarrow C_{kl} = r_{ij} \cdot d_{kl} \tag{3}$$

Variables:

$$w_{ik} = \begin{cases} If \; workstation \; i \; is \; assigned \; to \; location \; k \rightarrow 1 \\ Otherwise \rightarrow 0 \end{cases} \tag{4}$$

$$W_{jl} = \begin{cases} If workstation j is assigned to location l \rightarrow 1 \\ Otherwise \rightarrow 0 \end{cases} \tag{5}$$

Constraints:

$$\sum_i W_{ik} = 1 \tag{A}$$

$$\sum_k W_{ik} = 1 \tag{B}$$

$$\sum_j W_{jl} = 1 \tag{C}$$

$$\sum_l W_{jl} = 1 \tag{D}$$

$$W_{ik} = W_{jl}; \; if \; i = j \, and \, k = l \tag{E}$$

This is a nonlinear mixed integer programming formulation. The first part of the formulation (1) indicates the total flow cost between the workstations taking into account that workstation i is assigned to facility k and workstation j is assigned to facility l. The second part of the formulation forces the assignment of workstations, which r_{ij} is negative to the farthest possible location. On the other hand, the constraints (A), (B), (C) and (D) assure that each workstation is assigned to a location and that each location is assigned to just one workstation. The constraint (E) guarantees that both assignments are the same in both binary variables (Table 1).

Table 1. Relationship nomenclature (from Sule [4])

Closeness relationship associated cost	
Absolutely necessary	50
Primarily important	30
Important	20
Ordinarily important	10
Without importance	0
Not desirable	−10

4 Approach Proposed

4.1 Sule's Conventional Method

In order to generate a preliminary facility distribution, [4] suggests the completion of a series of steps: (the tables and figures shown at this point are the results obtained from the practical exercise). First, is required to establish the necessary content for each workstation, determining its area. The sum of these areas will determine the total required area. For this study, it is needed to make sure that the total required area does not exceed the total available area. Second, it has to be determined the amount of material to be moved between workstations or f_ij by using a single unit of measurement in order to handle raw materials in a generic way, as well as product in process and finished goods. For the case of study of this paper, the unit Kg/Hour is used. Later, the closeness can be determined by the flow of materials, personal needs in multiple workstations, communication requirements, security restrictions and any other aspect to be considered. Finally, a graphic display of the relationships table is generated. To do this, it is necessary to illustrate an initial arrangement of workstations by using a nodal diagram. Later, a grid or net representation in which the initial arrangement can be seen in the form of blocks is generated. The representation of the solutions is shown in a matrix, as follows (Table 2):

$$\begin{bmatrix} 8 & 5 & 3 \\ 6 & 1 & 2 \\ 4 & 9 & 7 \end{bmatrix}$$

The location of each working station can be represented through the matrix position. For instance, working station 8 is located in position 1.1 at the matrix and working station 7 is located in position 3.3 (Table 3).

Table 2. Available area vs required area.

Area requirement			
Facility	Length (m)	Width (m)	Area (m^2)
1	20	8	160
2	20	8	160
3	38	4	152
4	16	4	64
5	16	16	256
6	15	3	45
7	7	1.4	9.8
8	8.4	1.4	11.76
9	13	1	13
10	18	9	162
11	5	5	25
12	3	3	9
13	3	3	9
14	5	5	25
15	2	2	4
16	20	2	40
17	4	6	24
18	11	5	55
19	7	1.4	9.8
20	5	2	10
21	12	4	48
22	12	4	48
Total required área			1340.36
Total available área			1482

4.2 Genetic Algorithm

A Genetic Algorithm (GA) is a problem solving technique that uses the concepts of evolution and hereditary to produce good solutions to complex problems that typically

Table 3. Example of Sule's performance

have enormous search spaces and are therefore difficult to solve. Figure 1 shows a pseudocode that illustrates the general phases of a GA. The biggest difference with other meta-heuristics is that GA maintains a population of solutions rather than a unique current solution. Solutions are coded as finite-length strings called chromosomes and a measure of their adaptation (the fitness) is computed by an engine. Starting from an existing population representing the initial solution of the problem, a set of iterations generate new chromosomes (solutions) by applying crossover and mutation operators, according to a probability, to two chosen parents. The main advantage of GA is its intrinsic parallelism, which allows the exploration of a larger solution space.

Parent 1	4	10	6	2	1	3	9	5	8	7

Parent 2	1	2	3	4	5	6	7	8	9	10

Fig. 1. Illustration of parent's chromosomes

4.2.1 Solution Representation and Initial Population

In a broad way, the genetic algorithm presented here is an optimization procedure that seeks to minimize the total cost of facility layout design proposed. Once the values of decision variables are found, the total cost is computed by the procedure shown in Fig. 2.

The structure of each individual in the solution is a chain of chromosomes, each one giving the values of decisions variables (see problem formulation section) for a specific layout conditions. That is, we have a total of v chromosomes, each with m × z genes representing the location of each working station obtained by Sule's method.

Algorithm 1. Pseudocode of Genetic Algorithm Proposed
Begin
    Prompt parameters;
    Generate Initial Population;
    Evaluate Objective Function;
    While Stop_criteria=False
        Calculate Fitness Function;
        Selection operator;
        Crossover operator with probability=Pc;
        Mutation operator with probability= Pm;
        Evaluate Objective Function;
    End
    Print Best Individual Solution;
End

4.2.2 Selection, Crossover and Mutation

The selection procedure selects the best individuals to be considered for the next generation. In our procedure, the number of selected individuals is limited by the size of population and by the constraints of the problem (i.e., capacity constraints). The individuals with best values of the fitness function are selected.

Partial Mapped Crossover (PMX) is done over all individuals of the population and elitism procedure is carried out in order to choice the two best solutions that will generates the next solutions. In addition, with the PMX operator is possible registering the information (genetic code) of best solutions, with the aim to compare it with the next generation. This procedure is explained as follows:

First, the two best solutions -with higher value of elitism function- are selected from the set of initial solutions.

After, from the Parent 1, is selected a random section of consecutive genes and copied to Child 1 (Fig. 2). Next, for the positions of the same section in the Parent 2 is selected the values that have not been copied to Child 1 (Fig. 3):

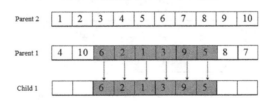

Fig. 2. First step of PMX crossover operator

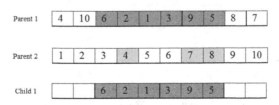

Fig. 3. Second step of PMX crossover operator

For each one of these values:

i. Is kept the index of this value in the Parent 2 and is searched the value of the Parent 1 in this same position (Fig. 4)
ii. Search this same value in the Parent 2 (Fig. 5)
iii. If the index of this value in the Parent 2 is part of the original section, go to I using this same value (Fig. 6)
iv. If the position is not part of the original section, assign the value in this position of Child 1 (Fig. 7)
v. Finally, is copied to Child 1 the rest of positions of Parent 2 (Fig. 8)

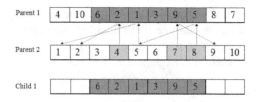

Fig. 4. Third step of PMX crossover operator

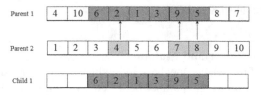

Fig. 5. Fourth step of PMX crossover operator

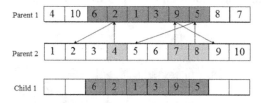

Fig. 6. Fifth step of PMX crossover operator

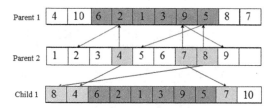

Fig. 7. Sixth step of PMX crossover operator

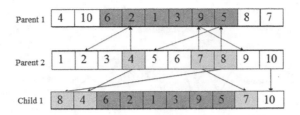

Fig. 8. Seventh step of PMX crossover operator

To generate other Child, this procedure is repeated; this time, the random section is taken from Parent 2. After crossover, it is necessary to verify that the resulting individuals correspond with a feasible solution.

On the other hand, Mutation operator is defined as an interchange operation. This means that two positions of the chromosome are interchanged as follows: the first position with the third one, the third position with the last one, etc. (see Fig. 3). Chromosomes to be mutated are those with the lowest value of the total cost (objective function) after the selection process. At the same way of crossover, after mutation, it is necessary to verify that the resulting individuals corresponds with a feasible solution (Fig. 9).

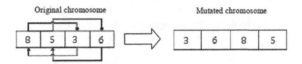

Fig. 9. Example of mutation operation

4.2.3   Fitness Function

The fitness function allows the algorithm to compare the quality of the individuals in the population (i.e., to evaluate the quality of the different solutions). Let $f(x1),\ldots, f(xn)$ be the values of the objective function for each individual. Since the objective is to minimize the total cost, chromosomes with the lowest probability are selected. That is, the lower the value of $f(xh)$. The higher the probability ph of being selected. Hence, individual xh will be selected for reproduction.

5   Experimental Environment, Parameter Settings and Results

This section first describes the process employed to setup the parameters of the GA. Second, the datasets employed for the extended experimental study. Finally, the analysis of results is presented, the convergence of the GA proposed as well as a comparison with an exact method which is the QAP formulation. The algorithm was coded using Visual Basic for Applications (VBA) ®. Experiments were carried out on a PC with processor Intel Pentium ® Dual Core with 2.40 GHz and 4.0 GB of RAM.

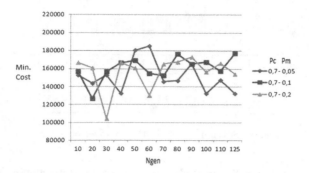

Fig. 10. Evolution of the total cost with PC = 0.7

In order to identify the effect of various factors on the response variables a 33 factorial experimental design was performed. The DOE was helpful to select, for these factors, the best option among multiple levels. The response variable was the total cost of the layout including flow and installation cost. Three factors were considered relevant for the pilot test relating to the quality of the solution. Those factors were crossover probability, mutation probability and amount of generations. Additionally, for each factor three levels were considered as follows (Table 4):

Table 4. Factor arrangements (for each level)

Factor	Level		
	−1	0	1
Crossover probability (Pc)	0.7	0.8	0.9
Mutation probability (Pm)	0.05	0.1	0.2
Number of generations (NGen)	5	10	20

5.1   Crossover Probability and Mutation Probability

Initially, no significant difference in the use of a level of probability (crossover or mutation) was found. Therefore, a review of recent literature was done in order to make a determine the crossover probability levels to be evaluated. The selected levels were 0,7; 0,8 y 0,9 and the mutation probabilities were 0,05; 0,1 y 0,2, just as described in [34]. These levels avoid the algorithm to converge prematurely to sub-optimal solutions.

On the other hand, from the insights given by these results, we next analyzed the behavior of the proposed genetic algorithm in terms of the objective function (total cost) in order to evaluate its convergence over the number of generations or iterations (see Figs. 11, 12 and 13). As shown in Fig. 11, there is not a clear convergence of the algorithm when the Pc = 0.7; the value of the total cost tends to improve when Pm = 0.2, and the initial solution is not improved at all in the case of Pm = 0.1. The value of the total cost moves within the range between USD 132467 and USD 177456, Pc = 0.7 with Pm = 0.05 and Pc = 0.7 with Pm = 0.1, respectively. Figure 11 shows

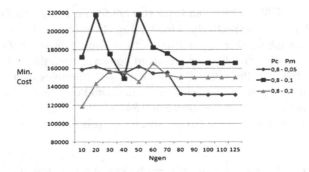

Fig. 11. Evolution of the total cost with Pc = 0.8

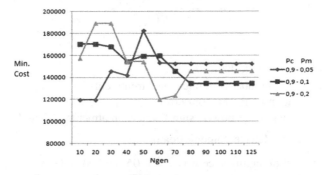

Fig. 12. Evolution of the total cost with Pc = 0.9

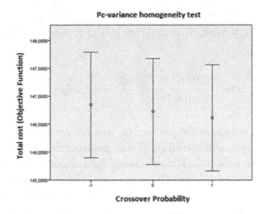

Fig. 13. Constant variance homogeneity

the evolution of the total cost value with Pc = 0.8 and different values of Pm. We observe a converge starting from the iteration number 80. For the case of Pc = 0.8 with Pm = 0.05, the solution value improves over the number of generations, while the opposite phenomenon occurs with Pc = 0.8 and Pm = 0.2. The value of the minimum cost moves between USD 131389 and USD 166099 for Pc = 0.8 with Pm = 0.05 and Pc = 0.8 with Pm = 0.1, respectively. Finally, Fig. 11 presents the evolution of the minimum cost over the number of iterations with crossover probability of 0.9. In this case the value of the objective function trends to converge from the 80th iteration. For the case of Pc = 0.9 with Pm = 0.1, the solution value improves over the number of generations, while the opposite phenomenon occurs with Pc = 0.9 and Pm = 0.05. The value of the minimum cost moves between 134500 and 152435 for Pc = 0.9 with Pm = 0.1 and Pc = 0.9 with Pm = 0.05, respectively (Fig. 10).

Sample Size:

According to [35] the sample size (n) can be calculated as follows:

$$n = \frac{2\left(t_{\left(\frac{\alpha}{2}, k \times n_0 - k\right)}\right)^2 \sigma^2}{dT^2} \tag{6}$$

Where:

k: number of treatments to be tested
n_0: proposal for the number of replicas per treatment
σ: standard deviation of the random error
$t_{\left(\frac{\alpha}{2}, k \times n_0 - k\right)}$: combination of power, significance, treatments and replicas.
dT: difference to be detected among treatments

The selected level of significance was $\alpha = 0.05$, and considering the 21 degrees of freedom it was obtained the following value of (T distribution of Student):

$$t_{\left(\frac{\alpha}{2}, k \times n_0 - k\right)} = 2,08 \tag{7}$$

Considering that $\sigma = 3030,70$ and dT = 1085,95; 68 is obtained the pertinent sample size:

$$n = \frac{2(2.08)^2 (3030.70)^2}{(1085.95)^2} = 67.37 \approx 68$$

In view of that the proposed sample size of was 68, an arrangement for two out of the three factors was done, and therefore, it was possible to determine the sample size in each level for the factors. In addition, taking into account that the relation was 9 treatments per factor, a total of 204 data were obtained per factor level; at the same way, considering that 3 factors were available, a total of 624 runs were done. Besides, since 624 is not a multiple of 27 (the orthogonal arrangement with size 3^3), it was rounded to the closest multiple: 621. Hence, 23 replicas were implemented for the 68 data per treatment (621/27 = 23).

5.2   Assumptions Verification and ANOVA

(a) **Normality:** A normality test was done in order to prove that data were generated based on a normal process, obtaining with a Kolmogorov-Smirnov test which was made with a significance level of 0.05, and it was possible to conclude that data does not behave in a normal way. For that reason, was necessary the use of a transformation of Response Variable [35]. In this case, the best transformation is the square root of this variable.

(b) **Constant variance:** The evaluation of the Levene test helped to prove that the treatments had a homogeneous variance, when data is distributed in a random way in a horizontal strip (no pattern). In the case of mutation probability, Levene's P-value was higher than 0.05. This result had consistency with the behavior observed on the graphics bellow. The graphic contains bars of the interactions between each factor and the objective function (Fig. 14).

(c) **Independence:** In the correlation analysis a significance level higher than 0.05 is evidenced for all the interactions. Consequently, it can be inferred that there is no value correlation in time. Since the model's assumptions are confirmed, a variance analysis or ANOVA is carried out, obtaining the following results (Fig. 15):

Square root Cost

Levene's Statistical	Freedom degree1	Freedom degree 2	P-value
0,007	2	618	0,993

Fig. 14. Levene's test

Pruebas de los efectos inter-sujetos

Variable dependiente: Raíz Costo Función Objetivo

Origen	Suma de cuadrados tipo III	gl	Media cuadrática	F	Sig.
Modelo corregido	56,663[a]	26	2,179	,044	1,000
Intersección	13369376	1	13369376	269097	,000
Pc	5,581	2	2,790	,056	,945
Pm	50,405	2	25,203	,507	,602
Ngen	,618	2	,309	,006	,994
Pc * Pm	,047	4	,012	,000	1,000
Pc * Ngen	,001	4	,000	,000	1,000
Pm * Ngen	,001	4	,000	,000	1,000
Pc * Pm * Ngen	,009	8	,001	,000	1,000
Error	29511,356	594	49,682		
Total	13398944	621			
Total corregida	29568,019	620			

a. R cuadrado = ,002 (R cuadrado corregida = -,042)

Fig. 15. Design of experiments ANOVA

The significance values of the interactions between 2 factors (for example Pc*Pm) or even between the three factors (Pc*Pm*NGen) is higher than 0.05, therefore, then it is possible to infer that any change in the factors alters the response variable in a significant way.

5.3 Comparison of the Proposed Model

A comparison with some instances is done. These have been the subject of multiple comparisons through the years. Specifically, the instances proposed by [39] are selected, to then carry out 10 runs with the proposed method for each instance.

Additionally, in order to maintain certain coherence in the experimental analysis, a relative deviation index, in percentage, was employed, as shown in the following equation, where F_x^{GA} corresponds to the averages values of the objective function (i.e., total cost) obtained using the proposed genetic algorithm (GA). Also, F_x^{TAI} corresponds to the best values of the objective function based on mentioned Taillard instances. These values are shown in Table 6. It is necessary to clarify that the instances that were chosen are the most comparable with the problem under study. Not all the instances accomplished with this.

$$\%dev = \frac{F_x^{GA} - F_x^{TAI}}{F_x^{TAI}} \times 100\% \tag{8}$$

At the other hand, a MINLP (Mixed Integer Nonlinear Programming) model was coded and running using AMPL® version 8.0 for MS Windows®. Because of computational complexity, it was run for instances with number of facilities less and equal to 22. Thus, optimal solutions were obtained for these small sub problems. For instances with higher number of facilities, to get a solution was not possible in a reasonable time. At the same way, the relative deviation index was calculated. Here, F_x^{QAP} was the value obtained with the MILP model:

$$\%dev = \frac{F_x^{GA} - F_x^{QAP}}{F_x^{QAP}} \times 100\% \tag{9}$$

From Tables 5 and 6 it can be noticed that the implemented genetic algorithm gave a good solution taking into account that its best solution deviated from the exact solution obtained with AMPL in 6.48 %. This is not a great deviation and the relevant aspect is the time spent with the genetic algorithm is about 7.5 s against 3958.1 with AMPL. This is equivalent to a time reduction of 99.95 %.

Table 5. Results in the study case

GA evaluation			
Computational time (QAP)	QAP solution	GA solution	Relative error
3958.1	97623	103949	6.48 %

Table 6. Results and QAPLIB instances comparisons

Instances vs proposed method (comparison)			
Instance	Best know solution	Best obtained solution	Relative error
tai10a	135028	137662	1.95 %
tai10b	118376	132277	11.74 %
tai12a	224416	225430	0.45 %
tai12b	394695	403852	2.33 %
tai15a	388214	390160	0.50 %
tai15b	576528	600987	4.02 %
tai17a	491812	508939	3.48 %
tai20a	703482	725155	3.08 %
tai20b	122459	130148	6.27 %

6 Conclusions

This work focused on a comparative analysis between a Genetic Algorithm and QAP formulation for Facility Layout Problem. The integration of the proximity measure in the proposed model turns out to be highly important: it forces the inclusion of important factors in the total cost. Besides flow and distance, it is also important to consider the need for distance between working stations due to risks in the security of the facility's operative personnel; or, closeness issues, due to the fact that a group of people are required to distribute their daily activities between several working stations.

The proposed algorithm generated solution that compared with instances from other methods, provides a sign that the proposed method can behave satisfactorily both with problems conceived randomly and with real-life problems. Additionally, it is important to point that such behavior can be even more evidenced in problems with less than 15 working stations.

For further research, several lines are still open. For example, many other issues of the problem under study could be included in the analysis in order to keep the problem much more realistic: probabilistic constraints, i.e. stochastic capacities in production plant. The procedure can be improved by implementing: different procedures to generate the initial population, other types of crossover or mutation strategies, or even other fitness function, etc. Other heuristic procedures could be employed to hybridize the genetic algorithm. Finally, because of the NP-completeness of the problem under study, researchers could be interested in analyzing the behavior of various meta-heuristic algorithms such as GRASP (Greedy Randomized Adaptive Search Procedure), Tabu Search, etc.

References

1. Niebles-Atencio, F., Solano-Charris, E.L., Montoya-Torres, J.R.: Ant colony optimization algorithm to minimize makespan and number of tardy jobs in flexible flowshop systems. In: Proceedings 2012 XXXVIII Conferencia Latinoamericana en Informática (CLEI 2012), Medellin, Colombia, pp. 1–10, 1–5 October 2012. doi:10.1109/CLEI.2012.6427154
2. Niebles-Atencio, F., Dionicio, N.R.: A Sule's method initiated genetic algorithm for solving QAP formulation in facility layout design: a real world application. J. Theor. Appl. Inf. Technol. **84**(2), 157–169 (2016)
3. Eldrandaly, K.A., Nawara, G.M., Shouman, M.A., Reyad, A.H.: Facility layout problem and intelligent techniques: a survey. In: 7th International Conference on Production Engineering, Design and Control, PEDAC, Alex, Egypt, pp. 409–422 (2001)
4. Sule, D.: Logistics of facility location and allocation (2001)
5. Koopmans, T., Beckmann, M.: Assignment problems and the location of economic activities. Econometrica **25**, 53–76 (1957)
6. Meller, R.D., Gau, K.Y.: The facility layout problem: recent and emerging trends and perspectives. J. Manufact. Syst. **15**, 351–366 (1996)
7. Azadivar, F., Wang, J.: Facility layout optimization using simulation and genetic algorithms. Int. J. Prod. Res. **38**(17), 4369–4383 (2000)
8. Coello, C.A.C., Lamont, G.B., van Veldhuizen, D.A.: Evolutionary Algorithms for Solving Multi-objective Problems, 2nd edn., pp. 145–156 (2007)
9. Kratica, J., Tošić, D., Filipović, V., Dugošija, Ð.A.: New genetic representation for quadratic assignment problem. Yugoslav J. Oper. Res. **21**(2), 225–238 (2011). doi:10.2298/ YJOR1102225K
10. Singh, S.P., Sharma, R.R.K.: A review of different approaches to the facility layout problems. Int. J. Adv. Manufact. Technol. **30**(5–6), 425–433 (2006)
11. Drira, A., Pierreval, H., Hajri-Gabouj, S.: Facility layout problems: a survey. Ann. Rev. Control **31**(2), 255–267 (2007)
12. Burkard, R.E., Çela, E., Pardalos, P.M., Pitsoulis, L.: The quadratic assignment problem. In: Du, D.Z., Pardalos, P.M. (eds.) Handbook of Combinatorial Optimization, vol. 3, pp. 241– 337. Kluwer, Dordrecht (1998)
13. Loiola, E.M., de Abreu, N.M.M., Boaventura-Netto, P.O., Hahn, P., Querido, T.: A survey for the quadratic assignment problem. Eur. J. Oper. Res. **176**, 657–690 (2007)
14. Voß, S.: Heuristics for nonlinear assignment problems. In: Pardalos, P.M., Pitsoulis, L. (eds.) Nonlinear Assignment Problems, pp. 175–215. Kluwer, Boston (2000). Armour and Buffa
15. Dakeroglu, T., Cosar, A.: A novel multistart hyper-heuristic algorithm on the grid for the quadratic assignment problem. Eng. Appl. Artif. Intell. **52**, 10–25 (2016)
16. Dorigo, M., Maniezzo, V., Colorni, A.: The ant system optimization by a colony of cooperating agents. IEEE Trans. Syst. Man Cybern.—Part B **26**, 29–41 (1996)
17. Stutzle, T., Dorigo, M.: ACO algorithms for the quadratic assignment problem. In: Corne, D., Dorigo, M., Glover, F. (eds.) New Ideas in Optimization. McGraw-Hill Ltd., Maidenhead (1999)
18. Nissen, V.: Solving the quadratic assignment problem with clues from nature. IEEE Trans. Neural. Netw. **5**, 66–72 (1994). Drezner 2003
19. Merz, P., Freisleben, B.: Fitness landscape analysis and memetic algorithms for the quadratic assignment problem. IEEE Trans. Evol. Comput. **4**, 337–352 (2000)
20. Misevicius, A.: An improved hybrid genetic algorithm: new results for the quadratic assignment problem. Knowl. Based Syst. **17**, 65–73 (2004)

21. Fleurent, C., Ferland, J.: Genetic hybrids for the quadratic assignment problem. DIMACS Ser. Math. Theor. Comput. Sci. **16**, 190–206 (1994)
22. Tate, D.M., Smith, A.E.: A genetic approach to the quadratic assignment problem. Comput. Oper. Res. **22**(1), 78–83 (1995)
23. Ahuja, K.R., Orlin, J.B., Tiwari, A.: A greedy algorithm for the quadratic assignment problem. Comput. Oper. Res. **27**, 917–934 (2000)
24. McLoughlin III, J.F., Cedeño, W.: The enhanced evolutionary tabu search and its application to the quadratic assignment problem. In: Beyer, H.-G., O'Reilly, U.-M. (eds.) Proceedings of the Genetic and Evolutionary Computation Conference—GECCO 2005, pp. 975–982. ACM Press, New York (2005)
25. Tseng, L., Liang, S.: A hybrid metaheuristic for the quadratic assignment problem. Comput. Optim. Appl. **34**, 85–113 (2005)
26. Xu, Y.-L., Lim, M.H., Ong, Y.S., Tang, J.: A GA-ACO-local search hybrid algorithm for solving quadratic assignment problem. In: Metal, K. (ed.) Proceedings of the Genetic and Evolutionary Computation Conference-GECCO 2006, vol. 1, pp. 599–605. ACM Press, New York (2006)
27. Stützle, T.: Iterated local search for the quadratic assignment problem. Eur. J. Oper. Res. **174**, 1519–1539 (2006)
28. Bölte, A., Thonemann, U.W.: Optimizing simulated annealing schedules with genetic programming. Eur. J. Oper. Res. **92**, 402–416 (1996)
29. Wilhelm, M.R., Ward, T.L.: Solving quadratic assignment problems by simulated annealing. IIE Trans. **19**, 107–119 (1987)
30. Battiti, R., Tecchiolli, G.: The reactive tabu search. ORSA J. Comput. **6**, 126–140 (1994)
31. Drezner, Z.: The extended concentric tabu for the quadratic assignment problem. Eur. J. Oper. Res. **160**, 416–422 (2005)
32. JamesT, R.C., Glover, F.: A cooperative parallel tabu search algorithm for the quadratic assignment problem. Eur. J. Oper. Res. **195**, 810–826 (2009)
33. Misevicius, A.: A tabu search algorithm for the quadratic assignment problem. Comput. Optim. Appl. **30**, 95–111 (2005)
34. Hernandez, H., Montoya, J.R., Niebles, F.: Design of multi-product/multi-period closed-loop reverse logistics network using a genetic algorithm. In: Proceedings 2014 IEEE Symposium on Computational Intelligence in Production and Logistics Systems, Orlando, USA (2014)
35. Montgomery, D.C.: Design and Analysis of Experiments. Wiley, Hoboken (2008)
36. Angel, E., Zissimopoulos, V.: On the hard ness of the quadratic assignment problem with metaheuristics. J. Heuristics **8**, 399–414 (2002)
37. Gambardella, L.M., Taillard, E.D., Dorigo, M.: Ant colonies for the quadratic assignment problems. J. Oper. Res. Soc. **50**, 167–176 (1999)
38. Maniezzo, V., Dorigo, M., Colorini, A.: The ant system applied to the quadratic assignment problem. Technical report IRIDIA/94-28. Universite Libre de Bruxelles, Belgium (1994)
39. Taillard, E.D.: Robust taboo search for the QAP. Parallel Comput. **17**, 443–455 (1991)

Machine Learning Methods

An Empirical Evaluation of Machine Learning Algorithms for Image Classification

Thembinkosi Nkonyana[(✉)] and Bhekisipho Twala

Department of Electrical and Electronics Engineering Science,
University of Johannesburg, P.O. Box 524 Auckland Park,
Johannesburg 2006, South Africa
{tnnkonyana,btwala}@uj.ac.za

Abstract. Image classification is an important aspect that needs techniques which can better predict or classify images as they become larger and complex to solve. Thus, the demand for research to find advanced algorithms and tools to solve problems experienced in classification, has shown great increase in interest over the years. The contribution of this paper is the evaluation of four machine learning techniques [multilayer perceptron (MLP), random forests (RF), k-Nearest Neighbor (k-NN), and the Naïve Bayes (NB)] in terms of classifying images. To this end, three industrial datasets are utilized against four performance measures (namely, precision, receiver operating characteristics, root mean squared error and mean absolute error). Experimental results show RF achieving higher accuracy while the NBC exhibits the worst performance.

Keywords: Machine learning · Image classification · Performance measures

1 Introduction

Image classification is an important aspect in special fields such as image analysis, remote sensing, and statistical pattern recognition. It is an important problem which appears in many application domain such as quality control, biometry (face recognition), medicine, office automation (character recognition), geology (soil type recognition) and so on. The type of image classification that we are concerned about in this paper is satellite remote sensing (RS) images. Satellite RS systems and aircraft applications can be used and applied in fields such as Land Use Land Cover, Oceanography, Meteorology, Climatology, Natural disasters, Water resources, Civil Engineering, and many more areas [10]. The nature and characteristics of images can be captured from a satellite in spatial, spectral, temporal, and multispectral resolutions [16, 23, 24]. This is mostly made possible by advanced sensors which are able to captured images that contain spatial information, which needs to be interpreted or analyzed into information that can assist in various planning and decision making.

Classification (as opposed to prediction) is one of the best ways to get information from an image, by assigning pixels to a class. All classification algorithms are based on the assumption that the image in question depicts one or more features (for example, spectral regions in the case of remote sensing). The classes may be specified a prior by an analyst (as in supervised classification) or automatically clustered (as in unsupervised

© Springer International Publishing Switzerland 2016
Y. Tan et al. (Eds.): ICSI 2016, Part II, LNCS 9713, pp. 79–88, 2016.
DOI: 10.1007/978-3-319-41009-8_8

classification) into sets of prototype classes, where the analyst simple specifies the number of desired categories. In this paper we follow the supervised classification approach.

Over the years there has been a growing demand for remotely sensed data. Specific objects of interest are being monitored with earth observation data. Examples include food security, environmental concerns, public safety and more. The improvement in the quality of remotely-sensed data does not guarantee more accurate feature extraction. The image classification techniques are very important for better accuracy. For example, let us say water is the feature that wants to be extracted from a high spatial resolution image but other pixels are mistakenly classified as water. This could have serious consequences in planning and decision making processes. This is the reason why correctly classification of images (according to its visual content) become important.

RS has attracted lots or research work in many institutions, as the technology is able to reach locations where a normal person cannot reach using their own physical efforts [20]. Various satellite sensor systems are in use to today such as IKONOS, LANDSAT, QuickBird, SPOT and many more. These enable us to capture areas which are too large to be reached by human beings or observed by a human eye. Also, as the technology advances, images are becoming more complex, hence, in part, the requirement for state-of-the-art techniques which can be able to accurately classify images. In previous years, more traditional classification techniques have been applied for the classification of images. However, most of these techniques have yielded incorrect and incomplete results due to their inability to deal with high dimensionality of the feature space [10]. Recently, the machine learning community have proposed and applied a lot of algorithms with very positive results. The strength of ML algorithms lies in their non-parametric nature as opposed to traditional methods.

A ML algorithm is used to fit a model to data. To create good predictive models in ML that are capable of generalizing, we believe that an empirical comparison (evaluation) of different ML algorithms against several performance measures and using three randomly selected datasets should suffice. To this end, the contribution of this paper is an empirical comparison of the effectiveness of four machine learning algorithms in classifying satellite images based on four performance measures. Three industrial datasets are utilized for this task. To the best of our knowledge few research studies have conducted such an extensive experiments.

The rest of the paper is organized as follows: Sect. 2 presents a brief description of the machine learning techniques used in the paper, followed by brief related work in Sect. 3. The experimentation protocols and results are presented in Sect. 4. We end the paper with conclusions and discussion and later future work in Sect. 5.

2   Machine Learning Techniques

2.1   Artificial Neural Network

In ML, artificial neural networks (ANNs) are family of models inspired by biological neural networks (the central nervous system of animals, in particular the brain) and are used to estimate or approximate functions that can depend on a large number of inputs

This type of a classifier, actually acts like a human brain. In other fields ANNs have been used to process features such as recognition of faces, speech [8, 9] while in other areas they have been very popular in remote sensing [27]. ANNs are typically organized in layers based on different kinds of nodes (input, hidden and output) and learning rules. The Multi-Layer Perceptron's makes use of three or more layers of neurons that has non-linear functions. The three layers consist of the input layer, hidden layer and output layer. This is a very popular ML technique and it is non-parametric in nature.

2.2 Decision Trees

A decision tree (DT) or classification tree is a decision support tool that uses a tree-like graph or model of decisions and their possible consequences, including chance event outcomes, resource costs and utility. The topmost decision node in a tree is called root node. Each internal node represents a "test" on an attribute while each branch represents the outcome of the test and each leaf node represents a class label (i.e. a decision taken after computing all attributes). In various research projects and specialty, this type of technique has been used by many and is very popular [2, 3]. DTS are non-parametric in their nature, and they are very simple to apply. A DT can also be easily transformed to a set of rules by mapping from the root node to the leaf nodes one by one. Random Forests belong to the family of DTs. RFs are a combination of tree predictors such that each tree depends on the values of a random vector sampled independently and with the same distribution for all tress in the forest.

2.3 k-NN

In pattern recognition and statistical estimation, the k-nearest neighbor (k-NN) is a non-parametric method use for classification and regression. All available instances are stored and new instances are classified based on similarity measure (for example, distance functions like the Euclidean or Hamming). For continuous attributes, the Euclidean function is used while for categorical instances the Hamming distance must be used. An instance is classified by a majority vote of its neighbors, with the instance being assigned to a class most common amongst its k nearest neighbors measured by a distance function [24]. If the value of k is equal to 1, then the instance is simple assigned to the class of its nearest neighbor. In remote sensing, this type of classifier focuses on the value of k, by looking at pixels which are close to each other, and assumes that they belong to each other in the feature space [24, 26].

2.4 Naïve Bayes

The Naïve Bayes classifier (NBC) is based on the Baye's theorem with independent assumptions between attributes. A NBC is easy to build with no complicated iterative parameter estimation with makes it particularly useful for very large datasets. It provides a way of calculating the posterior probability of the class label which is given a feature vector that represents the instances based on Bayes theorem [25]. The NBC utilizes the training data, and the conditional probability of each attribute A, given class label C, and can be able to handle an arbitrary number of independent attributes whether continuous or categorical as described in the next sentence. The posterior probability is calculated by first, constructing a frequency table for each attribute against the target and then transforming the frequency tables and finally use the Naive Bayes to calculate the posterior probability for each class. The class with the highest posterior probability is the outcome of prediction.

3 Related Work

Given the increasing amount of images that are currently available there has been considerable interest in the remote sensing community to leverage this data to learn processing models. Thus, image classification has become an important problem in ML and RS. Much work has explored the use of supervised learning classifiers such as support vector machines (SVM), DT, NBC, k-NN and ANN on land use classification with very promising results [1–3, 4–7]. To improve image classification accuracy, [19] developed a Bayesian hybrid approach which was compared to a single classifier, which showed the former as more effective. The boosting approach was proposed and evaluated by [19] as means of improving image classification accuracy. [18] addressed the image segmentation problem in remotely sensed satellite images by following an unsupervised classification approach. Traditional methods such as principal component analysis and simultaneous diagonalization of variance matrices have also been proposed but results have been poor [21]. Recently the use of extreme learning machine classifier has been proposed for purposes of classifying an image, using different types of kernels [22].

From the above related research studies, it appears that no extensive evaluation of ML methods against several performance measures and several datasets has been carried out for image classification prediction. Most of the studies have either looked at one or two ML techniques versus one performance measure and/or one image dataset. Since, a core of a learner is to generalize (generalization in this context is the ability of a learning machine to perform accurately on new/unseen examples after having experiences a learning dataset) from its experience, our approach in this paper is to use as many ML techniques and performance measures as possible. By so-doing we are addressing the generalization problem which previous studies have failed to do or taken into account in their respective experiments.

4 Experiments

4.1 Experimental Set-Up

The objective of the experiments is to show the effectiveness of four ML algorithms in terms of classifying images. The three image datasets were obtained from the machine learning repository UCI [11]; these are the urban land cover, image segmentation, and wildt. The urban land cover datasets is purposely made for classification of urban land cover using high resolution aerial imagery [12, 13]. It consists of 168 instances, 148 attributes and 9 classes. The classes are namely made up of trees, grass, soil, concrete, asphalt, buildings, cars, pools, and shadows. This datasets consists of training and testing data all, which is all for classifying a high resolution aerial image into 9 types of urban land cover. The image segmentation dataset is made up of 2310 instances, 19 attributes and 7 classes [11]. The instances on this dataset were randomly drawn from a database of 7 outdoor images. The 7 classes which are in this datasets consists of brickface, window, cement, foliage, sky, path and grass. The wildt dataset is a Quickbird high-resolution remote sensing dataset [11]. It is made up of a small training samples of diseased trees and a large number of other land cover. The dataset is made up of testing and testing data for detecting diseased trees in Quickbird imagery, and it consists of 4889 instances, 6 attributes and 2 classes [14].

Generally, we applied the ANN, DT, k-NN and RF algorithms to each image dataset in the training set and evaluate the resulting model using the dataset in the test set. The cross validation methods involves partitioning each dataset randomly into 10-fold. We use one partition as a testing set and the remaining partitions to form the training set. We repeat this process ten (10) times, each of the partitions as the testing dataset and the remaining partitions to form a training set. Four evaluation measures are used to test the effectiveness of the ML methods. All the ML algorithms were created from WEKA with default settings [15].

In total, 240 experiments were carried out, i.e., 4 (ML techniques) × 4 (performance measures) × 3 (datasets) × 10 (replications). For statistical significance tests, a two-way mixed effects model analysis of variance was used; the fixed effects being the ML methods and performance measures and the random effect being the three image datasets.

4.2 Experimental Results

Figure 1 shows RF achieving a higher percentage in correctly classifying the image, followed by MLP and NB. The worst performance is by k-NN. However, there appears to be no significant difference in terms of performance between RF and MLP (on the

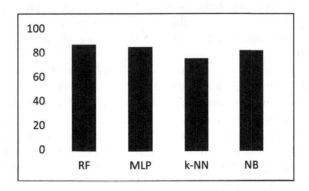

Fig. 1. Results for correctly classified instances (landsat satellite problem)

one hand) while there is a significant difference in performance between RF and NB; RF and k-NN, MLP and NB; MLP and k-NN and NB and k-NN (on the other hand). This is the case at the 5 % level of significance.

For the landsat satellite image problem, RF is the best performing technique across three of the performance measures with NB achieving the worst performance in all the four performance measures. k-NN achieves the least mean absolute error while RF drops to third for the same measure (Table 1).

Table 1. Landsat satellite image performance measure results

Methods	MAE	RMSE	Precision	ROC
RF	0.019	**0.077**	**0.979**	**0.999**
MLP	0.015	0.097	0.962	0.994
k-NN	**0.011**	0.099	0.966	0.980
NB	0.058	0.230	0.819	0.971

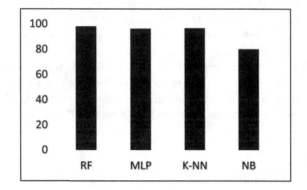

Fig. 2. Results for correctly classified instances (image segmentation problem)

The results presented in Fig. 2 (once again) show RF with the highest accuracy in terms of correctly classified instances, followed by MLP, k-NN, and NB, respectively. The performance of k-NN slightly improves for this kind of dataset (compared with the landsat image problem where it was the worst performing method). There are only three significant pairs between methods at the 5 % level, i.e., RF and NB; MLP and NB and k-NN and NB.

Table 2 presents results of the image segmentation problem, which shows RF outperforming all the methods across all the performance measures, with fierce competion from MLP. Overall, the worst performing method is the k-NN followed by NB. In fact, NB is the worst performing method in terms of the mean absolute error.

The results summarised in Fig. 3 are similar to the landsat satellite problem in terms of correctly classified instances. Once again, RF exhibits superior performance, followed by MLP, NB and k-NN, respectively. Also, there appears to be significant difference between different pairs of methods at the 5 % level.

Table 2. Segmentation image performance measure results

Methods	MAE	RMSE	Precision	ROC
RF	**0.162**	**58.084**	**0.900**	**0.945**
MLP	0.304	82.411	0.797	0.858
k-NN	0.357	123.03	0.649	0.628
NB	0.381	106.66	0.730	0.816

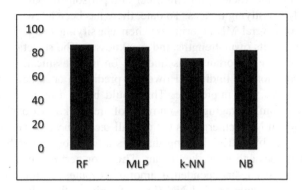

Fig. 3. Results for wildt correctly classified instances

Table 3. Wildt image performance measure results

Methods	MAE	RMSE	Precision	ROC
RF	0.070	**0.168**	**0.858**	**0.977**
MLP	**0.040**	0.174	0.838	0.968
k-NN	0.060	0.237	0.745	0.852
NB	0.043	0.205	0.813	0.964

Table 4. Overall average performance measure results

Methods	MAE	RMSE	Precision	ROC
RF	**0.084**	**0.175**	**0.912**	**0.974**
MLP	0.120	0.223	0.866	0.940
k-NN	0.143	0.310	0.787	0.820
NB	0.161	0.317	0.787	0.917

Table 3 shows RF achieving the highest accuracy rates across three performance measures while NB (surprisingly) outperforms RF in terms of the mean absolute error. This is the only dataset where RF achieves the worst performance in terms of the mean absolute area.

In terms of overall performance (across the three datasets), RF exhibits the highest accuracy rates (Table 4), closely followed by MLP. However, MLP achieves the third best performance in terms of the mean absolute area while NB exhibits the second best performance. The worst performance is by k-NN which achieves the highest errors across 50 % of the performance measures although there is no significant difference in performance between k-NN and NB in terms of the precision measure.

5 Conclusion

The contribution of this article is an empirical comparison of four ML algorithms in terms of accurately classifying images. To date, there has been a few studies examining the effectiveness of several ML algorithms when classifying images. Each algorithm used in this paper has its own strengths and weaknesses. The selection of a particular ML algorithm can have important consequences on the classification of images and subsequent interpretations of findings. However, prediction accuracy rates of an "averaging" like RF were very impressive. This could be as a result of a reduction in variance resulting from averaging the number of trees like is done in bagging or boosting. Even though MLP emerges as the overall second best of the four techniques, it has come under fire by critics claiming that it is a black box whereby one does not understand what it does and what they mean. The poor performance of NB could be attributed to it very strong independence among attributes assumption while determining the number of neighbors in k-NN (i.e. determining the value of k) could have contributed to its poor performance. So far, we have restricted our experiments to four ML methods and only three datasets. It would be interesting to carry out a comparative study of more than four methods (including SVM,) against more than three image datasets. Furthermore, it would be possible to explore the different patterns and levels of missing values.

Acknowledgements. I would like to thank the University of Johannesburg for funding me and affording me the opportunity to utilize the resources to complete this work.

References

1. Yasmin, M., Sharif, M., Irum, I., Mohsin, S.: An efficient content based retrieval using EI classification and colour features. J. Appl. Res. Technol. **12**(5), 877–885 (2014)
2. Xuerong, L., Qianguo, X., Lingyan, K.: Remote sensing image method based on evidence theory and decision tree. Proc. SPIE **7857**, 7857Y-1 (2010)
3. Yang, C.C., Prasher, S.O., Enright, P., Madramootoo, C., Burgess, M., Goel, P.K., Callum, I.: Application of decision tree technology for image classification using remote sensing data. Agric. Syst. **76**(3), 1101–1117 (2003)
4. Lu, K.C., Yang, D.L.: Image processing and image mining using decision trees. J. Inf. Sci. Eng. **25**, 989–1003 (2009)
5. Ghose, M.K., Pradhan, R., Ghose, S.S.: Decision tree classification of remotely sensed satellite data using spectral separability matrix. Int. J. Adv. Comput. Sci. Appl. (IJACSA) **1**(5), 93–101 (2010)

6. Pal, M., Mather, P.M.: Decision tree based classification of remotely sensed data. In: Centre for Remote Imaging, Sensing and Processing (CRISP), National University of Singapore, Singapore Institute of Surveyors and Valuers (SISV), Asian Association on Remote Sensing (AARS) (2001)

7. Thai, H.L., Hai, T.S., Thuy, N.T.: Image classification using support machine and neural network. Int. J. Inf. Technol. Comput. Sci. **5**, 32–38 (2012)

8. Liu, Z.K., Xiao, J.Y.: Classification of remotely sensed image data using artificial neural networks. Int. J. Remote Sens. **12**(11), 2433–2438 (1991)

9. Rashmi, S., Mandar, S.: Textural feature based image classification using artificial neural network. In: Unnikrishnan, S., Surve, S., Bhoir, D. (eds.) ICAC3 2011. CCIS, vol. 125, pp. 62–69. Springer, Heidelberg (2011)

10. Xu, X., Zhong, Y., Zhang, L.: Adaptive subpixel mapping based on a multiagent system for remote-sensing imagery. IEEE Trans. Geosci. Remote Sens. **52**(2), 0196–2892 (2014)

11. Lichman, M.: Machine Learning Repository, University of California, School of Information and Computer Science, Irvine, CA (2013). http://archive.ics.uci.edu/ml

12. Johnson, B., Xie, Z.: Classifying a high resolution image of an urban area using super-object information. ISPRS J. Photogram. Remote Sens. **83**, 40–49 (2013)

13. Johnson, B.: High resolution urban land cover classification using a competitive multi-scale object based approach. Remote Sens. Lett. **4**(2), 131–140 (2013)

14. Johnson, B., Tateishi, R., Hoan, N.: A hybrid pansharpening approach and multiscale object-based image analysis for mapping diseased pine and oak trees. Int. J. Remote Sens. **34**(20), 6969–6982 (2013)

15. Hall, M., Frank, E., Holmes, G., Pfahringer, B., Reutmann, P., Witten, I.H.: The WEKA data mining software: an update. SIGKDD Explor. **11**(1), 10–18 (2009)

16. Yu, Q., Gong, P., Clinton, N., Biging, G., Kelly, M., Schirokauer, D.: Object-based detailed vegetation classification with airborne high spatial resolution remote sensing imagery. Photogram. Eng. Remote Sens. **72**(7), 799–811 (2006)

17. Pradhan, R., Ghose, M.K., Jeyaram, A.: Land cover classification of remotely sensed satellite data using Bayesian and hybrid classifier. Int. J. Comput. Appl. **7**(11), 0975–8887 (2010)

18. Mitra, P., Shankar, B., Pal, S.K.: Segmentation of multispectral remote sensing images using active support vector machines. Pattern Recogn. Lett. **25**, 1067–1074 (2004)

19. Xiaohc, Z., Liang, Z., Jixian, Z., Huiyong, S.: An object-oriented classification method of high resolution imagery based on improved ADA tree. In: International Symposium on Remote Sensing of Environments: Earth and Environmental Science, vol. 17 (2014). doi:10.1088/1755-1315/17/1/012212

20. Blaschke, T.: Object based image analysis for remote sensing. ISPRS J. Photogram. Remote Sens. **65**, 2–16 (2009)

21. Subramanian, S., Gat, N., Sheffiel, M., Barhen, J., Toomarian, N.: Methodology for hyperspectral image classification using novel neural network. In: Algorithms for Multispectral and Hyperspectral Imagery III, SPIE, vol. 3071 (1997)

22. Kavitha, K., Arivazhagan, S., Suriya, B.: Histogram binning and morphology based image classification. Int. J. Curr. Res. Acad. Rev. **2**(6), 56–66 (2014)

23. Gupta, G.R., Kamalapur, S.M.: Study of classification of remote sensing images using particles swarm optimization based approach. Int. J. Appl. Innov. Eng. Manag. **3**(10), 165–169 (2014). ISSN 2319-4847

24. Twala, B., Nkonyana, T.: Extracting supervised learning classifiers from possibly incomplete remotely sensed data. In: 1st Brics Countries on Computational Intelligence, Brics-CCI 2013, pp. 476–482 (2013)

25. John, G.H., Langley, P.: Estimating continuous in Bayesian classifiers. In: Proceedings of the Eleventh Conference on Uncertainty in Artificial Intelligence, pp. 338–345. Morgan Kaufmann, San Francisco (1995)
26. Richards, J.A., Jia, X.: Remote Sensing Digital Image Analysis. Springer, Berlin (1999)
27. Arafi, A., Safi, Y., Fajr, R., Bouromi, A.: Classification of mammographic images using artificial neural networks. Appl. Math. Sci. **7**(89), 4415–4423 (2013)

An Improved Ensemble Extreme Learning Machine Based on ARPSO and Tournament-Selection

Ya-Qi Wu[1(✉)], Fei Han[1], and Qing-Hua Ling[1,2]

[1] School of Computer Science and Communication Engineering,
Jiangsu University, Zhenjiang, Jiangsu, China
`js.rg.wyq@163.com`
[2] School of Computer Science and Engineering,
Jiangsu University of Science and Technology, Zhenjiang, Jiangsu, China

Abstract. Extreme learning machine (ELM) performs more effectively than other learning algorithm in many cases, it has fast learning speed, good generalization performance and simple setting. However, how to select and cluster the candidate are still the most important issues. In this paper, KGA-ARPSOELM, an improved ensemble of ELMs based on K-means, tournament-selection and attractive and repulsive particle swarm optimization (ARPSO) strategy is proposed to obtain better candidates of the ensemble system. To improve classification and selection ability in the ensemble system, K-means is applied to cluster the ELMs efficiently while tournament- selection is used to choose the optimal base ELMs with higher fitness value in proposed method. Moreover, experiment results verify that the proposed method has the advantage of being more convenient to get better convergence performance than the traditional algorithms.

Keywords: Extreme learning machine · Attractive and repulsive particle swarm optimization · K-means · Tournament-selection

1 Introduction

Extreme learning machine (ELM) [1, 2] is proposed for single-hidden layer feedforward neural networks (SLFNs). Different from BP algorithm and other learning algorithm, it can simply set number and type of hidden layer neurons randomly without the iterative solution. Moreover, ELM does not need to adjust input parameters in the network, fast learning speed and better generalization performance are features and advantages of the ELM. Thus ELM has received increasing attention in recent years.

Some studies show that extreme learning machine optimization using particle swarm algorithm (PSO) is very effective [3]. However, traditional PSO still has the disadvantages of premature convergence and easily fall into local minima [4]. To reduce the impact of this problem, attractive and repulsive particle swarm optimization (ARPSO) [5] was proposed to ensure the diversity of swarm effectively in the search process, so it could get better convergence accuracy. In the E-ARPSOELM [6], by considering the classification accuracy and diversity of the ensemble system, ARPSO was applied to select the base ELMs from the initial ELM pool.

© Springer International Publishing Switzerland 2016
Y. Tan et al. (Eds.): ICSI 2016, Part II, LNCS 9713, pp. 89–96, 2016.
DOI: 10.1007/978-3-319-41009-8_9

How to find suitable ensemble ELMs with more concise and compact methods are the jobs of this study. K-means algorithm [7] is still the one of the most widely used partitioning clustering algorithm in different areas of science. In this paper, K-means method is used to cluster the base ELMs in the first phase, which greatly increases the diversity between the different classes of samples. Then, tournament-selection is used to choose better members in each cluster, which further improves the classification accuracy of K-means method. Finally, we use ARPSO to search the ELMs by considering both the classification performance and diversity of the ensemble system.

2   Preliminaries

2.1   Extreme Learning Machine

For N arbitrary distinct samples (x_i, t_i), where $x_i = [x_{i1}, x_{i2}, \ldots, x_{in}] \in R^n$, $t_i = [t_{i1}, t_{i2}, \ldots, t_{im}]^T \in R^m$, the standard SLFN with $\tilde{N}$ hidden neurons can be expressed in matrix form as

$$H\beta = T \tag{1}$$

$$
H(w_1, w_2, \cdots, w_{\tilde{N}}, b_1, b_2, \cdots, b_{\tilde{N}}, x_1, x_2, \cdots, x_N)
$$
$$
= \begin{bmatrix} g(w_1 \cdot x_1 + b_1) & \cdots & g(w_{\tilde{N}} \cdot x_1 + b_{\tilde{N}}) \\ \vdots & \cdots & \vdots \\ g(w_1 \cdot x_N + b_1) & \cdots & g(w_{\tilde{N}} \cdot x_N + b_{\tilde{N}}) \end{bmatrix}_{N \times \tilde{N}} \quad \beta = \begin{bmatrix} \beta_1^T \\ \cdot \\ \cdot \\ \beta_{\tilde{N}}^T \end{bmatrix}_{\tilde{N} \times m} \quad T = \begin{bmatrix} t_1^T \\ \cdot \\ \cdot \\ t_N^T \end{bmatrix}_{N \times m}
$$

where $w_i = [w_{i1}, w_{i2}, \ldots, w_{in}]^T$ is the weight vector connecting the *i-th* hidden neuron and the input neurons, and $\beta_i = [\beta_{i1}, \beta_{i2}, \ldots, \beta_{im}]^T$ is the weight vector connecting the *i-th* hidden neuron and the output neurons and b_i is the bias of the *i-th* hidden neuron. Simultaneously, H is called the hidden layer output matrix [8].

Due to the characteristics of ELM algorithm, training a SLFN is simply equivalent to find a least-squares solution $\hat{\beta}$ of the linear system $H\beta = T$ as follows:

$$\hat{\beta} = H^+ T \tag{2}$$

In conclusion, since the solution is obtained by an analytical method and all the parameters of SLFN need not be adjusted, ELM has good generalization performance than other gradient-based algorithms.

2.2   Attractive and Repulsive Particle Swarm Optimization

Despite the global search ability of the PSO algorithm has a great advantage, but it is easy to lose the diversity of the swarm [9]. To avoid the problem of premature convergence, attractive and repulsive particle swarm optimizer (ARPSO) [5] was proposed, which guarantees the diversity of the swarm. The position of the *i-th* particle is assumed to be $X_i = (x_{i1}, x_{i2}, \ldots, x_{iD})$, $i = 1, 2, \ldots, N$ and the velocity of the *i-th* particle can be expressed as $V_i = (v_{i1}, v_{i2}, \ldots, v_{iD})$; $P_i = (p_{i1}, p_{i2}, \ldots, p_{iD})$ is denoted as the best historical position of the *i-th* particle which is called *pbest*; $P_g = (p_{g1}; p_{g2}; \ldots; p_{gD})$ as the best solution of the swarm is called *gbest* [9]. The velocity and position of the particles in the algorithm are updated as follows:

$$V_i(t+1) = W * V_i(t) + dir * (c_1 * r_1 * (P_i - X_i(t)) + c_2 * r_2 * (P_g - X_i(t))) \quad (3)$$

$$X_i(t+1) = X_i(t) + V_i(t+1) \quad (4)$$

$$dir = \begin{cases} -1, & dir > 0, diversity < d_{low} \\ 1, & dir < 0, diversity > d_{high} \end{cases} \quad (5)$$

The value of c_1, c_2 is usually between 0 and 2; r_1, r_2 are the generated uniformly random numbers in the range of $(0,1)$; w is the inertia weight. The function of the diversity of the populations is:

$$diversity(S) = \frac{1}{|S| \bullet |L|} \bullet \sum_{i=1}^{|S|} \sqrt{\sum_{j=1}^{D} (p_{ij} - \overline{p}_j)^2} \quad (6)$$

$|S|$ is the number of all the particles, $|L|$ is the length of the radius in the search space, D is the dimension of the problem, p_{ij} is the *j-th* value of the *i-th* particle. When the diversity reaches the upper bound, d_{high}, in the attraction phase ($dir = 1$), the swarm is attractive. If the diversity drops below the critical point, the swarm is the repulsion phase. Due to these characteristics makes the algorithm a better search capability [10].

2.3 Tournament-Selection

Genetic algorithm (GA) is very important in modern calculation method for a global optimization method. Selection operator of GA has important influence on the performance of the algorithm process such as roulette wheel selection and tournament selection [11]. The strategy of roulette wheel selection is the process of roulette wheel simulation. The basic idea of the strategy is the probability of each individual appearing in offspring which based on the individual fitness value. However, when solving minimization problem, it needs to transform adaptive value, then the issue turns into a maximization problem. Thus, the algorithm steps of roulette wheel selection strategy are more complicated.

Tournament selection strategy takes out a certain amount of individual from the population each time. Then according to the fitness value choice one of the best into the offspring population, until the new population size reaches the size of the original population by repeating this operation. Obviously, this selection strategy makes the individual of better fitness value has higher probability of survival. Also, it does not need to consider the problem of transition between minimizing and maximization. So, tournament selection is superior in precision and speed of solving, this article uses the tournament selection to realize the choice operation, which merit in the candidate by selecting candidates randomly. The selection strategy not only increases the diversity of the population, but also avoids the premature convergence phenomenon.

3 The Proposed Algorithm

The critical steps to build an ensemble system include how to select the optimal base classifiers. And diversity between base ELMs can greatly affect the accuracy and stability of the ensemble system [12, 13]. In this study, the ELMs for ensemble are determined adaptively by ARPSO with K-means method and tournament-selection. The detailed steps of the proposed ensemble algorithm are described as follows:

Step 1: The dataset is divided into the training and testing datasets. On the training datasets, bagging method is used to randomly assign different sub-training sets with the same size [6]. With a sub-training set, a corresponding ELM is randomly generated to train a SLFN. It can be generated m alternative ELM by repeating m times.

Step 2: Cluster the base ELMs by K-means method. Use Euclidean Distance as the similarity measurement of the proposed algorithm by typing variable of input weight, bias of hidden neurons and output weight.

Step 3: Select ELMs by tournament selection.

(a) Each cluster is determined to choose the number of R, which expressed as a percentage of the population with the number of individuals.
(b) Select R individuals in each class of the population randomly. The probability of each individual selected is same. The actual category of generated ELMs compared with the expected category of the training datasets in step 1. We can obtain deviation as accuracy of each generated ELMs. According to the accuracy of each individual as fitness, select the highest fitness of the individual into the offspring population.
(c) Repeat sub-step (b) by percentage, and get the individual to become a new generation of population, which used to be optimized.

Step 4: Initial the swarm by randomly initialize the position. Optimize the new population by ARPSO [6].

The above optimization process is repeated until the goal is met or the maximum optimization epochs are completed. Finally, the optimal ensemble of ELMs is obtained, and it is applied to the test dataset.

4 Experiment Results

We choose five datasets from UCI Repository including Diabetes, Satellite Image, Wine, Image Segmentation and Zoo. To verify the effectiveness of the proposed algorithm, it is compared with E-ELM, EOS-ELM, E-PSOELM and E-ARPSOELM. The specifications of the five data are presented in Table 1. Moreover, the training sets and validation sets occupy 70 % and 30 % of the whole training sets respectively. In the proposed algorithm, all experiments are carried out in a MATLAB 2012b environment running in an Intel Core 2 Duo Processor with 2.93 GHz and 2 GB RAM.

Table 1. The description of the two data

Datasets	Training sets	Testing sets	Categories	Attributes
Diabetes	576	192	2	8
Satellite image	4435	2000	6	36
Wine	120	58	3	13
Image segmentation	1500	810	7	19
Zoo	70	31	7	17

In the experiments, the number of the hidden node in all ELMs is same. As for ARPSO in the E-ARPSOELM, KGA-ARPSOELM on all datasets, the inertia weights w_{max} and w_{min} are set as 0.9 and 0.1, respectively; the diversity controlling parameters d_{low} and d_{high} are taken as $5 * 10^{-6}$ and 0.25; $c1$ and $c2$ are equal to 2; the population size is 50; the size of the initial ELM pool is 100. These parameters are determined by trial and error. Each experiment is conduct 20 trials, and the mean results are calculated.

Table 2 shows the mean classification accuracy of the five algorithms on the five data. It can be found that the ensemble systems based on ARPSO including the E-ARPSOELM and KGA-ARPSOELM obtains higher test accuracy than other algorithms on all data. The KGA-ARPSOELM has better test accuracy and the generalization performance than the E-ARPSOELM on all cases. Moreover, the KGA-ARPSOELM achieves the least standard test deviation among all algorithms, which indicates the proposed algorithm is the most stable.

Figure 1 shows the diversity of the ensemble systems in the E-ELM, EOS-ELM, E-PSOELM, E-ARPSOELM and KGA-ARPSOELM on all data. The diversity of the ensemble systems obtained by the KGA-ARPSOELM is greater than that in other ensemble of ELMs on the diabetes and satellite image dataset. However, the E-PSOELM has obvious advantage on the segment, wine and zoo dataset. Due to choosing the higher diversity of the ensemble system in the KGA-ARPSOELM, and improve the system structure further, the KGA-ARPSOELM is second only to it in the center position. These results still verified that the KGA-ARPSOELM could improve the diversity of the ensemble system.

Figure 2(a) shows the testing accuracy with different number of basic ELMs in KGA-ARPSOELM. While the number of initial ELMs is 40, the K-means and tournament selection have no impact on the testing accuracy of the ensemble system. Therefore, the accuracy of 40 initial ELMs is significantly lower than that of 80 primal ELMs on all the dataset. Meanwhile, there is a clear downward trend on diabetes and image segmentation dataset, while the number of initial ELMs is more than 80 and 100. Under the comprehensive consideration, choose 100 as the number of basic ELMs is more suitable for the ensemble system.

In the step 4 of KGA-ARPSOELM, the selected ELMs obtained by tournament selection strategy are optimized by ARPSO. Figure 2(b) shows the test accuracy between two approaches on the five datasets with different number of selected ELMs. Two approaches easily achieve higher test accuracy while the number of ELMs is 40, and the test accuracy of KGA-ARPSOELM is higher than or equal to that of the other approach in each trial. Therefore, it is necessary to classify and select the ELMs obtained by the bagging method.

Table 2. Classification accuracy of five algorithms on five data

Datasets	Algorithm	Training accuracy	Testing accuracy	Standard deviation
Diabetes	ELM	0.7850	0.8146	0.0121
	OS-ELM	0.7836	0.8182	0.0126
	E-PSOELM	0.8176	0.8316	0.0085
	E-ARPSOELM	0.8223	0.8359	0.0077
	KGA-ARPSOELM	0.8236	0.8438	0.0075
Wine	ELM	0.9887	0.9759	0.0189
	EOS-ELM	0.9975	0.9886	0.0087
	E-PSOELM	0.997	0.9888	0.0083
	E-ARPSOELM	1	0.9914	0.0088
	KGA-ARPSOELM	1	1	0
Image segmentation	ELM	0.9725	0.9513	0.0039
	EOS-ELM	0.9736	0.9523	0.0035
	E-PSOELM	0.9772	0.9530	0.0030
	E-ARPSOELM	0.9828	0.9562	0.0025
	KGA-ARPSOELM	0.9829	0.9627	0.0030
Zoo	ELM	0.9957	0.9387	0.0346
	EOS-ELM	0.9863	0.9435	0.0329
	E-PSOELM	1	0.9613	0.0307
	E-ARPSOELM	1	0.9765	0.0275
	KGA-ARPSOELM	1	0.9884	0.0192
Satellite image	ELM	0.9209	0.8913	0.0028
	EOS-ELM	0.9215	0.8923	0.0026
	E-PSOELM	0.9211	0.8914	0.0033
	E-ARPSOELM	0.9316	0.8976	0.0023
	KGA-ARPSOELM	0.9319	0.9015	0.0021

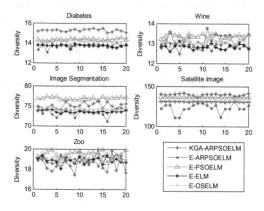

Fig. 1. The diversity of the ensemble systems in the five algorithms with 20 trials (Color figure online)

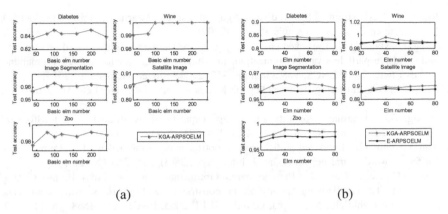

(a)   (b)

Fig. 2. (a) The testing accuracy on five data with different number of the initial ELMs (b) The testing accuracy on five data with different number of selected ELMs

5 Conclusions

In this paper, KGA-ARPSOELM was proposed to cluster and select the basic ELMs to improve accuracy of the ensemble system. In the first phase, K-means was used to improve the diversity of ensemble system. In the second phase, the fitness was determined by the accuracy of each ELM while tournament selection was used to select base ELMs. Finally, the selected ELMs were optimized by the optimization strategy of ARPSO. Experimental results on the five data verified that the proposed approach outperforms the original ELM algorithm. Future work will include how to establish an effective and simple ensemble system of ELMs and apply to other classification and selection.

Acknowledgements. This work was supported by the National Natural Science Foundation of China (Nos. 61271385, 61572241) and the Initial Foundation of Science Research of Jiangsu University (No. 07JDG033).

References

1. Huang, G.B., Zhu, Q.Y., Siew, C.K.: Extreme learning machine: a new learning scheme of feedforward neural networks. In: IEEE International Joint Conference on Neural Networks, vol. 2, pp. 985–990. IEEE Press (2004)
2. Huang, G.B., Zhu, Q.Y., Siew, C.K.: Extreme learning machine: theory and applications. Neurocomputing **70**(1), 489–501 (2006)
3. Xu, Y., Shu, Y.: Evolutionary extreme learning machine – based on particle swarm optimization. In: Wang, J., Yi, Z., Żurada, J.M., Lu, B.-L., Yin, H. (eds.) ISNN 2006. LNCS, vol. 3971, pp. 644–652. Springer, Heidelberg (2006)
4. Kennedy, J., Eberhart, R.: Particle swarm optimization. In: IEEE International of First Conference on Neural Networks (1995)

5. Riget, J., Vesterstrøm, J.S.: A diversity-guided particle swarm optimizer-the ARPSO. Department of Computer, Technical report (2002)
6. Yang, D., Han, F.: An improved ensemble of extreme learning machine based on attractive and repulsive particle swarm optimization. In: Huang, D.-S., Bevilacqua, V., Premaratne, P. (eds.) ICIC 2014. LNCS, vol. 8588, pp. 213–220. Springer, Heidelberg (2014)
7. Anil, K.J.: Data clustering: 50 years beyond K-means. Pattern Recogn. Lett. **31**(8), 651–666 (2010)
8. Huang, G.B.: Learning capability and storage capacity of two-hidden-layer feedforward networks. IEEE Trans. Neural Netw. **14**(2), 274–281 (2003)
9. Han, F., Zhu, J.S.: Improved particle swarm optimization combined with backpropagation for feedforward neural networks. Int. J. Intell. Syst. **28**(3), 271–288 (2013)
10. Zeng, J.C., Jie, J., Cui, Z.H.: Particle Swarm Optimization. Science Press, Beijing (2004)
11. Frühwirth, T.: Parallelizing union-find in constraint handling rules using confluence analysis. In: Gabbrielli, M., Gupta, G. (eds.) ICLP 2005. LNCS, vol. 3668, pp. 113–127. Springer, Heidelberg (2005)
12. Wang, D., Alhamdoosh, M.: Evolutionary extreme learning machine ensembles with size control. Neurocomputing **102**(2), 98–110 (2013)
13. Soria Olivas, E., Gómez Sanchis, J., Martín, J.D., Vila Francés, J., Martínez, M., et al.: BELM: Bayesian extreme learning machine. IEEE Trans. Neural Netw. **22**, 505–509 (2011)

An Improved LMDS Algorithm

Taiguo Qu[✉] and Zixing Cai

School of Information Science and Engineering, Central South University,
Changsha 410083, Hunan, China
qutaiguo88@sohu.com

Abstract. Classical multidimensional scaling (CMDS) is a widely used method
for dimensionality reduction and data visualization, but it's very slow. Landmark MDS (LMDS) is a fast algorithm of CMDS. In LMDS, some data points
are designated as landmark points. When the intrinsic dimension of the landmark points is less than the intrinsic dimension of the data set, the embedding
recovered by LMDS is not consistent with that of classical multidimensional
scaling. A selection algorithm of landmark points is put forward in this paper to
ensure the intrinsic dimension of the landmarks to be equal to that of the data
set, so as to ensure that the embedding recovered by LMDS is the same as that
of CMDS. By introducing the selection algorithm into the original LMDS, an
improved LMDS algorithm called iLMDS is presented in this paper. The
experimental results verify the consistency of iLMDS and CMDS.

Keywords: Classical multidimensional scaling · Landmark points · Intrinsic
dimension · Selection algorithm

1 Introduction

Given the distances between the data points, the classical multidimensional scaling
(CMDS) attempts to embed them in a low-dimensional Euclidean space, such that the
distances in the space match, as well as possible, the original distances [1]. CMDS is a
common method for dimensionality reduction and data visualization [2–8] and it's
contained in MATLAB, SAS, SPSS, STATISTICA, S-Plus, etc.

CMDS has a closed-form solution, but it requires $\Theta(N^3)$ time, where N is the data
size. To improve the speed of CMDS, many algorithms [9–14] have been proposed.
Based on spring-mass model, [9–11] tries to find the embedding of the data set by
minimizing the cost function. Chalmers's algorithm [9] requires $\Theta(N^2)$ time, which is
further reduced to $\Theta(NlgN)$ by Morrison's algorithm [10]. But the latter can only give
the two-dimensional embedding. Williams et al. [11] improved Chalmers's algorithm
on that it can be applied to higher dimensional and larger data sets. These algorithms
are iterative and they are easy to fall into local minimum. There are some other kind of
fast algorithms such as FastMap [12]. In FastMap, the samples are projected onto a set
of orthogonal pivots. All the above-mentioned algorithms are faster than CMDS, but
they are approximate solutions. Therefore, how to obtain the same solution as that of
CMDS with high speed is still to be further studied.

Landmark MDS [13] designated a set of landmark points firstly, then apply CMDS
to find an embedding of the landmarks, and finally apply distance-based triangulation to

© Springer International Publishing Switzerland 2016
Y. Tan et al. (Eds.): ICSI 2016, Part II, LNCS 9713, pp. 97–104, 2016.
DOI: 10.1007/978-3-319-41009-8_10

find the embedding of the data set. LMDS is much faster than CMDS, therefore, it has been widely used [14–16].

In LMDS, the landmark points play a key role. When the intrinsic dimension of the landmarks is equal to the intrinsic dimension of the data set, the solution of LMDS is the same as that of CMDS, while the intrinsic dimension of the landmark points is less than that of the data set, the solution of LMDS is just the orthogonal projection of the data set onto the subspace spanned by the landmarks.

Random choice and MaxMin [13] are two common methods for landmark selection. By MaxMin, landmark points are chosen one at a time, and each new landmark maximizes, over all unused data points, the minimum distance to any of the existing landmarks. The first point is chosen arbitrarily. MaxMin is a little expensive, which is not in the spirit of LMDS. Some other selection methods are proposed in [17–19]. But all the above-mentioned methods can't guarantee that the intrinsic dimension of the landmarks to be equal to that of the data set.

A selection criterion of landmarks is proposed in this paper to ensure the solution of LMDS is the same as that of CMDS.

2   Intrinsic Dimension

Definition. The intrinsic dimension of a data set is the dimension of the smallest Euclidean space containing such points [20].

Consider a set of data points with coordinates x_i, $i = 1,...,N$, where N is the data size, the intrinsic dimension of the set is as follows:

$$m = rank(X_{N-1} - x_N 1_{N-1}^T) = rank(\, x_1 - x_N \quad \cdots \quad x_{N-1} - x_N \,)$$

where m denotes the intrinsic dimension of the data set, $X_{N-1} = (\, x_1 \quad \cdots \quad x_{N-1} \,)$, 1_{N-1} denotes a $(N-1)$-dimensional column vector of ones.

Let $H_N = I - \frac{1}{N} 1_N 1_N^T$, where I denotes the identity matrix. From this it follows

$$X_N H_N = (\, x_1 - \mu_X \quad \cdots \quad x_{N-1} - \mu_X \quad x_N - \mu_X \,)$$

where $\mu_X = \frac{1}{N} \sum_{i=1}^{N} x_i$ is the mean of the data points.

For any $i = 1,\ldots,N-1$,

$$x_i - x_N = (x_i - \mu_X) - (x_N - \mu_X) \tag{1}$$

For any $i = 1,\ldots,N$,

$$x_i - \mu_X = (x_i - x_N) - (\mu_X - x_N) = (x_i - x_N) - (\frac{1}{N} \sum_{j=1}^{N} x_j - x_N)$$

$$= (x_i - x_N) - \frac{1}{N} \sum_{j=1}^{N-1} (x_j - x_N) \tag{2}$$

It can be seen from (1) and (2) that the vectors in $X_N H_N$ and those in $X_{N-1} - x_N 1_{N-1}^T$ can be mutually linearly expressed, so

$$rank(X_N H_N) = rank(X_{N-1} - x_N 1_{N-1}^T) = m$$

Let $d_{ij} = \sqrt{(x_j - x_i)^T (x_j - x_i)}$ be the distance between points i and j, $D = (d_{ij})$, let

$$A = (a_{ij}) = (-\frac{1}{2} d_{ij}^2) \tag{3}$$

$$B = H_N A H_N \tag{4}$$

It follows at once that

$$B = (XH)^T (XH)$$

therefore, $m = rank(XH) = rank(B)$.

3   CMDS Algorithm and LMDS Algorithm

CMDS algorithm can be summarized as follows (Table 1):

Table 1. CMDS algorithm

$Y^{CMDS} = CMDS(D, m)$

Step1: Find matrix A according to (3).

Step2: Find matrix B according to (4).

Step3: Find the positive eigenvalues $\lambda_1 \geq ... \geq \lambda_m > 0$ and the associated eigenvectors

$v_1, ..., v_m$ such that $v_i^T v_j = \delta_{ij}$, $i, j = 1, ..., m$.

Step4: The coordinates of the data points in the m dimensional Euclidean space are given

by $y_{ij}^{CMDS} = \sqrt{\lambda_j} v_{ji}$ ($i = 1, ..., N$, $j = 1, ..., m$).

By using the iterative method based on QR decomposition to carry out the spectral decomposition, CMDS requires $O(N^3)$ time, so it's very slow. Many fast methods are put forward, one of which is LMDS. LMDS consists of four steps (Table 2).

According to the literature [13], the embedding vector z_a gives the components of x_a when projected onto the principal axes of the landmark points. When the intrinsic dimension of the landmarks is equal to that of the data set, the LMDS coordinates are in agreement with the CMDS coordinates. But, if the landmarks span a lower-dimensional affine subspace, then LMDS recovers the orthogonal projections of the points onto this subspace.

Table 2. LMDS algorithm

$Y^{LMDS} = LMDS(\boldsymbol{D}, m)$

Step 1. Select n landmark points (randomly), without loss of generality, assume the first n points are selected as the landmarks.

Step 2. Apply classical MDS on landmarks

  2.1. Find matrix $\boldsymbol{A}_n = (\boldsymbol{\delta}_1, ..., \boldsymbol{\delta}_n)$, where $[\boldsymbol{A}_n]_{ij} = -d_{ij}^2 / 2$, $i, j=1, ..., n$.

  2.2. Find matrix $\boldsymbol{B}_n = \boldsymbol{H}_n \boldsymbol{A}_n \boldsymbol{H}_n$, where $[\boldsymbol{H}_n]_{ij} = \delta_{ij} - 1/n$.

  2.3. Find the positive eigenvalues $\xi_1 \geq ... \geq \xi_{m_1} > 0$ and associated eigenvectors

   $\boldsymbol{u}_1, ..., \boldsymbol{u}_{m_1}$ such that $\boldsymbol{u}_i^T \boldsymbol{u}_j = \delta_{ij}$, $i, j = 1, ..., m_1$, where $m_1 = rank(\boldsymbol{B}_n)$ is the intrinsic

   dimension of the landmarks.

Step 3. The embedding vector z_a of data point a is given by the formula:

$$z_a = -\frac{1}{2} L^\#(\delta_a - \delta_\mu), \text{ where } L^\# = \begin{pmatrix} \boldsymbol{u}_1^T / \sqrt{\xi_1} \\ ... \\ \boldsymbol{u}_{m_1}^T / \sqrt{\xi_{m_1}} \end{pmatrix}, \delta_\mu = \frac{1}{n}\sum_{i=1}^{n}\delta_i, \ \delta_a \text{ is the squared}$$

  distances between a and the landmarks.

Step 4. Return the PCA-normalized coordinates Y^{LMDS} of the embedding vectors $z_1, ..., z_N$.

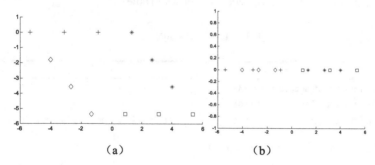

(a)   (b)

Fig. 1. (a) The coordinates of the data set; (b) The embedding recovered by LMDS.

In Fig. 1, the three points represented by "+" are selected as the landmarks, the subspace spanned by them is 1-dimensional (a line), the embedding recovered by LMDS are just the projections of the data set onto a line.

According to the literature [13], LMDS requires $O(mnN + n^3)$ time and $O(nN)$ space, while CMDS requires $O(N^3)$ time and $O(N^2)$ space.

4   The Selection Algorithm of Landmark Points

To begin with, let m_1, m denote the intrinsic dimensions of the landmarks and the data set respectively. As mentioned before, the selection of "proper" landmarks is the key to the successful recovering by LMDS. To ensure m_1 to be equal to m, a landmark selection algorithm (Table 3) is proposed. Without loss of generality, the first n points are designated as landmarks. n is initialized to be m. If the dimension of the landmarks is less than m, one more point is added until the dimension of the landmarks is equal to m.

By introducing the selection algorithm into LMDS, an improved LMDS algorithm called iLMDS is put forward in this paper. In iLMDS, the number of landmarks is automatically increased until the dimension of the landmarks is equal to that of the data set, so the embedding recovered by iLMDS is the same as that of CMDS. Figure 2 is the embedding recovered by iLMDS of the data set in Fig. 1.

The time spent in the selection of landmarks is usually far less than the time of LMDS, iLMDS preserves the linear time complexity of LMDS.

Table 3. The selection algorithm of landmark points

$n = LandmarkSelection(\boldsymbol{D}, m)$
Step1. Let $m_1 = 0$, $n = m$;
Step2. while $m_1 < m$
$\quad\quad\quad n++$;
$\quad\quad\quad$ Find matrix $\boldsymbol{\varDelta}_n$, where $[\boldsymbol{\varDelta}_n]_{ij} = -d_{ij}^2 / 2$, $i, j = 1, ..., n$.
$\quad\quad\quad$ Find matrix $\boldsymbol{B}_n = \boldsymbol{H}_n \boldsymbol{\varDelta}_n \boldsymbol{H}_n$, where $[\boldsymbol{H}_n]_{ij} = \delta_{ij} - 1/n$
$\quad\quad\quad m_1 = rank(\boldsymbol{B}_n)$;
$\quad\quad$ end
Step 3. Return n.

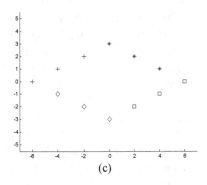

(c)

Fig. 2. The embedding recovered by iLMDS

5   Experiments

The experiments are carried on Acer TM4330. In the first experiment, the performance of iLMDS and CMDS are compared. In the second experiment, the relationship between the running time and the data size is studied.

5.1   The Comparison of iLMDS and CMDS

The experiment is carried out on USPS [21] and some data sets from UCI [22]. Limited by the performance of the computer, for those data sets with more than 4000 samples, only 4000 samples are taken from each of them. The experimental results are shown in Table 4, where N, m and n denote data size, intrinsic dimension and the number of landmarks respectively. tiLMDS and tCMDS denote time of iLMDS and CMDS.

To quantitatively analyze the difference between the two algorithms, we define:

$$\varepsilon = \max_{i,j} \frac{\left| y_{i,j}^{iLMDS} - y_{i,j}^{CMDS} \right|}{\sigma_j}$$

where $y_{i,j}^{iLMDS}$, $y_{i,j}^{CMDS}$ is the j-th component of y_i^{iLMDS}, y_i^{CMDS} respectively, σ_j is the standard deviation of the j-th component of $\{y_{1j}^{CMDS}, \ldots, y_{N,j}^{CMDS}\}$.

From Table 4 we can see: (1) For most data sets, n is equal to or slightly greater than $m + 1$, but there are also some data sets, where n is much larger than $m + 1$, which leads to an increase in time; (2) iLMDS is much faster than CMDS. For most sets, the time of iLMDS is less than 1 s; (3) The time of iLMDS increases with N and m, which validates the theoretical analysis in Sect. 4; (4) For each set, the difference between the two algorithms is very small. The difference is caused by rounding errors of the computer. This fully verifies that the solution of iLMDS is in agreement with that of CMDS.

Table 4. The comparison of iLMDS and CMDS

Data set	N	m	n	tiLMDS(s)	tCMDS(s)	ε
Iris	150	4	6	0.0019	0.0370	3.8×10^{-9}
Connectionist bench	208	60	61	0.0292	0.0540	4.4×10^{-9}
Seeds	210	6	7	0.0538	0.0636	9.9×10^{-8}
Glass	214	5	6	0.0335	0.0625	1.2×10^{-7}
User knowledge	258	5	6	0.0738	0.136	1.4×10^{-9}
Wholesale	440	6	7	0.0025	0.913	1.9×10^{-9}
Wine quality (red)	1599	9	11	0.0331	58.2	3.3×10^{-9}
Texture	1599	33	34	0.0923	59.6	2.0×10^{-8}
Margin	1600	45	46	0.0632	57.5	8.1×10^{-7}
Shape	1600	64	65	0.1029	56.0	4.7×10^{-9}
Wine quality (white)	4000	9	14	0.0575	901	8.7×10^{-6}
Letter recognition	4000	16	17	0.0596	890	3.4×10^{-10}
Waveform version 1	4000	21	22	0.0785	890	7.3×10^{-9}
Waveform version 2	4000	40	41	0.2221	896	4.7×10^{-8}
USPS	4000	256	499	132.0	812	1.1×10^{-8}

5.2   The Relationship Between the Runtime of ILMDS and the Data Size

In order to quantitatively analyze the relationship between the data size and the time of iLMDS and CMDS, MATLAB is used to generate 40 20-dimensional random data sets, with $N = 100,200,\ldots,4000$. The time of dcMDS and CMDS are shown in Fig. 3:

It can be seen from Fig. 3 that iLMDS has a substantial increase in speed, for example, tiLMDS(4000) = 0.0555 s, tCMDS(4000) = 1021.8 s. The time of iLMDS falls between $N/45000$ and $N/150000$, while CMDS falls between $2.5 \times 10^{-8} \times N^3$ and $9.2 \times 10^{-9} \times N^3$, which confirms the previous theoretical analysis.

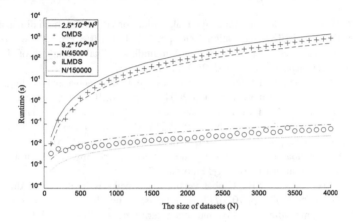

Fig. 3. Runtime of iLMDS and CMDS vs. the size of datasets

6   Conclusion

Dimensionality reduction and data visualization are commonly encountered problems in the fields of statistics, machine learning, and knowledge discovery. A selection criterion of landmarks is proposed in this paper to ensure that the intrinsic dimension of the landmarks is equal to the intrinsic dimension of the data set. The same solution as that of CMDS can be guaranteed by the improved LMDS with little extra time. Therefore, the improved LMDS can be applied to larger data sets.

References

1. Trevor, F.C., Michael, A.A.C.: Multidimensional Scaling. Chapman & Hall/CRC, London (2001)
2. Ingram, S., Munzner, T.: Dimensionality reduction for documents with nearest neighbor queries. Neurocomputing **150**, 557–569 (2015)
3. Mjolsness, E., DeCoste, D.: Machine learning for science: state of the art and future prospects. Science **293**(5537), 2051–2055 (2001)

4. Patel, A.P., Tirosh, I., Trombetta, J.J., et al.: Single-cell RNA-seq highlights intratumoral heterogeneity in primary glioblastoma. Science **344**(6190), 1396–1401 (2014)
5. Wang, S., Zhuo, Z., Yang, H., et al.: An approach to facial expression recognition integrating radial basis function kernel and multidimensional scaling analysis. Soft. Comput. **18**(7), 1363–1371 (2014)
6. Vogelstein, J.T., Park, Y., Ohyama, T., et al.: Discovery of brainwide neural-behavioral maps via multiscale unsupervised structure learning. Science **344**(6182), 386–392 (2014)
7. Arkin, A., Shen, P., Ross, J.: A test case of correlation metric construction of a reaction pathway from measurements. Science **277**(5330), 1275–1279 (1997)
8. Li, S.S.-M., Wang, C.C.L., Kin, C.H.: Bending-Invariant correspondence matching on 3-D human bodies for feature point extraction. IEEE Trans. Autom. Sci. Eng. **8**(4), 805–814 (2011)
9. Chalmers, M.: A linear iteration time layout algorithm for visualizing high dimensional data. In: Proceedings of IEEE Visualization 1996, pp. 127–132. IEEE Press, New York (1996)
10. Morrison, A., Ross, G., Chalmers, M.: Fast multidimensional scaling through sampling, springs and interpolation. Inf. Visual. **2**(1), 68–77 (2003)
11. Williams, M., Munzner, T.: Steerable, progressive multidimensional scaling. In: IEEE Symposium on Information Visualization 2004, pp. 57–64. IEEE Press, Austin (2004)
12. Faloutsos, C., Lin, K.: FastMap: a fast algorithm for indexing, data-mining, and visualization. In: Proceedings of the 1995 ACM SIGMOD International Conference on Management of Data, pp. 163—174. ACM Press, Washington (1995)
13. Silva, V.D., Tenenbaum, J.B.: Sparse multidimensional scaling using landmark points. Technical report, Stanford University, Palo Alto (2004)
14. Silva, V.D., Tenenbaum, J.B.: Global versus local methods in nonlinear dimensionality reduction. In: NIPS 2002, pp. 705–712. MIT Press, Cambridge, Massachusettes (2003)
15. Platt, J.C.: Fast embedding of sparse music similarity graphs. In: NIPS 2003, pp. 571–578. MIT Press, Cambridge, Massachusettes (2004)
16. Brandes, U., Pich, C.: An experimental study on distance-based graph drawing. In: Tollis, I. G., Patrignani, M. (eds.) GD 2008. LNCS, vol. 5417, pp. 218–229. Springer, Heidelberg (2009)
17. Silva, J., Marques, J., Lemos, J.: Selecting landmark points for sparse manifold learning. In: NIPS 2005, pp. 1241–1248. MIT Press, Cambridge, Massachusettes (2006)
18. Gu, R.J., Xu, W.B.: An improved manifold learning algorithm for data visualization. In: IEEE International Conference on Machine Learning and Cybernetics 2006, pp. 1170–1173. IEEE Press, Piscataway, N.J. (2006)
19. Li, J., Hao, P.: Finding representative landmarks of data on manifolds. Pattern Recogn. **42**(11), 2335–2352 (2009)
20. Young, G., Householder, A.S.: Discussion of a set of points in terms of their mutual distances. Psychometrika **3**(1), 19–22 (1938)
21. Data for MATLAB hackers. http://www.cs.toronto.edu/~roweis/data.html
22. UCI Machine Learning Repository. http://archive.ics.uci.edu/ml/datasets.html

Clustering Algorithm

An Improved K-means Clustering Algorithm Based on the Voronoi Diagram Method

Jiuyuan Huo[✉] and Honglei Zhang

College of Electronic and Information Engineering,
Lanzhou Jiaotong University, 730070 Lanzhou, China
{huojy,zhanghonglei}@mail.lzjtu.cn

Abstract. To solve the problem that the K-means clustering algorithm is over dependent on the K value and the clustering center, we proposed an improved K-means clustering algorithm, VK-means algorithm in this paper. In the initial stage, the Voronoi diagram was adapted in the K-means algorithm to get a better K value and clustering center. By means of weighted average of K-means algorithm, the results of the criterion function is improved. This method could fast convergence and improve performance of the algorithm. The superiority of the improved algorithm was verified by experiments on Weka platform.

Keywords: K-means · Clustering center · Criterion function · Voronoi diagram · Weight

1 Introduction

K-means clustering algorithm is a typical and most popular clustering algorithm that based on partitioning method. Since it was proposed in 1967, with the characters of reliability theory, simple and fast convergence, the algorithm has been the subject of widespread concern and has been widely used. At present, the clustering algorithms based on K-means have become one of the most important research directions in the field of data mining clustering technology. In recent years, using clustering analysis, especially the K-means clustering algorithm for mining and analyzing data have spread to all areas of finance, health care and banking, and its development has important practical significance for various fields.

2 K-means Clustering Algorithm

Clustering analysis is a data mining technology that based on the different thought. It could be divided into many types, such as partitioning algorithms, hierarchical algorithms, algorithms based on density, algorithms based on grid and algorithms based on model. Among them, the K means clustering algorithm is one of the most popular methods based on partitioning.

© Springer International Publishing Switzerland 2016
Y. Tan et al. (Eds.): ICSI 2016, Part II, LNCS 9713, pp. 107–114, 2016.
DOI: 10.1007/978-3-319-41009-8_11

2.1   Researches of K-means Clustering Algorithm

The K-means clustering algorithm was proposed by J.B. MacQueen in 1967. Its simplicity and efficiency make it become a popular cross disciplinary clustering method. After many years' development, the clustering method has made remarkable achievements, and many of which are based on K-means to make up the deficiency of K-means clustering algorithm.

For examples, in the literature [1], a new K-means algorithm for dynamic selection of the initial cluster center point was proposed. The minimum spanning tree is constructed in the algorithm to obtain the initial cluster centers. The literature [2] presented an initialization method based on hierarchy. The method can find better initial clustering centers to short the iteration time and get an excellent performance. A K-means algorithm based on distributed architecture is proposed in the paper [3] that is feasible to apply the new matrix pruning algorithm to the very large data set, and it can improve the performance of clustering results. A new clustering algorithm based on improved artificial bee colony (ABC) is proposed in the paper [4]. The two strategies (reverse learning initialization strategy and nonlinear selection strategy) improved the original ABC algorithm and combined it with the K-means clustering algorithm, thus, it improves the robustness of the algorithm and the clustering effect, and obtains higher stability.

2.2   Theory of K-means Algorithm

The main thought of K-means algorithm [5, 6]: first specify the number of clusters K, and randomly select the K initial clustering centers from the data set X that has a given size N; then according to the similar features of the data object and the initial cluster centers to assign all remaining data objects to the K initial cluster centers based on similarity to the closed clustering center, then the K disjoint stable cluster set is generated; then adjust the division of initial K clusters by updating the clusters' center, and recalculate the cluster center of each cluster; compare the new clustering center and the last time clustering center is the same, if they are different, then this process will be repeated until each cluster center does not change. And when the overall difference function converges, then end the process.

Basic Steps of K-means Algorithm. Given a data set X which containing N data objects, the number of clusters of K and the initial clustering center $c_1, c_2, \ldots, c_k$, the K clusters were represented by $C_1, C_2, \ldots, C_k$, and N_i represents the number of data objects in C_i. The basic steps of K-means clustering algorithm is described as follows:

   Input: a data set X containing N data objects, the cluster number K.
   Output: K clusters $C_1, C_2, \ldots, C_k$.

(1) Random select the K initial cluster centers from data set X;
(2) According to the initial cluster center $c_1, c_2, \ldots, c_k$ to divide the data set, and calculate the Euclidean distance of the remaining data object to the initial cluster centers, and clustering by the distance;

(3) According to the formula (3) to re-calculate the centroid of the cluster and defines it as $c_1^*, c_2^*, c_3^*, \ldots, c_k^*$;

(4) If for $\forall i \in \{i = 1, 2, \ldots, k\}$, there are $c_i = c_i^*$, then the algorithm ends, the final formation of the cluster is $c_1^*, c_2^*, c_3^*, \ldots, c_k^*$, otherwise, the algorithm return step (2);

(5) Output the clustering results.

Deficiencies of K-means Algorithm. K-means algorithm shows good performance in clustering, and its defects are also very significant, as described in the Sect. 2.1. Its main problems are: (1) the selection of K value of the algorithm has a great impact on the results of clustering. If the selected value of K is not ideal, it will produce a result that is not ideal or even almost useless. (2) The selection of the initial cluster centers will directly affect the merits of the algorithm clustering. If the center points were not appropriately selected, it will easy to obtain the local optimal solution, and will generate the problem that the algorithm iteration cannot be finished. (3) The performance of the algorithm for large data volume is not ideal. The K-means clustering algorithm has the limitation for the amount of data. In the iteration cycle, the algorithm updates the data objects in the cluster and calculates the cluster centers, thus, it will lower the efficiency of the search strategy in the case of large number of data, and increase the time cost of the algorithm. (4) The clustering result is easily affected by the isolated points. (5) The clustering results are not ideal for the uneven shape cluster.

3 Improvement of K-means Clustering Algorithm

According to the difficult to determine the K value and the clustering center points of the K-means clustering algorithm, thus, the unsuitable selection will lead the algorithm easy to the problem of falling into the local optimum. Firstly, through adding Voronoi diagram in the initial phase of the algorithm to preprocess data, and to adaptive select the K value and the clustering centers; improve the criterion function by the weighted average method to get the global optimal solution. And at last, the improved clustering algorithm was analyzed and compared through the Weka software to [7] verify its superiority. It contains tools for data pre-processing, classification, regression, clustering, association rules, and visualization.

3.1 Voronoi Diagram

Voronoi diagram is a basic concept in computational geometry, and is also called the Tyson polygon or Dirichlet diagram [8]. It is composed of a set of continuous polygons consisting of vertical split lines connected to two adjacent points in a straight line. According to the N points which are not similar on the plane, the region is divided into areas by the principle of the nearest neighbor; each point correlates with its closest region.

There are many construction algorithms for the construction of Voronoi diagram. In this paper, we choose the simple incremental method. At the same time, the construction of the Voronoi diagram is carried out in the two-dimensional space plane, thus, we need to transform the data into two-dimensional.

In this paper, we first make the two-dimensional data sets to the two-dimensional closed region of the whole data, and then construct the Voronoi diagram in the two-dimension. Then, the area of the polygons of Voronoi diagram can be calculated through the polyarea function in Matlab software.

3.2   Criterion Function Based on the Weighted Average Method

The idea of the improved criterion function is to calculate the weighted average value of each object in the cluster, and divide each object in the data set X to the cluster which has the highest similarity, and repeat the operation. Finally, the criterion function J converges, and the global optimal solution is obtained.

Given a data set X containing N data objects, the number of clusters K and initial clustering centers $c_1, c_2, \ldots, c_k$. The K cluster is represented as $C_1, C_2, \ldots, C_k$, N_i represents the number of data objects in C_i. Then, the weights assigned to each data object as shown in the formula (1):

$$\lambda_j = \frac{\lambda_j^*}{\sum\limits_{j=1}^{N} \lambda_j^*}. \tag{1}$$

Among them, $\lambda_j^* = \dfrac{\sum\limits_{j=1}^{N} d(x_j, y_j)}{N}$, N represents the total number of data objects in the data set X, $d(x_j, y_j)$ represents the distance between the data object x_j and data object y_j. From the formula, we can know that the smaller weight means the higher degree of similarity, and the greater value means the greater difference.

The weighted average value of each object in the cluster is defined as the formula (2):

$$c_i^* = \frac{1}{N_i} \sum\limits_{\substack{j=1 \\ x_j \in C_i}}^{N} \lambda_j x_j. \tag{2}$$

Among them, N_i represents the number of data objects in C_i, λ_j represents the weight of the data objects in the cluster, and x_j represents a data object in the cluster.

In order to improve the clustering quality, and ensure the criterion function converges and reaches the global optimum. A weight coefficient δ_i was adopted in the traditional criterion to make the data objects tend to divide to the clusters that have a small number of data objects. The improved criterion formula as shown in formula (3):

$$J^* = \sum_{i=1}^{K} \delta_i \sum_{\substack{j=1 \\ x_j \in C_i}}^{N} distance(x_j, c_i^*)^2. \tag{3}$$

Among them, $\delta_i = \frac{N_i}{N}$, and $distance(x_j, c_i^*)$ represents the Euclidean distance between the x_j and c_i^*.

4 An Improved K-means Algorithm - VK-means Algorithm

4.1 VK-means Algorithm

Input: a data set X containing N data objects, the number of the initial Voronoi units m and density threshold η_0.

Output: K cluster $C_1, C_2, \ldots, C_k$.

(1) Preprocess Data: generating a set of data points D in the two-dimensional plane region.
(2) Generate the number of cluster K.
  (a) Generate Voronoi diagram: construct a closed planar rectangular area of the data, randomly generate the initial Voronoi mother points and generate the Voronoi diagram in a rectangular area.
  (b) Calculate density gain η_i: calculate data point density of the rectangular area; calculate the data point density of each Voronoi diagram; calculate the density coefficient η_i of each Voronoi diagram;
  (c) Judge whether the inequation $\forall \eta_i, \exists \eta_i >= \eta_0$ is satisfied? If it was established, output the m value which is the number of clusters K; conversely, removing the j-th Voronoi diagram unit which not established; calculate the centroid of the remaining established X-j Voronoi diagram units as the new Voronoi mother point, and repeat the step (1) and step (2).
(3) According to the formula (6) to calculate the centroid of data objects in each unit, and re-divide the data set. Repeat this process, and calculate the value of the target criterion function based on the formula (7) to gradually reduce the value, until the convergence is no longer change.
(4) Output the clustering result.

The flow chart of the improved algorithm VK-means is shown in Fig. 1.

4.2 Experiments Analysis

We have taken the experiments of the K-means clustering algorithm and the VK-means clustering algorithms on the Weka platform for the campus card consumption data of Lanzhou Jiaotong University. The experiments intercepted the students' consumption transaction flow in campus canteens from March 2014 to June 2014. The improved VK-means algorithm was integrated into the Weka platform for the experiments.

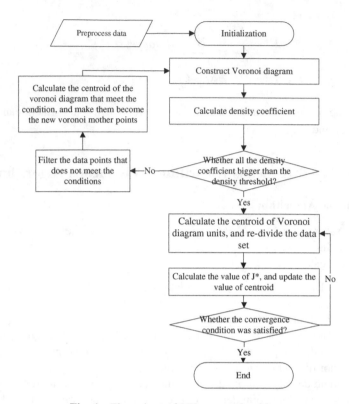

Fig. 1. Flow chart of VK-means algorithm.

The initialization parameters of the algorithm were set as follows: the initial Voronoi unit number $m = 50$ (initial cluster number k), the initial Voronoi mother points randomly generated $seed = 100$, as well as the density threshold $\eta_0 = 0.25$.

In this paper, the experiments were mainly carried on from two aspects: the relationship between the testing times and the clustering accuracy, the relationship between the amount of data and operation time. The comparing results of experiments were shown in Figs. 2 and 3.

Under the same parameters and data amount, it can be seen from the Fig. 2 that the clustering accuracy of VK-means clustering algorithm is obviously more stable and more accurate than the K-means clustering algorithm. The average clustering accuracy of VK-means algorithm is about 86.86, and K-means clustering algorithm's average clustering accuracy is 74.096. And it can be drawn in Fig. 3 that although with the increase of the amount of data, the execution time of the two kinds of algorithms are all in a rapid increase, but the increasing trend of VK-means clustering algorithm is more slowly than the K-means clustering algorithm which means the improved algorithm has the better data processing performance data the original algorithm.

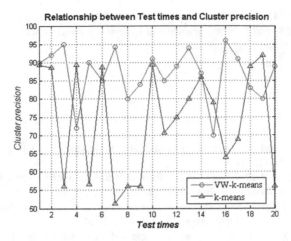

Fig. 2. Relationship between the clustering accuracy and the test times.

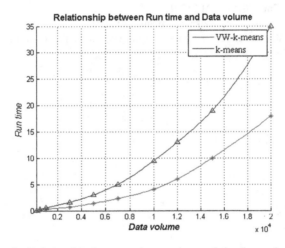

Fig. 3. Relationship between the run time and the data volume.

5 Conclusions

The traditional K-means clustering algorithm has a strong dependence to the K value and initial center points. If the selection is not ideal, it will lead to the worse clustering results. The improved algorithm VK-means algorithm could overcome the above deficiencies by the Voronoi diagram successive screening mechanism to choose the optimal K value and the cluster center points. At the same time, the weighted average criterion function was used to make the algorithm tend to converge more quickly, and to optimize the performance of the algorithm.

Acknowledgments. This work is supported by National Nature Science Foundation of China (Grant No. 61462058) and Lanzhou Science and technology project (2014-1-127).

References

1. Feng, B., Hao, W.N., Chen, G.: Optimization to K-means initial cluster centers. Comput. Eng. Appl. **14**, 182–185 (2013)
2. Ismkhan, H.: A novel fast heuristic to handle large-scale shape clustering. J. Stat. Comput. Simul. **86**, 160–169 (2015)
3. Tsapanos, N., Tefas, A., Nikolaidis, N.: A distributed framework for trimmed kernel k -means clustering. Pattern Recogn. **48**, 2685–2698 (2015)
4. Cao, Y.C., Cai, Z.Q., Shao, Y.B.: An improved artificial bee colony clustering algorithm based on the K-means. Comput. Appl. **34**, 204–207 (2014)
5. Zhang, J., Zhu, Y.Y.: The improvement and application of a K-means clustering algorithm. Electron. Technol. Appl. **41**, 125–128 (2015)
6. Xie, J.Y., Wang, Y.: K-means algorithm based on minimum deviation initialized clustering centers. Comput. Eng. **40**, 205–211 (2014)
7. Weka platform. http://www.cs.waikato.ac.nz/ml/weka/
8. Sabo, K., Scitovski, R.: Interpretation and optimization of the K-means algorithm. Appl. Math. **59**, 391–406 (2014)

Brain Storm Optimization with Agglomerative Hierarchical Clustering Analysis

Junfeng Chen[1]([⊠]), Jingyu Wang[1], Shi Cheng[2], and Yuhui Shi[3]

[1] College of IOT Engineering, Hohai University, Changzhou, China
chen-1997@163.com
[2] Division of Computer Science, University of Nottingham Ningbo, Ningbo, China
shi.cheng@live.com
[3] Department of Electrical and Electronic Engineering,
Xi'an Jiaotong-Liverpool University, Ningbo, China
yuhui.shi@xjtlu.edu.cn

Abstract. Brain storm optimization (BSO) is a relatively new swarm intelligence algorithm, which simulates the problem-solving process of human brainstorming. In General, BSO employs flat clustering which has a number of drawbacks. In this paper, the agglomerative hierarchical clustering is introduced into BSO and its impact on the performance of the creating operator is then analyzed. The proposed algorithm is applied to numerical optimization problems in comparison with the BSO with k-means Clustering. Experimental results show that the proposed algorithm achieves satisfactory results and guarantees a high coverage rate.

Keywords: Brain storm optimization · Agglomerative hierarchical clustering · k-means clustering

1 Introduction

Brain Storm Optimization (BSO) is a newly developed swarm intelligence algorithm, which is inspired by the problem-solving process of human brainstorming. Since its introduction [8–10], BSO has been applied to solving practical problems, such as electric power dispatch problems [5,6], design problems in aeronautics field [4,7], wireless sensor networks [1] and optimization problems in finance [11]. In general, BSO employs three operators: clustering, creating and selecting. The clustering strategy is a sticking point, having a great effect on global optimization performance of BSO. k-means-type clustering is employed in most of BSO variants [3]. Besides, Zhan et al. [12] proposed a simple grouping method with a hope to reducing the computational burden. Chen et al. [2] introduced affinity propagation clustering into BSO by analyzing the clusters' variations over iterations. However, all these clustering methods fall under the category of flat clustering, which has a number of drawbacks, such as requiring a prespecified number of clusters as input and being nondeterministic. In this paper, the hierarchical clustering is employed by BSO and its impact on the performance of

© Springer International Publishing Switzerland 2016
Y. Tan et al. (Eds.): ICSI 2016, Part II, LNCS 9713, pp. 115–122, 2016.
DOI: 10.1007/978-3-319-41009-8_12

the creating operator is carefully analyzed. Introduced affinity propagation clustering into BSO by analyzing the clusters' variations over iterations. However, all these clustering methods fall under the category of flat clustering, which has a number of drawbacks, such as requiring a prespecified number of clusters as input and being nondeterministic. In this paper, the hierarchical clustering is employed by BSO and its impact on the performance of the creating operator is carefully analyzed.

The paper is organized as follows. Section 2 is a brief introduction to fundamentals of BSO. Section 3 presents the proposed BSO with hierarchical clustering algorithm. Comparative performance study on benchmark functions is presented in Sect. 4, followed by conclusion in Sect. 5.

2   Fundamentals of Brain Storm Optimization

The BSO employs three operators: clustering, creating and selecting. The three operation strategies are all connected and have a great impact on the optimization performance. Meanwhile, clustering is a sticking point, which has a big influence on the results of the follow-up creating operator indirectly. The basic creating operator generates new ideas in four patterns. Figure 1 shows the creating patterns of candidate solutions in three clusters.

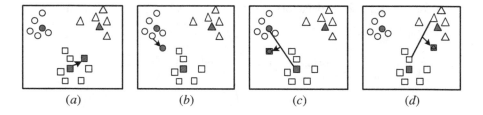

Fig. 1. The creating patterns of candidate solutions. (Color figure online)

As shown in Fig. 1, the various individuals are grouped into three clusters and represented by circular, triangle and squares symbols, respectively. The red symbols are the cluster centers and the blue symbols designate the generated individuals. If the cluster centers are viewed as special individuals, the updating mechanisms of BSO can be summarized as the *inter-cluster creating pattern* and the *intra-cluster creating pattern*. Specifically, the intra-cluster creating pattern (see the creating pattern 1 in Fig. 1(a) and pattern 2 in Fig. 1(b)) are the local search strategies which can yield small variation, while the inter-cluster creating pattern (see the creating pattern 3 in Fig. 1(c) and pattern 4 in Fig. 1(d)) are the global search strategies that create large variation. Thus, it can be said that the basic BSO takes the advantages of fine-grained search and coarse-grained search.

3   Brain Storm Optimization Based on Hierarchical Clustering

3.1   Agglomerative Hierarchical Clustering

In data mining and statistics, hierarchical clustering is a method of cluster analysis which seeks to build a hierarchy of clusters. Strategies for hierarchical clustering generally fall into two types: agglomerative and divisive. In the general case, the complexity of divisive clustering is worse than that of agglomerative clustering. So the agglomerative hierarchical clustering is appropriate for the BSO algorithm. It not only does not require a prespecified number of clusters, but also helps the creating operator to enhance search performance of the exploration and exploitation.

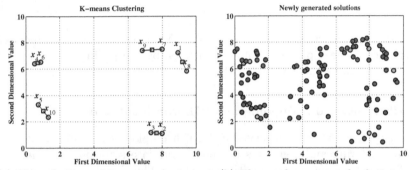

(a) The solutions of k-means clustering. (b) The newly generated solutions.

Fig. 2. The solutions distributions based on k-means clustering. (Color figure online)

Here, an illustration is presented to compare the impact on the creating performance by different clustering methods. We randomly generate candidate solutions $\{x_1, x_2, \cdots, x_{10}\}$ in a limited area. Figure 2 shows the solutions distributions based on k-means clustering and Fig. 3 shows the solutions distributions based on agglomerative hierarchical clustering.

The solutions are grouped into 5 clusters by k-means clustering in Fig. 2(a) and the red and blue circles represent the solutions generated by intra-cluster creating pattern and inter-cluster creating pattern, respectively, as shown in Fig. 2(b). The solutions are merged as a tree with linkage levels by using agglomerative hierarchical clustering as shown in Fig. 3(a). As shown in Fig. 3(b), the red circles represent the generated solutions based on the singleton cluster, which is similar to the intra-cluster creating pattern. The blue and yellow circles represent the solutions generated based on two clusters and three clusters, respectively. It should note that the generated pattern of blue and yellow solutions is analogous to inter-cluster creating pattern.

From Figs. 2(b) and 3(b), we can see that the solutions distribution based on agglomerative hierarchical clustering is more scattered and disperse than

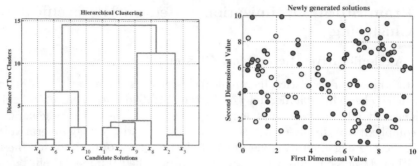

(a) The solutions of hierarchical clustering. (b) The newly generated solutions.

Fig. 3. The solutions distributions based on agglomerative hierarchical clustering. (Color figure online)

those solutions based on k-means clustering. Moreover, we can enhance search performance of the exploration and exploitation by controlling solution numbers of one cluster and multi-clusters.

3.2 BSO Based on Agglomerative Hierarchical Clustering

In this paper, we propose a new BSO, which employs the agglomerative hierarchical clustering and generates the new candidate solutions by synthesizing the information from one or multiple clusters. The detailed procedure of the proposed BSO is listed as Algorithm 1.

As can be seen in Algorithm 1, the proposed BSO algorithm takes the advantages of exploration and exploitation, and it can automatically switch creating mode between the inter-cluster pattern and intra-cluster pattern. Specifically, if the newly generated candidate solutions based on the singleton cluster information have a better overall property, then the proposed BSO algorithm will increase the coefficient p_1 and decrease the coefficients p_2 and p_3, which are probabilities for controlling the creating size of integrating one, two and three clusters, respectively.

4   Numerical Experiments and Performance Analysis

4.1   Description of Functions and Parameters

Numerical Experiments are carried out to test the performance of the proposed algorithm. Table 1 illustrates the functions' mathematical expressions, dimensions, domain, function minimum and name. The first four functions $f_1 \sim f_4$ are unimodal functions and the last four functions $f_5 \sim f_8$ are multimodal functions.

We select two versions of BSOs to perform the simulation experiments, i.e., Brain Storm Optimization based on k-means Clustering (BSO-KC) and Brain

Algorithm 1. Procedure of BSO with agglomerative hierarchical clustering

---

Input: initial solution set $X = (x_1, x_2, \cdots, x_N)$, maximum iteration (MaxIter)

Output: new generated candidate solution x_i, $i = 1, 2, \ldots, n$

1 Initialize the solution set $X = (x_1, x_2, \cdots, x_N)$ randomly;

2 Evaluate the N candidate solutions and we have $f(x_i)$, $i = 1, 2, \cdots, N$;

3 Set the coefficients $p_1 = p_2 = p_3 = 1$;

4 **repeat**

5   Built the hierarchical structure with linkage levels (K) by using the agglomerative hierarchical clustering;

6   create the $p_1 \cdot N$ candidate solutions based on the structure information of singleton cluster at level 0;

7   create the $p_2 \cdot N$ candidate solutions based on the structure information of two clusters at level $round(K/2)$;

8   create the $p_3 \cdot N$ candidate solutions based on the structure information of three clusters at level $round(K/4)$;

9   evaluate the new candidate solutions and obtain the best candidate solution $\hat{x}$;

10   calculate the average fitness values for $\bar{f}(x_i)$, $i = 1, \cdots, p_1 \cdot N$, $\bar{f}(x_j)$, $j = 1, \cdots, p_2 \cdot N$ and $\bar{f}(x_k)$, $k = 1, \cdots, p_3 \cdot N$;

11   **if** $\max\{\bar{f}(x_i), \bar{f}(x_j), \bar{f}(x_k)\} == \bar{f}(x_i)$ **then**

12   set the factors $p_1 = p_1 + 2 \times Iter/MaxIter$; $p_2 = p_2 - Iter/MaxIter$; $p_3 = p_3 - Iter/MaxIter$;

13   **if** $\max\{\bar{f}(x_i), \bar{f}(x_j), \bar{f}(x_k)\} == \bar{f}(x_j)$ **then**

14   set the factors $p_1 = p_1 - Iter/MaxIter$; $p_2 = p_2 + 2 \times Iter/MaxIter$; $p_3 = p_3 - Iter/MaxIter$;

15   **if** $\max\{\bar{f}(x_i), \bar{f}(x_j), \bar{f}(x_k)\} == \bar{f}(x_k)$ **then**

16   set the factors $p_1 = p_1 - Iter/MaxIter$; $p_2 = p_2 - Iter/MaxIter$; $p_3 = p_3 + 2 \times Iter/MaxIter$;

17   select the best N solutions to form a new solution set;

18 **until** *meet the maximum iterations or solution precision criteria*;

19 **return** *the optimal solution and its evaluation value* $\{\hat{x}, f(\hat{x})\}$.

Table 1. Benchmark functions.

Benchmark function	Dim. (D)	Domain	Min. f_{min}	Name
$f_1 = \sum_{i=1}^{D} x_i^2$	30	$(-100, 100)^D$	0	Sphere
$f_2 = \sum_{i=1}^{D} (i \cdot x_i^2)$	30	$(-100, 100)^D$	0	Sum squares
$f_3 = \exp\left(0.5\sum_{i=1}^{D} x_i^2\right) - 1$	30	$(-1, 1)^D$	0	Hartmann
$f_4 = \sum_{i=1}^{D} (x_i + 0.5)^2$	30	$(-100, 100)^D$	0	Step
$f_5 = \sum_{i=1}^{D} \left[x_i^2 - 10\cos(2\pi x_i) + 10\right]$	30	$(-5.12, 5.12)^D$	0	Rastrigin
$f_6 = -\sum_{i=1}^{D} \left[x_i^2 - 10\cos(2\pi x_i) + 10\right] + D \cdot$ 418.98291	30	$(-500, 500)^D$	0	Schwefel 2.26
$f_7 = -\frac{1}{4000}\sum_{i=1}^{D} x_i^2 - \prod_{i=1}^{D} \cos\left(\frac{x_i}{\sqrt{i}}\right) + 1$	30	$(-600, 600)^D$	0	Griewank
$f_8 = -20\exp\left[-0.2\sqrt{\frac{1}{D}\sum_{i=1}^{D} x_i^2}\right] -$ $\exp\left[\sqrt{\frac{1}{D}\sum_{i=1}^{D} \cos(2\pi x_i)}\right] + e + 20$	30	$(-32, 32)^D$	0	Ackley 1

Storm Optimization based on Agglomerative Hierarchical Clustering (BSO-AHC). In all the following experiments, the number of population is 50 and the maximum number of iterations is 500. For the BSO-KC algorithm, the number of clusters is 5 i.e., $k = 5$. It employs Gaussian random function with mean μ and standard derivation σ, which are set to 0 and 1, respectively. The slope of the logsig function (K) is 20. Weight values of the two selected solutions are set to 0.5, i.e., $\omega_1 = \omega_2 = 0.5$. Probability of choosing a cluster or a cluster center is set to 0.5. Probability of selecting two clusters or the centers of the two clusters is set to 0.5. For the proposed BSO-AHC algorithm, it adopts centroid linkage and average distance for the agglomerative hierarchical clustering.

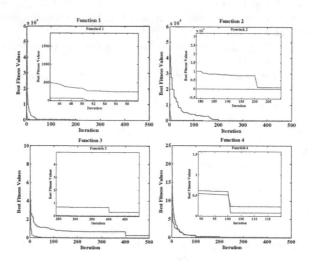

Fig. 4. The simulation results of unimodal functions. (Color figure online)

4.2 Experimental Results and Analysis

The simulation results of the unimodal functions are illustrated as Fig. 4, where blue line is the convergence curve of the BSO-KC algorithm and red line is the convergence curve of the BSO-AHC algorithm. It can easily be observed from Fig. 4 that the BSO-AHC is much faster than the BSO-KC algorithm when approaching the optimum solution of each unimodal function. Additionally, for function f_4, both algorithms nearly have the same convergence speed, but the BSO-AHC algorithm is in high precision.

The results of multimodal functions are given in Table 2, where "Best" indicates the best fitness values found in the generation, "Mean" and "Dev" stand for the mean value and standard deviation, respectively. From Table 2, it can be seen that BSO-AHC also obtains satisfactory solutions on the optimization of complex multimodal problems. BSO-AHC performs better than the BSO-KC algorithm according to three criteria (Best, Mean and Dev.) for all multi-modal functions.

Table 2. Comparison results of multimodal functions.

Function	f_5			f_6		
	Best	Mean	Dev	Best	Mean	Dev
BSO-KC	$3.46E - 05$	$1.99E - 04$	$8.32E - 03$	$6.57E - 02$	0.14	1.38
BSO-AHC	0	0	0	$7.78E - 08$	$2.32E - 06$	$3.67E - 06$
Function	f_7			f_8		
	Best	Mean	Dev	Best	Mean	Dev
BSO-KC	$3.28E - 05$	$1.09E - 03$	$6.98E - 03$	$8.61E - 03$	$4.14E - 02$	0.79
BSO-AHC	0	0	0	$1.22E - 06$	$3.14E - 05$	$4.22E - 05$

5 Conclusion

In this paper, the hierarchical clustering was adopted in this study. Meanwhile, the impact on the performance of the creating operator was carefully discussed. A new BSO with hierarchical clustering was then proposed. The agglomerative hierarchical clustering not only does not require a prespecified number of clusters, but also helps the creating operator to enhance search performance of the exploration and exploitation. Finally, the proposed algorithm was applied to numerical optimization problems in comparison with the BSO with k-means Clustering. The convergence curves and statistical results show that the proposed method can identify the regions with high quality solutions in the search space quickly and obtains satisfactory solutions on the optimization of multimodal problems according to Best, Mean and Dev. criteria.

Acknowledgments. The research work reported in this paper was partially supported by the National Natural Science Foundation of China under Grant Number 61273367 and 61403121 and the Fundamental Research Funds for the Central Universities under Grant Number 2015B20214.

References

1. Chen, J., Cheng, S., Chen, Y., Xie, Y., Shi, Y.: Enhanced brain storm optimization algorithm for wireless sensor networks deployment. In: Tan, Y., Shi, Y., Buarque, F., Gelbukh, A., Das, S., Engelbrecht, A. (eds.) ICSI-CCI 2015. LNCS, vol. 9140, pp. 373–381. Springer, Heidelberg (2015)
2. Chen, J., Xie, Y., Ni, J.: Brain storm optimization model based on uncertainty information. In: 2014 Tenth International Conference on Computational Intelligence and Security, pp. 99–103, November 2014
3. Cheng, S., Qin, Q., Chen, J., Shi, Y.: Brain storm optimization algorithm: a review. Artif. Intell. Rev. (2016) (in press)
4. Duan, H., Li, C.: Quantum-behaved brain storm optimization approach to solving loney's solenoid problem. IEEE Trans. Magn. **51**(1), 1–7 (2015)

5. Jadhav, H., Sharma, U., Patel, J., Roy, R.: Brain storm optimization algorithm based economic dispatch considering wind power. In: Proceedings of the 2012 IEEE International Conference on Power and Energy (PECon 2012), Kota Kinabalu, Malaysia, pp. 588–593. December 2012

6. Jordehi, A.R.: Brainstorm optimisation algorithm (BSOA): an efficient algorithm for finding optimal location and setting of facts devices in electric power systems. Electr. Power Energy Syst. **69**, 48–57 (2015)

7. Li, J., Duan, H.: Simplified brain storm optimization approach to control parameter optimization in F/A-18 automatic carrier landing system. Aerosp. Sci. Technol. **42**, 187–195 (2015)

8. Shi, Y.: Brain storm optimization algorithm. In: Tan, Y., Shi, Y., Chai, Y., Wang, G. (eds.) ICSI 2011, Part I. LNCS, vol. 6728, pp. 303–309. Springer, Heidelberg (2011)

9. Shi, Y.: An optimization algorithm based on brainstorming process. Int. J. Swarm Intell. Res. (IJSIR) **2**(4), 35–62 (2011)

10. Shi, Y.: Developmental swarm intelligence: developmental learning perspective of swarm intelligence algorithms. Int. J. Swarm Intell. Res. (IJSIR) **5**(1), 36–54 (2014)

11. Sun, Y.: A hybrid approach by integrating brain storm optimization algorithm with grey neural network for stock index forecasting. Abstr. Appl. Anal. **2014**, 1–10 (2014)

12. Zhan, Z.h., Zhang, J., Shi, Y.h., Liu, H.l.: A modified brain storm optimization. In: Proceedings of the 2012 IEEE Congress on Evolutionary Computation (CEC 2012), pp. 1–8, June 2012

Discovering Alias for Chemical Material with NGD

Ching Yi Chen[1], Ping-Yu Hsu[1], Ming Shien Cheng[1,2(✉)],
Jui Yi Chung[1], and Ming Chia Hsu[1]

[1] Department of Business Administration, National Central University, No. 300,
Jhongda Road, Jhongli City, Taoyuan County 32001, Taiwan, ROC
984401019@cc.ncu.edu.tw, mscheng@mail.mcut.edu.tw
[2] Department of Industrial Engineering and Management, Ming Chi University
of Technology, No. 84, Gongzhuan Road, Taishan District,
New Taipei City 24301, Taiwan, ROC

Abstract. The complexity of chemical substance names makes it difficult to fully describe chemical substances using just several keywords. We usually find related information through search engines or look up an online chemical dictionary. However, the chemical material names used in academy usually translated from English, and the same chemicals often have many different aliases. This English Chinese translation creates many problems when querying information for chemicals. Recent studies have proposed to use Normalized Google Distance (NGD) to determine semantic relevance between two words. Therefore, this study proposes to find alias based on NGD with two methods, namely, novel and category affixed methods. The Experimental results show that the latter method can derive better result.

Keywords: Normalized google distance (NGD) · Text mining · Chemical material name

1 Introduction

The complexity of chemical substance names makes it difficult to fully describe chemical substances using just several keywords. When the average user, who has no professional knowledge of chemistry, comes across an unfamiliar name of a chemical substance, search engines or online chemistry dictionaries can provide a vast amount of information; however, the scientific names of chemical substances are mostly translations from English into Chinese; as a result of different translations, the same chemical substance will often have several aliases. Different industries, such as the drug industry, refer to the chemical substances by different aliases such as chemical names, common names, and trade names. For example, the chemical name of aspirin is o-acetylsalicylic acid, which is also commonly called a panacea.

The Chemical Abstracts Service (CAS) of the American Chemical Society [1] allocated a set of unique identification numbers called CAS Registry Numbers or CAS Numbers to every kind of substance appearing in the literature to avoid the trouble

Y. Tan et al. (Eds.): ICSI 2016, Part II, LNCS 9713, pp. 123–131, 2016.
DOI: 10.1007/978-3-319-41009-8_13

caused by having many different aliases for chemical substances; however, English names are mainly used on the Internet, as corresponding Chinese names are lacking, and the average user is not aware that the names can be retrieved through CAS Numbers.

According to the Normalized Google Distance (NGD) proposed by Chen and Lin [2], the Google search engine can be used as a cloud database. After users input the keywords into the Google search engine, they use the number of returned search results to calculate the degree of correlation between two keywords. In 2005, the number of web pages in Google search engine exceeded 80 billion. The huge volume of data can provide more accurate results, and the return rate is quite fast.

The main research objective is to enable the average user to use the alias of a chemical substance to find the correct name. This study therefore proposed two methods: "simple method" and "categorical affixation method" to do a one-word search in real time based on the keywords users input and to add category names to the words to conduct the query. After obtaining the search results, the NGD between the name of the chemical substance and its alias was calculated, and the results were sorted to find the name of the chemical substance that was most relevant to the keyword input by the user and to calculate the average distance of the correct answer. Finally, a comparison was made to find the best method.

The research thesis framework is divided into five parts: (1) Introduction: This part explains the research motivation and objective. (2) Literature Review: This part is the literature on NGD and text mining in recent years. (3) Experimental methods: This part describes the experimental design and experimental methods. (4) Experimental results: This part contains the analysis and discussion of experimental results. (5) Conclusion: This part introduces the research conclusions.

2   Related Work

Google search engine has become one of the main sources of obtaining information and retrieving information. When users wish to understand an unfamiliar keyword, they can input the keyword into Google search engine and infer the meaning of the keyword from the other words that also appear on the same page. Users can also judge what correlation exists between the keyword and the other words. In 2007, Cilibrasi and Vitanyi [3] found that Google search engine could be used to detect the correlation between two words.

Having an indicator that measured the correlation between two words, Cilibrasi and Vitanyi developed a kind of statistical indicator based on the found data records to show the logical distance of a group of words, and called it the Normalized Google Distance (NGD). The closer the NGD value is to 0, the more likely these two words will simultaneously appear in the same document each time, and it can be inferred that the correlation between the two keywords is higher. The bigger the NGD value is, the lower the chances of these two words appearing at the same time, and the correlation is weaker.

The computational formula of NGD is as follows:

$$\text{NGD}(x, y) = \frac{G(x, y) - \min(G(x), G(y))}{\max(G(x), G(y))} = \frac{\max(logf(x), logf(y)) - logf(x, y)}{logN - \min(logf(x), logf(y))}$$

f(x) : **representing the number of search results containing keyword x.**

f(y) : **representing the number of search results containing keyword y**

f(x, y) : **representing the number of search results containing keywords x and y**

N : **representing the total number of webpages included in the Google search engine.**

Cilibrasi and Vitanyi proposed that the main objective of NGD is to resolve the situation where there is an inability to engage in file data collection preparation. Thus Google's large volume of data is a cross-field and cross-language corpus, and Google search engine can be used to timely retrieve the number of search results and timely determine the meaning and degree of relevance of the meaning of two words. In 2010, Chen [2] proposed using the calculated NGD distance between words to build the word relationship network, and the analysis of distribution of terms can timely extract the words in the documents in order of importance. In 2014, Liu et al. [5] proposed a three-phase prediction (TPP) algorithm to enable the average user with no legal professional background to directly use everyday words as the keywords for a query to describe their problem; NGD calculations were done to find the relevant legal terminology and find the regulations related to the user's problem.

However, the number of search results returned by the Google search engine changes frequently, creating huge differences in NGD value at different points in time. If there are many potential keywords that need the similarity degree between two words to be mutually calculated, the Google search engine need to search a large volume. Therefore, Chen's research proposed the concept of Google core distance (GCD); a modification was made to the equation to reduce the number of times a query is made to the Google search engine to reduce the degree of influence of errors. The equation is as follows:

$$0 \leq \text{GCD}(x, y) \leq \sqrt{(R - r1)^2 + (R - r2)^2}$$

r1:representing the number of search results containing keyword x

r2:representing the number of search results containing keyword y

R : representing the total number of webpages contained in the Google search engine.

Figure 1 shows the concept of GCD. Suppose the results returned for x and y are perpendicular to each other; the vectors on both sides can then be used to calculate the length of the third side, that is, the concept distance between two words, x and y. The radical sign added to the equation has the function of reducing the error size, thereby reducing the error's degree of influence.

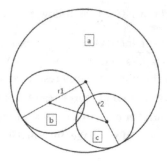

Fig. 1. Distance data of Google using similarity distance as the basis. Source: [3]

3   System Design

Based on the normalized Google distance (NGD) proposed by Vitanya and Cilibrasi, this study proposed two methods: "simple method" and "categorical affixation method" to timely query the keywords that users input into the Google search engine, obtain the number of search results to calculate NGD, find the relevance between two words, and thus sort the results calculated and find the name of the chemical substance related to the keyword. However, the number of search results returned by the Google search engine frequently changes, causing the NGD value to have great differences that may have been caused by queries being made at different time points and preventing the correct name from being sorted to the top. Therefore, this study used the average distance of the correct chemical substance name found to compare which of the two methods is better.

3.1   Data Sources

It was found that most of the terminology of online professional chemistry dictionaries was in English or simplified Chinese. In Taiwan, only the "Bilingual Glossary for Academic Terms Informational Website" published by National Academy for Educational Research provides queries for various academic terms and free data downloads. But the queried results only have noun translations and interpretation of nouns in the dictionary. There are fewer data that can be queried, and the different aliases of the same chemical substance are not integrated. Therefore, this study used the chemical substances in the Chinese MSDS library in the chemistry foundation database of China's ChemNet as the experimental data.

The Material Safety Data Sheet (MSDS), also known as the identification card of chemical substances, provides data on the important safety and health features of chemical substances and is an important reference data of the occupational health and safety environment. The objective is to provide to operating personnel and emergency personnel a safe method of handling this kinds of substances. The MSDS format contents chemical substance composition identification information such as Chinese and English chemical substance names and other names, CAD number, physical and chemical properties such as meting point, boiling point, impact on health, protective equipment, leak handling, etc. This study only used Chinese names and aliases of the chemistry substances as the experimental terms. Chinese chemical substance names were the terms of training data, and the aliases were randomly sampled as the terms of test data. In addition, based on Wikipedia's categorization of chemical substance names [6], we manually tagged each term in the training data with a category name. To make a comparison with the "simple method," category names are added to the queried terms to see whether the results increased in accuracy.

3.2   Data Classification

This study thus used Chinese chemical substance names as the training data terms. Based on the classification of Wikipedia's chemical substance names, the training data was used to do classification of chemical substances, which are shown in Fig. 2.

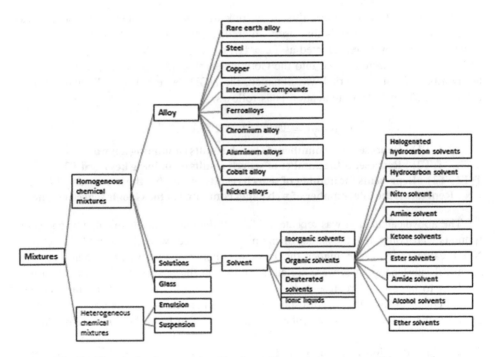

Fig. 2. Chemical substance classification table – mixture, source: this research

Chemical substances can be classified as compounds and mixtures. The figure's classifications table is the classification of mixtures.

Fig. 3. Experimental flow chart of the study

3.3   Experimental Methods

Figure 3 is experimental flow chart of this study, and $\{t_1, t_2, ..., t_{98}\}$ is the study's training date. After the user entered the keywords, the keywords and each term in the training date were used to calculate the NGD of two words. Finally, each term of the training date was sorted from smallest to largest NGD size, which was represented as $\{n_1, n_2, ..., n_{98}\}$. The higher up the ranks a name is, the higher the relevance of the keyword.

3.3.1   Simple Method

Based on the flowchart process, this study took the aliases of chemical substances as the test data of single words that users entered, represented as T1, and all chemical substance names were represented as $T2 = \{t_1, t_2, ..., t_{98}\}$. Thus, "T1" and "T2" and "T1" AND "T2" were entered into the Google search engine to do a query, and after the number of results queried was obtained, the NGD value between T1 and T2 was calculated. The calculation equation is as follows:

$$NGD(x, y) = \frac{max(logf(T1), logf(T2)) - logf(T1, T2)}{logN - min(logf(T1), logf(T2))}$$

f(T1) : **Represents the number of search results containing keyword T1.**
f(T2) : **Represents the number of search results containing keyword T2.**
f(T1, T2) : **Represents the number of search results containing the keywords T1 & T2.**
N : **Represents the total number of web pages contained in the Google search engine.**

The above calculation was repeated. After all the names of chemical substances in t_1 and t_2 were computed, the names of chemical substances were then sorted. The closer NGD value is to 0, the higher the relevance between two words. Therefore, the chemical substance names were sorted according to the size of NGD value from the smallest to largest and represented as $\{n_1, n_2, ..., n_{98}\}$. Then, find the rank of the correct chemical substance t_n corresponding to the chemical substance alias t_1 inputted, giving it the rank order -1 to calculate the distance of tn from the correct answer.

3.3.2   Classification Affixation Method

This study used the aliases of chemical substance as the test data of single words that users input, represented as T1. Unlike the simple method, this method added the category names to the chemical substance names, represented as $T2 = \{t_1, t_2, ..., t_{98}\}$. Thus "T1" and "T2" AND "category name of tn" and "T1" AND "T2" AND "category of tn" were entered into the Google search engine to conduct queries, and after the query results were obtained, the NGD of T1 and T2 was calculated.

The above calculation was repeated. After all chemical substance names in T1 and T2 were computed; the chemical substance names had to be sorted. The closer the NGD value is to 0, the higher the relevance there is between two terms. Therefore, the chemical substance names were sorted according to the NGD size from smallest to largest, represented as $N2 = \{n_1, n_2, ..., n_{98}\}$. Then, find the rank of the correct chemical substance t_n corresponding to the chemical substance alias T1 inputted, giving it the rank order -1 to calculate the distance of t_n from the correct answer.

4 Experimental Results

4.1 Simple Method

Using the queried "alcohol" as example, T1 = "alcohol"; the correct chemical substance name is "ethanol"; and t_1 = "ethanol"; respectively, the three groups of terms are T1 = "alcohol"; t_1 = "ethanol"; and "alcohol" AND "ethanol." The three groups of terms were entered into the Google search engine to do queries, and the results were f (T1) = 17,200,000, $f(t_1)$ = 2,300,000, and $f(T1, t_1)$ = 683, 000. According to the statistical data in 2014, the total number of web pages Google search engine established was N = 67,000,000,000 [4]. After calculations, NGD = 0.313843828. The above calculation was repeated. After the queried terms "alcohol" and other 97 chemical substance names in T2 were used to calculate NGD, the chemical substance names were sorted according to the NGD size from smallest to largest. "Ethanol" ranked 22nd, and its distance from the correct answer was $22 - 1 = 21$.

4.2 Classification Affixation Method

Similarly, using the "alcohol" query as an example, T1 = "alcohol". The correct chemical substance name was "ethanol" and the chemical classification name was "alcohol solvent"; respectively, the three groups of terms were T1 = "alcohol"; t_1 = " ethanol" AND "alcohol solvents"; and "alcohol" AND "ethanol" AND "alcohol solvent"; the three groups of terms were entered into the Google search engine to conduct the queries, and the respective results were obtained: f(T1) = 17,200,000; f (t_1) = 5,700,000; f(T1, t_1) = 1,520,000; N = 67,000,000,000. After calculations the NGD = 0.258877911. The queried word "alcohol" and other 97 chemical substance names in T2 and the category names were all used to calculate NGD, and then the chemical substance names were sorted according to the NGD size from smallest to largest. "Ethanol" ranked 4th, and its distance from the correct answer was $4 - 1 = 3$.

4.3 Comparison Result of Two Methods

This study randomly selected 50 chemical substance aliases from the test data and respectively used the simple method and classification affixation method to calculate the distances of 50 chemical substance names from the correct answer. The average distances from the correct answer, standard deviations, and T-test of the two methods are as follows (Table 1):

Table 1. The comparison table of experimental results

	Simple method	Classification affixation method
Average distance	38.58	20.74
Standard deviation	24.68448095	20.06669878
T-test	0.0000285450***	

Source: Compiled by this study.

The biggest difference of the simple method and classification affixation method is that the T2 in the classification affixation method had chemical substance classifications added to do the queries. From the experiment we found that the classification affixation method calculated the average distance of the correct answer to be 20.74, which is smaller than the result of 38.58 in the simple method. This represents that adding the chemical substance category in the Google search engine can obtain a more accurate number of results and can find the correct chemical name more effectively. The simple method only used two single words to do queries, resulting in the average distance from the correct answer being farther away. This may be because chemical substance names lack unified naming rules for entering two terms. The T-test = $0.000028545 < 0.05$, which represents that these two methods had significant differences. The experiment also found that the Google search engine's numbers of returned search results had a huge difference because the queries were made at different time points. Also, if the order of terms is different, it can also obtain different search results.

5   Conclusion and Future Research

5.1   Conclusion

The particularity of the chemistry field has created complicated chemical substance names, making it difficult to use several keywords for a full description. There are many new compounds or newly discovered chemical substances that must use chemical structure description, and many chemical substances have an isomeric structure. The average text retrieval method may not be able to do a full chemistry field-related retrieval. This study proposed two methods based on the Normalized Google Distance (NGD) proposed by Cilibrasi and Vitanyi and timely calculated the NGD between chemical substance names and aliases according to the keywords entered by users and further determined whether two words had the same meaning. The simple method directly used the chemical substance names and aliases to calculate the NGD of two words, while the classification affixation method add the category name and alias to the already known chemical substance name to do the calculation. The experimental results show that the classification affixation method of adding classification names in the

Google search engine can obtain a more accurate number of results, and the average distance from the correct answer was shorter.

5.2   Future Research

This study proposed two methods: the "simple method" and "classification affixation method." Although the "classification affixation method" produced better results, the calculated results were still not accurate enough, leaving huge room for improvement. In the future, the direction can be strengthened by choosing different words groups such as chemical structural formulae, molecular formulae, and so on. WordNet dictionary can provide a more complete knowledge base of lexical and semantic relationships.

References

1. Chemical Abstracts Service. https://www.cas.org/content/chemical-substances/faqs
2. Chen, P.I., Lin, S.J.: Automatic keyword prediction using Google similarity distance. Expert Syst. Appl. **37**(3), 1928–1938 (2010)
3. Cilibrasi, R.L., Vitanyi, P.M.B.: The Google similarity distance. IEEE Trans. Knowl. Data Eng. **19**, 370–383 (2007)
4. Google indexed number. http://www.statisticbrain.com/total-number-of-pages-indexed-by-google/
5. Yi-Hung, L., Yen-Liang, C., Wu-Liang, H.: Predicting associated statutes for legal problems. Inf. Process Manag. **51**(1), 194–211 (2015)
6. Wikipedia, Category Name of Chemical Substances. http://zh.wikipedia.org/wiki/. Category: %E5 %8C%96 %E5 %AD%A6 %E7 %89 %A9 %E8 %B4 %A8

Estimate the Kinematics with EMG Signal Using Fuzzy Wavelet Neural Network for Biomechanical Leg Application

Weiwei Yu[1(✉)], Yangyang Feng[1], Weiyu Liang[1], Runxiao Wang[1], and Kurosh Madani[2]

[1] School of Mechanical Engineering, Northwestern Polytechnical University, Youyi xilu 127 hao, Xi'an, People's Republic of China
yuweiwei@nwpu.edu.cn
[2] Images, Signals and Intelligence Systems Laboratory (LISSI/EA 3956), UPEC, Senart-Fontainebleau Institute of Technology, Bât.A, 77127 Lieusaint, France

Abstract. Several linear and nonlinear models were proposed to predict the forward relationship between EMG signals and kinematics for biomechanical limbs, which is meaningful for EMG-based control. Although using nonlinear model to predict the kinematics is able to represent rational complex relationship between EMG signals and desired outputs, there exists high risk for overfitting models to training data and calculating burden because of the multi-channel variation EMG signals. Inspired by the hypothesis that CNS modulates muscle synergies to simplify the motor control and learning of coordinating variation of redundant joints, this paper proposed to extract the synergies to reduce the dimension of EMG-based control. Furthermore, the fuzzy wavelet neural network was developed to generate velocity–adapted gait by the reference gaits only with the limited set of experimental trials. The experimental results show the efficiency and robust of this approach.

Keywords: EMG · Muscle synergy · Fuzzy wavelet neural network

1 Introduction

The myoelectric signal is the electrical manifestation of muscular contractions, which reflexes the plentiful neural control information [1]. EMG based control has advantages for achieving intuitive simultaneous and proportional control of exoskeletons, myoelectric prostheses and bio-robot, and moreover for the development of novel diagnostic tools and rehabilitation approaches.

In past few years, several contributions were proposed to predict the forward relationship between EMG signals and kinematics for biomechanical limb. Such as the approaches based on the linear models. A linear 'mixing matrix' was built to provides an explicit expression of the EMG in terms of force functions, which correspond to the wrist intended activations of physiological DOFs of natural movement [2]. In their later study, muscle synergies of both DOF were extracted at once, and three synergies can achieve simultaneous control when applying separately on each DOF [3].

© Springer International Publishing Switzerland 2016
Y. Tan et al. (Eds.): ICSI 2016, Part II, LNCS 9713, pp. 132–140, 2016.
DOI: 10.1007/978-3-319-41009-8_14

Nonlinear model based approach are not as reliant on robust features as linear models, and therefore are able to represent rational complex relationship between synergies and desired outputs. J.M. Hahne systematically compare linear and nonlinear regression techniques including linear regression, mixture of linear experts, multilayer-perceptron and kernel ridge regression, for an independent, simultaneous and proportional myoelectric control of wrist movements. And got the results that the kernel ridge regression outperformed the other methods, but with higher computational costs [4]. Muceli S et al. adopted MLP artificial neural networks to estimate kinematics of the hand wrist from EMG signals of the contralateral limb during mirrored bilateral movements in free space [5]. A Hybrid time-delayed artificial neural network was investigated to predict clenching movements during mastication from EMG signals. Actual jaw motions and EMG signals from the masticatory muscles were recorded and used as output and input respectively [6].

However, there may cause two problems if using nonlinear model to predict the kinematics of biomechanical leg with EEG signals. Firstly, the multi-channel EMG signals vary greatly with time, which brings the computing burden for training non-linear model. Secondly, there exists high risk for overfitting models to training data, resulting in sensitivity to transient EMG changes and frequent retraining [7]. Thus, several approaches were developed to reduce the overfitting [8] and the model sensitivity to EMG transient changes [9].

A prominent hypothesis suggests that the biological system underlie muscle contractions during movement execution in a modular fashion by the CNS. These modularities have been observed in forms of muscle synergies. Evidences observed in many cases and species show that regularities of the muscle synergies appear to be very similar across subjects and motor tasks [10, 11]. Inspired by this hypothesis, this paper proposes to apply the muscle synergies to reduce the dimension of EMG-based control of biomechanical leg. Furthermore, a fuzzy wavelet neural network was developed to generalize the flexible gait only with limited sets of experimental data, which avoid the overfitting models to training data.

2   Extract Synergies from EMG Signal

It is hypothesis that the CNS coordinates groups of muscles to simplify the generation of intricate movements. These building modules, known as muscle synergies, can be used as a small number of co-activation patterns to imitate the performance of movement [12]. This paper proposed using Nonnegative Matrix Factorization to extract synergies. Given a set of time-variable n channel EMG data vectors, the vectors are placed in the columns of an $n \times m$ matrix X, in which m is the number of samplings in each channel of EMG data set. This matrix is then approximately factorized into an $n \times k$ matrix W and an $k \times m$ matrix C, such as:

$$X = WC \tag{1}$$

For example, if choose the structure of generators, which are combined to generate command EEG signal across tasks, as the form of spatial dimensionality. For K generators:

$$\mathbf{x}^r(t) = \sum_{k=1}^{K} c_k^r(t) \mathbf{w}_k \tag{2}$$

Where $\mathbf{x}^r(t)$ are the set of signals ($t = 1 \cdots m$) for task condition r. $c_k^r(t)$ is condition dependent, time-varying combination coefficient for the k th generator. $\mathbf{w}_k$ is the condition-independent, time-invariant k th spatial generator. Fundamentally, a muscle synergy consists of a time-invariant weighing coefficient $\mathbf{w}_k$ and a time-varying activation coefficient $c_k^r(t)$. The weighing coefficients within a synergy determine the number of muscles along with the extent of their activation, while the activation coefficient captures when the muscles are active during a task.

3 Estimation Kinematics Using Fuzzy Wavelet Neural Network

3.1 Wavelet Neural Network

As it is analyzed in Sect. 2, the muscle synergies can explain the muscle co-activation patterns of leg movement. The wavelet neural network (WNN) is adopted as the system identification method to estimate the relationship between the muscles co-activation patterns and coordinated variation of joint angles.

The WNN is provided with the features of wavelets theory and feed-forward neural network, with one hidden layer, whose activation functions are drawn form an orthonormal wavelet family. The wavelet and the neural network processing can be performed separately. For a multi-input multi-output nonlinear indentification system, the output can be defined by wavelons as:

$$y_\vartheta(x) = \sum_{i=1}^{M} w_i \sqrt{\lambda_i} \phi(\lambda_i x - t_i) \tag{3}$$

in which λ and t are the dilation and translation parameters respectively. The parameters $\bar{y}$, w_i, t_i and λ_i can be grouped into a parameter vector ϑ. The objective function to be minimized is defined as:

$$e(\vartheta) = \frac{1}{2} E\left[(y_\vartheta(x) - f(x))^2 \right] \tag{4}$$

The minimization of above function is performed using stochastic gradient algorithm. This recursively modifies ϑ, after each sample pair $\{x_k, f(x_k)\}$, the objective function is written as:

$$b(\vartheta, x_k, f(x_k)) = \frac{1}{2}(y_\vartheta(x_k) - f(x_k))^2 \tag{5}$$

The parameters t_i and λ_i can be fixed at initialization of the network. w_i, which is the only parameter in θ needed to be adjusted, is modified in the opposite direction of $e(\vartheta, x_k, f(x_k))$, and the gradient can be computed by the partial derivatives as follows:

$$\frac{\partial e}{\partial w_i} = e_k \sqrt{\lambda_i} \phi(z_{ki}) \tag{6}$$

in which,

$$e_k = y_\vartheta(x_k) - f(x_k) \tag{7}$$

$$z_{ki} = \lambda_i x_k - t_i \tag{8}$$

A multi-inputs and 3-outputs WNN is constructed to predict the coordinating variation of three joint angles with muscle synergies, taking the muscle activation coefficient as input signals. The outputs are the coordinated variation of hip, knee and ankle angles.

3.2   Fuzzy Inference System

The proposed fuzzy inference system is based on the well know Takagi-Sageno fuzzy inference system. The TS-FIS is described by a set of $R_k (k = 1 \ldots N_k)$ fuzzy rules presented in Eq. (9). $x_i (i = 1 \ldots N_i)$ are the inputs of the FIS with N_i dimension input space, and $A_i^j (j = 1 \ldots N_j)$ are linguistic terns, which numerically defined by membership functions distributed in the universe of discourse for each input x_i. Each output rule y_k is a linear combination of input variables.

$$if \; x_1 \; is \; A_1^j \ldots and \ldots x_i \; is \; A_i^j \; then \; y_k = (x_1, \ldots x_{Ni}) \tag{9}$$

Each reference gait has been memorized, which records the relationship between muscle co-activation coefficient and joint coordinated variation. Based on the experiment results, totally five reference gaits are chosen and memorized. The movement velocity is modeled by five fuzzy sets ("VeryFast", "Fast", "Medium", "Slow", "VerySlow") with triangle membership functions. The generalized joint angles Y is estimated by the weighted average of reference gait, which is calculated using

$$Y = \sum_k \bar{u}_k O_k(C) \tag{10}$$

with $\bar{u}_k$ is given by Eq. (11):

$$\bar{u}_k = u_k / \sum_{k=1}^{N_r} u_k \tag{11}$$

and u_k is computed with the membership function according to Eq. (12):

$$u_k = \mu_1^j \mu_2^j \cdots\cdots\cdots \mu_{Ni}^j \tag{12}$$

3.3 Fuzzy Wavelet Neural Network

The final desired trajectory of hip, knee and ankle joint angles are generalized by the FIS and wavelet neural network architecture, on the basis of predefined membership functions. The output Y is carried out in two stages:

Firstly, take the extracted co-activation coefficient as the inputs of the WNN. The coordinating variation of hip, knee and ankle joint angles $O_k(C)$ can be identified according to Eq. (3), indicating each reference gait pattern with corresponding muscle co-activation coefficient.

Secondly, the generalized joint angles Y is estimated using the weighted average of all wavelets neural networks, which is computed by Eqs. (10), (11) and (12).

By using the Fuzzification of several outputs of wavelets neural network, it is possible to obtain a global generalization, which allows to decrease the memory size and computing cost using only a small set of identification system.

4   Experiment Results and Analysis

4.1   Experimental Protocol

Three dimensional kinematic data were collected using VICON 10-camera motion capture system at frequency of 500 Hz. The BioVision 8 channel EMG device was used to record muscle activity of seven muscles: tibialis anterior, gastrocnemius, soleus, vastus medialis, rectus femoris, glusteus maximus and biceps femoris. The eighth channel of EMG equipment was used for synchronization.

After removing DC offset of the EMG signal for each channel, rectification of the EMG signal should be done firstly. Then, a band-pass filter is constructed for a cut of frequency between 10 Hz and 30 Hz. In this experiment, the sampling frequency of EMG signal is 1000 Hz, thus, the 5th order Butterworth filter is adopted.

A volunteer subject was trained for vertical hopping with uniform velocity before the experiment. The experiment contained 20 trials and each trial with 10 hops. The subject was asked for jumping with relative constant velocity during each trial (Fig. 1) and the movements were analyzed with the range of rhythm between 470 ms and 630 ms per jump after all the trials finished.

4.2   Results and Analysis

Variance Accounted For (VAF) is usually used to calculate the percentage of variability in the EMG dataset that is accounted for by the extracted synergies. The VAF of five trials of all 20 trials is below 90 % if choosing the number of synergies as three, which means that that extracted synergy set do not sufficiently explain the EMG variance. This can be also explained in Fig. 2(a), in which all the trials have different muscle coordination patterns.

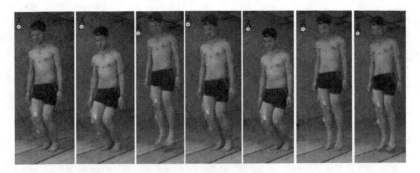

Fig. 1. The subject was doing vertical hopping within the experimental environment.

If increasing the number of synergies to four, the results show that VAF of all trials is above 90 %, which indicates that the recorded EMGs are well reconstructed by the extracted synergies. The synergies weighting coefficients in Fig. 2(b) indicate that the four co-activation patterns are (vastus medialis, rectus femoris, glusteus maximus), (gastrocnemius, soleus), (tibialis anterior) and (biceps femoris) respectively.

A 4-inputs and 3-outputs WNN is constructed to predict the coordinating variation of three joint angles with muscle synergies, taking the muscle activation coefficients as input signals. The output joint angles of WNN are presented in Fig. 3, in which $y1$, $y2$ and $y3$ stands for the hip, knee and ankle angles. If the neural network uses the last four steps to estimate the output joint angles, the fit to working data for output $y1$, $y2$ and $y3$ are 77.89 %, 84.56 % and 84.20 % respectively, as shown in Fig. 3(a). If use the last three steps to estimate the outputs, the fit to working data for output $y1$, $y2$ and $y3$ are 83.31 %, 82.90 % and 84.83 % respectively, refers to Fig. 3(b), in which blue lines are the actual joint angle and grey lines are the outputs of wavelet neural network.

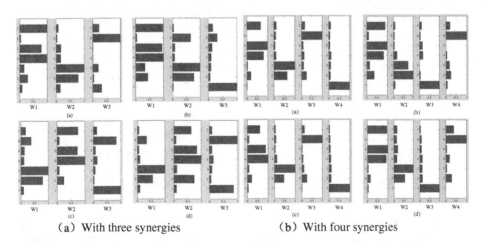

(a) With three synergies   (b) With four synergies

Fig. 2. Synergies weighing coefficients of vertical hopping with different synergies

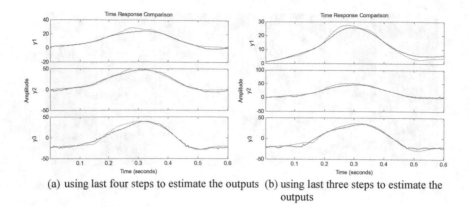

(a) using last four steps to estimate the outputs (b) using last three steps to estimate the outputs

Fig. 3. Estimated joint angles with wavelet neural network

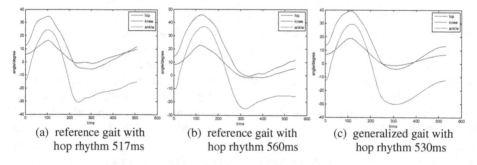

(a) reference gait with   (b) reference gait with   (c) generalized gait with
hop rhythm 517ms   hop rhythm 560ms   hop rhythm 530ms

Fig. 4. Generalized gait with fuzzy inference system

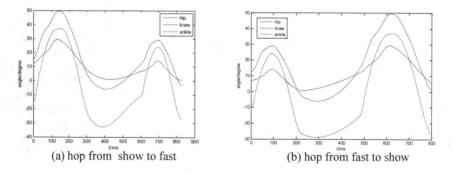

(a) hop from show to fast   (b) hop from fast to show

Fig. 5. Generalized gait of continuous hopping movement

Five triangle membership functions were set according to the different hopping velocity grouped based on the experimental data. Figure 4(c) presents the generalized gait with hoping rhythm of 530 ms after the fuzzy inference system, which refers to the reference gaits with rhythm of 517 ms and 560 ms (Fig. 4(a) and (b)).

Figure 5(a) represents joint angle trajectories with the hopping velocity changing from slow to fast. While the joint angle trajectories from fast hopping to slow are shown in Fig. 5(b). It should be noticed that in the case of show hopping, all of the hip, knee and ankle angles are more compressed than the fast hopping. This also follows the regularities observed from the subject hopping experiment. The generalized gait pattern is compliant in the condition of variable velocity, which increases the robust of EMG-based control for biomechanical leg application.

5 Conclusion

To generate more compliant gait pattern with EMG signals is meaningful for achieving intuitive simultaneous and proportional control of exoskeletons, myoelectric prostheses and bio-mechanical leg. Based on the hypothesis that CNS modulates muscle synergies to simplify the motor control and learning of coordinating variation of redundant joints, this paper proposed to extract the synergies to reduce the dimension of EMG-based control. The identification system built on wavelet neural network is used for predicting the coordinating variation of joint angles with synergies weighting coefficients, which remembered as reference gait patterns. The Fuzzy Inference System is adopted to merge these reference motions in order to built more complex gaits. The simulation results based on the experiments of hopping indicate that the generalizing gaits are compliance and can self-adapted to variable velocity.

Acknowledgement. This research has been made possible by "111 project" (Grant No. B13044), National Science Foundation of China with grant number 51475373 and Natural Science Basic Research Plan of Shaanxi Province with grant number 2016JQ6009.

References

1. Jing, N., Vujaklija, I., Rehbaum, H., Graimann, B., Farina, D.: Is accurate mapping of EMG signals on kinematics needed for precise online myoelectric control? IEEE Trans. Neural Syst. Rehabil. Eng. 22(5), 49–558 (2014)
2. Jiang, N., Englehart, K.B., Parker, P.A.: Extracting simultaneous and proportional neural control information for multiple-DOF prostheses from the surface electromyographic signal. IEEE Trans. Biomed. Eng. 56, 1070–1080 (2009)
3. Muceli, S., Jiang, N., Farina, D.: Extracting signals robust to electrode number and shift for online simultaneous and proportional myoelectric control by factorization algorithms. IEEE Trans. Neural Syst. Rehabil. Eng. 22, 623–633 (2014)
4. Hahne, J.M., Biessmann, F., Jiang, N., Rehbaum, H., et al.: Linear and nonlinear regression techniques for simultaneous and proportional nyoelectric control. IEEE Trans. Neural Syst. Rehabil. Eng. 22(2), 269–279 (2014)
5. Muceli, S., Farina, D.: Simultaneous and proportional estimation of hand kinematics from EMG during mirrored movements at multiple degrees-of-freedom. IEEE Trans. Neural Syst. Rehabil. Eng. 20, 371–378 (2012)

6. Kalani, H., Moghimi, S., Akbarzadeh, A.: SEMG-based prediction of masticatory kinematics in rhythmic clenching movements. Biomed. Signal Process. Control **20**, 24–34 (2015)
7. Ison, M., Artemiadis, P.: The role of muscle synergies in myoelectric control: trends and challenges for simultaneous multifunction control. J. Neural Eng. **11**, 051001 (2014)
8. Clancy, E.A., Liu, L., Liu, P., Moyer, D.V.Z.: Identification of constant-posture EMG-torque relationship about the elbow using nonlinear dynamic models. IEEE Trans. Biomed. Eng. **59**, 205–212 (2012)
9. Gijsberts, A., Bohra, R., Sierra, G.D., Werner, A., Nowak, M., Caputo, B., Roa, M.A., Castellini, C.: Stable myoelectric control of a hand prosthesis using non-linear incremental learning. Front. Neurorobot **8**, 8 (2014)
10. d' Avella, A., Saltiel, P., Bizzi, E.: Combinations of muscle synergies in the construction of a natural motor behavior. Nat. Neurosci. **6**(3), 300–308 (2003)
11. d' Avella, A., Bizzi, E.: Shared and specific muscle synergies in natural motor behaviors. Proc. Natl. Acad. Sci. U.S.A. **102**, 3076–3081 (2005)
12. Alessandro, C., Deils, I., Nori, F., Panzeri, S., Berret, B.: Muscle synergies in neuroscience and robotics: from input-space to task-space perspectives. Front. Comput. Neurosci. **7**(43), 1 (2013)

A *Physarum*-Based General Computational Framework for Community Mining

Mingxin Liang[1], Xianghua Li[1(✉)], and Zili Zhang[1,2(✉)]

[1] School of Computer and Information Science,
Southwest University, Chongqing 400715, China
li_xianghua@163.com, zhangzl@swu.edu.cn
[2] School of Information Technology, Deakin University,
Locked Bag 20000, Geelong, VIC 3220, Australia

Abstract. Community mining is a crucial and essential problem in complex networks analysis. Many algorithms have been proposed for solving such problem. However, the weaker robustness and lower accuracy still limit their efficiency. Aiming to overcome those shortcomings, this paper proposes a general *Physarum*-based computational framework for community mining. The proposed framework takes advantages of a unique characteristic of a *Physarum*-inspired network mathematical model, which can differentiate inter-community edges from intra-community edges in different type of networks and improve the efficiency of original detection algorithms. Some typical algorithms (e.g., genetic algorithm, ant colony optimization algorithm, and Markov clustering algorithm) and six real-world datasets have been used to estimate the efficiency of our proposed computational framework. Experiments show that the algorithms optimized by *Physarum*-inspired network mathematical model perform better than the original ones for community mining, in terms of robustness and accuracy. Moreover, a computational complexity analysis verifies the scalability of proposed framework.

Keywords: Community mining · General computational framework · *Physarum* network mathematical model

1 Introduction

Many complex systems in the real world can be formulated as a complex network, such as social networks, internet and biological networks [1]. And the community structure is one of the essential topological properties of a network, where vertexes across communities are connected sparsely, and vertexes within a community are connected densely. More importantly, the community structure can provide both the sketches of a complex network and some insights into the properties of vertexes [2].

Due to the importance of community structure, many algorithms have been proposed for detecting communities [1,3]. Generally speaking, there are two typical kinds of algorithms: optimization algorithms and model-based algorithms.

© Springer International Publishing Switzerland 2016
Y. Tan et al. (Eds.): ICSI 2016, Part II, LNCS 9713, pp. 141–149, 2016.
DOI: 10.1007/978-3-319-41009-8_15

Concerning the optimization algorithms, such as Genetic Algorithm (GA) [4] and Ant Colony Optimization algorithm (ACO) [5], community mining is to find a community division of network with the maximum objective function values. Currently, the most popular objective function is the modularity [6]. In term of model-based algorithms, a representative algorithm, Markov clustering algorithm (MCL), is used to highlight their mechanisms. MCL [3] is based on the flow simulation in which high-flow regions are clustered together. However, some shortcomings, such as the weaker robustness and lower accuracy, still limit their performances.

Currently, *Physarum*, which is a unicellular and multi-headed slime mold, shows an intelligence of path finding and network designing in biological experiments [7,8]. Moreover, inspired by the intelligence of *Physarum*, a mathematic model is proposed [9], which has shown an ability to optimize some nature-inspired algorithms (e.g., GA [10] and ACO [11]) in terms of efficiency and robustness. Given the aforementioned observation and works, the following two questions are raised:

- Does the *Physarum*-based mathematical model (PM) have a potential to recognize community structures in complex networks?
- Can the intelligence of PM optimize traditional algorithms to overcome the shortcomings of community detection?

Considering the questions above mentioned, the main contributions of this paper are twofold. First, inspired by PM, a *Physarum* network mathematical model (PNM) is proposed, which has an ability to recognize the inter-community edges coarsely. Second, utilizing the proposed PNM, a *Physarum*-based general computational framework for community mining is proposed to overcome the weaker robustness and lower accuracy of traditional detection methods. And some experiments are used to estimate the efficiency of proposed framework.

The remaining parts of this paper are organized as follows: Sect. 2 formulates the community mining and introduces the basic ideas of traditional algorithms for community detection. Section 3 proposes the PNM and a *Physarum*-based general computational framework for community mining. Then, some typical algorithms (e.g., GA, ACO and MCL) and six classical real-world datasets are used for estimating the efficiency and scalability of proposed framework in Sect. 4. Final, Sect. 5 concludes this paper.

2 Related Works

2.1 Formulation of Community Mining

A complex network can be formulated as a graph $G(V, E)$, where V and E stand for the sets of vertexes and edges, respectively. And, a community is denoted as c which contains a subset of V with the common certain features. A community division is denoted as $C = \{c_i | c_i \neq \emptyset, c_i \neq c_j, c_i \cap c_j = \emptyset, 1 \leq i, j \leq N_c\}$, where N_c stands for the number of communities in C. The main goal of community detection is to find a community division which maximizes a particular quality metrics function f. Therefore, the formulation of community detection can be represented as Eq. (1).

$$\arg \max_{C} f(G, C) \tag{1}$$

In this paper, two quality metrics, i.e., modularity and normalized mutual information (NMI), are used to provide more comprehensive comparisons. The modularity, denoted as Q, is measured by the inherent characteristics of a network, i.e., the heterogeneous relationships among edges between dense intra-community connections and sparse inter-community connections [6]. While NMI, as the object function of community detection, is widely used to quantify the similarity between detected communities and standard communities based on a certain feature in the real world [12].

2.2 Algorithms for Community Mining

Given the practical significance of community detection, many algorithms have been proposed to detect communities ranging from optimization algorithms to model-based algorithms

Concerning the perspective of optimization algorithms, a basic hypothesis for community detection is that the quality of community division can be evaluated by a network-based objective function, such as the modularity (Q). And, the object of such algorithms is to find community divisions with the maximal values of object function. In the following, we take two typical optimization algorithms (i.e., ACO and GA) as examples to describe the process of community mining of such algorithms.

- ACO [5] is a typical optimization algorithm, which searches the optimized community division based on the cooperation of ants. Each ant first finds a community division with a value of object independently. Then, each ant informs the quality of its community division to others based on a pheromone matrix by updating local pheromone. After all ants finish the searching process, a global pheromone is updated based on the results of elitist ants. With the iteration going on, the pheromone leads ants to find better community divisions.
- GA [4] is also an optimization algorithm, in which divisions need to be coded as integer strings, called chromosomes. And the integers in chromosomes are called genes. Every chromosome has a fitness value, which reflects the corresponding object function value. In each iteration step, some chromosomes with a higher fitness are searched by exchanging genes among chromosomes (i.e., crossover operation) and changing genes randomly (i.e., mutation operation). And only chromosomes with a higher fitness will survive with the increment of iteration steps. Final, the chromosome with the highest fitness will be output as the best community division.

In terms of the model-based algorithms, the basic hypothesis is the dynamic process taking on a network, which can be used to reveal community structures. Usually, this dynamic process is described by a model, such as the Markov chains. And communities emerge in the wake of the dynamic process. Here, we take MCL as an example to introduce the community detection of such algorithms.

- MCL [3] is a model-based algorithm, which is based on a flow simulation. In MCL, every vertex has a quantity of fluxes, and the distribution of fluxes is represented by a matrix, called the flow matrix. Moreover, the flowing of fluxes is described by a Markov chain with a transfer probability matrix. In such situation, a positive feedback operator is emerged through the fluxes flowing based on the markov chain. With the iterations going on, fluxes of vertexes in a tightly-linked group will flow together, and those vertexes are clustered as a community.

3   *Physarum* Computational Framework for Community Mining

3.1   A *Physarum* Mathematic Network Model for Community Mining

Inspired by *Physarum*, a mathematical model is designed by Tero, et al. [9], which has shown the abilities of path finding [9], network designing [8] and algorithms optimizing [11]. In this paper, a *Physarum*-based network mathematical model (PNM) is designed to distinguish inter-community edges from intra-community ones, based on PM. The core mechanism of PNM is the feedback system between the cytoplasmic fluxes and conductivities of tubes in PM. This feedback system has two main processes. First, $Q^t_{i,j}$, $D^t_{i,j}$ and $L_{i,j}$ stand for the flux, conductivity and length of $e_{i,j}$, respectively. p^t_i indicates the pressure of v_i. And the relationship among the flux, conductivity, length and pressure are formulated as Eq. (2). According to the Kirchhoff's law as shown in Eq. (3), the pressures and fluxes can be obtained at each time step. And then, $Q^t_{i,j}$ feeds back to $D^t_{i,j}$ based on Eq. (4). After that, a time step finishes. In the wake of such feedback process, a high efficient network is emerged.

$$Q^t_{i,j} = \frac{D^{t-1}_{i,j}}{L_{i,j}} |p^t_i - p^t_j| \tag{2}$$

$$\sum_i Q^{t-1}_{ij} = \begin{cases} I_0, \; if \; v_j \; is \; an \; inlet \\ -I_0, \; if \; v_j \; is \; an \; oulet \\ 0, \; others \end{cases} \tag{3}$$

$$D^t_{i,j} = \frac{Q^t_{i,j} + D^{t-1}_{i,j}}{k} \tag{4}$$

The major modification of PNM is the scheme of inlets/outlets choosing. In PNM, when a vertex is chosen as an inlet, the others are chosen as outlets.

In other words, Eq. (3) is modified as Eq. (5). And, in each time step of PNM, every vertex is chosen as the inlet once. When v_i is chosen as the inlet, a local conductivity matrix, denoted as $D^t(i)$, is calculated based on the feedback system. Finally, at the end of each time step, the global conductivity matrix is updated by the average of local conductivity matrixes based on Eq. (6). As time steps increase, the inter-community edges tend to have a larger conductivity.

$$\sum_i \frac{D^{t-1}_{i,j}}{L_{i,j}} |p^t_i - p^t_j| = \begin{cases} I_0, & \text{if } v_j \text{ is an inlet} \\ \frac{-I_0}{|V|-1}, & \text{others} \end{cases} \tag{5}$$

$$D^t = \frac{1}{|V|} \sum_{i=1}^{|V|} D^t(i) \tag{6}$$

3.2 *Physarum*-Based General Computational Framework for Community Mining

Utilizing the ability of PNM, a general computational framework for community mining is proposed, which aims to overcome the lower accuracy and the weaker robustness through optimizing the initialization or dynamic process of traditional algorithms. In the following, we take the optimized initialization scheme of such framework to introduce the details.

Based on PNM, the optimized initialization scheme detects communities through distinguishing inter-community edges from intra-community ones. A matrix DA is used to denote the property of $e_{i,j}$, in which $da_{i,j} = 1$ if and only if $e_{i,j}$ is an intra-community edges. The optimized initialization based on PNM is summarized as follows. First, a conductivity matrix D is obtained based on PNM. And then, all edges are supposed to be inter-community, i.e., $DA = -A$. Thereafter, some vertexes are chosen randomly. Meanwhile, the edges joining those chosen vertexes are marked as intra-community edges, except for the edges with top conductivities. After that, DA could be used to denote communities.

Based on DA, the vertexes connected by intra-communities edges are identified as a community. In other works, the connected components of a network composed of intra-community edges, are identified as communities of such network. After that, a community division emerges, and the optimized initialization of community mining is completed.

4 Experiments

4.1 Datasets

Six real-world networks[1] and three typical kinds of algorithms (i.e., GA, ACO, MCL) are used for estimating the efficiency and scalability of proposed framework. The basic information of those networks is shown in Table 1.

---

[1] http://www-personal.umich.edu/~mejn/netdata/.

Table 1. Basic information of used networks. The columns of clusters show the number of communities in standard community divisions, in which "-" means that standard division is nonexistent.

Network	Node	Edges	Clusters	Network	Node	Edges	Clusters
Karate	34	78	2	Dolphins	62	160	2
Polbooks	105	411	3	Football	115	613	12
Netscience	1589	2742	-	Polblogs	1490	19025	2

For a clear expression, a prefix (i.e., PNM-) is added to the original names of optimized algorithms for distinguishing. For example, the optimized algorithm of GA is denoted as PNM-GA. All experiments are implemented with the same parameters setting and running environment. Moreover, for some random algorithms (i.e., GA and ACO), the results are averaged over 20 repeated runnings for eliminating fluctuation and evaluating the robustness of algorithms.

4.2 Accuracy Comparison

We take NGACD [4] as an example to estimate the efficiency of our proposed framework for GA. Figure 1 plots the box chart of results returned by NGACD and PNM-NGACD. Due to the randomness of maximum and minimum, efficiency comparison is mainly based on the quartiles and means. Results show that the first and third quartiles, median and means of Q returned by PNM-NGACD are higher than that of NGACD on all datasets, which means that PNM-NGACD has a stronger exploring ability. Moreover, the lengths of boxes of PNM-NGACD are shorter than that of NGACD, which verifies that the robustness of PNM-NGACD is stronger than that of NGACD.

For ACO, we take ANCC [5] as an example to estimate the efficiency of our proposed framework. Figure 2(a) shows a comparison between ANTCC and

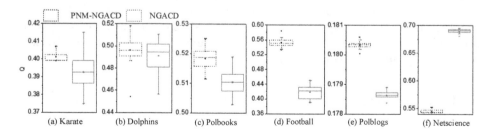

Fig. 1. Box charts of results returned by PNM-NGACD and NGACD on six networks in term of Q, in which the bottom and top of box are the first and third quartiles respectively, and the band inside the box denotes the median. The ends of whiskers represent the minimum and maximum of Q. Moreover, the small quadrates in boxes stand for the means of Q.

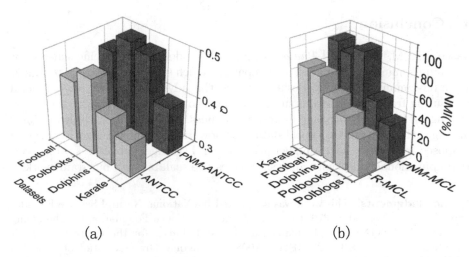

Fig. 2. Comparison of results. (a) shows the results returned by PNM-ANTCC, ANTCC, and (b) reports the results returned by PNM-MCL, R-MCL, in term of Q.

PNM-ANTCC in term of Q. The result shows that the Q values returned by PNM-ANTCC have an obvious improvement comparing with that of ANTCC.

Because MCL is a typical model-based algorithm, and INM is a more reasonable metrics for such algorithm. The comparison on networks with standard division, between a representational MCL (i.e., R-MCL [3]) and PNM-MCL for accuracy, is based on INM. As is shown in Fig. 2(b), the NMI of PNM-MCL is higher than that of R-MCL. Meanwhile, the NMI values of PNM-MCL on karate and dolphins network are 1, which means that the communities found by PNM-MCL are same as the known communities.

4.3 Computation Complexity Analysis

Using T, N to stand for the maximal iterative step and the number of nodes in network, respectively. The computation complexities of *Physarum*-based initialization and dynamic process are analyzed as follows.

***Physarum*-based initialization:** For *Physarum*-based initialization, at every iterative step, every node should be chosen as the inlet once. When a node is chosen, a corresponding system of equations needs to be solved. In other words, N systems of equations have to be solved as every iterative step. The worst computation complexity of solving a system of equations is $O(N^3)$. With a empirical setting (i.e., $T = 1$), the total computation complexity of *Physarum*-based initialization is $O(N^4)$.

***Physarum*-based dynamic process:** Generally speaking, optimizing the feedback system based on the *Physarum*-based dynamic process, dose not change the operation sequence. For example, the computation complexity of PNM-MCL is $O(T \times N^2)$, which is the same as that of R-MCL.

5   Conclusion

Based on PM, a modified *Physarum* network model with a specific scheme for community mining is proposed in this paper, which shows an ability of recognizing inter-community edges. Besides, taking the advantages of PNM, a general computing framework for community mining is proposed to overcome the shortcomings of lower accuracy and weaker robustness. Experiments with three different algorithms on six real-world datasets show that optimized algorithms with proposed framework have a higher accuracy and stronger robustness. Moreover, a computational complexity analysis verifies the scalability of our framework.

Acknowledgments. This work was supported by National Natural Science Foundation of China (Nos. 61402379, 61403315), Natural Science Foundation of Chongqing (No. cstc20 13jcyjA40022), Fundamental Research Funds for the Central Universities (Nos. XDJK2016D053, XDJK2016A008), Chongqing Graduate Student Research Innovation Project, and Specialized Research Fund for the Doctoral Program of Higher Education (No. 20120182120016). Prof. Zili Zhang and Dr. Xianghua Li are the corresponding authors of this paper.

References

1. Tremblay, N., Borgnat, P.: Graph wavelets for multiscale community mining. IEEE Trans. Sig. Process. **62**(20), 5227–5239 (2014)
2. Nematzadeh, A., Ferrara, E., Flammini, A., Ahn, Y.Y.: Optimal network modularity for information diffusion. Phys. Rev. Lett. **113**, 088701 (2014)
3. Satuluri, V., Parthasarathy, S.: Scalable graph clustering using stochastic flows: applications to community discovery. In: Proceedings of the 15th International Conference on Knowledge Discovery and Data Mining, pp. 737–746. ACM (2009)
4. Li, X., Gao, C., Pu, R.: A community clustering algorithm based on genetic algorithm with novel coding scheme. In: The 10th International Conference on Natural Computation, pp. 486–491. IEEE (2014)
5. Jina, W.: Ant aolony algorithms based community clustering research. Master's thesis, Sun Yat-sen University (2009)
6. Girvan, M., Newman, M.E.: Community structure in social and biological networks. Proc. Nat. Acad. Sci. **99**(12), 7821–7826 (2002)
7. Nakagaki, T., Yamada, H., Tóth, Á.: Intelligence: maze-solving by an amoeboid organism. Nature **407**(6803), 470–470 (2000)
8. Tero, A., Takagi, S., Saigusa, T., Ito, K., Bebber, D.P., Fricker, M.D., Yumiki, K., Kobayashi, R., Nakagaki, T.: Rules for biologically inspired adaptive network design. Science **327**(5964), 439–442 (2010)
9. Tero, A., Kobayashi, R., Nakagaki, T.: A mathematical model for adaptive transport network in path finding by true slime mold. J. Theor. Biol. **244**(4), 553–564 (2007)
10. Liang, M., Gao, C., Liu, Y., Tao, L., Zhang, Z.: A new Physarum network based genetic algorithm for bandwidth-delay constrained least-cost multicast routing. In: Tan, Y., Shi, Y., Buarque, F., Gelbukh, A., Das, S., Engelbrecht, A. (eds.) ICSI-CCI 2015. LNCS, vol. 9141, pp. 273–280. Springer, Heidelberg (2015)

11. Liu, Y., Gao, C., Zhang, Z., Lu, Y., Chen, S., Liang, M., Tao, L.: Solving np-hard problems with Physarum-based ant colony system. IEEEE/ACM Trans. Comput. Biol. Bioinform. (2015). doi:10.1109/TCBB.2015.2462349
12. Jin, D., Chen, Z., He, D., Zhang, W.: Modeling with node degree preservation can accurately find communities. In: 29th AAAI Conference on Artificial Intelligence, pp. 160–167. AAAI (2015)

Rank-Based Nondomination Set Identification with Preprocessing

Vikas Palakonda and Rammohan Mallipeddi[✉]

School of Electronics Engineering, Kyungpook National University,
Bukgu, 702 701 Daegu, Republic of Korea
vikas.11475@gmail.com, mallipeddi.ram@gmail.com

Abstract. In multi-objective optimization, finding a nondomination set is a computationally expensive process and the complexity grows with the population size. In this paper, we propose a nondomination set identification method with (a) rank-based preprocessing step where the obvious dominant solutions are eliminated and (b) better order of comparison based on the average rank so that the number of comparisons can be significantly reduced. In preprocessing, the maximum rank information of the solutions that are best in each individual objectives is used. In addition, during nondomination set identification process to check if a solution is nondominant the solution is compared only with solutions that are better in terms of average rank. The experiment results demonstrate the effectiveness of the proposed method in identifying the nondominant set in less number of comparisons.

Keywords: Multi-objective evolutionary algorithms · Nondomination set identification · Preprocessing step · Rank-based sorting

1 Introduction

Most real world problems are associated with the multiple objectives resulting in a set of optimal solutions which arise due to trade-off between the conflicting objectives and to choose a particular over the other requires some problem specific knowledge and a number of problem-related factors [1]. However, most Multi-Objective Evolutionary Algorithms [MOEAs] have adopted the Pareto-based approach where the comparisons between the solutions will be done based on the qualities of Pareto dominance [2]. In the literature it has been proved that most of the MOEAs have concentrated on sorting the population into various fronts that is nondomination sorting whereas some algorithms have concentrated only on finding the nondominated set of solutions that is Pareto front. In this paper, we mainly concentrate on finding the Pareto front rather than sorting entire population into fronts.

A solution $x^{(1)}$ is said to dominate the other solution $x^{(2)}$, if both conditions 1 and 2 are true: (1) solution $x^{(1)}$ is no worse than $x^{(2)}$ in all objectives and (2) solution $x^{(1)}$ is strictly better than $x^{(2)}$ in at least one objective [1, 3].

Due to the importance of nondomination, significant work has been reported with aim of reducing the number of comparisons and thus the computational complexity such as Naïve method [1], Kung's method [1] and Jun du's algorithm. In this paper, we

© Springer International Publishing Switzerland 2016
Y. Tan et al. (Eds.): ICSI 2016, Part II, LNCS 9713, pp. 150–157, 2016.
DOI: 10.1007/978-3-319-41009-8_16

proposed a novel method with two steps: (a) rank-based preprocessing step where the obvious dominant solutions are eliminated and (b) better sequence or order of comparison based on the average ranks so that the number of comparisons can be significantly reduced. In the proposed method, initially, each individual in the population is assigned a set of ranks where the nth rank of the individual is assigned based on the objective value of the individual in the nth objective. The ranking will be given as follows: in a minimization problem the solution which has least value in an objective will be the assigned with the integer rank "one" whereas for a maximization problem, the solution with highest value will be assigned with integer rank "one".

After assigning the ranks, the solutions that have at least one integer rank "one" will be considered as "obvious" nondominated solutions. Then, we introduce a preprocessing step which will get rid of the "obvious" dominated solutions by using the rank information of the "obvious" nondominant solutions. The removal of the obvious dominant solutions reduces the number of comparisons to be done to identify the nondomination set. After the preprocessing step, the obvious nondominant solutions will be temporarily removed and the other remaining solutions will be assigned a specific order based on their average ranks. It is based on the assumption that a solution can be dominated only by individuals that are better in terms of average rank compared to it.

In the remainder of the paper, Sect. 2 briefly describes the importance and issues in finding the nondomination set while solving multi-objective optimization problems. In Sect. 3, we explain our proposed rank based sorting method to eliminate the solutions through preprocessing and to find the nondomination set. In Sect. 4, we describe our experimental set up and simulation results to highlight the effectiveness of the proposed algorithm in reducing the computationally complexity while finding a Pareto optimal set. Finally, in Sect. 5 we outline the conclusions of this paper.

2 Concept of Nondomination: Importance and Issues

When solving a multi-objective optimization problem, there exists a set of solutions which are superior to rest of the solutions in the search space when all the objectives are considered but are inferior to some other solutions in the space in one or more objectives. The set of solutions that satisfy the above criteria are referred to as ***Pareto-optimal*** or ***non-dominated*** solutions while the rest of the solutions are known as the dominated solutions [1, 3, 4]. Therefore, in multi-objective optimization, the main goals to be achieved during the optimization are: (1) convergence to the Pareto optimal set and (2) maintenance of diversity in the solutions of the Pareto optimal set. To achieve convergence to Pareto optimal set, most MOEAs employ nondominated sorting algorithms [5]. Nondominated sorting is a technique in which the solutions present in a population are assigned to different fronts depending on their dominance relationships [6]. However, the process of the finding nondominated set is time-consuming and therefore designing an efficient nondominated set identification algorithm plays a crucial part in improving the performance of MOEAs [5].

In multi-objective optimization, to determine nondominant solutions among a set of solutions it is essential to compare each solution with all the remaining solutions present in the population with respect to all the objectives. This is referred to as Naïve approach where the number of comparisons to be done to determine the set of non-dominated solutions increases with increase in the population resulting in computational complexity. The computational complexity of Naïve method is $O(MN^2)$ [1]. To reduce the time complexity associated with nondomination sorting, it is essential to minimize the number of comparisons. In literature, various nondomination set identification methods were proposed to minimize the runtimes. The most popular one is Kung's method which has a computational complexity of $O(MN\log N)$ for two and three objectives whereas for objectives more than three it will be $O(MN(\log N)^{(M-2)})$ [1, 7]. In Kung's method all the solutions are sorted with respect to the first objective and adopt a divide and conquer approach in which the solutions will be divided into two halves. The solutions in the bottom half will be compared with the solutions in the top half and the nondominated solutions will be merged with the top half to form a merged population M [1, 7, 8]. Recently, in [2] a sorting based algorithm to find the non-dominated set of solutions which has time complexity of $O(MN\log N)$ in the best cases whereas in the worst cases it will be $O(MN^2)$. In this algorithm, the solutions are ranked on each objective and the summation of the ranks is obtained based on which the order of comparisons depend.

3   Proposed Method

In this paper, we propose a novel nondomination set identification method where the solutions are ranked on each of the objectives under consideration. In the current work, a prescreening process helps in eliminating the obvious dominated solutions from the current population which in turn minimizes the number of comparisons needed to identify the nondomination set. First, each solution will be assigned an integer rank with respect to the each objective and the solutions that have least rank with respect to the objectives are considered as obvious nondominated solutions. In other words, the solutions that have an integer rank of "one" in any particular objective will be considered as nondominated. The maximum rank of these solutions will be taken and mean will be calculated and is employed to identify and thus eliminate the obvious dominant solutions. In other words, the solutions whose minimum rank is greater than the mean of the maximum ranks of the obvious nondominant solutions (found in the previous set) will be eliminated. To make our preprocessing step more comprehensive, we demonstrate the process with an example multi-objective problem where minimization of all the objectives is considered. In Table 1, we present a set of 10 solutions (S1, S2 ... S10) with 3 different objective values (F1, F2 and F3). In Table 1, the value entry 0.60 (5) represents that individual S1 has an objective value of 0.6 on F1 and is ranked 5th according to F1. The last row Table 1 presents the average of the ranks for each solution.

From Table 1, we observe that solutions S3, S1 and S9 are the solutions with integer rank "one" on objectives F1, F2 and F3, respectively. Therefore, solutions S3, S1 and S9 are obvious nondominant solutions. After that, we find the mean of the maximum ranks of these obvious nondominant solutions (S3, S1 and S9). From Table 1, the maximum ranks of S3, S1 and S9 are 10, 5 and 4 respectively. The mean of these maximum ranks of S3, S1 and S9 is 6.3 which is obtained as (5 + 10 + 4)/3. Now, the solutions which have the minimum rank greater than the mean of the maximum ranks (6.3) will be

Table 1. Objective function values and ranks of 10 population members

	S1	S2	S3	S4	S5	S6	S7	S8	S9	S10
F1	0.60	0.91	0.33	0.85	0.80	0.37	0.72	0.76	0.55	0.4 (3)
	(5)	(10)	(1)	(9)	(8)	(2)	(6)	(7)	(4)	
F2	21 (1)	28 (2)	44 (3)	87 (8)	93 (9)	67 (6)	98	75 (7)	51 (4)	64 (5)
							(10)			
F3	297	475 (6)	943	508	676	786	419	416	112	118
	(3)		(10)	(7)	(8)	(9)	(5)	(4)	(1)	(2)
AR*	3	6	4.6	8	8.3	5.6	7	6	3	3.3

*AR – average rank

eliminated. From the Table 1, we observe that the solutions S4 and S5 have a minimum rank greater than the mean value 6.3 and are eliminated. Thus, the preprocessing step gets rid of the obvious dominant solutions (S4 and S5).

After preprocessing, it is evident that solutions S3, S1 and S9 are obvious nondominant solutions while solutions S4 and S5 are obvious dominant. In the next step, the obvious nondominant solutions are removed temporarily and the remaining individuals should be compared in an appropriate order by sorting in the ascending order with respect to the mean rank. In other words, the solutions are ordered as S10, S6, S2, S8 and S7 based on the average ranks 3.3, 5.6, 6, 6 and 7, respectively.

Once the solutions are sorted, a solution will be compared for nondominance with solutions that are better (in terms of average rank). This is based on the assumption that: a solution A can dominate solution B only when the mean rank of A is less than that of B. In other words, solutions C and D with mean ranks greater than B can be nondominant with respect to B but cannot dominate B because their mean ranks are not less than that of B. In other words, solution S2 will only be compared with solutions S10 and S6 which have better average rank than S2 for nondominance. And S8 and S9 can never dominate S2 since their average rank is greater than that of S2.

Based on the above intuition, we start comparing solutions as shown below for nondominance. We start with S6 and is compared only with S10. S2 is compared with only S10, S6 and S2. Now when S8 is compared with S10, S6 and S2, S8 will be eliminated because it is dominated by S10. Since S8 is eliminated S7 does not have to be compared with S8 thus eliminating few comparisons. In other words, if a solution X is completed dominated by any solutions which come before it, then solution X will be removed. Thus, unnecessary comparisons between the solutions can be avoided when the solutions ranked before it are being compared for nondominance. Even though, some extra comparisons are needed for sorting the individuals according to the ranking but the remaining process will offset it.

4 Experimental Setup and Simulation Results

In this section, we have examined the effect of the preprocessing step and then compared our rank based sorting with and without preprocessing step with nondomination set identification algorithms namely Naïve [1], Deb's [1] and Du's methods [2]. All simulations reported are conducted on a PC with a 3.40 GHz Intel Core i7-2600QM CPU and Windows 7 SP1 64-bit operating system. In the first setup, randomly generated populations (N) varying from 500 to 5000 solutions were taken and the number of objectives (M) varying from 2 to 10. To highlight the effect of preprocessing, we reported the mean of number of solutions eliminated for 100 runs for different objectives (M) and populations sizes (N) in Table 2.

Table 2. Mean number of solutions eliminated due to preprocessing

	2	3	4	5	6	7–10
500	144.71	28.64	4.57	0.81	0.02	0
1000	289.54	55.15	8.32	0.66	0.10	0
2000	547.56	132.33	17.80	2.03	0.23	0
3000	840.83	159.22	28.98	3.61	0.13	0
4000	1.30E + 03	244.22	28.81	4.59	0.45	0
5000	1.83E + 03	291.39	35.26	4.71	0.46	0

Table 3. Performance comparison of various algorithms in terms of average number of comparisons required (Objectives 2, 5 and 10)

N	Algorithm	Number of objectives		
		2	5	10
1000	Naïve	2000000	5000000	10000000
	Kung	96130.667	941384.67	3936507.7
	Jun du	1314346.7	2042507.3	3214097.7
	RBS	12849.2	162042.67	**3175667.3**
	RBSWP	**11262.354**	**162036.48**	**3175667.3**
2000	Naïve	8000000	20000000	40000000
	Kung	322929.39	3672582.6	13742789
	Jun du	1443197.5	4302527	10674327
	RBS	27969.047	380933.19	**10599146**
	RBSWP	**24540.69**	**380906.42**	**10599146**
3000	Naïve	18000000	45000000	90000000
	Kung	667060.37	8048619.8	28378485
	Jun du	1706122.4	7300922.2	21309845
	RBS	43868.158	648442.52	**21197993**
	RBSWP	**39846.688**	**648426.77**	**21197993**
4000	Naïve	32000000	80000000	160000000
	Kung	1182924.9	13744621	46376980
	Jun du	2137145.1	11189783	33654935
	RBS	59990.853	924062.27	**33521808**
	RBSWP	**54872.467**	**924043.9**	**33521808**
5000	Naïve	50000000	125000000	250000000
	Kung	1746514.9	21497116	68285585
	Jun du	2531164.6	16656192	48661076
	RBS	74242.017	1152838.4	**48501137**
	RBSWP	**66782.779**	**1152798.6**	**48501137**

From the results presented in Table 2, it has been observed that the number of solutions that are eliminated by the preprocessing step increases with the increase in the number of solutions or population size (N). In addition, it can also be observed the number of solution eliminated due to the preprocessing step decrease with the increase in the number of objectives. From Table 2, it can be concluded that for $M \geq 7$, no solutions are eliminated by the preprocessing step. However, for multi-objective problems with number of objectives less than 5 the preprocessing step is observed to be eliminating solutions that result in significant reduction in the number of comparisons that need to be done to obtain the set of nondominant solutions.

The performance of various nondomination set identification algorithms on randomly generated populations of various sizes and various objectives is reported in Table 3. In Table 3, RBS refers to the proposed algorithm with no preprocessing step involved while RBSWP is the proposed algorithm with preprocessing step. All the algorithms were run on the same set of randomly generated populations and the results presented are average results of 100 runs. The best results are highlighted. From the highlighted results it can be observed that RBSWP outperforms all the other algorithms on each objectives for all population sizes. However, as the number of objectives increases the performance of RBSWP becomes similar to that of RBS. This is due to the fact that as the number of objectives increases the number of solutions that can be eliminated due to preprocessing are virtually zero.

In Figs. 1, 2 and 3, we summarize the performance of Kung's, Jun du's and the proposed RBSWP algorithms on 2-, 5- and 10-objectives respectively. From Fig. 1, it is clear that RBSWP outperforms the other algorithms by finding the nondomination set in comparatively less number of comparisons. As the number of objectives increases the performance of Jun du's algorithm outperforms Kung's method as observed in Figs. 1, 2 and 3. In addition, as the number of objectives increases further to 10, the performance of Jun du's algorithm and the proposed RBSWP becomes nearly

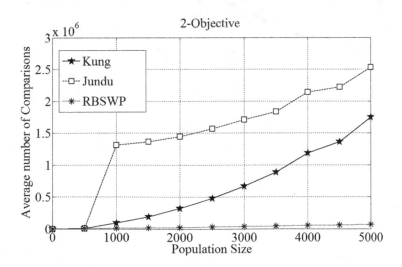

Fig. 1. Comparison of different algorithms in 2-objective case

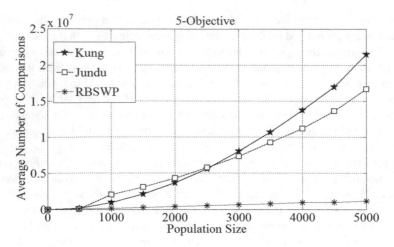

Fig. 2. Comparison of different algorithms in 5-objective case

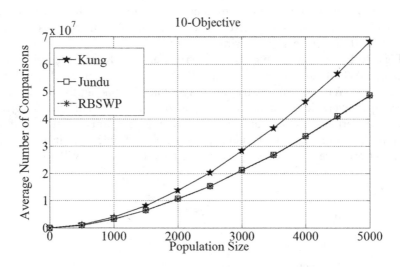

Fig. 3. Comparison of different algorithms in 10-objective case

identical according to Fig. 3. However, the slightly improved performance of proposed RBSWP compared to Jun du's algorithm in terms of average number of comparisons can be observed from Table 3.

5   Conclusion

In this paper, a novel nondomination set identification method referred to as rank based sorting with preprocessing is proposed, in which each solution in the population is assigned an integer rank with respect to each objective. At the beginning of the optimization of the problem, a preprocessing step is introduced which would reduce the number of solutions to be checked for nondominance by eliminating the obvious dominant solutions and then efficiently comparing the remaining solutions. Thus, most of the unnecessary comparisons can be avoided which results in the reduction of the computational complexity. The proposed preprocessing step being general can be applied in conjunction with Naïve, Kung's and the Jun du's methods. In addition, the comparison on individuals in an order that is determined based on the average rank of the individuals saves a considerable number of comparisons thus reducing the computational complexity further.

Acknowledgement. This study was supported by the BK21 Plus project funded by the Ministry of Education, Korea (21A20131600011).

References

1. Deb, K.: Multi-objective Optimization using Evolutionary Algorithms, vol. 16. Wiley, Hoboken (2001)
2. Du, J., Cai, Z.: A sorting based algorithm for finding a non-dominated set in multi-objective optimization. In: 3rd International Conference on Natural Computation, pp. 436–440 (2007)
3. Srinivas, N., Deb, K.: Muiltiobjective optimization using nondominated sorting in genetic algorithms. Evol. Comput. **2**, 221–248 (1994)
4. Deb, K., Pratap, A., Agarwal, S., Meyarivan, T.: A fast and elitist multiobjective genetic algorithm: NSGA-II. IEEE Trans. Evol. Comput. **6**, 182–197 (2002)
5. Shi, C., Chen, M., Shi, Z.: A fast nondominated sorting algorithm. In: International Conference on Neural Networks and Brain, pp. 1605–1610 (2005)
6. Zhang, X., Tian, Y., Cheng, R., Jin, Y.: An efficient approach to nondominated sorting for evolutionary multiobjective optimization. IEEE Trans. Evol. Comput. **19**, 201–213 (2015)
7. McClymont, K., Keedwell, E.: Deductive sort and climbing sort: new methods for non-dominated sorting. Evol. Comput. **20**, 1–26 (2012)
8. Fang, H., Wang, Q., Tu, Y.-C., Horstemeyer, M.F.: An efficient non-dominated sorting method for evolutionary algorithms. Evol. Comput. **16**, 355–384 (2008)

Spiking Simplicial P Systems with Membrane Coefficients and Applications in Document Clustering

Jie Xue$^{(\boxtimes)}$ and Xiyu Liu

School of Management Science and Engineering, East Road of Wenhua, Jinan
250014, Shandong, China
{xiaozhuzhu1113, sdxyliu}@163.com

Abstract. The purpose of this paper is to propose a new kind of P systems on simplicial complexes with membrane coefficients (SS P system). We extend the basic membrane structures on complexes. Four classes of communication rules based on chain structure and edge operation are also provided. The computational completeness of the new P system is proved by simulation of register machine. A document clustering problem is used to verify the effectiveness of SS P systems. It also provides a new thought to the clustering of documents. Since document clustering has great significance in document summarization and computer science. Comparison result between SS P systems and other P systems shows comparative advantage of our new P system.

Keywords: SS P system · Membrane coefficient · Document clustering

1 Introduction

Membrane computing is a new branch of natural computing which is initiated by Păun at the end of 1998 [1]. In recent years, one of the main research fields on P systems is new classes of P systems. Păun provided numerical P systems adding integers to multisets of objects [2]. Aman presented a real-time extension of P systems in which each membrane and each object has a lifetime attached to it [3]. Adorna gave evolution communication P systems, which propose a way to measure the communication costs by means of quanta of energy [4]. Dragomir describe a P system with object processing rules and structure changing rules [5].

However, only limited attention has been paid to the problem of membrane structure and design of rules. Păun [6] presented cell-like P systems, which are the first membrane computing models. Cell-like P systems are hierarchical membrane structure with rewriting and communication rules. Objects are rewritten in membrane and communicated between lower and higher membranes. Martín-Vide provided tissue-like P systems, which are based on graph [7]. In tissue-like P systems, several one-membrane cells are considered as evolving in a common environment. Communication rules implement the transportation of strings among channels. Division rules [8, 9] can divide one membrane into double. Ionescu [16] proposed spiking neural P systems with firing and forgetting rules. Firing rules can send spikes to other neurons within d steps.

© Springer International Publishing Switzerland 2016
Y. Tan et al. (Eds.): ICSI 2016, Part II, LNCS 9713, pp. 158–168, 2016.
DOI: 10.1007/978-3-319-41009-8_17

Forgetting rules delete spikes. Miguel investigated spiking neural P systems with axons. Recently, Bogdan considered mobile membranes with objects on surface. Sburlanv introduces a P system that uses vectors of rules to describe a causal dependence relation between the executions of the rules [10]. We proposed lattice based communication P systems in [11]. But it is flat structure with two types of rules. We also presented simplicial membrane structure in [12] with two classes of rules.

So far, presented papers are engaged in searching computational power of membrane computing, solving NP problems, solving arithmetic problems, producing languages or strings. Nearly, none of them combine P systems with practical applications. Yang uses P system in image edge detection [13]. Metta combine SN P systems with Petri net [14]. The comparison of applications by P systems with exist methods has not been provided.

Motivated by researches above, we continue to research the structure of membranes. We propose simplicial structure with coefficients on membranes. Objects are only spikes. Up to the authors' knowledge, this is the first paper to describe coefficients on membranes, which give a more complex execution of rules, objects and membranes. The new P system is still equal to Turing machine. We prove it by simulating the register machine. The computational power of SS P systems is compared with other P systems. We verify the feasibility of SS P system by solving document clustering problem. It is a new combination of document clustering and membrane computing. Further studies on SS P systems will be summarized in our next study.

2   Spiking Simplex P System with Membrane Coefficients

Definition 2.1. Assume $a_0, a_1, \ldots, a_q$ are q + 1 points possessing the widest position in Euclidean space Rn, $q \leq n$. Let $x = \sum_{i=0}^{q} \lambda_i a^i, \lambda_0, \lambda_1, \ldots, \lambda_q$ meets the condition: $\lambda_0 + \lambda_1 + \ldots + \lambda_q = 1$, $\lambda_0 \geq 0$, $\lambda_1 \geq 0$, $\lambda_q \geq 0$, then, x is called a q dimensional simplex, denoted as sq.

Definition 2.2. K is a simplicial complex that satisfies the following conditions: (1) Any face of a simplex s from K also belongs to K; (2) The intersection of any two simplices of s is either empty or a face of both.

Definition 2.3. Assume a simplex $s^q = (a^0, a^1, \ldots, a^q)$, $a^{i_0}, a^{i_1}, \ldots .a^{i_k}$ are arbitrary k + 1 points, they possess the widest position, thus, $s^k = (a^{i_0}, a^{i_1}, \ldots, a^{i_k})$ is a r dimensional simplex, which is face of sq, denoted as $s^k \prec s^q$, and if k = q−1, sq is called the parent of sk. There are two relationships between sq and sk, denoted as $[s^q : s^k] = 1, -1$.

Definition 2.4. $s^{q-1} = (a^{i_0}, a^{i_1}, \ldots, a^{i_{q-1}})$, assume that are k−1 dimensional face of sq, then, $x_q = g_0 s_0^{q-1} + g_1 s_1^{q-1} + \ldots + g_n s_q^{q-1}$ is a chain of sq. For two chain xq and yq, $x_q + y_q = (g_1 + h_1) s_1^q + (g_2 + h_2) s_2^q + \ldots + (g_{\alpha_q} + h_{\alpha_q}) s_{\alpha_q}^q$.

Definition 2.5. A $(q - 1)$ dimensional chain ∂s^q is called an edge chain, ∂ is edge operator. $\partial s^0 = 0$, $\partial s^q = \sum_{j=1}^{\alpha_{q-1}} [s_i^q : s_j^{q-1}] s_j^{q-1} = \sum_{i=0}^{q} (-1) a^0, \cdots, a^i \ldots a^q$.

2.1 Spiking Simplex P System with Membrane Coefficients

A spiking simplex P system with membrane coefficient, denoted as SSP system. SSP system is a construct:

$$\prod = (\alpha, p_i, g_i, E, m_s, m_c, R_{pc}, R_e, R_c, R_{sf}, i_0)$$

Where, α is spike, pi is polarization of spiking in membrane i, $p \in \{+, -\}$; gi is the coefficient of membrane i, gi is a function f(x). E is the set of objects α with limited multiplicity in the environment; ms is the structure of simplex membranes; mc is the structure of chain membranes; $\{R_{pc}, R_e, R_c, R_{sf}\}$ are rules of SSP systems; i_0 is the label of output membrane.

Membrane structures are studied in our previous paper [12]. Besides, if $s^k \prec s^q$ is a face of sq and $k = q - 1$, then we define sq as the parent membrane of sk. If $[s^q : s^k] = 1$, we say that sk is a forward child membrane of sq, otherwise, if $[s^q : s^k] = -1$, sk is a reverse child membrane of sq.

For a simplex membrane sq, $\partial s^q = \sum_{j=1}^{\alpha_{q-1}} [s_i^q : s_j^{q-1}] s_j^{q-1} = \sum_{i=0}^{q} (-1) a^0 \ldots a^i \ldots a^q$ is a chain membrane of sq. For a simplex membrane $a^1 a^2 a^3$, its child membranes $a^1 a^2, a^2 a^3, -a^1 a^3$ consist of the chain membrane $a^1 a^2 \rightarrow a^2 a^3 \rightarrow -a^1 a^3$.

In SSP systems, according to forwards and reverse faces, we define membranes with spikes and anti-spikes. Besides, membranes have coefficients (functions) which perform in rules.

Different from conventional P systems which use an alphabet with multiple objects, SN P system just use one object: the spike a, which reduce the complexity of P system. However, SN P systems just have firing rules and forgetting rules. To do better communication in P system and save multisets, we introduce rules on simplex and chain with spikes.

Rules Rpc among parent-children membranes may have the following forms:

$$[\partial s^q, s^q] : [(\{p_i g_i \alpha^i\}, asc), (\{p_j g_j \alpha^j, ent_j\})] \rightarrow [(p_m \alpha^m, des), (p_k \alpha^k, ent)]$$

$\{p_0 g_0 \alpha^{i0}, p_1 g_1 \alpha^{i1}, p_2 g_2 \alpha^{i2}, \ldots, p_q g_q \alpha^{iq}\}$ from membrane $\{s_0^{q-1}, s_1^{q-1}, s_2^{q-1}, \ldots, s_q^{q-1}\}$ transform into a^k and ascend to their common face s^q; α^j divides into $\{p_0 g_0 \alpha^{i0}, p_1 g_1 \alpha^{i1}, p_2 g_2 \alpha^{i2}, \ldots, p_q g_q \alpha^{iq}\}$ and enter into $\{s_0^{q-1}, s_1^{q-1}, s_2^{q-1}, \ldots, s_q^{q-1}\}$ respectively at the same time.

Rules Re correspond to the edge operations may have the following forms:

$$[s^q \rightarrow \partial s^q] : [(\alpha^j, des) \rightarrow (p_j g_j \alpha^j, ent_j)]$$

α^j divides into $p_0 g_0 \alpha^{j0}, p_1 g_1 \alpha^{j1}, p_2 g_2 \alpha^{j2}, \ldots, p_q g_q \alpha^{jq}$ and descend into generated membranes $\{s_0^{q-1}, s_1^{q-1}, s_2^{q-1}, \ldots, s_q^{q-1}\}$ respectively and membrane s^q dissolves simultaneously.

Rules Rc among chain membranes may have the following forms:

$$[(\partial s_i^q, \partial s_j^q, \partial s^q)] : [(\{p_{im} g_{im} \alpha^{im}\}, \{p_{jn} g_{jn} \alpha^{jn}\}, mer) \rightarrow \{p_l g_l \alpha^k\}]$$

$\{p_{i0} g_{i0} \alpha^{i0}, p_{i1} g_{i1} \alpha^{i1}, p_{i2} g_{i2} \alpha^{i2}, \ldots, p_{iq} g_{iq} \alpha^{iq}\}$ in chain membrane ∂s_i^q and $\{p_{j0} g_{j0} \alpha^{j0}, p_{j1} g_{j1} \alpha^{j1}, p_{j2} g_{j2} \alpha^{j2}, \ldots, p_{jq} g_{jq} \alpha^{jq}\}$ in membrane ∂s_j^q merge into $\{p_0 g_0 \alpha^{k0}, p_1 g_1 \alpha^{k1}, p_2 g_2 \alpha^{k2}, \ldots, p_q g_q \alpha^{kq}\}$ according to $p_0 g_0 \alpha^{k0} = p_{i0} g_{i0} \alpha^{i0} + p_{j0} g_{j0} \alpha^{j0}$ and chain membrane ∂s_i^q and ∂s_j^q merge into a new chain membrane ∂s^q.

Rules Rsf among simplexes and their common face may have the following forms:

$$[s_i^q, s_j^q, s^{q-1}] : [(p_i g_i \alpha^i, p_j g_j \alpha^j, \lambda)] \rightarrow [(p_i g_i \alpha^m, p_j g_j \alpha^n, g(p_i \alpha^i + p_j \alpha^j))]$$

$p_i g_i \alpha^i$ from ∂s_i^q and $p_j g_j \alpha^j$ from ∂s_j^q transform into $p_i g_i \alpha^n$ and $p_j g_j \alpha^m$ in ∂s_i^q and ∂s_j^q; the sum $g(p_i \alpha^i + p_j \alpha^j)$ of spikes with coefficient of membrane s^{q-1} is stored into their common child membrane s^{q-1} simultaneously. s^{q-1} is empty in initial configuration. Especially, $\alpha^+ \alpha^- \rightarrow \varepsilon$.

2.2 Configuration and Computation

The configuration and computation of spiking simplex P system with membrane coefficient are described as follows. The spikes $\{a^i | i \in alphabet\}$ consist of the initial configuration of the system. Computation is the change of system, which is caused by the application of the four types of rules $\{R_{pc}, R_e, R_c, R_{sf}\}$. Spikes change in two classes of membranes.

Rules make spikes evolving through membranes. The computation begins with the $\{a^i | i \in alphabet\}$ in the n cells. In each time unit, rules are used in simplex membranes or chain membranes in SSP systems. If there are no rules being applicable for a membrane, then no spikes changes. The system is synchronously evolving for all membranes. When the system reaches a configuration that no rules are applicable, the computation halts. A configuration is stable when the execution of rules does not make any spikes change in membrane. The computation is successful if and only if it halts/stable. The result of a SS P system is spikes presented in membrane i_0 in the halting configuration.

3   Document Clustering by SS P System

In this section, we use SS P systems to implement a simple document clustering to verify the effectiveness of our P systems. Documents are represented using the vector space model (VSM) [18]. Dimensions of vectors correspond to features of each document, the feature item refers to various units constituting document. Weights of features reflect the importance of them in text content. For a document D, let $T = (t_1, t_2, \ldots, t_n)$ represent all the terms occurring in a document D. ω_{ij} is the weight of t_j in d_i. A document d_i is denoted as:

$$v(d_i) = ((t_1, \omega_{i1}), (t_2, \omega_{i2}), \ldots, (t_n, \omega_{in})), \omega_{ij} = tf_{ij} \times idf_j = tf_{ij} \times \log(N/df_j)$$

tf_{ij} is the occurrence of word(term) t_j in document d_i; df_j is the number of documents containing t_j [19].

Preprocessing data may be high dimensional, which will add the complexity of computation. Thus, dimensionality reduction technique is used to extract strong related words to constitute the feature set. We use document frequency (df) and terms contribution (tc) to do feature selection.

(1) Calculate document frequency (df) after automatic segmentation, filter out those words with df too low or too high
(2) Calculate terms contribution (tc), $tc(t) = \sum \omega_{it} \times \omega_{jt}$, choose the first L highest score features, put them from T to a new set S (initial S is empty), delete d with chosen features from D. The process will continue until there are no documents in D.

A weighted graph model is built with documents to do the clustering. Each document is represented by a node on a graph. Edges on a document graph are weighted with a value representing the similarity between documents. The weight is a cosine similarity between the vectors of two documents.

$$sim(d_i, d_j) = \sum_{k=1}^{n} (\omega_{ik} \times \omega_{jk}) / \sqrt{\sum_{k=1}^{n} (\omega_{ik}^2 \times \omega_{jk}^2)}$$

In the proceeding of document clustering, we adopt the idea of finding minimum path for every vertex.

SSP system designed for DC is a construct:

$$\prod = (\alpha, p_i, g_i, E, m_s, m_c, R_{pc}, R_e, R_c, R_{sf}, i_0)$$

Where, $\alpha = \{\alpha^0, \alpha^{t_j}, \alpha^i \alpha^{t_j} \alpha^0 \ldots \alpha^{t_k}\}$, the function of α^0 is like space in sentence, which separates spiking. Besides, α^0 cannot be changed by rules. $p_i = +, g_i = 1$, membrane structure is shown in Fig. 1. D is a membrane chain $\{d_1, d_2, \ldots, d_m\}$.

A document $d_i : \alpha^i \alpha^{ht_j} \alpha^0 \ldots \alpha^{qt_k}, 1 \leq i \leq m, 1 \leq j, k \leq n$. Words (terms) tj occur in a document D: $\alpha^{t_j} \alpha^0$.

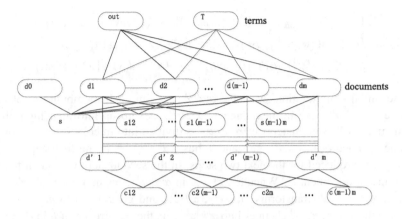

Fig. 1. Membrane structure design for 30 documents clustering

Rules are designed as below:

$R_1 = [D, T] : [(\{\alpha^i \alpha^{ht_j} \ldots \alpha^{qt_k}\}, asc), (\{\alpha^{t_1} \ldots \alpha^{t_n}, ent_i\})] \rightarrow [(\alpha^{t_1} \ldots \alpha^{t_n}, des), (\{\alpha^i \alpha^{ht_j} \ldots \alpha^{qt_k}\}, ent)]$

$R_2 = [D, T] : [(\{\alpha^i \alpha^{ht_j} \ldots \alpha^{qt_k}\}, asc), (\{\alpha^{df_{ij}}\}, ent_i)] \rightarrow [(\{\alpha^i \alpha^{ht_j} \ldots \alpha^{qt_k} \alpha^{t_1} \ldots \alpha^{t_n}\}, des), (\alpha^s, ent)]$

$R_3 = [T \rightarrow D] : [(\alpha^s \alpha^i \alpha^{ht_j} \ldots \alpha^{qt_k}, des) \rightarrow (\{\alpha^{htf_{ij}} \ldots \alpha^{qtf_{ik}}\}, ent_i)]$

$R_4 = [D, d_1] : [(\{\alpha^{htf_{ij}} \ldots \alpha^{qtf_{ik}}\}\{\alpha^{df_{ij}} \ldots \alpha^{df_{kj}}\}, mer) \rightarrow \{\log(N/x)\alpha^{xdf_{ij}} \alpha^{htf_{ij}} \ldots \alpha^{qtf_{ik}}\}]$

$\delta_{\min} \leq df_j \leq \delta_{\max}, \alpha^{\delta_{\min}df_j} \leq \alpha^{xdf_j} \leq \alpha^{\delta_{\max}df_j}$

$R_5 = [d_1, d_0, s] : [(\alpha^{xdf_j}\{\alpha^{htf_{ik}} \ldots \alpha^{qtf_{ik}}\}, \lambda, \lambda)] \rightarrow [\alpha^{xdf_j}\{\alpha^{htf_{ik}} \ldots \alpha^{qtf_{ik}}\}, \lambda, \alpha^{xdf_j}\{\alpha^{htf_{ij}} \ldots \alpha^{qtf_{ik}}\})]$

$R_6 = [D, d_1] : [(\lambda, asc), (\{(\log(N/x) \times h)\alpha^{\omega_{ij}}, ent_i\})] \rightarrow$
$[(\{\log(N/x)\alpha^{xdf_{ij}} \alpha^{htf_{ij}}\}, des), (\{\log(N/x_1) \times h_1)\alpha^{\omega_{1j}}\}, ent)]$

denote $\log(N/x) \times x \times h$ as $g_{\omega ij}$.

$R_7 = [d_i, d_j, v_{ij}] : [(g_{\omega ik}\alpha^{\omega_{ik}}, g_{\omega jk}\alpha^{\omega_{jk}}, \alpha)] \rightarrow [(\lambda, \lambda, g_{sim(d_i, d_j)}\alpha^{s_{ij}} g_{\omega ik} \times g_{\omega jk}\alpha^{\omega_k})], g_{\omega ik} \times g_{\omega jk} \neq 0$

$R_8 = [\{s_{ij}\}, s] : [(\{g_{sim(d_i, d_j)}\alpha^{s_{ij}} g_{\omega ik} \times g_{\omega jk}\alpha^{\omega_k}\}, mer) \rightarrow g_{TC(t_k)}\alpha^{\omega_k} g_{sim(d_i, d_j)}\alpha^{s_{ij}}], g_{sim(d_i, d_j)} \geq \beta_s$

$R_9 = [d_1, d_0, s] : [(\lambda, \lambda, \alpha^{xdf_j}\{\alpha^{htf_{ij}} \ldots \alpha^{qtf_{ik}}\}g_{TC(t_k)})] \rightarrow [\alpha^{xdf_j}\{\alpha^{htf_{ik}} \ldots \alpha^{qtf_{ik}}\}, \lambda, \lambda)], g_{TC(t_k)} \geq \beta$

$R_{10} = [d_i, d_j, s] : [(g_{\omega ik}\alpha^{\omega_k}, g_{\omega jk}\alpha^{\omega_k}, \lambda)] \rightarrow [(g_{sim(d_i, d_j)}\alpha^{s_{ij}}, g_{sim(d_i, d_j)}\alpha^{s_{ij}}, \lambda)]$

$R_{11} = [d_k, D'] : [(\lambda, asc), \{(g_{sim(d_i, d_j)}\alpha^{s_{ij}}, ent_j)\}, (-g_{sim(d_i, d_k)}\alpha^{s_{ik}}, ent_k)] \rightarrow$
$[(\{g_{sim(d_i, d_j)}\alpha^{s_{ij}}\}, des), (\{g_{sim(d_i, d_j)}\alpha^{s_{ij}}\}, ent)]$

$D' = \{j\}, 1 \leq j \leq m, j \neq i$, d_k is a document, where k is a number belonging to $\{1 \leq j \leq m, j \neq i\}$, $k \neq i \neq j$. k is chosen randomly.

$R_{12} = [d_i, \{c_{jk}\}] : [(\{g_{simjk}\alpha^{s_{jk}}\}, -\{g_{simj'k}\alpha^{s_{j'k}}\}, asc), (\lambda, ent_{jk})]$
$\rightarrow [(\lambda, des), (g_{sim(d_i, d_{j'})}\alpha^{s_{ij'}}, ent)].$

This process will continue until there is no negative spiking produced, then,

$$R_{13} = [d_i, \{c_{jk}\}] : [(\{g_{simjk}\alpha^{s_{jk}}\}, asc), (\lambda, ent_{jk})] \rightarrow [(\lambda, des), (\alpha^{ik}\alpha^0, ent)]$$

$$R_{14} = [d_i, \{c_{jk}\}] : [(\{0 \times \alpha^0\}, asc), (\lambda, ent_{jk})] \rightarrow [(\lambda, des), (\{\alpha^{ij}\alpha^0\}, ent)]$$

$$R_{15} = [d_i, I_{out}] : [(\{\alpha^{ik}\alpha^0\}, asc), \{(\lambda, ent_i)\}] \rightarrow [(\alpha^{halt}, des), (\{\alpha^{ik}\alpha^0\}, ent)], ik \leq t$$

In the initial configuration, we have $\alpha^i\alpha^{t_j}\alpha^0\ldots\alpha^{t_k}$ in each membrane d_i of chain membrane D and $\alpha^{t_1}\ldots\alpha^{t_n}$ in membrane T. $\alpha^i\alpha^{t_j}\alpha^0\ldots\alpha^{t_k}$ go up into their high dimensional membrane T. Simultaneously, $\alpha^{t_1}\ldots\alpha^{t_n}$ is divided into m copies and go down to membrane d_i. Each d_i has a $\alpha^{t_1}\ldots\alpha^{t_n}$ and T has $\alpha^i\alpha^{t_j}\alpha^0\ldots\alpha^{t_k}$ in the first step. $\alpha^{df_{ij}}$ and $\alpha^{htf_{ij}}\ldots\alpha^{qtf_{ik}}$ are produced for the next two steps; In the second step, we search for term α^{t_j} appearing in document d_i. When $\alpha^{t_j}, \alpha^{t_k}, \ldots$ in $\alpha^i\alpha^{t_j}\alpha^0\ldots\alpha^{t_k}$ in T meet $\alpha^{t_j}, \alpha^{t_k}, \ldots$ in $\alpha^{t_j}, \alpha^{t_k}, \ldots, \alpha^i\alpha^{t_j}\alpha^0\ldots\alpha^{t_k}$ transforms into $\alpha^{df_{ij}}, \alpha^{df_{ik}}\ldots$ and goes down to d_i at the same time. In the third step, $\alpha^{ht_j}\alpha^0$ changes into $\alpha^{htf_{ij}}\alpha^0$, h is the occurrence of t_j in d_i. Each tf goes into their membrane d_i at the same time. Membrane T dissolves. In the help of α^s, rule in step two and three will not execute in reverse order, which ensure the right computation of our algorithm.

Next step, we calculate the factor TF*IDF by rules, $\alpha^{df_{ij}}\ldots\alpha^{df_{kj}}$ merges into α^{xdf_j} by chain rule. In this step, coefficient g comes into work. $g = \log(N/x)$ in this rule. Here, membrane d_1 contains all tf and idf. In the fifth step, one backup of $\alpha^{xdf_j}\{\alpha^{htf_{ij}}\ldots\alpha^{qtf_{ik}}\}$ is stored in membrane s, where $\delta_{\min} \leq df_j \leq \delta_{\max}$. This step select feature initially. All weight $\omega_{ij} = \log(N/x_i) \times x_i \times h_i$ are calculated among coefficient g by rules and put into membrane d_i respectively. $\{\log(N/x)\alpha^{xdf_j}\alpha^0\alpha^{htf_{ij}}\}$ in membrane d_i are changed into empty in the following step. In the seventh step, $g_{\omega ik}\alpha^{\omega_{ik}}$ in membrane d_i is transformed into empty and representing similarity between documents d_i and d_j is produced. $g_{\omega ik} \times g_{\omega jk}\alpha^0$ is also produced for calculating terms contribution (tc), which is used for feature selection. For those $g_{sim(d_i,d_j)} \geq \beta_s$, $g_{TC(t_k)}\alpha^0$ is merged by $g_{\omega ik} \times g_{\omega jk}\alpha^0$ and those $\alpha^{t_j}, \alpha^{t_k}, \ldots$ are selected by $g_{TC(t_k)} \geq \beta$ in the following two steps. Because of new features are chosen, for the next two steps, $g_{\omega_{ij}}$ and $g_{sim(d_i,d_j)}$ are calculated afresh. So far, we complete steps of feature selection.

From step 12, the process of clustering begins. D' is a chain membrane $\{d'_1, d'_2, \ldots, d'_m\}$. $g_{sim(d_i,d_j)}\alpha^0$ are put down to membrane $d_{j'}$ respectively. Among all $\{g_{sim(d_i,d_j)}\alpha^0\}$, one is transformed into negative $\{-g_{sim(d_i,d_k)}\alpha^0\}$. Therefore, (m-1) positive $\{g_{sim(d_i,d_j)}\alpha^0\}$ and one negative $\{-g_{sim(d_i,d_k)}\alpha^0\}$ appear in membranes of chain $\{d'_1, d'_2, \ldots, d'_m\}$. Next step, addition goes on among membrane $d_{k'}$ and membranes $\{d'_1, d'_2, \ldots, d'_m\}$ without $d_{k'}$. $\pm|g_{sim(d_i,d_j)} - g_{sim(d_i,d_k)}|$ is produced and put into membrane c_{jk}. Rough compared results are turned out among $sim(d_i, d_k)$ and all other similarity of d_i in step 14. g_{simjk} means the sum is positive. That is to say $sim(d_i, d_j)$ is larger than $sim(d_i, d_k)$. Thus, $g_{simjk}\alpha^0$ becomes empty. Otherwise, $-g_{simj'k}\alpha^0$ stands for $sim(d_i, d_{j'})$ is larger than $sim(d_i, d_k)$. $g_{sim(d_i,d_{j'})}\alpha^0$ is produced and put up to d_i for the next comparison. From step 15, steps 12–14 will continues until no negative spiking appearing in step 14. $g_{simj'k}\alpha^0$ in this moment is the maximum one. The minimum similarities are output α^{ik} through membrane out at last. α^{ik} with same i or k is gathered into a same cluster. The process of document summarization by SS P system is done. All steps are 16. All membranes are $2(m + 2 + Cm2)$.

4   Example and Discussion

In this section, we choose three documents in CNN in Tables below to the calculation of SS P system (Tables 1, 2 and 3).

59 terms are used to represent all the 30 documents as Table 4. Result of the 30 documents is {1,2,3,4,5,6,7,8,9,10},{11,12,13,14,15,16,17,18,19,20}, {21,22, 23,24, 25,26,27,28,29,30}. It meets with the content of documents. The correct rate is 100 %.

Table 1. Nelson Mandela 2013/12/6 CNN (http://edition.cnn.com/) Africa

N	Documents
1	Nelson Mandela, the prisoner-turned president who reconciled South Africa
2	That was Nelson Mandela, who emerged from prison after 27 years to lead his country out of decades of apartheid
3	April 27, 1994, was the crowning moment in the life of Nelson Mandela
4	Mandela is woven into the fabric of the country and the world
5	He was South Africa's president
6	He was banned from reading newspapers in prison
7	South Africa: Following Nelson Mandela
8	In 1988 at age 70, Mandela was hospitalized with tuberculosis a disease whose effects plagued him until the day he died
9	Only three of Mandela's children are still alive
10	In 1996, he did not seek a second term as president, keeping his promise to serve only one term

Table 2. McDonnell Douglas DC-10 2013/12/6 CNN (http://edition.cnn.com/ Asia

N	Documents
1	Now it's the McDonnell Douglas DC-10's turn
2	The final flight honor goes to Bangladesh Biman Airlines, operator of the world's last passenger DC-10
3	The DC-10 won't completely disappear from the skies.
4	Biman has since announced one last chance to experience the DC-10 in action
5	McDonnell Douglas's famous jetliner logged its maiden voyage for passengers on August 5, 1971
6	The DC-10 is a reliable airplane, fun to fly, roomy and quiet
7	The DC-10 had its share of high-profile accidents
8	The aircraft would be hampered by tragic accidents
9	The aircraft will then remain grounded in standby mode while awaiting the arrival of its successor
10	The airline says will be making its final scheduled flight on December 7

Table 3. Hobbit2013/12/6 CNN (http://edition.cnn.com/) Travel

N	Documents
1	"Hobbit" mania is in full swing in New Zealand this week
2	New 'Hobbit' art flies at NZ airport
3	Wellington International Airport has unveiled a new installation featuring two realistic "great eagles"
4	"The Hobbit" and "The Lord of the Rings" trilogies.
5	The giant eagles appeared at the end of the previous Hobbit movie
6	Air New Zealand's new Hobbit livery
7	it used to transport the actors back to New Zealand
8	Air New Zealand has offered Hobbit-themed flights
9	Air New Zealand also recently released a new Hobbit-themed commercial
10	The sculpture is suspended above the airport's food court

Table 4. Document description by terms

d1:{t1,t2,t3,t4, t5}	d2:{t1,t2,t6, t7,t8,t9}	d3:{t1,t2,t10}	d4:{t2,t8,t11, t12}	d5:{t3,t4,t5}
d6:{t6,t13}	d7:{t1,t2,t4, t5}	d8:{t2,t14, t15}	d9:{t2,t16, t17}	d10:{t3,t18,t19, t20}
d11:{t21,t22, t23}	d12:{t12,t23, t24,t25 ,t26,t27,t28}	d13:{t23,t29}	d14:{t23,t26, t30,t31}	d15:{t21,t22,t28, t32,t33}
d16:{t23, t34}	d17:{t23, t35}	d18:{t35, t36}	d19 {t36,t37, t38,t39}	d20:{t24,t27}
d21:{t40,t41, t42,t43}	d22:{t40,t44, t45}	d23:{t45,t46, t47,t48}	d24:{t40,t49, t50,t51}	d25:{t40,t48,t52}
d26:{t40,t42, t53,t54}	d27:{t42,t55}	d28:{t24,t42, t53}	d29:{t42,t53, t56}	d30:{t45,t57,t58, t59}

Besides, by SS P systems, steps are 14. The time complexity is O(n). Rules like (2) demonstrate the capability of parallel computing of SS P system fully.

In step 11, rule $[d_k, D'] : [(\lambda, asc), \{(g_{sim(d_i,d_j)}\alpha^{s_{ij}}, ent_j)\}, (-g_{sim(d_i,d_k)}\alpha^{s_{ik}}, ent_k)] \rightarrow [(\{g_{sim(d_i,d_j)}\alpha^{s_{ij}}\}, des), (\{g_{sim(d_i,d_j)}\alpha^{s_{ij}}\}, ent)]$ choose positive spike randomly. In the calculation of this example, we adopt the best situation choose the maximum spike as the positive one. However, in worst case, the maximum spike will be chosen at last. will execute repeat until the maximum spike is selected. 14 steps and $24 + 2C_N^2$ rules are consumed in the best situation. 26 steps and $38 + 2C_N^2$ rules are used in worst situation. Generally, calculation steps are between the best and worst situation.

5   Conclusion

In this paper, a new P system with membrane coefficients on complex is proposed which has four types of rules. All rules are based on edge operations and chain structure. Through complex structure, objects can be transformed among higher and lower dimensions. A membrane can communication with all its face membranes simultaneously also. Membrane dissolves by edge operations. Membranes of different chain structure communicate at the same time. Middle data can be stored in common face of two membranes with same dimension. Furthermore, we use positive and negative spikes in the process, which expand using of P system. Besides all of these, we add coefficients on membranes, which make membrane deal with functions. The computation completeness of SS P system has been proved. We simulate the ADD instruction by 3 membranes and 1 rule. The subtraction instruction consumes 3 membranes and 2 rules. Combinations of document clustering and SS P system are also proposed in Sect. 5. The clustering algorithm is based on similarity. 15 rules, $2(m + 2 + C_m^2)$ membranes and 16 steps are consumed in solving the problem. An example of 30 documents is clustered in 13 steps, which shows the effectiveness of SS P system. SS P systems are different in many aspects from the P systems considered before, which extends membrane computing in some degree.

Acknowledgments. Research is supported by the Natural Science Foundation of China (No.61170038, 61472231), the Natural Science Foundation of Shandong Province (No. ZR2011FM001), Humanities and Social Sciences project Supported by Chinese Ministry of Education (12YJA630152).

References

1. Păun, G.: Turku center for computer science-TUCS report (1998). http://www.tucs.fi. Information on http://www.weld.labs.gov.cn
2. Pabreve, G., Pabreve, R.: Membrane computing and economics: numerical P systems. Fundamenta Informaticae **73**(1), 213–227 (2006)
3. Aman, B., Ciobanu, G.: Behavioural equivalences in real-time P systems. In: 14th International Conference on Membrane Computing, p. 49
4. Adorna, H., Păun, G., Pérez-Jiménez, M.J.: On communication complexity in evolution communication P systems. Rom. J. Inf. Sci. Technol. **13**(2), 113–130 (2010)
5. Dragomir, C., Ipate, F., Konur, S., Lefticaru, R., Mierla, L.: Model checking kernel P systems. In: 14th International Conference on Membrane Computing, p. 131
6. Păun, G.: Computing with membranes. J. Comput. Syst. Sci. **61**(1), 108–143 (2000)
7. Martín-Vide, C., Păun, G., Pazos, J., Rodrıguez-Patón, A.: Tissue P systems. Theor. Comput. Sci. **296**(2), 295–326 (2003)
8. Díaz-Pernil, D., Gutiérrez-Naranjo, M.A., Pérez-Jiménez, M.J., Riscos-Núñez, A.: A uniform family of tissue P systems with cell division solving 3-col in a linear time. Theor. Comput. Sci. **404**(1), 76–87 (2008)
9. Pan, L., Ishdorj, T.-O.: P systems with active membranes and separation rules. J. UCS **10**(5), 630–649 (2004)
10. Sburlan, D.: P systems with chained rules. In: Gheorghe, M., Păun, G., Rozenberg, G., Salomaa, A., Verlan, S. (eds.) CMC 2011. LNCS, vol. 7184, pp. 359–370. Springer, Heidelberg (2012)
11. Xue, J., Liu, X.: Lattice based communication P systems with applications in cluster analysis. Soft Comput. **18**, 1–16 (2013)

12. Liu, X., Xue, A.: Communication P systems on simplicial complexes with applications in cluster analysis. Discrete Dyn. Nat. Soc., Article ID: 415242 (2012)
13. Yang, S., Wang, N.: A P systems based hybrid optimization algorithm for parameter estimation of FCCU reactor-regenerator model. Chem. Eng. J. **211**, 508–518 (2012)
14. Metta, V.P., Krithivasan, K., Garg, D.: Modeling spiking neural P systems using timedpetri nets. In: World Congress on Nature and Biologically Inspired Computing, 2009, NaBIC 2009, pp. 25–30. IEEE (2009)
15. Păun, G., Rozenberg, G., Salomaa, A.: The Oxford Handbook of Membrane Computing. Oxford University Press, Inc., Oxford (2010)
16. Ionescu, M., Pabreve, G., Yokomori, T.: Spiking neural P systems. Fundamenta informaticae **71**(2), 279–308 (2006)
17. Zhang, X., Wang, S., Niu, Y., Pan, L.: Tissue P systems with cell separation: attacking the partition problem. Sci. China Inf. Sci. **54**(2), 293–304 (2011)
18. Lloret, E., Palomar, M.: A gradual combination of features for building automatic summarisation systems. In: Matoušek, V., Mautner, P. (eds.) TSD 2009. LNCS, vol. 5729, pp. 16–23. Springer, Heidelberg (2009)
19. Ferreira, R., Lins, R.D., Freitas, F., Cavalcanti, G.D.C., Lima, R., Simske, S.J., Favaro, L., et al.: Assessing sentence scoring techniques for extractive text summarization. Expert Syst. Appl. **40**, 5755–5764 (2013)

Classification

Crop Classification Using Artificial Bee Colony (ABC) Algorithm

Roberto A. Vazquez[1](✉) and Beatriz A. Garro[2]

[1] Intelligent Systems Group, Facultad de Ingeniería, Universidad La Salle,
Benjamin Franklin 47, Condesa, 06140 Mexico, D.F., Mexico
`ravem@lasallistas.org.mx`
[2] IIMAS-UNAM, Ciudad Universitaria, Mexico, D.F., Mexico
`beatriz.garro@iimas.unam.mx`

Abstract. Identifying which crop is growing in certain areas is important to many national and multinational agricultural agencies for forecasting grain supplies, monitoring farming activity, facilitating crop rotation records, etc. In order to achieve that, the agencies require to schedule censuses on a regular basis. Recently, different techniques based on remote sensing have been applied to collect the information and perform a crop classification task. In this paper, we described a methodology to perform a crop classification task based on the Gray Level Co-Occurrence Matrix (GLCM) and the artificial bee colony (ABC) algorithm. The proposed methodology selects the set of features from the GLCM that allow classify the crops with a good accuracy using the ABC algorithm in terms of a distance classifier. The accuracy of the proposed methodology was tested over a specific region of Mexico and compared against different distance classifiers.

Keywords: Crop classification · ABC algorithm · Feature selection

1 Introduction

Remote sensing is a field in which aerial sensors are employed to obtain information from the surface of the earth. Once the information is acquired, a numerous amount of techniques can be used to give solution to a vast range of problems, including crop classification.

In several countries, it is necessary to collect actual information about which crop is growing in relevant regions. A census provides a data set which allows to know several variables concerning to the production volume, crop identification, location, etc. in order to characterize the agricultural sector structure and performance, forecast grain supplies, monitor farming activity, facilitate crop rotation records, etc. However, the huge quantity of information, combined with the cost and periodicity for obtain the information, have leaded to applied solutions based on remote sensing. Satellite image processing, as a method for crop

Y. Tan et al. (Eds.): ICSI 2016, Part II, LNCS 9713, pp. 171–178, 2016.
DOI: 10.1007/978-3-319-41009-8_18

classification using remote sensing, is a good option considering cost and data actualization frequency.

Several researches related to crop classification have been published in the last years [10]. Crop classification systems can be separated into three essential parts: the selection of the information source, the feature extraction process and the classification process.

Different commercial companies provide satellite images obtained from different sensors such as Synthetic Aperture Radar (SAR), which captures dielectric characteristics of objects such as their structure (size, shape and orientation). Multispectral and hyperspectral images capture the image at specific frequencies across the electromagnetic spectrum [2,11]. Concerning to the feature extraction stage, texture descriptors have been widely used as a mechanism to identify to which crop one given pixel or region belongs. Perhaps one of most popular texture descriptor is the Gray Level Co-Occurrence Matrix (GLCM), see for example [8]. Once the set of features is extracted, it is possible to apply different approaches based on parametric and non-parametric classifier such as [4], as well as artificial neural networks [1] for classifying the crops.

Despite of the advances achieved in this field, there are several concerns that should be analyzed to increase the efficiency of these systems. One of this concerns is the selection of the best set of features to guaranty an acceptable accuracy during the classification. Several approaches has been applied to select the set of features such as random projection [6], genetic algorithms [3,9].

In this paper, we present an approach which focuses on computing the Gray Level Co-Occurrence Matrix (GLCM) from satellite images obtained from the visible electromagnetic spectrum, select the best set of features by means of the Artificial Bee Colony (ABC) algorithm, and then train different distance metrics for classifying the crops. Due to there is not a binary version of ABC, the representation of the individual is binarized to select the set of features used to classify the crops. Then, the features selected are classified by distance metrics such as Euclidean and Manhattan distances. The accuracy of the proposal is tested over a region of Mexico where five different crops were detected.

2   Artificial Bee Colony (ABC) Algorithm

The Artificial Bee Colony (ABC) algorithm is a very popular heuristic method based on the metaphor of bees foraging behavior. This algorithm is composed of a population of solutions NB (bees) $\mathbf{x}_i \in \mathbb{R}^n, i = 1, \ldots, NB$ that represents the positions of the food sources [7]. This algorithm proposed three types of bees that play an important role to achieve the convergence near to the optimal solution in the search space: *employed bees*, *onlooker bees* and *scout bees*.

The *employed bees* search for new neighborfood source near of their hive that are compared against the old one using Eq. 1. Then, the best food source is saved in their memory.

$$v_i^j = x_i^j + \phi_i^j \left(x_i^j - x_k^j \right).$$

(1)

where $k \in \{1, 2, ..., NB\}$ and $j \in \{1, 2, ..., n\}$ are randomly chosen indexes and $k \neq i$. ϕ_i^j is a random number between $[-a, a]$.

After that, the bees evaluate the quality of each food source in terms of the amount of the nectar (the information) using a fitness function. Finally, they return to the dancing area in the hive to share the information with the *onlooker bees*.

The *onlooker bees* observe the dance of the *employed bees* to get information such as where the food source can be found, if the nectar is of high quality, as well as the size of the food source. The *onlooker bees* choose probabilistically a food source depending on the amount of nectar shown by each employed bee, see Eq. 2.

$$p_i = \frac{fit_i}{\sum\limits_{k=1}^{NB} fit_k} . \tag{2}$$

where fit_i is the fitness value of the solution i and NB is the number of food sources which are equal to the number of *employed bees*.

Finally, the *scout bees* help the colony to randomly create new solutions when a food source cannot be improved anymore ("limit" or "abandonment criteria") by using Eq. 3.

$$x_i^j = x_{\min}^j + rand\,(0, 1)\left(x_{\max}^j - x_{\min}^j\right). \tag{3}$$

Using these models of bees, the authors in [7] introduce the pseudo-code of the ABC Algorithm described in Algorithm 1.

Algorithm 1. Pseudo-code ABC algorithm.

```
1:  Initialize the population of solutions xᵢ∀ᵢ, i = 1, ..., NB.
2:  Evaluate the population xᵢ∀ᵢ, i = 1, ..., NB.
3:  for cycle = 1 to maximum cycle number MCN do
4:      Produce and evaluate new solutions vᵢ using Eq. 1.
5:      Apply the greedy selection process.
6:      Calculate the probability values pᵢ for the solutions xᵢ by Eq. 2.
7:      Produce and evaluate new solutions vᵢ for the solutions xᵢ selected
        depending on pᵢ.
8:      Apply the greedy selection process.
9:      Replace the abandoned solutions with a new one xᵢ by using Eq. 3.
10:     Memorize the best solution achieved so far.
11:     cycle = cycle + 1
12: end for
```

---

3 Co-occurrence Matrix

The Gray Level Co-Occurrence Matrix (GLCM) is a textural measure that describes some properties about the spatial distribution of the gray levels in an image [5]. Furthermore, this matrix measure how often a pixel value known as the reference pixel with the intensity value i occurs in a specific relationship to a pixel value known as the neighbor pixel with the intensity value j. The spatial relationship between two pixels can be specified with different offsets and angles.

In order to compute the GLCM from a image or from an specific region, it is necessary to define a neighborhood relationship and window size for creating a two dimensional histogram (a squared matrix with the quantification of the image as length). Each cell of this histogram is filled with the occurrence count of the given pixel relationship. Once the histogram is calculated, a normalization is computed using by (4).

$$\mathbf{P}_{ij} = \mathbf{V}_{ij} / \sum_{i,j=0}^{N-1} \mathbf{V}_{ij}. \tag{4}$$

where each element of matrix $\mathbf{V}_{ij}$ is the number of occurrences that the relationship between pixels with value i and pixels with value j occurs.

After computing the probability, it is possible to calculate a set of 8 properties describing characteristics of the evaluated region such as contrast (c_1), dissimilarity (c_2), homogeneity (c_3), angular second moment (ASM) (c_4), entropy (c_5), energy, average and standard deviation, see Eqs. (5) and (6).

$$c_1 = \sum_{i,j=0}^{N-1} \mathbf{P}_{ij} (i-j)^2 , c_2 = \sum_{i,j=0}^{N-1} \mathbf{P}_{ij} |i-j|^2 , c_3 = \sum_{i,j=0}^{N-1} \frac{\mathbf{P}_{ij}}{1+(i-j)^2}. \tag{5}$$

$$c_4 = \sum_{i,j=0}^{N-1} \mathbf{P}_{ij}^2 , c_5 = \sum_{i,j=0}^{N-1} \mathbf{P}_{ij} (-In\mathbf{P}_{ij}). \tag{6}$$

4   Proposed Methodology

In contrast with traditional research projects where satellite images are obtained with multi-spectral, hyper-spectral or Synthetic Aperture Radar devices, the proposed methodology was adapted for performing with satellite images obtained from the Internet maps service Google Earth. These images only contain data from the visible part of the electromagnetic spectrum. The image obtained from the maps service was manually segmented, by visual inspection, into different crop classes defining several polygons in the image. Each one of these polygons belong to a single class, making easier the feature extraction process. For obtaining the features that describe the crops in the image, the well-known GLCM method was used. The GLCM method provides a set of 8 features from each color channel that define the set of feature patterns.

Following the methodology described in [3], we applied the ABC algorithm to select the best set of features computed from the GLCM over three color channels. In order to do that, the problem to be solved was defined as follows: Giving a set of input patterns $\mathbf{X} = \{\mathbf{x}_1, \ldots, \mathbf{x}_p\}, \mathbf{x_i} \in \mathbb{R}^n, i = 1 \ldots, p$ and a set of desired classes $\mathbf{d} = \{d_1, ..., d_p\}, d \in \mathbb{N}$, find a set of features $G \in \{0,1\}^n$ such that a function defined by $\min (F(\mathbf{X}|_G, \mathbf{d}))$ is minimized.

The solution of the problem is represented in terms of a set of features and is defined with an array $I \in \mathbb{R}^n$ that contains the food source's position. Each solution $I_q, q = 1, \ldots, NB$ is binarized by means of Eq. 7 using a threshold level th in order to select the best set of features defined as $G^k = T_{th}(I^k), k = 1, \ldots, n$; values whose component is set to 1, indicates that this feature will be selected to make up the set of features.

$$T_{th}(x) = \begin{Bmatrix} 0, x < th \\ 1, x \geq th \end{Bmatrix}. \tag{7}$$

The aptitude of an individual is computed by means of the classification error (CER) function, defined in Eq. 8, that measures how many crop samples have been incorrectly classified.

$$F(\mathbf{X}|_G, \mathbf{d}) = \frac{\sum\limits_{i=1}^{p} \left(\left| \underset{k=1}{\overset{K}{\arg\min}} \left(D(\mathbf{x}_i|_G, \mathbf{c}^k) \right) - d_i \right| \right)}{tng}. \tag{8}$$

where tng is the total number of samples of a crop to be classified, D is a distance measure, K is the number of classes and $\mathbf{c}$ is the center of each class. Once computed the best set of features, the classification process consist in applying the distance classifiers using only the set of features that minimize the classification error during the training phase.

5 Experimental Results

To evaluate the accuracy of the proposed methodology, a testing image (8000 × 8000 pixels) from the northwest region of Mexico was obtained. From this image, five classes were identified and a set of polygons over each of these classes were drawn, see Fig. 1. From this polygons, we built a dataset composed of 2752 patterns with 24 features (eight features for each space color channel).

In order to validate statistically the experimental results, the proposed methodology was executed 30 runs for each Euclidean and Manhattan distance. The information of the dataset was normalized and randomly splited into two subsets: 50 % of the samples for the training subset, and the remain for the testing subset. Furthermore, the experiments take into account the threshold for binarizing each individual. To evaluate the threshold value and knowing how much affect, the threshold value was changed in five different configurations: 0.1, 0.3, 0.5, 0.7, 0.9. The parameters of ABC algorithm for all the experimentation were: population size $(NB = 40)$, maximum number of cycles $MNC = 1000$, limit $l = 100$ and food sources $NB/2$.

Figure 2 shows the learning error evolution during the 1000 iterations for five configurations and two distance measures.

Table 1 summarize the experimental results obtained with the Euclidean distance classifier. In average, during testing phase the best results were obtained with $th = 0.5$ getting a 78.5 % of accuracy. Using this configuration, the best accuracy obtained during testing phase was 79.7 % selecting 10 features from each color channel and GCLM feature: RF2, RF4, RF5, RF6, RF7, RF8, GF6, BF4, BF5, BF6. It is important to notice that most of the features selected for the euclidean distance came from the red color channel.

Table 2 summarize the experimental results obtained with the Manhattan distance classifier. In average, during testing phase the best results were obtained with $th = 0.5$ getting a 78.7 % of accuracy. However, the best accuracy was obtained with $th = 0.9$ reaching an accuracy of 80.2 % during testing phase. It is important to notice that only four features (RF4, RF7, BF5, BF8) were selected for achieving that accuracy with the Manhattan distance classifier from the red and blue color channel.

In addition to these experiments, the accuracy of the proposed methodology was compared against the distance classifiers without using the feature selection stage. Whereas the Euclidean distance classifier achieved an accuracy 75.3 %, the Manhattan distance classifier achieved an accuracy of 73.87 %. From this results, we could say that the results obtained with the proposed methodology were better than those obtained using only the distance classifier.

Table 1. Behavior of the proposed methodology using an Euclidean distance classifier.

th	Average accuracy		Average # of features	Average # of iter.	Best accuracy		# of features	# of iter.
	Tr. cl.	Te. cl.			Tr. cl.	Te. cl.		
0.1	0.772 ± 0.008	0.770 ± 0.006	15.5	1000	0.787	0.783	16	1000
0.3	0.783 ± 0.008	0.777 ± 0.009	11.3	1000	0.799	0.794	10	1000
0.5	0.791 ± 0.008	0.785 ± 0.009	9.1	1000	0.808	0.797	10	1000
0.7	0.797 ± 0.009	0.780 ± 0.013	7.2	1000	0.808	0.795	8	1000
0.9	0.792 ± 0.006	0.782 ± 0.010	6.3	1000	0.807	0.795	8	1000

Tr. cl. = Training classification rate, Te. cl. = Testing classification rate.

Table 2. Behavior of the proposed methodology using a Manhattan distance classifier.

th	Average accuracy		Average # of features	Average # of iter.	Best accuracy		# of features	# of iter.
	Tr. cl.	Te. cl.			Tr. cl.	Te. cl.		
0.1	0.777 ± 0.009	0.770 ± 0.011	14.6	1000	0.791	0.784	14	1000
0.3	0.789 ± 0.008	0.776 ± 0.009	11.5	1000	0.802	0.791	11	1000
0.5	0.798 ± 0.008	0.787 ± 0.007	8.8	1000	0.807	0.797	9	1000
0.7	0.797 ± 0.007	0.784 ± 0.009	7.0	1000	0.805	0.800	7	1000
0.9	0.794 ± 0.009	0.781 ± 0.010	4.9	1000	0.807	0.802	4	1000

Tr. cl. = Training classification rate, Te. cl. = Testing classification rate.

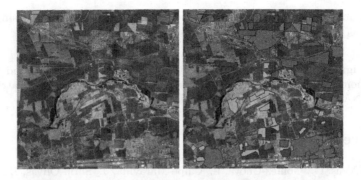

Fig. 1. Scaled image used for the experiments shown in the present study. (Color figure online)

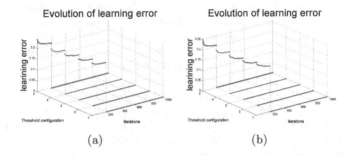

Fig. 2. Learning error evolution using different distance classifiers. *th* values are labeled with numbers from 1 to 5. 2(a) Learning error evolution for Euclidean distance measure. 2(b) Learning error evolution for Manhattan distance measure.

6 Conclusions

In this paper we described an approach to select relevant features from a GLCM obtained from a RGB color space for performing a crop classification task using ABC algorithm and distances classifiers. The features selection task was carried out by means of Artificial Bee Colony (ABC) algorithm using different configurations to binarize each individual or solution to obtained the best set of features from the GLCM to contribute to solve a problem.

Concerning to the learning error evolution, the experimental results show that both distance achieved the best results with few iterations. The best accuracy was obtained with $th = 0.5$ and $th = 0.9$ for Euclidean and Manhattan classifiers, respectively. Using this parameters, only 10 and 4 features were used with the euclidean and Manhattan classifiers, improving the results obtained without the selection stage. This results support that the ABC algorithm is an excellent tool for performing a feature selection task.

Although the proposed methodology provide acceptable results, nowadays, we are testing different classifier such as artificial neural networks, including spiking neural networks to improve the classification accuracy using ABC algorithm as well as different color spaces.

Acknowledgments. The authors thank DGAPA, UNAM and Universidad La Salle for the economic support under grants number I-61/12 and NEC-03/15. Beatriz Garro thanks CONACYT for the postdoctoral scholarship.

References

1. Camps-Valls, G., Gómez-Chova, L., Calpe-Maravilla, J., Soria-Olivas, E., Martín-Guerrero, J.D., Moreno, J.: Support vector machines for crop classification using hyperspectral data. In: Perales, F.J., Campilho, A.J.C., de la Blanca, N.P., Sanfeliu, A. (eds.) Pattern Recognition and Image Analysis. LNCS, vol. 2652, pp. 134–141. Springer, Heidelberg (2003). doi:10.1007/978-3-540-44871-6_16
2. Damodaran, B., et al.: Assessment of the impact of dimensionality reduction methods on information classes and classifiers for hyperspectral image classification by multiple classifier system. Adv. Space Res. **53**(12), 1720–1734 (2014). http://www.sciencedirect.com/science/article/pii/S0273117713007308
3. Garro, B.A., Rodríguez, K., Vázquez, R.A.: Classification of DNA microarrays using artificial neural networks and ABC algorithm. Appl. Soft Comput. **38**, 548–560 (2016). http://www.sciencedirect.com/science/article/pii/S1568494615006171
4. Gomez-Chova, L., et al.: Feature selection of hyperspectral data through local correlation and SFFS for crop classification. In: Proceedings of 2003 IEEE International Conference on IGARSS 2003, vol. 1, pp. 555–557, July 2003
5. Haralick, R., Shanmugam, K., Dinstein, I.: Textural features for image classification. IEEE Trans. Syst. Man Cybern. (SMC) **3**(6), 610–621 (1973)
6. Hariharan, S., et al.: Polarimetric SAR decomposition parameter subset selection and their optimal dynamic range evaluation for urban area classification using random forest. Int. J. Appl. Earth Obs. Geoinf. **44**, 144–158 (2016). http://www.sciencedirect.com/science/article/pii/S0303243415300192
7. Karaboga, D.: An idea based on honey bee swarm for numerical optimization. Technical report, Computer Engineering Department, Engineering Faculty, Erciyes University (2005)
8. Sandoval, G., Vazquez, R.A., Garcia, P., Ambrosio, J.: Crop classification using different color spaces and RBF neural networks. In: Rutkowski, L., Korytkowski, M., Scherer, R., Tadeusiewicz, R., Zadeh, L.A., Zurada, J.M. (eds.) ICAISC 2014, Part I. LNCS, vol. 8467, pp. 598–609. Springer, Heidelberg (2014)
9. Stavrakoudis, D., et al.: A boosted genetic fuzzy classifier for land cover classification of remote sensing imagery. J. Photogramm. Remote Sens. **66**(4), 529–544 (2011). http://www.sciencedirect.com/science/article/pii/S0924271611000438
10. Tatsumi, K., et al.: Crop classification of upland fields using random forest of time-series landsat 7 ETM+ data. Comput. Electron. Agric. **115**, 171–179 (2015). http://www.sciencedirect.com/science/article/pii/S0168169915001234
11. Zhang, E., et al.: Weighted multifeature hyperspectral image classification via kernel joint sparse representation. Neurocomputing **178**, 71–86 (2016). http://www.sciencedirect.com/science/article/pii/S0925231215016136

Classification of Distorted Handwritten Digits by Swarming an Affine Transform Space

Somnuk Phon-Amnuaisuk[1(✉)] and Soo-Young Lee[2]

[1] Media Informatics Special Interest Group, School of Computing & Informatics,
Universiti Teknologi Brunei, Bandar Seri Begawan, Brunei Darussalam
somnuk.phonamnuaisuk@utb.edu.bn
[2] Brain Science Research Center, Korea Advanced Institute of Science and
Technology, Daejeon, Korea
sylee@kaist.ac.kr

Abstract. Given an affine transform image having a distorted appearance, if a transform function is known, then an inverse transform function can be applied to the image to produce the undistorted original image. However, if the transform function is not known, can we estimate its values by searching through this large affine transform space? Here, an unknown affine transform function of a given digit is estimated by searching through the affine transform space using the Particle Swarm Optimization (PSO) approach. In this paper, we present important concepts of the proposed approach, describe the experimental design and discuss our results which favorably support the potential of the approach. We successfully demonstrate the potential of this novel approach that could be used to classify a large set of unseen distorted affine transform digits with only a small set of digit prototypes.

Keywords: Searching affine-transform space · Particle swarm optimization

1 Background

All pixel appearance-based image recognition techniques must deal with a large set of appearance variations of the same object[1]. The issue has been tackled using various plausible approaches: (i) build a large look-up table for all possible variations of an object's appearance; (ii) find a good way to compactly abstract essential variations of an object; and (iii) find a transformation model that could successfully transform an object back to its original pose before matching it to a smaller set of prototypes in the original pose.

There are pros and cons for these plausible approaches. The look-up table is commonly constructed using the machine learning approach and it is computationally expensive and might not be practical to create a model (i.e., a large

---

[1] In contrast to the pixel appearance-based approach, other approaches may possess 3D information of an object of interest.

© Springer International Publishing Switzerland 2016
Y. Tan et al. (Eds.): ICSI 2016, Part II, LNCS 9713, pp. 179–186, 2016.
DOI: 10.1007/978-3-319-41009-8_19

table look-up) of all possible variations of an object's appearance from different views. Recently, with the progress in deep learning research [1], it has been shown that a large amount of examples covering standard digits and their affine transformed versions can be successfully learned by a sophisticated deep ANN model[2]. However, it is unlikely that the model should be able to generalize more than what is in the training examples. This could be an issue in some domains where it might not be possible to have access to all variations of the object's appearance.

Another approach is to abstract essential appearance features. In this approach, the computer program finds unique patterns of interest points that could identify the same image presented under a different translation, scaling and rotation. It would be nice if an object viewed from different viewpoints could share some highly abstract features that are unique to each object. This is known as invariant representation and researchers have explored this concept with some success [2–4].

Here we propose the third approach, motivated by the need to find ways to handle a large amount of possible variations of handwritten digits without the need to obtain all the examples for the training process. Since a given handwritten digit could have a large amount of affine-transformed variations and it is a big challenge to construct a comprehensive classification model from all variations, we propose to handle this situation by searching through a transform space to find the right transformation matrix that would successfully transform an input affine transformed digit back to its original pose. This tactic would allow us to handle a large amount of variations with a small set of prototypes.

The rest of the paper is organized into the following sections: Sect. 2 discusses our proposed concept and gives the details of the techniques behind it; Sect. 3 provides the experimental design; Sect. 4 provides a critical discussion of the output from the proposed approach; and finally, the conclusion and further research are presented in Sect. 4.

2   Searching the Affine-Transformed Space Using PSO

In this experiment, we limit our domain to only handwritten digits of zero to nine. We resort to the MNIST handwritten digit database[3] where 60,000 digits (roughly 6,000 different handwritten examples for each digit) are prepared as a training set and 10,000 digits (roughly 1,000 different handwritten examples for each digit) are prepared as a testing set.

2.1   Affine Transformed Digits

Let I be an input handwritten digit that has been geometrically transformed in a 2D space. The transform image I could appear to be distorted and become

---

[2] Lecun's state of the art LeNet-5 can be viewed online at http://yann.lecun.com/exdb/lenet/index.html.

[3] Available from http://yann.lecun.com/exdb/mnist/.

quite different from the original version, D. It would be a great challenge for a classification model to correctly classify it if the model is not trained with the distorted examples. The following affine transformation is considered in this work: translation (T), rotation (R), scaling (S_c), and shearing (S_h), where

$$T = \begin{bmatrix} 1 & 0 & t_x \\ 0 & 1 & t_y \\ 0 & 0 & 1 \end{bmatrix} ; R = \begin{bmatrix} cos\theta & -sin\theta & 0 \\ sin\theta & cos\theta & 0 \\ 0 & 0 & 1 \end{bmatrix} ; S_c = \begin{bmatrix} s_x & 0 & 0 \\ 0 & s_y & 0 \\ 0 & 0 & 1 \end{bmatrix} ; S_h = \begin{bmatrix} 1 & h_x & 0 \\ h_y & 1 & 0 \\ 0 & 0 & 1 \end{bmatrix}$$

The overall transformation function $A : D \mapsto I$, translates the pixel information at a coordinate $D(h', w')$ to a new coordinate $I(h, w)$. The transformation matrix can be expressed as:

$$T \times R \times S_c \times S_h = \begin{bmatrix} (s_x cos\theta - h_y s_y sin\theta) & (h_x s_x cos\theta - s_y sin\theta) & t_x \\ (s_x sin\theta + h_y s_y cos\theta) & (h_x s_x sin\theta + h_y cos\theta) & t_y \\ 0 & 0 & 1 \end{bmatrix} \quad (1)$$

2.2 Representing Transform Parameters in PSO

Particle Swarm Optimization (PSO) is a computation technique developed by [5]. Given a distorted digit input, we decide to transform each reference digit (our prototypes) to fit the distorted input. Once the transformation parameters that can transform a prototype digit to produce similar appearances as the distorted input digit has been estimated, the inverse matrix is computed and employed to transform the input digit back to its original position. Here, PSO searches for a transformation matrix that will produce the best matched appearance. From the transformation function presented in Eq. 1, the following parameters form the elements in a particle: $\mathbf{p} = (\theta, s_x, s_y, h_x, h_y, t_x, t_y)$. Formally, let $\mathbf{p}(t) = (x_{11}, ..., x_{ij})$ represent a particle's parameters at time t, the particle's parameters at time $t + 1$ is calculated as follow:

$$x_{ij}(t + 1) = x_{ij}(t) + v_{ij}(t + 1) \quad (2)$$

where v_{ij} is the velocity of the parameter j of the particle i.

$$v_{ij}(t + 1) = wv_{ij}(t) + c_1 r_{1j}(p_{ij}(t) - x_{ij}(t)) + c_2 r_{2j}(g_{ij}(t) - x_{ij}(t)) \quad (3)$$

where w is the inertia weight for PSO, v_{ij} is the velocity of the parameter j of the particle i. The $c_1 r_{1j}$ and $c_2 r_{2j}$ are random weight parameters. The p_{ij} and g_{ij} are the local best and global best particles respectively. Local best particle (Pbest) refers to a particular particle instance with the best performance observed through its lifetime. Global best particle (Gbest) refers to a particular particle instance with the best Pbest observed from the whole swarm. The setting of PSO search control parameters are empirically determined as according to [6].

2.3   Designing Matching Evaluation Measures

Each particle described above represents a transformation matrix and will be used to transform a representative prototype of each digit class. The performance of each particle is determined by whether it can transform each prototype digit to fit the input digit [7,8]. We have decided to use the following measures: (i) mutual information, (ii) ink intersection, and (iii) non-intersected ink. The combination of these measures are used to rank the particles and control the way the affine-space is explored. Each of these measures is explained below:

Mutual Information. Mutual information is calculated using the Shannon entropy (H) [9,10]. Given an image A, its entropy is calculated using $H(A) = -\sum p_A log(p_A)$. Mutual information of two images A and B, $MI(A, B)$, is the summation of the entropy $H(A)$ and $H(B)$ less the joint entropy H(A,B). The joint entropy and mutual information can be expressed as follows:

$$H(A, B) = -\sum_{a,b} p_{A,B}(a, b) log[p_{A,B}(a, b)] \tag{4}$$

$$MI(A, B) = H(A) + H(B) - H(A, B) \tag{5}$$

Mutual information is one of the popular metrics employed in a medical image registration process [8].

Information Extracted from Ink. Intuitively, if two digits are of the same class, then their ink should intersect. Hence, ink intersection provides a good measurement of whether the two digits are the same. We have devised two ink intersection measures, namely: global intersection (GI) and global non-intersection (GN). It is named as such because GI and GN are calculated from the whole images (of digits). Let A be a reference digit, then GI and GN of two digits A & B are computed as follow:

$$GI(A, B) = \frac{\sum_{i,j} min(A_{ij}, B_{ij})}{\sum_{i,j} A_{ij}} \tag{6}$$

$$GN(A, B) = e^{-b}; b = \left| \frac{\sum_{i,j} A_{ij} - \sum_{i,j} B_{ij}}{\sum_{i,j} A_{ij}} \right| \tag{7}$$

Furthermore, two identical digits might be translated to different positions. Hence we have extended the GI and GN to handle this translation case and they are named local intersection (LI) and local balance (LN) respectively. Local features are calculated in the same way as global features but the local area covers only 25 % of the total area around the centroid of a digit. The centroid c_i and c_j are computed using the equations below:

$$M_{00} = \sum_i \sum_j Iij; \quad M_{10} = \sum_i \sum_j i \times Iij; \quad M_{01} = \sum_i \sum_j j \times Iij \tag{8}$$

$$c_i = \frac{M_{10}}{M_{00}}; \quad c_j = \frac{M_{01}}{M_{00}} \tag{9}$$

where M_{00} is known as the zeroth moment, M_{10} and M_{01} are the first moments of i and j respectively [11]. The fitness of each particle is evaluated using a combination of these 5 measures: MI + GI + GN + LI + LN.

3 Experimental Design & Results

For each digit class in the MNIST database, the training examples were clustered into 10 clusters and the prototype digit was the average of these 10 cluster centers. To evaluate our proposed approach, 500 digit examples were randomly selected from the 10,000 testing examples. Each digit was randomly transformed according to the transformation function in Eq. 1.

PSO then searched for an optimal affine transformation parameters that transformed the prototype digits such that the result would best match the input digit i.e., maximize mutual information between the two images, maximize intersected ink and minimize non-intersected ink.

Since there were ten digit prototypes, each unknown input digit was matched against all 10 prototypes. We decided to create ten sub-swarms, one for each prototype digit. Hence, ten possible transformation matrices were obtained from all PSO subswarms in each run. Among these ten transformation matrices, we expected the subswarm of the corresponding digit to give the best match, i.e., if the unknown input digit was *five*, then the subswarm matching the prototype digit *five* should give the best matching. This implied that (i) the matching score could be used as a classification score and PSO acted like a classifier in this case; and (ii) if the PSO could locate the optimal transformation matrix, it was expected that the inverse transformation matrix should transform the distorted input digit back to its original position. Hence, the PSO performed a kind of an invariant appearance generative function that transformed different invariant appearances back to its representative prototype.

Classifying Input Digit from the Gbest Score. Figure 1 displays 100 plots of Gbest and the mean of Pbest results of the 10 subswarms (columns) of 10 input digit (rows). For each subswarm, the mean value of the Pbest of the subswarm from iterations 0 to 40 is displayed in a (red) dash line and the value of the Gbest of the subswarm is displayed in a (black) solid line. In the overall picture, it is clear that the Gbest score provides a good indicator and can be used as a classifier.

Although the average Gbest score shows a clear classification result, the accuracy rate for each individual run is, of course, not 100 %. From the experimental results, it is observed that the correct subswarm may come in at the second or third places many times with a small difference behind the winner but they would usually be the first with a substantial margin over the second place. This is why the average score shows a clear classification result.

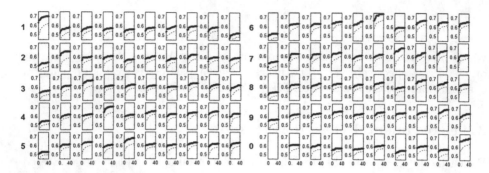

Fig. 1. Summary of performance from all sub-swarms. The Gbest and Pbest values are averaged over 50 runs (i.e., 500 test digits). The average Gbest scores indicate correct digit classes. That is, for distorted input digit 'one' (row #1), the Gbest score of column #1 is the highest. (Color figure online)

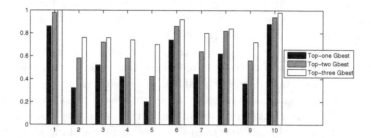

Fig. 2. Summary of the performance calculated from the top Gbest, the top-two Gbest, and the top-three Gbest.

Figure 2 summarises the classification results taken from Gbest score with (i) first place, (ii) first or second places and (iii) first or second or third places. The results show that the classification results of digits can be improved if the results from other competing digits are included in the process. This can be extended on in the future.

Our experiment shows that the Gbest score might be applied as a classification score. That is, given an unknown distorted digit, if the subswarm number three has the best Gbest score, then the unknown input digit can be classified as the digit three. Although in a conceptual sense, the proposed measures appear to be expressive enough to correctly identify the correct digit using the best score obtained from the subswarm as the class indicator, the performance of the individual run can still be improved. From the experiment, we foresee that the tuning of the evaluation function will help improve the accuracy of the classifier (i.e., to improve the case of second and third places).

Estimated Transformation Function by PSO. The Gbest of each swarm is the estimated transformation function obtained from the PSO search through the affine transform space. The information from the Gbest particle is decoded back to the transform function $\mathbf{p}(t) = (x_{11}, ..., x_{ij}) \Rightarrow (\theta, s_x, s_y, h_x, h_y, t_x, t_y)$.

Figure 3 shows examples of distorted input digits and transformed output digits from the system. Two examples are shown here (rows # 1,2,3 and rows #4,5,6). In each example, the distorted input digits are presented then examples of good transform and examples of not so successful transform are presented.

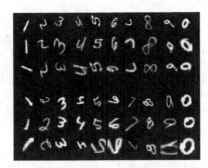

Fig. 3. Examples of distorted input digits (row#1 and #4) and their computed original pose generated using estimated transform functions from PSO (rows #2,3,5,6).

4   Discussion & Conclusion

Attempts to classify handwritten digits using pixel information can be categorized as either generative e.g., [12] or discriminative e.g., [13–15]. The generative approach finds generative models while the discriminative approach finds classification models where the classification models are learned using supervised learning approach. Research activities are dominated by the discriminative approach due to advances in classification techniques. The merit of the generative approach is that it only requires a small set of prototype digits (as its training data).

In this paper, we investigate the feasibility of searching for appropriate affine-transformation parameter of any handwritten digits using particle swarm optimization technique (PSO). Our approach is generative since it finds transform parameters that generate the original pose of a distorted digit.

The experimental results show that the current measurement metrics, derived from mutual information and ink intersections, can guide the search to find appropriate unknown affine transform functions of given distorted digits with some success (see Fig. 3). However, there is still a lot of room for improvement. It is also observed that the Gbest score can be a good classification score (see Figs. 1, 2). Although, the average Gbest score can successfully indicate the digit class their accuracies of each individual run can still be improved.

It was observed that the average Gbest score perfectly indicates the correct class and the Gbest scores of the correct class were always very competitive if they are not the winner. From this observation, we are convinced that further tuning of the matching evaluation measures will improve the performance of the approach future and this will be one of the priority areas in our future work.

Acknowledgments. We wish to thank anonymous reviewers for their comments that have helped improve this paper. We would like to thank the GSR office for their partial financial support given to this research.

References

1. Lecun, Y., Bengio, Y., Hinton, G.: Deep learning. Nature **521**, 436–444 (2015)
2. Lowe, D.G.: Distinctive image features from scale-invariant keypoints. Int. J. Comput. Vis. **60**(2), 91–110 (2004)
3. Mikolajczyk, K., Schmid, C.: Scale and affine invariant interest point detectors. Int. J. Comput. Vis. **60**(1), 63–86 (2004)
4. Sohn, K., Lee, H.: Learning invariant representations with local transformations. In: Proceedings of the 29th International Conference on Machine Learning, ICML 2012, Edinburgh, Scotland (2012)
5. Eberhart, R.C., Kennedy, J.: A new optimizer using particle swarm theory. In: Proceedings of the Sixth International Symposium on Micromachine and Human Science, Nagoya, Japan, pp. 39–43 (1995)
6. Phon-Amnuaisuk, S.: Investigating a hybrid of Tone-Model and particle swarm optimization techniques in transcribing polyphonic guitar sound. Appl. Soft Comput. **29**, 211–220 (2015)
7. Jain, A.K.: Representation and recognition of handwritten digits using deformable templates. IEEE Trans. pattern Anal. Mach. Intell. **19**(12), 1386–1391 (1997)
8. Li, Q., Ji, H.B.: Medical image registration based on maximization of mutual information and particle swarm optimization. In: Proceedings of the 5th International Conference on Photonics and Imaging in Biology and Medicine, vol. 6534, 65342M (2007)
9. Shannon, C.E.: A mathematical theory of communication. The Bell Syst. Tech. J. **27**(379–423), 623–656 (1948)
10. MacKay, D.J.C.: Information Theory, Inference, and Learning Algorithms. Cambridge University Press, Cambridge (2003)
11. Gonzalez, R.C., Woods, R.E.: Digital Image Processing. Prentice Hall, Upper Saddle River (2001)
12. Revow, M., Williams, C.K.I., Hinton, G.E.: Using generative models for handwritten digit recognition. IEEE Trans. Pattern Anal. Mach. Intell. **18**(6), 592–606 (1996)
13. Bottou, L., Cortes, C., Denker, J.S., Drucker, H., Guyon, I., Jackel, L.D., LeCun, Y., Müller, U.A., Säckinger, E., Simard, P., Vapnik, V.: Comparison of classifier methods: a case study in handwritten digit recognition. In: Proceedings of the 12th IAPR International Conference on Pattern Recognition. Conference B: Computer Vision & Image Processing, vol. 2, pp 77–82 (1994)
14. Liu, C.H., Nakashima, K., Sako, H., Fujisawa, H.: Handwritten digit recognition: benchmarking of state-of-the-art techniques. Pattern Recogn. **36**(2003), 2271–2285 (2003)
15. Ciresan, D.C., Meier, U., Masci, J., Gambardella, L.M., Schmidhuber, J.: Flexible, high performance convolutional neural networks for image classification. In: Proceedings of the Twenty-Second International Joint Conference on Artificial Intelligence (IJCAI 2011), pp, 1237–1242 (2011)

DKDD_C: A Clustering- Based Approach
for Distributed Knowledge Discovery

Marwa Bouraoui$^{(\boxtimes)}$, Houssem Bezzezi$^{(\boxtimes)}$,
and Amel Grissa Touzi$^{(\boxtimes)}$

Signal, Image and Technology of Information Laboratory,
National Engineering School of Tunis, Tunis El Manar University,
BP 37, Le Belvedere, 1002 Tunis, Tunisia
bourawimarwa@gmail.com, hbezzezi@gmail.com,
grissa.touzi@topnet.tn

Abstract. In this paper, we address the problem of knowledge discovery. Several approaches have been proposed in this field. However, existing approaches generate a huge number of association rules that are difficult to exploit and assimilate. Moreover, they have not been proven themselves in a distributed context. As contribution, we propose, in this paper, DKDD_C, a new Distributed Knowledge Discovery approach. Exploiting, KDD based on data classification, we propose to give the choice to the user, either to generate Meta-Rules (rules between classes arising of preliminary data classification), or to generate classical Rules between distributed data. DKDD_C took place in both local and global processes. We prove that our solution minimizes the number of distributed generated association rules and then, offer a better interpretation of the data and optimization of the execution time. This approach has been validated by the implementation of a user-friendly platform as an extension of the Weka platform for the support of Distributed KDD.

Keywords: Distributed knowledge discovery · Mining association rules · Distributed database · Clustering · Weka plateform extension

1 Introduction

Nowadays, our ability to collect and store data from any type exceeds our possibilities of analysis, synthesis and Knowledge Discovery in Data (KDD). However, the performance of conventional centralized approaches degrade when the size of the processed data increases, in terms of execution time and memory space, hence we note the emergence towards the Distributed Knowledge Discovery (DKDD).

Several approaches and tools have been proposed in this context. Through our study, we found that these theoretical and practical approaches have different limits:

- Theoretically, DKDD algorithms generate a huge number of association rules that are difficult to exploit and assimilate.
- Practically, existing tools (1) support only some KDD algorithm that generates a large number of association rules that are difficult to assimilate (2) tools have not

© Springer International Publishing Switzerland 2016
Y. Tan et al. (Eds.): ICSI 2016, Part II, LNCS 9713, pp. 187–197, 2016.
DOI: 10.1007/978-3-319-41009-8_20

been proven themselves in a distributed context. (3) Are applied only to one restricted type of data.

We propose, in this paper, DKDD_C, a distributed knowledge discovery approach based on classification, which minimizes the number of distributed generated association rules and then offer a better interpretation of the data and optimized both the space memory and the execution time. By exploiting, KDD based on data classification, we propose to give the choice to the user, either to generate Meta-Rules (rules between classes arising of preliminary data classification), or to generate Rules between distributed data without preliminary classification. This approach has been validated by the implementation of a user-friendly plat-form as an extension of the Weka platform for the support of DKDD.

This paper is organized as follows: Section 2 presents some basic concepts for the DKDD, Sect. 3 presents our motivation, Sect. 4 provides a description of our proposed approach, and Sect. 5 presents the implementation and validation of our approach. We end with a conclusion and some perspectives.

2   Basic Concepts

In this section we present the basic concepts related to our research.

2.1   Mining Distributed Association Rules

Association Rules Mining is an important problem in the field of knowledge discovery aims at finding meaningful relationships between the attributes of databases in order to identify the groups of items that are most frequently purchased together. The first effective algorithm is Apriori [5]. Other algorithms have been proposed to improve performances such as CHARM [9] and Closet [10].

However, given the increase of the size of the processed data which lengthen the execution time and fill the memory space, the performances of centralized conventional approaches are deteriorating increasingly. Thus, new algorithms are opting for the parallelization and the distribution of this research problem.

Several approaches have been proposed in the literature. The first proposed algorithm is CD [6], which present a simple distribution of Apriori [5] algorithm. Other algorithms have been proposed later to propose more effective solutions. These algorithms include FDM [8] which introduce powerful pruning techniques called global and local pruning that minimize the size of candidates. ODAM [11] which reduce average size of transactions by eliminating all infrequent items from DB transactions after the first pass and therefore efficiently generate candidate support counts in latter passes. EDFIM [2] extended the ODAM algorithm to all passes and then offer better results. We can cite also the L-Matrix algorithm [7] that reduces the time of scan of database partition by using LMatrix to find support counts instead of scanning the databases partition time after time, which will save a lot of memory. In [3], DDRM algorithm is proposed that is a dynamic extension of Prefix-based [4] algorithm and has used a lattice-theoretic approach for mining association rules.

2.2   Clustering

Clustering is an important task in data mining that aims to organize data into groups of similar observations. Clustering approaches can be classified into two categories, a crisp approaches that consist to associate each object to a single cluster (such as K-means [13]), and fuzzy clustering (such as Fuzzy K-means [12]), that deals with overlapping data clusters. The main advantage of this type of data clustering is that it gives the flexibility to express that data points can belong to more than one cluster.

2.3   Panorama of Existing Platforms

The growing interest of the Data Mining method was accompanied with the appearance of many tools. Among them, we can mention:

- WEKA: This tool includes a set of training algorithms of Data Mining covering some supervised and not supervised methods of classification.
- TANAGRA: Free Data Mining software for academic and research purposes. It proposes several data mining methods for exploratory data analysis, statistical learning, machine learning and databases area.
- ORANGE: Open source data visualization and analysis for novice and experts. Data mining through visual programming or Python scripting. Components for machine learning.

Through our study, we see a height convergence to the Weka tool, and this is for several reasons including: (1) the diversity of its implemented data mining algorithms compared to other tools (2) it implements two type of clustering unlike other tools that don't implement the fuzzy clustering (3) it is very portable as it is completely implemented in Java and therefore runs on all modern platforms (4) it is a free tool.

3   Motivation

Several approaches and tools have been proposed for the DKDD. Through our study, we found that these approaches have different limits:

- Theoretically, DKDD algorithms generate a huge number of association rules, which causes (1) this requires a large memory space for data modeling, and data structures required by these algorithms such as trees or graphs or trellis. (2) The execution time for the management of these data structures is important. (3) Generated rules from these data are generally redundant rules. (4) Algorithms of mining of association rules give a very large number of rules that are difficult to assimilate.
- Practically, existing tools (1) support only some KDD algorithm that generates a large number of association rules that are difficult to use (2) tools have not been proven themselves to a distributed context. (3) Are applied only to one restricted type of data (4) they produce the output on text mode, so there are no the visualization of association rules (5) do not provide the calculation of the execution time.

We propose, in this paper, DKDD_C, a distributed knowledge discovery approach based on classification, which minimizes the number of distributed generated association rules and then offer a better interpretation of the data and optimized both the space memory and the execution time.

This approach has been implemented under the centralized data mining tool Weka, that we have extended it to support DKDD. The user interface is a modified Weka Explorer environment that supports not only the execution of both local and global data mining tasks, but also present solutions for practical existing tools limits cited above, such as graphic visualization of association rules, calculation of execution time, loading data from various source and different type of data.

4 New Approach

In this section, we present DKDD_C, our new approach of Distributed KDD.

4.1 General Principe

Our approach provide a solution for Distributed KDD and more precisely an effective generation of distributed association rules. The DKDD_C process takes place in two main phases:

- **Data collection phase** which consist in the preparation and collection of necessary information for the distributed knowledge extraction phase.
- **Distributed knowledge extraction phase**, which is divided into two distinct phases according to the user's choice:
  - Local DKDD phase: for knowledge discovery from a specific site of the DDB.
  - Global DKDD phase: for knowledge discovery from all sites of the DDB.

In the knowledge discovery phase, we added a classification step as a preliminary step to the mining of association rules. By exploiting, KDD based on data classification, we propose to give the choice to the user, either to generate Meta-Rules (rules between classes arising of preliminary data classification), or to generate Rules between distributed data without preliminary classification.

The Meta-Knowledge concept (Meta-Rules) models a certain abstraction of the data that is fundamental when the number of data is enormous. Moreover, the number of generated rules is smaller with this concept. Indeed, theoretically, while classifying data, we construct homogeneous groups of data. These groups, called clusters, have the same properties, so defining rules between these clusters implies that all the data elements belonging to those clusters will be necessarily dependent on these same rules. Thus the number of generated rules is smaller since one processes the extraction of the knowledge on the clusters which number is relatively lower compared to the initial data elements.

4.2   Architecture of the New Approach

In this section, we present the general architecture of our approach as shown in Fig. 1.

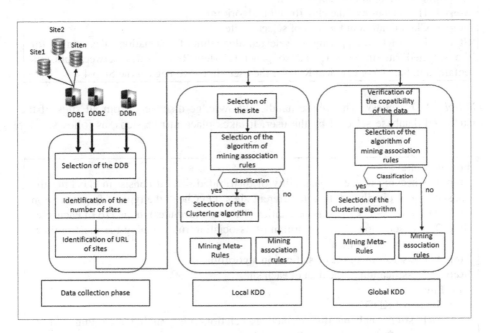

Fig. 1. General architecture of DKDD_C

Data collection phase. The data collection phase is the first phase of our knowledge discovery process. It consist in the preparation and collection of necessary information for the distributed knowledge discovery phase. It takes place in three steps as follows:

Step 1: The selection of the DDB on which we wish to apply the DKDD process.
Step 2: The identification of the site number of the DDB to define the data location in each remote site.
Step 3: The identification of the URL of each database to enable the remote access to databases of remote sites and import data on which we wish extract knowledge.

Distributed Knowledge Extraction phase. This phase is the main phase of our approach that is divided into two distinct processes according to the user's choice: Local and Global DKDD process. We detail in the following sections these two processes.

Local KDD process. This phase handles extracting knowledge from distributed data from a specific site selected by the user. It takes place in five steps, as follows:

Step 1: The selection of the site on which we want to apply the KDD process.
Step 2: The selection of the algorithm of association rules mining the most optimal according to the field of data.
Step 3: The selection of the classification algorithm
Step 4: Classification of the data of selected site.
Step 5: Local KDD: Applying the selected algorithm of association rules mining for the selected site in Step 1, and so generate Meta-Rules (rules between clusters obtained in Step 4) or classical Rules if the user choose to neglect the Step 4.

Global KDD process. This phase handles knowledge discovery from all sites of the distributed database selected by the user. It takes place in five steps, as follows:

For each site:
Step 1: Checking of the compatibility of distributed data: it consist in checking the consistency of distributed data in different sites: (1) Check the type of fragmentation of the DDB. (2) Check the various entities (Tables / Attributes) in the DB of each site.
Step 2: The selection of the algorithm of association rules mining the most optimal according to the field of data.
Step 3: The selection of the classification algorithm
Step 4: Classification of the data of each site
End For
Step 5: Global KDD
For each site: Applying the selected association rules mining algorithm and so generate Meta-Rules or classical Rules if the user choose to neglect the Step 4.
End For

The classification step in the knowledge extraction phase, Step 3 for both Local and Global process, is depending on the user' choice. This classification phase is proposed in Grissa [1] for centralized databases. We then have added this model as a step in our Distributed KDD approach (Fig. 2).

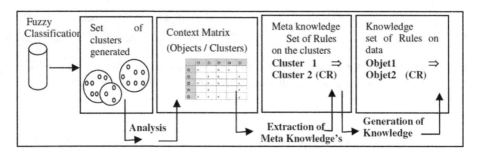

Fig. 2. The KDD approach based on classification

The different steps of this model are conducted as follows:

Step 1: Enter the datasets (any type of data).
Step 2: Apply a fuzzy clustering algorithm to organize data into groups (or clusters).
Step 3: Determine the fuzzy formal context of the matrix obtained in step 2.
Step 4: Apply on the matrix obtained an algorithm of association rule generating for the KDD process in the form of association rules between clusters.
Step 5: Generate knowledge of the dataset as association rules.

---

5 Implementation and Validation of the New Approach

5.1 Implementation.

For the implementation of our approach, we have worked with the Weka platform Version 3.6.12. We detail below the functioning of our solution:

Data collection and preparation phase. For managing distributed data, we have extended the original interface of Weka in which we added the option "Distributed Data" (Fig. 3). The user can consult existing DDB information: the number of distribution sites and their URLs (Fig. 4).

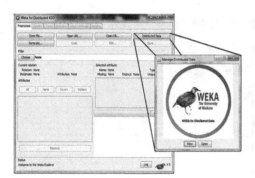
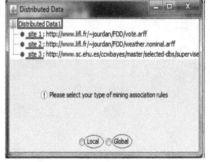

Fig. 3. Access to distributed process   **Fig. 4.** Selecting the DDB

Our solution works in two main phases, namely Local and Global DKDD and this according to the user needs (Fig. 4).

Local DKDD. This phase consist in managing the distributed data by site, and this in order to visualize the result of data mining of each site separately. By selecting a specific site, our solution imports its data through its URL. Figure 5 shows the loading of data of site 2 in the Weka data preprocessing panel.

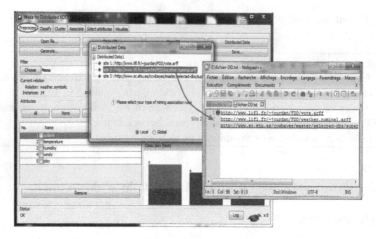

Fig. 5. Loading of sites of the selected DDB

Thus the user selects the classification algorithm and that of association rule mining (algorithms already implemented in Weka) that we have operated for the generation of Meta-Rules (Fig. 6). When achieving Local DKDD process, Association Meta-Rules are displayed in the "Out Put" panel of Weka (Fig. 7)

Fig. 6. Selection of rules mining algorithm   **Fig. 7.** Local mining of association rules

Global DKDD. This is the principal phase of our approach, our solution supports distributed data retrieval site by site just by giving the user the possibility to confirm the import of data from sites. Figures 8, 9 and 10.

By specifying the algorithms of classification and association rules extraction, our solution performs the classification process according to the selected algorithm, then generates the Meta-Rules between clusters obtained according to the selected algorithm of association rules mining (Fig. 11).

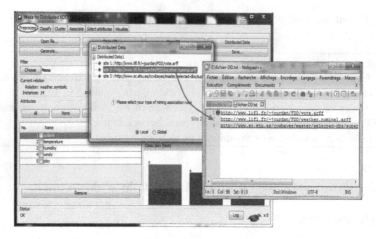

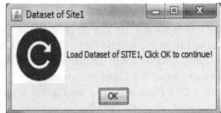

Fig. 8. Loading of sites   **Fig. 9.** Loading of data of site1

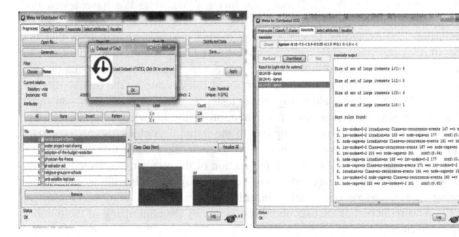

Fig. 10. Global execution   **Fig. 11.** Global execution

5.2 Validation

The proposed approach presents a number of advantages which can be summarized as follows:

- **The concept of "Global" and "Local" KDD:** The concept of "Global" KDD enables a global view of the DDB dataset and the concept of "Local" KDD allows a detailed view of initial database dataset on each remote site.
- **The definition of KDD approach based on classification (definition of Meta-Knowledge concept):** The Meta-Knowledge concept is very important in the case of very voluminous dataset because it gives a global view of this data set and models a certain abstraction of the data that is fundamental when the number of data is enormous. Moreover, the number of generated rules is smaller with this concept.
- **The extensibility of this approach:** In our approach, the step of generating association rules can be applied with any KDD algorithm. Studies have shown that, the fact that an algorithm is more optimum than another is according to the field of

used data. This means that we can apply the most optimum method according to the field of the dataset. In addition, the data classification step in our KDD process can be applied with any fuzzy clustering algorithm to classify the initial data.

The performance of our approach was evaluated on two datasets, "Car" which is derived from the simple hierarchical decisional model, and "Mushrooms" which is a set of dense data on fungi describing their characteristics. We have installed these datasets at each site of the DDB (site number = 3). As classification algorithm, we have tested with FCM (fuzziness = 2 and accuracy = 0.002). As algorithm of association rules extraction, we have tested with Apriori and Close algorithms. The parameters for these two algorithms are, confidence = 10 % and support = 0.5 % with "Car" datasets and confidence = 70 % or 50 % and support = 10 % with "Mushrooms" datasets.

To test performances according to the number of generated association rules (NAR) we have used both "Car" and "Mushrooms" datasets. We have tested our approach with different number of cluster (NC), we show here tests with 10 and 15 clusters (Fig. 13). According to the tests we carried out, we can conclude that the number of rules generated while applying our DKDD_C approach is reduced relative to the number of rules generated in the case of a direct application of an association rule mining algorithm. This reduction of the number of rules is due by the intervention of the classification step (fuzzy clustering in our tests). Indeed, the number of generated rules is smaller since one processes the extraction of the knowledge on the clusters which number is relatively lower compared to the initial data elements.

For execution time test, we have used "Car" datasets. Tests show that the execution times of these algorithms are increasing in function of the number of selected clusters, and they are lower than the execution times generated by these algorithms without the clustering step (using conventional approaches) for a reduced number of clusters (Fig. 12). This reduction of the execution time is due by the minimizing of the number of generated rules.

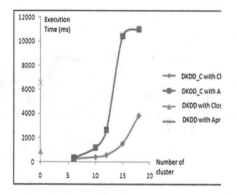

Fig. 12. Execution time

Fig. 13. Number of generated rules

Datasets MinSupp	MinConf	Apriori			Close		
		DKDD	DKDD_C		DKDD	DKDD_C	
		NAR	NC	NAR	NAR	NC	NAR
Mushroom 10%	70%	>110000	10	1185	48401	10	371
			15	12711		15	3144
	50%	>110000	10	1808	92651	10	581
			15	16533		15	4212
Car 0.5%	10%	42315	10	3794	35234	10	139
			15	23128		15	3753

6 Conclusion

We proposed, in this paper, DKDD_C, a distributed knowledge discovery approach based on classification, which minimizes the number of distributed generated association rules. Indeed, based on, KDD based on data classification, we propose to give the choice to the user, either to generate meta-rules, that is to say, generate rules between classes arising of preliminary classification on distributed data (Meta-Rules), or to generate rules between data distributed without preliminary classification. This approach has been implemented under the centralized data mining tool Weka, that we have extended it to support DKDD from different type of distributed data.

As perspective, we propose (1) to extend our approach to dynamic databases with a variable number of sites (2) to test and adapt our approach for large distributed databases.

References

1. Grissa, A.: Towards a discovering knowledge comprehensible and exploitable by the end-user. In: 2nd International Conference on Advances in Databases, Knowledge, and Data Applications, pp. 126–134 (2010)
2. Adelpoor, A., Abadeh, M.S.: An efficient frequent itemsets mining algorithm for distributed databases. Int. J. Comput. Sci. Electron. Eng. (IJCSEE) **1**(1) (2013)
3. Wessel, T.: Parallel mining of association rules using a lattice based approach. Nova Southeastern University (2009)
4. Zaki, M.J.: Hierarchical parallel algorithms for association mining. In: Kargupta, H., Chan, P. (eds.) Advances in Distributed and Parallel Knowledge Discovery, pp. 336–339. MIT Press, Cambridge, MA (2000)
5. Agrawal, R., Skirant, R.: Fast algorithms for mining association rules. In: Proceedings of the 20th International Conference on Very Large Databases, pp. 478–499 (1994)
6. Agrawal, R., Shafer, J.C.: Parallel mining of association rules. IEEE Trans. Knowl. Data Eng. **8**(6), 962–969 (1996)
7. Arokia Renjit, J., Shunmuganathan, K.L.: Mining the data from distributed database using an improved mining algorithm. Int. J. Comput. Sci. Inf. Secur. (IJCSIS) **7**(3), 116–121 (2010)
8. Cheung, D.W., Han, J., Ng, V.T., Fu, A.W., Fu, Y.: A fast distributed algorithm for mining association rules. In: Proceedings of the Fourth International Conference on Parallel and Distributed Information Systems, pp. 31–42 (1996)
9. Zaki, M.J., Hsiao, C.J.: CHARM: An efficient algorithm for closed itemset mining. In: The 2nd International Conference on Data Mining, Arlington, pp. 34–43 (2002)
10. Pei, J., Han, J., Mao, R., Nishio, S., Tang, S., Yang, D.: CLOSET: an efficient algorithm for mining frequent closed itemsets. In: Proceedings of the ACM SIGMOD DMKD 2000, Dallas, TX, pp. 21–30 (2002)
11. Ashrafi, M.Z., Taniar, D., Smith, K.: ODAM: an optimized distributed association rule mining algorithm. IEEE Comput. Soc. **5**(3), 1541–4922 (2004). IEEE Distributed Systems
12. Bezdek, J.: Fuzzy mathematics in pattern classification. Ph.D. Dissertation, Cornell University (1973)
13. McMueen, J.: Some methods for classification and analysis of multivariate observations. In: The Fifth Berkeley Symposium on Mathematical Statistics and Probability, pp. 281–297 (1967)

Fuzzy Rule-Based Classifier Design with Co-operation of Biology Related Algorithms

Shakhnaz Akhmedova$^{(\boxtimes)}$, Eugene Semenkin, and Vladimir Stanovov

Siberian State Aerospace University, Krasnoyarsk, Russia
shahnaz@inbox.ru, {eugenesemenkin,
vladimirstanovov}@yandex.ru

Abstract. A meta-heuristic called Co-Operation of Biology Related Algorithms (COBRA) is applied to the design of a fuzzy rule-based classifier. The basic idea consists in the representation of a fuzzy classifier rule base as a binary string and the use of the binary modification of COBRA with a biogeography migration operator for the selection of the fuzzy classifier rule base. The parameters of the membership functions of the fuzzy classifier, represented as a string of real-valued variables, are adjusted with the original version of COBRA. Two medical diagnostic problems are solved with this approach. Experiments showed that the modification of COBRA demonstrates high performance and reliability in spite of the complexity of the optimization problems solved. Fuzzy classifiers developed in this way outperform many alternative methods at the given classification problems. The workability and usefulness of the proposed algorithms are confirmed.

Keywords: Bionic algorithms · Biogeography · Optimization · Fuzzy systems · Classification

1 Introduction

Classification problems consist in identifying to which of a set of categories a new instance from a given data set belongs. Classification is a fundamental problem that has to be solved to approximate the human functions of recognizing sounds, images or other inputs. Classification is important for automation in today's commercial and military arenas and has many applications: computer vision, speech recognition, document classification, credit scoring, biological classification, etc.

Currently various algorithms for solving classification problems are being developed: the naive Bayes classifier, decision trees, quadratic classifiers, support vector machines, neural networks, etc. Moreover new and improved versions of the above-mentioned approaches are proposed. Researchers frequently use classifiers based on fuzzy logic (fuzzy rule-based classifiers or FRBC) for categorization. There are various works in which this method has been used and it has been established that generally it is efficient and works successfully (for example, [1]).

© Springer International Publishing Switzerland 2016
Y. Tan et al. (Eds.): ICSI 2016, Part II, LNCS 9713, pp. 198–205, 2016.
DOI: 10.1007/978-3-319-41009-8_21

Nowadays researchers propose different optimization techniques for the design of fuzzy rule-based classifiers which demonstrate good results (for example, [2]). However, in this study the meta-heuristic called Co-Operation of Biology Related Algorithms (COBRA) [3] and its modification for solving optimization problems with binary variables which uses a biogeography migration operator were applied for the design of fuzzy rule-based classifiers (for adjusting the parameters of membership functions and for the optimal selection of the rule base respectively). The COBRA meta-heuristic is based on the cooperation of five nature-inspired algorithms (Particle Swarm Optimization (PSO) [4], Wolf Pack Search (WPS) [5], the Firefly Algorithm (FFA) [6], the Cuckoo Search Algorithm (CSA) [7] and the Bat Algorithm (BA) [8]. The workability and reliability of COBRA for optimization problems with real-valued variables was shown in [3] on a set of benchmark functions with up to 50 variables and later confirmed in [9] on neural networks' weight coefficient adjustment with up to 110 real-valued variables and structure selection with 100 binary variables.

The aim of this paper is to improve the binary modification of COBRA by using a biogeography migration operator and demonstrate the workability and usefulness of the developed meta-heuristic on harder optimization problems related to the design of the fuzzy classifier rule base and the adjustment of membership function parameters.

The rest of the paper is organized as follows. In Sect. 2 the problem statement is presented. Then in Sect. 3 we describe the optimization technique COBRA and its binary modification with a biogeography migration operator. In Sect. 4 the workability of the meta-heuristic is demonstrated with fuzzy rule-based classifier design for two of medical diagnostics problems. The Conclusion contains a discussion of the results and consideration of further research directions.

2 Problem Statement

Let $L = \{l_1, \ldots, l_c\}$ be a set of class labels and $x = [x_1, \ldots, x_n]^T \in R^n$ be a vector describing an object. Each component of x expresses the value of a feature, thus a classifier is any mapping $C : R^n \to L$. A classifier is considered as a black box at the input of which x is submitted and at the output the values of c functions $f_1(x), \ldots, f_c(x)$, which express the support for the respective classes, are obtained. The maximum membership rule assigns x to the class with the highest support.

In fuzzy rule-based classifiers the features are associated with linguistic labels. Fuzzy systems are meant to be transparent models implementing logical reasoning, presumably understandable to the end-user of the system. A class of such systems employs if-then rules and an inference mechanism which, ideally, should correspond to the expert knowledge and decision-making process for a given problem.

Thus solving classification problems using fuzzy systems requires two problems to be solved: rule base selection and membership function tuning. These problems can be considered as optimization tasks: the selection of the classifier rule base can be described as an optimization problem with binary variables and the parameter tuning of membership functions as an optimization problem with real-valued variables.

In this study there are three Gaussian membership functions for each feature or attribute of a given input vector. So there are 2 parameters for each function and therefore $2 \times 3 \times n$ real-valued parameters that have to be tuned. As a result each data feature or attribute is represented by 2 bits: "00" means the feature is not used in a given rule, "01" means that for a given feature the 1st membership function is used, "10" means that the feature uses the 2nd membership function and "11" means the feature uses the 3rd membership function. Let m be the number of rules and consider classification problems with 2 classes (so 1 bit for class label), thus each rule base can be presented by a binary string with length equal to $(2 \times n + 1) \times m$.

Consequently the binary version of COBRA with a biogeography migration operator is used for finding the best rule base and the original COBRA is used for the membership function parameter adjustment of every rule base.

3   Optimization Techniques

The meta-heuristic approach Co-Operation of Biology Related Algorithms (COBRA) originally developed for solving optimization problems in continuous variable space was introduced in [3]. The proposed approach consists in generating 5 populations which are then executed in parallel cooperating with each other.

The proposed algorithm is a self-tuning meta-heuristic, so there is no need to choose the population size for each algorithm. The number of individuals in the population of each algorithm can increase or decrease depending on whether the fitness value improves: if the fitness value does not improve over a given number of generations, then the size of all populations increases and vice versa. Besides, on each generation a "winner algorithm" is determined: the algorithm with the best population's average fitness value. The population of the winner algorithm "grows" by accepting individuals removed from other populations. The migration operator of the suggested approach consists in the exchange of individuals between populations in such a way that a part of the worst individuals of each population is replaced by the best individuals of others.

However, frequently applied problems are defined in discrete valued spaces where the domain of the variables is finite. Thus a binary modification of the COBRA called COBRA-b was developed [9]. Namely its component-algorithms were adapted to search in binary spaces by applying a sigmoid transformation [10] to the velocity components (PSO, BA) or coordinates (WPS, FFA, CSA) to squash them into a range [0, 1] and force the component values of the positions of individuals to be 0's or 1's with an inverse exponential function [10].

So the binarization of individuals in algorithms is conducted using the calculated value of the sigmoid function. After that a random number $rand$ from the range [0, 1] is generated and the corresponding component value of the position of the individual is 1 if $rand$ is smaller than $s(v)$ and 0 otherwise.

An experiment showed that the COBRA algorithm and its modification COBRA-b work successfully and that they are reliable. Moreover, it was established that the meta-heuristics COBRA and COBRA-b outperform their component-algorithms. Yet in some cases COBRA-b requires too many calculations.

Biogeography-based optimization (BBO) is an evolutionary algorithm that translates the natural distribution of species into a general problem solution [11]. Habitat, in biogeography, is a particular type of local environment occupied by an organism, where the island is any area of suitable habitat. Each island represents one solution, where a good problem solution means that the island has lots of good biotic and abiotic factors, and attracts more species than the other islands.

In BBO the number of species on an island is based on the dynamic between new immigrated species onto an island and the extinct species from that island. Let's consider a migration model with a linear immigration rate λ and emigration rate μ, where they can be plotted as a logistic, exponential or any proper function [12]. The maximum immigration rate I occurs when the island is empty of any species and thus it offers a maximum opportunity to the species on the other islands for settling on it; whereas if the arrivals on that island increase, the opportunity for settling and λ will also decrease. And, as λ decreases, the species density increases, thus μ will increase and reach its maximum value E when λ reaches its minimum value. So the purpose of the migration process is to use "good" islands as a source of modification to share their features with "bad" islands, so the poor solutions can be probabilistically enhanced and may become better than those good solutions.

This migration operator was applied to COBRA-b. Now populations do not exchange individuals in such a way that a part of the worst individuals of each population is replaced by the best individuals of other populations. In the new version of COBRA-b the individuals of each population can be updated by the individuals of the other populations. However a certain number of individuals with the highest fitness value will not be changed but can be used for updating other individuals.

A set of real-parameter test optimization problems, whose number of variables varied from 2 to 4, were used for the preliminary investigation of COBRA-b and its

Table 1. Summary of results obtained by COBRA-b and its modification

F	D	ANFE1	AFV1	STD1	ANFE2	AFV2	STD2
1	2	740	0.000182069	0.00024851	116	1.10078e-5	3.05451e-5
	3	3473	0.000188191	0.00068965	435	0.000162824	0.001195637
	4	6730	0.00579879	0.0242297	555	4.12793e-5	0.000133099
2	2	567	0.000236274	0.00026535	103	0.000260617	2.32735e-5
	3	775	0.000150127	0.00016824	298	4.1847e-5	4.90123e-5
	4	916	0.000355086	0.00029257	419	0.000658365	9.9126e-5
3	2	1439	0.00019874	0.00033005	1041	0.00275265	0.11245036
	3	2046	0.00150713	0.00245315	1861	0.00150873	0.3043927
	4	3030	0.00126295	0.00281119	2580	0.00283	0.465772932
4	2	931	0.000209168	0.00026854	116	4.0362e-5	7.43944e-5
	3	868	0.000191162	0.00023388	327	0.000330693	5.33858e-5
	4	1710	0.000347666	0.00025729	518	0.000734619	0.000147079
5	2	899	0.00032841	0.00046868	157	9.87119e-5	1.85875e-5
	3	1332	0.000506847	0.00140048	527	0.000644101	0.000376341
	4	2258	0.00411721	0.158903	1082	0.00030445	6.7481e-5
6	2	1734	180.0002	0.00018536	1554	180.002	0.029217289
	3	3294	169.801	0.169149	2567	170.015	0.084792983
	4	5462	159.2	0.279294	4583	160.023	0.122501061

modification: the Rosenbrock, Sphere, Ackley, Griewank, Hyper-Ellipsoidal and Rastrigin functions [13]. Each real-valued variable was represented by a binary string with a length equal to 10. Thus the number of variables while solving the given test problems varied from 20 to 40. For each problem and dimension the number of program runs was equal to 100 so the results in Table 1 were averaged by that number. The maximum number of function evaluations was equal to 10000; but if the obtained result differed from the known optimum by less than 0.001, calculations were stopped. The following abbreviations were used in Table 1: ANFE is the average number of function evaluations, AFV is the average function value, STD is the standard deviation. The number "1" in the abbreviations signifies the results for COBRA-b and the number "2" for its modification.

As Table 1 shows, the modification of COBRA-b with a BBO migration operator allows better solutions to be found with a smaller number of calculations. The obtained results demonstrate that the new version of the algorithm outperforms the original COBRA-b both by the average number of function evaluations and by the best function value achieved during the work of the algorithm, averaged over 100 program runs.

4   Experimental Results

Two medical diagnostic problems were solved with the proposed fuzzy rule-based classifiers: Breast Cancer Wisconsin and Pima Indians Diabetes. For Breast Cancer Wisconsin one has 10 attributes (a patient's ID that was not used for calculations and 9 categorical attributes which possess values from 1 to 10), 2 classes, 458 records of patients with benign cancer and 241 records with malignant cancer. For Pima Indians Diabetes one has 8 attributes (all numeric-valued), 2 classes, 500 patients that tested negative for diabetes and 268 patients that tested positive.

From the viewpoint of optimization, fuzzy rule-based classifiers for these problems have from 170 to 190 binary variables for the rule base and from 48 to 54 real-valued variables for the membership function parameters. For the final parameter adjustment of membership functions (for the best obtained rule base) the maximum number of function evaluations was equal to 15000. Alternative algorithms for comparison are taken from [14, 15]. The obtained results are demonstrated in Tables 2 and 3 where the portion of correctly classified instances from testing sets (%) is presented. The authors' results are averaged over 15 algorithm executions.

Table 2. Performance comparison of classifiers for Breast Cancer Wisconsin problem

Author	Method	Accuracy (%)
This study (2016)	FRBC+COBRA	98.84
Polat et al. (2007)	LS-SVM	98.53
Guijarro-Berdias et al. (2007)	LLS	96.00
Karabatak et al. (2009)	AR+NN	97.40
Peng et al. (2009)	CFW	99.50
Pena-Reyes et al. (1999)	Fuzzy-GA1	97.36

Table 3. Performance comparison of classifiers for Pima Indian Diabetes problem

Author	Method	Accuracy (%)
This study (2016)	FRBC+COBRA	79.78
S.M. Kamruzzaman et al. (2005)	FCNN+PA	77.34
H. Temurtas et al. (2009)	PNN	78.13
H. Temurtas et al. (2009)	MLNN+LM	82.37
K. Kayaer et al. (2003)	GRNN	80.21
M.R. Bozkurt et al. (2012)	DTDN	76.00

Examples of the rule base for the Breast Cancer Wisconsin and Pima Indians Diabetes problems obtained during one of the program runs are presented in Tables 4 and 5. The following denotations are used: DC – feature is not used, 1, 2 or 3 – the first, the second or the third membership function for a given feature is used, B or M – benign or malignant cancer, P or N – patient has or does not have diabetes.

Figure 1 demonstrates examples of the membership functions of rules obtained for the last feature of the Breast Cancer Wisconsin problem and for the seventh feature of the Pima Indians Diabetes problem during one of the program runs:

Table 4. Example of the rule base for the Breast Cancer Wisconsin problem

3	2	DC	1	DC	DC	3	2	1	B
1	2	2	2	DC	DC	1	DC	2	B
3	3	1	3	1	3	DC	3	3	B
DC	2	2	3	2	1	2	3	3	M
2	1	2	3	3	1	DC	DC	1	M
2	DC	3	3	DC	DC	DC	DC	3	B
DC	1	DC	2	1	3	1	2	DC	B
DC	DC	3	2	DC	1	3	2	1	M
1	1	3	2	2	DC	3	1	2	M
DC	1	1	3	2	3	3	1	DC	B

Table 5. Example of the rule base for the Pima Indian Diabetes problem

1	3	3	3	DC	DC	3	2	P
1	1	3	1	3	3	3	2	N
3	3	1	1	3	1	DC	2	P
DC	2	DC	3	2	3	2	3	N
3	DC	3	1	DC	3	2	1	N
DC	2	DC	1	3	3	3	3	N
2	3	2	1	DC	2	DC	2	P
DC	3	DC	3	3	3	1	3	N
3	DC	2	2	1	3	1	2	P
3	3	3	DC	3	1	3	1	N

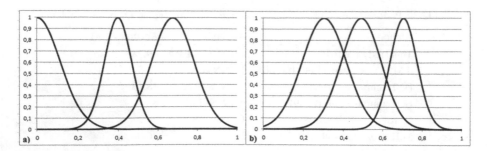

Fig. 1. Examples of the membership functions of rules for a feature: (a) Breast Cancer Wisconsin, (b) Pima Indian Diabetes

Consequently the inference should be drawn that the suggested algorithms successfully solved all the problems of designing classifiers with competitive performance. Thus the study results can be considered as confirming the reliability, workability and usefulness of the algorithms in solving real world optimization problems.

5   Conclusion

In this paper a new modification of an earlier proposed meta-heuristic called COBRA-b is introduced for solving optimization problems with binary variables. The modification of the algorithm COBRA-b consisted in the implementation of a migration operator from a BBO algorithm instead of the simple exchange of individuals between populations. The main purpose was to lessen the number of function evaluations required for solving an optimization problem.

The new version of the COBRA-b approach was tested by using a set of 6 benchmarks with differing numbers of variables. Experiments showed that the proposed algorithm works successfully, is reliable and demonstrates competitive behaviour. Moreover, the new optimization technique demonstrated better results than the original COBRA-b, so it outperformed not only the component algorithms but also the meta-heuristic COBRA-b.

Then the described optimization methods were used for the automated design of fuzzy rule-based classifiers. The modification of COBRA-b with a BBO migration operator was used for the rule base optimization of the classifier and the original COBRA was used for the parameter adjustment of membership functions. This approach was applied to 2 real-world classification problems, the solving of which is equivalent to solving big optimization problems where the objective functions have many variables and are given in the form of a computational program. The obtained results confirmed the reliability, workability and usefulness of the algorithms in solving real world optimization problems.

Directions for future research are heterogeneous: improvement of the cooperation and competition scheme within the approach, addition of other algorithms in cooperation, development of a modification for mixed optimization problems, etc. Also the application of the biogeography migration operator to real-parameter constrained and unconstrained versions of the meta-heuristic COBRA may improve its workability.

Acknowledgement. Research is performed with the financial support of the Ministry of Education and Science of the Russian Federation within the State Assignment for the Siberian State Aerospace University, project 2.1889.2014/K.

References

1. Angelov, P., Zhou, X., Klawonn, F.: Evolving fuzzy rule-based classifiers. In: IEEE Symposium on Computational Intelligence in Image and Signal Processing, pp. 220–225 (2007)
2. Sanz, J., Fernandez, A., Bustince, H., Herrera, F.: A genetic tuning to improve the performance of fuzzy rule-based classification systems with interval-valued fuzzy sets: degree of ignorance and lateral position. Int. J. Approximate Reasoning **55**(6), 751–766 (2011)
3. Akhmedova, Sh., Semenkin, E.: Co-operation of biology related algorithms. In: IEEE Congress on Evolutionary Computation, pp. 2207–2214 (2013)
4. Kennedy, J., Eberhart, R.: Particle swarm optimization. In: IEEE International Conference on Neural Networks, pp. 1942–1948 (1995)
5. Yang, Ch., Tu, X., Chen, J.: Algorithm of marriage in honey bees optimization based on the wolf pack search. In: International Conference on Intelligent Pervasive Computing, pp. 462–467 (2007)
6. Yang, X.S.: Firefly algorithms for multimodal optimization. In: 5th Symposium on Stochastic Algorithms, Foundations and Applications, pp. 169–178 (2009)
7. Yang, X.S., Deb, S.: Cuckoo search via levy flights. In: World Congress on Nature and Biologically Inspired Computing, pp. 210–214. IEEE Publications, USA (2009)
8. Yang, X.-S.: A new metaheuristic bat-inspired algorithm. In: González, J.R., Pelta, D.A., Cruz, C., Terrazas, G., Krasnogor, N. (eds.) NICSO 2010. SCI, vol. 284, pp. 65–74. Springer, Heidelberg (2010)
9. Akhmedova, Sh., Semenkin, E.: Co-operation of biology related algorithms meta-heuristic in ANN-based classifiers design. In: IEEE World Congress on Computational Intelligence, pp. 867–873 (2014)
10. Kennedy, J., Eberhart, R.: A discrete binary version of the particle swarm algorithm. In: World Multiconference on Systemics, Cybernetics and Informatics, pp. 4104–4109 (1997)
11. Simon, D.: Biogeography-based optimization. IEEE Trans. Evol. Comput. **12**(6), 702–713 (2008)
12. Simon, D.: A probabilistic analysis of a simplified biogeography-based optimization algorithm. Evol. Comput. **19**(2), 167–188 (2011)
13. Molga, M., Smutnicki, Cz.: Test Functions for Optimization Need (2005)
14. Marcano-Cedeno, A., Quintanilla-Dominguez, J., Andina, D.: WBCD breast cancer database classification applying artificial metaplasticity neural network. Expert Syst. Appl. **38**(8), 9573–9579 (2011)
15. Temurtas, H., Yumusak, N., Temurtas, F.: A comparative study on diabetes disease diagnosis using neural networks. Expert Syst. Appl. **36**(4), 8610–8615 (2009)

Identifying Protein Short Linear Motifs by Position-Specific Scoring Matrix

Chun Fang[1]([✉]), Tamotsu Noguchi[2], Hayato Yamana[3], and Fuzhen Sun[1]

[1] Shandong University of Technology, Shandong 255049, China
fangchun0409@163.com
[2] Meiji Pharmaceutical University, Tokyo 204-0004, Japan
noguchi-amotsu@aist.go.jp
[3] Waseda University, Tokyo 169-8555, Japan
yamana@yama.info.waseda.ac.jp

Abstract. Short linear motifs (SLiMs) play a central role in several biological functions, such as cell regulation, scaffolding, cell signaling, post-translational modification, and cleavage. Identifying SLiMs is an important step for understanding their functions and mechanism. Due to their short length and particular properties, discovery of SLiMs in proteins is a challenge both experimentally and computationally. So far, many existing computational methods adopted many predicted sequence or structures features as input for prediction, there is no report about using position-specific scoring matrix (PSSM) profiles of proteins directly for SLiMs prediction. In this study, we describe a simple method, named as PSSMpred, which only use the evolutionary information generated in form of PSSM profiles of protein sequences for SLiMs prediction. When comparing with other methods tested on the same datasets, PSSM-pred achieves the best performances: (1) achieving 0.03–0.1 higher AUC than other methods when tested on HumanTest151; (2) achieving 0.03–0.05 and 0.03–0.06 higher AUC than other methods when tested on ANCHOR-short and ANCHOR-long respectively.

Keywords: Short linear motifs · Protein · Prediction

1 Introduction

Short linear motifs (SLiMs) are short binding regions which are typically not more than 11 consecutive residues in length with less than 5 defined positions. Many positions are degenerate thus offering flexibility in terms of the amino acid types allowed at those positions [1]. SLiMs often evolve convergently and are typically appear in intrinsically disordered protein regions (IDRs), however, they differ from Molecular Recognition Features (MoRFs) which are also short binding segments (usually 5–25 residues in length) located within longer IDRs. Besides their length difference, they are defined from different perspectives, SLiMs are defined on the basis of a sequence pattern while MoRFs are defined as interaction-prone disordered segments that easily undergo disorder-to-order

© Springer International Publishing Switzerland 2016
Y. Tan et al. (Eds.): ICSI 2016, Part II, LNCS 9713, pp. 206–214, 2016.
DOI: 10.1007/978-3-319-41009-8_22

transitions upon binding to their partners. The MoRFs may themselves contain SLiMs. SLiMs play critical roles in several biological functions including protein scaffolding, cell regulation, cell signaling, post-translational modification, and cleavage. It is estimated that up to 1/3 of the eukaryotes proteins contain long disordered segments [2]. Within these disordered regions, there may be millions of SLiMs existing although few of them have been discovered and experimentally validated so far [3].

Their short length and the attribute of a weak binding affinity render SLiMs difficult to be detected experimentally, which makes computational methods indispensable for guiding experimental analysis. So far, many algorithms and tools have been developed for identifying SLiMs. SLiMPrints [4] combines relative local conservation (RLC) statistics and disorder predictions to identify putative SLiMs in the input sequence. GLAM2 [5] identify DNA or protein motifs using gapped local alignment. MEME [6] identify DNA or protein motifs using deterministic optimization which is based on Expectation Maximization (EM). ANCHOR [7] is a well-known web tool for the prediction of disordered binding regions, which relies on the pairwise energy estimation approach that is the basis of IUPred [8], a previous general disorder prediction method. SLiMPred [9] used a variety of predicted features including predicted disorder probabilities, predicted solvent accessibility, and predicted secondary structure propensities as input. These predicted features not only easy to result in high-dimensional feature space, but also greatly increase the complexity of algorithms. Furthermore, the performances of all these predictors are still not well and need to be improved continuously.

In order to develop more simple and higher efficient method for identifying SLiMs, we have tried various features for SLiMs prediction. What greatly surprised us is that directly using the position-specific scoring matrixes generated from proteins sequences seems to be the most effective method for prediction. Here, we describe a method which only adopts evolutionary information generated in form of PSSM profiles of protein sequences for SLiMs prediction. All the input features are extracted from PSSM only. Support vector machine (SVM) is used to build the classifiers.

2   Methods

2.1   Data Sets

Two groups of datasets were used in this study.

Dataset group 1: This group of datasets includes 602 SLiMs -contained human protein chains, which were extracted from the Eukaryotic Linear Motif database [10] (http://elm.eu.org/). The sequence identity between them is <30 %. 451 chains of them selected randomly were used as the training dataset, containing 4760 positive samples (SLiMs residues) and 104,235 negative samples (non-SLiMs residues), named as HumanTrain451. All the positive samples and the equal number of negative samples selected randomly were used for training the prediction model. The retained 151 chains, containing 1667 SLiMs residues and 126,521 non-SLiMs residues, were used as test dataset, named as HumanTest151.

Dataset group 2: This group of datasets is the same with the datasets used in the study of SLiMFpred [9]. 300 SLiM-contained protein chains were used as the training dataset, named as ELM300, containing 2,461 positive samples (SLiMs residues) and 223,437 negative samples (non-SLiMs residues). All the positive samples and the equal number of negative samples selected randomly were used for training the prediction model. Two other datasets — the ANCHOR-short and the ANCHOR-long — that contain 18 and 28 chains respectively are used as test datasets, they shares up to 30 % sequence identity with the ELM300.

2.2   Continuous Binding Residues Analysis

We have calculated the continuous binding residues length of the 602 SLiM-contained proteins in the Dataset group 1. The statistics in Fig. 1 demonstrate that most SLiMs are consecutive amino acids with length between 3 to 10 residues.

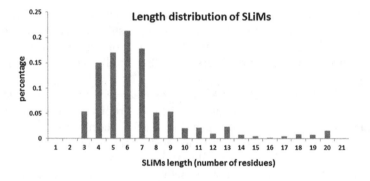

Fig. 1. Length distribution of SliMs

2.3   Physicochemical Properties Analysis of SLiMs and Their Flanking Regions

In protein evolution, different conservation patterns will display different physicochemical properties tendencies in different positions of the sequence. We next analyzed the SLiMs with respect to ten physicochemical properties, namely: hydrophobic, polar, small, proline, tiny, aliphatic, aromatic, positive, negative and charged. Patterns that contain SLiMs were extracted from the sequences. Regions flanking on either side of SLiM residue with 20-residue long were analyzed. For example, XXXXXOXXXX, where X is the flanking residue and O is a SLiM residue. The flanking length is the distance between X and O. Physicochemical property differences were calculated according to Formula (1), and the corresponding result is shown in Fig. 2. As shown in Fig. 2, within 20-residue long flanking regions, the differences of many physicochemical properties fluctuate significantly, while in the outlying regions, all property differences trend to 0. The SLiMs show slightly opposite property trends compared with their flanking regions. This phenomenon illustrates that SLiMs have their particular conservation patterns which are different from the non-SLiM regions.

$$property\ difference = \frac{frequence\ i}{\sum_{i=1}^{10} frequence\ i} - \frac{frequence\ j}{\sum_{j=1}^{10} frequence\ j} \qquad (1)$$

where i is one of the 10 properties appeared at a position within the SLiMs or their 20-residue long flanking regions, and j is the related property appeared in the whole sequences.

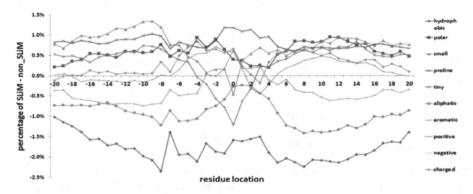

Fig. 2. Differences in physicochemical properties between SLiMs and their flanking regions. The values at position 0 on the horizontal axis mean the average values for SLiM residues, L means the left flanking side and R means the right flanking side. (Color figure online)

2.4   Evolutionary Information (PSSM)

Evolutionary information was obtained from PSSMs, generated by PSI-BLAST [11] searching against NCBI non-redundant (nr) database [12] by three times iteration with an e-value of 0.001.

2.5   SVM

Prediction of SLiMs can be addressed as a two- classification problem, namely, determining whether a given residue belongs to a SLiMs or not. Our prediction model was trained using the LIBSVM software package [13]. Here, the Radial Basis Function (RBF kernel) was selected as the kernel function.

3   Evaluation Criteria

Performances of the predictors were measured by the area under the corresponding ROC curve (AUC). ACC [14] calculated by the average of sensitivity and
specificity was also used as an evaluating indicator. The ROC plots with the
AUC values were created using the R statistical package [15]. The sensitivity,
specificity, true positive rate (TPR), false positive rate (FPR), and ACC are
defined as follows:

$$Specificity = \frac{TN}{TN + FP} \tag{2}$$

$$TPR = Sensitivity = \frac{TP}{TP + FN} \tag{3}$$

$$FPR = 1 - Specificity = \frac{TN}{TN + FP} \tag{4}$$

$$ACC = \frac{1}{2}(Sensitivity + Specificity) \tag{5}$$

where TP, TN, FP, and FN represent the number of true positives, true negatives,
false positives, and false negatives respectively.

4   Results and Discussion

4.1   Optimizing Window Size

To develop the predictor, a sliding-window size is needed to ultimately determine
the dimensions of the feature vectors. The training dataset HumanTrain451 in
Dataset group 1 was analyzed as an example. PSSMpred models with different
sliding-window sizes were tested by 5-cross-validation. The corresponding cross-
validation accuracies according to different sliding-window sizes are shown in
Fig. 3. Therefore, we chose the relatively best size 25 as the sliding-window size
for our models.

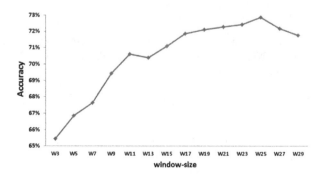

Fig. 3. Cross-validation accuracy of PSSMpred with different sliding-window sizes. W
indicates the length of windows, i.e., W3 means 3-residue long.

4.2   Performance Tested on Dataset Group1 and Comparison with Other Feature-Based Methods

In order to verify whether incorporating contextual information of residues facilitate the prediction. PSSMPred was compared with other four methods which were used in our previous paper [16]: (1) the masked-PSSM method, the smoothed-PSSM method, and the MFSPSSMpred model (please refer the details in the paper [16]). For a fair comparison, all the methods employed the same outside sliding-window size of 25. ROC plots of the three methods tested on HumanTest151 are shown in Fig. 4 which shows that the PSSMPred achieved the best performance.

We speculate the reasons is that: (1) some residues in SLiMs involve strongly independently and show slightly opposite property trends compared with their flanking regions (see Fig. 2); (2) The masking or smoothing steps adopted in the related methods did not strengthen but, on the contrary, weaken the independent feature of SLiMs residues, and therefore impede the prediction; and (3) every value in the PSSMs is calculated based on the assumption that the position of each value in the matrix is independent of the others, So PSSM is just suitable to represent the independent feature of SLiMs, and therefore facilitate the prediction.

It is noteworthy that, the masked-smoothed-PSSM method has been well used for identifying MoRFs in our previous study [16], while it performs not well for identifying SLiMs, meaning that they have different conservation patterns.

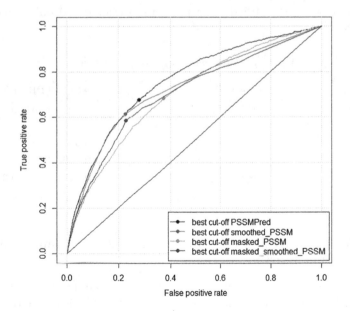

Fig. 4. ROC plots of PSSMPred compared with other feature-based tested on Human-Test151. (Color figure online)

We also compared the PSSMPred with another notable predictor Anchor [7] on the dataset HumanTest151. Result is shown is Fig. 5. The detailed comparisons of ACC, TPR, FPR and AUC of all the methods are listed in Table 1. Figure 5 and Table 1 demonstrate that, PSSMPred performed much better than Anchor, achieving significantly higher ACC and AUC than Anchor.

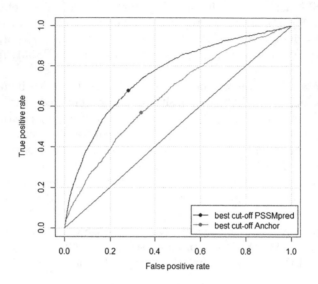

Fig. 5. ROC plots of PSSMPred compared with Anchor tested on HumanTest151. (Color figure online)

Table 1. Performance of the five methods.

Test dataset	Method	ACC	TPR	FPR	AUC
HumanTest151 (Dataset group1)	PSSMPred	0.409	0.494	0.323	0.758
	ANCHOR [7]	0.384	0.337	0.431	0.658
	smoothed_PSSM [16]	0.306	0.227	0.386	0.727
	masked_PSSM [16]	0.345	0.372	0.318	0.705
	masked_smoothed_PSSM [16]	0.322	0.230	0.415	0.705

4.3   Performance Tested on Dataset Group2 and Comparison with Other Existing Methods

In order to further verify the validity of our method, we next compared PSSMPred with other existing methods on the Dataset group2. Here, the training and testing datasets are the same with the datasets used in the research of SLiMPred [9]. ROC plots of PSSMpred tested on ANCHOR-short and ANCHOR-long are shown in Fig. 6. Details of ROC and ACC for all the predictors are shown in

Table 2. The results of other classifiers are quoted from the paper [9]. Figure 6 and Table 2 demonstrate that, PSSMpred outperformed the other predictors with respect to ACC and AUC on both the ANCHOR-short and ANCHOR-long datasets. Figure 6 also suggests that PSSMpred is more competent in identifying short SLiMs than identifying long SLiMs.

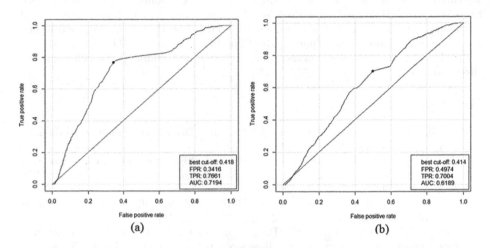

(a)   (b)

Fig. 6. ROC plots of PSSMpred tested on ANCHOR-short (a) and ANCHOR-long (b).

Table 2. Performance of the three methods.

Test dataset	Method	ACC	TPR	FPR	AUC
ANCHOR-short (Dataset group2)	PSSMPred	**0.71**	**0.77**	**0.34**	**0.72**
	Anchor [7]	0.37	0.31	0.57	0.67
	SLiMPred [9]	0.61	0.42	0.20	0.69
ANCHOR-long (Dataset group 2)	PSSMPred	**0.60**	**0.70**	**0.50**	**0.62**
	Anchor [7]	0.41	0.39	0.57	0.59
	SLiMPread [9]	0.52	0.29	0.25	0.56

5   Conclusions

In this study, we propose a PSSM-based method for SLiMs prediction. Our method employs no predicted results as input, and all input features are extracted only from the PSSMs of sequences. When comparing with other existing methods on the same datasets, PSSMpred outperformed them all, achieving 0.03–0.1 higher AUC than other methods when tested on HumanTest151; achieving 0.03–0.05 and 0.03–0.06 higher AUC than other methods when tested on ANCHOR-short and ANCHOR-long respectively. This study suggests that, conservation information included in PSSM of 25 amino acid patterns are more predictive than other features for SLiMs prediction, and incorporating contextual information of residues would not facilitate the prediction. The related Web server is freely available at: http://centos.sdutacm.org/fang/SLiMPed.php.

Acknowledgment. We gratefully thank Dr. Catherine Mooney for providing us with the datasets of SLiMPred, and Professor Xiaohong Liu of Shandong University of Technology for providing a lot of convenience for our study. This work was supported by a grant from Natural Science Foundation of Shandong Province (NO. ZR2014FQ028). We also thank the anonymous reviewers for his/her helpful comments, which improved the manuscript.

References

1. Pallab, B., Mainak, G., Peter, T.: Intrinsically Disordered Proteins Studied by NMR Spectroscopy. Chapter 9 (2014)
2. Ward, J.J., Sodhi, J.S., McGuffin, L.J., Buxton, B.F., Jones, D.T.: Prediction and functional analysis of native disorder in proteins from the three kingdoms of life. J. Mol. Biol. **337**(3), 635–645 (2004)
3. Tompa, P., Norman, E.D., Toby, J.G., Mabu, M.M.: A million peptide motifs for the molecular biologist. Mol. Cell **55**(2), 161–169 (2014)
4. Davey, N.E., Cowan, J.L., Shields, D.C.: Conservation-based discovery of functional motif fingerprints in intrinsically disordered protein regions. Nucleic Acids Res. **40**(21), 10628–10641 (2012)
5. Martin, C.F., Neil, F.W., Bostjan, K., Timothy, L.B.: Discovering sequence motifs with arbitrary insertions and deletions. PLoS Comput. Biol. **4**(5), e1000071 (2008)
6. Timothy, L.B., Nadya, W., Chris, M., Wilfred, W.L.: MEME: discovering and analyzing DNA and protein sequence motifs. Nucleic Acids Res. **34**(Suppl 2), W369–W373 (2006)
7. Dosztanyi, Z., Mszros, S.I.: ANCHOR: web server for predicting protein binding regions in disordered proteins. Bioinformatics **25**(20), 2745–2746 (2009)
8. Dosztanyi, Z., Csizmok, V., Tompa, P., Istvn, S.: IUPred: web server for the prediction of intrinsically unstructured regions of proteins based on estimated energy content. Bioinformatics **21**(16), 3433–3434 (2005)
9. Mooney, C., Pollastri, G., Shields, D.C., Haslam, N.J.: Prediction of short linear protein binding regions. J. Mol. Biol. **415**, 193–204 (2012)
10. Dinkel, H., Van, R.K., Michael, S., et al.: The eukaryotic linear motif resource ELM: 10 years and counting. Nucleic Acids Res. **42**, D259–D266 (2014)
11. Altschul, S.F., Madden, T.L., Schaffer, A.A., et al.: Gapped BLAST and PSI-BLAST: a new generation of protein database search programs. Nucleic Acids Res. **25**, 3389–3402 (1997)
12. NR. ftp://ftp.ncbi.nih.gov/blast/db/fasta/nr.gz
13. Chang, C.C., Lin, C.J.: LIBSVM: a library for support vector machines. ACM Trans. Intell. Syst. Technol. **2**(27), 1–27 (2011)
14. CASP10.   http://predictioncenter.org/casp10/doc/presentations/CASP10_DR_KF.pdf
15. R statistical package. http://www.r-project.org/
16. Fang, C., Noguchi, T., Yamana, H.: MFSPSSMpred: identifying short disorder-to-order binding regions in disordered proteins based on contextual local evolutionary conservation. BMC Bioinform. **14**, 300 (2013)

An Intelligent Identification Model
for Classifying Trembling Patterns
of Parkinson's Disease

Yo-Ping Huang[✉] and Chih-Hang Chuang

Department of Electrical Engineering, National Taipei University of Technology,
Taipei 10608, Taiwan
yphuang@ntut.edu.tw, a791167@gmail.com

Abstract. According to a recent survey that death caused by fall over was among the top three accidental deaths in elderly people aged over 65. Fall over not only affected the elderly mental and physical fitness, social functions, and living quality, but also placed heavy burden on caregivers. Parkinson's disease patients faced walking unbalance and tremor on hands and feet that made them hold a walking stick or umbrella as an assistive device. In this study we proposed an intelligent identification model to classify trembling patterns from signals detected by a sensor device attached to cane or umbrella hold by Parkinson's disease patients during walking. Alternatively, this paper was aimed to develop a systematic method that measured the trembling directions and severity and long-term monitored whether the tremor has been gradually improved after proper medication or rehabilitation. Experimental results were presented to verify the classification accuracy.

Keywords: Parkinson's disease · Smart cane · Self-organizing map · Intelligent model · Sensor device

1 Introduction

Parkinson's disease (PD) is one of the common diseases suffered from elderly people. Clinical symptoms in PD patients include resting tremor, bradykinesias, akinesia, rigidity and postural instability that will cause the risk of fall over. Among the induced disorders, trembling and walking difficulty affects the patients the most. Some of the side effects induced from PD patients include the higher risk of fall over and feeling unsafety to the ambient environment. Due to the fear of fall over again, they gradually lose self-mobility to interact with others and require others help or need to stay in the hospital since then. Tremor is the most common condition on physical imbalance for PD subjects. Current diagnostic procedure is based on clinical examination that was difficult to capture some tiny tremors. Rigas et al. [1] embedded a set of accelerometers on different parts of the body to measure tremor. They extracted features from acceleration signals to develop the hidden Markov models to evaluate trembling patterns and their severity on idle and moving postures. Yoneyama et al. [2, 3] exploited accelerometers to evaluate the severity of PD, proposed a method to detect gait patterns

© Springer International Publishing Switzerland 2016
Y. Tan et al. (Eds.): ICSI 2016, Part II, LNCS 9713, pp. 215–222, 2016.
DOI: 10.1007/978-3-319-41009-8_23

and presented a novel method to quantify the walking activities. Milosevic et al. [4] applied Self-Organizing Maps (SOM) visualization technique to classify complex mutual coordination between muscle activities.

Gait pattern is a prospective bio-feature for personal identification that can be extracted from a series of non-intrusive acceleration signals. Most current methods on gait pattern identification were obtained from acceleration signals measured at crisp gait periods. They may fail if the crisp periods were not detected accurately or there was a phase shift across from one period to the next. Zhang et al. [5] proposed a sparse representation method that was clustered by features from acceleration signals to identify the gait patterns. This can remove the risk of failure from using accelerometer for gait detection. Hsu et al. [6] used accelerometer and gyroscope to analyze gait patterns and balance for PD patients. Kurihara et al. [7] placed the accelerometer on the waist and used the transformation between the maximum intake oxygen and acceleration to find a regression line that related the walking speed to the intensity of exercise.

There was no handy tool to quantify the condition of improvement on tremor from using a cane. Not to say there was an effective method to monitor if patients hold the cane properly during walking after leaving the hospital. Mercado et al. [8] installed force sensors, accelerometer and alarm on single pole quadruple cane. The purpose was to exert partial body weight on the cane to protect the injured leg.

Bhatlawande et al. [9] designed an electronic cane for the blind people. An ultrasonic sensor was installed inside the cane that detected obstacles ahead and returned a voice or vibration signal to notify the user from colliding with the obstacles.

This paper was aimed on devising a sensor device equipped with accelerometer that was embedded to a smart cane to measure the hand trembling patterns of PD subjects. PD might be cured by medication use or surgery to delay its deterioration. If it can be diagnosed at an early stage, then PD might be cured and rehabilitated earlier. Consequently, the detection of gait patterns and hand trembling is crucial to the early prevention and cure. Besides that, an assistive device is an alternative for PD patients and can help alleviate the mental and physical burden of caregivers. Various auxiliary tools have been developed to detect fall over for elderly people but for the convenience and ease of use for elderly people either cane or umbrella is the most suitable assistive device for them.

Another application of the proposed system is to quantitatively measure the improvement after proper medication on PD subjects. The collected data can provide medical staff valuable information about the progress of rehabilitation. Besides that, the proposed system can automatically send an alarm signal to notify caregiver or medical staff if the fall over was detected so that an emergency handling can be taken to minimize the hurt to the subjects.

This paper is organized as follows. The proposed method was described in Sect. 2. Intelligent modeling was discussed in Sect. 3. Experimental results are presented in Sect. 4. Conclusions and future work are given in the final section.

2 The Proposed Sensor Device

The gait data were collected from subjects who used the designed smart cane to walk straight for 30 s and each step was about 30 cm. The sensor device was constituted by ATMEGA328 board, ADXL345 accelerometer, Arduino Bluetooth module and two rechargeable batteries (3.7 V, 1200 mAh) as shown in Fig. 1(a). The size of the prototype sensor device is 64 mm × 64 mm × 26 mm. A mobile phone was used to receive accelerometer signals and the gait patterns were analyzed by the developed APP. The devised smart cane was shown in Fig. 1(b) where the sensor device was attached to the cane at 75 cm above the ground.

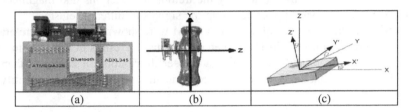

Fig. 1. The prototype sensor device and the smart cane.

To evaluate the hand trembling patterns we measured the signal variations from accelerometer. Through numerical transformation we can obtain the hand trembling trajectories in the three dimensional space as shown in Fig. 1(c) where $X' = X_g$, $Y' = Y_g$, $Z' = Z_g$ and X_g, Y_g, Z_g represent X-, Y- and Z-axis accelerometer variations, respectively. To analyze the hand trembling variations on the two-dimensional plane, the acceleration to the ground X_g is not considered in this study. The hand trembling trajectory on the plane is shown in Fig. 1(b). When the subject held the smart cane, the acceleration of left-right direction trembling is Z_g and the forward-backward direction trembling is Y_g. We approached from the variations of acceleration trembling on Z_g and Y_g to determine the trembling directions and severity. The sampling rate in our App is set to 3.333 Hz.

The sensor device continued to record the accelerometer signals that were transmitted to the smart phone for trembling patterns' identification by Bluetooth. To analyze the trembling patterns from the sensor device attached to a cane or umbrella, both x- and y-axis data from triaxial accelerometers were plotted on 2-dimensonal plane.

Before we apply fuzzy c-means (FCM) clustering method to group the measured signals some of the abnormal data were filtered by data clustering technique. In this study a variety of clusters, such as 3, 5 and 7 clusters, were considered in the FCM model. To simplify the illustration 3 clusters in FCM were given here. A linear regression model is established from the centroids of FCM to roughly identify the direction of tremor. Then, centroids of FCM, slope from linear regression model, and distance between two outmost centroids were used as inputs to the self-organizing feature map (SOM) to categorize the trembling patterns.

3   Intelligent Identification Model for Tremor

In this study we investigated intelligent models to identify trembling patterns of Parkinson's disease patients during walking. The proposed system has simulated on combinations of trembling patterns from left-right, forward-backward, 30°, 45° and 60°, and found that trembling patterns can be effectively identified by the intelligent models such as FCM and SOM.

Participants were instructed to use the functions provided by the designed smart cane. They followed the designated route to walk that allowed us to quantify the directions and magnitude of tremor. A mobile phone was used to store data transmitted from the sensor device through Bluetooth. Then, sensed data were plotted on a plane to display the trembling signals. To identify the trembling directions and magnitudes, a fuzzy c-means (FCM) was applied to group the signals to different clusters.

Clustering results from trembling data by FCM were shown in Fig. 2 where orange dots represented centroids of clusters. When comparing the results in Fig. 2(a) to those in Fig. 2(b) it is obvious that the former had closer distribution of centroids than the latter. This indicates that the distribution of centroids has potential to help quantify the

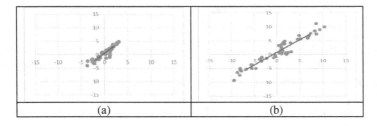

Fig. 2. Clustering results from trembling patterns. (a) mild; (b) severe.

degree of tremor, especially under the similar trembling directions.

After finding the clusters' centroids, we can use linear regression method to find a straight line that represents the principal trembling direction. Assuming p is the slope and d is the intercept of the linear regression line, then we have:

$$Y = pX + d + U, \tag{1}$$

where U is the error between Y and the linear regression model. Then we can use the least mean squared error, $Q(p, d)$, to optimize the regression line as follows:

$$Q(p,d) = \frac{1}{n}\sum_{i=1}^{n}(y_i - px_i - d)^2 = \frac{1}{n}\sum_{i=1}^{n}U_i^2. \tag{2}$$

Based on the three centroids shown in Fig. 2(a) we found the regression line as follows:

$$y = 1.3159x + 0.5693. \tag{3}$$

From the regression line (plotted in red) shown in Fig. 2(a) it had a slope of $p = 1.3159$ which corresponded to an angle of 52° with respective to the horizontal line. Besides that, an intercept of $d = 0.5693$ implied that the subject had a forward trembling pattern.

For comparison, using the three centroids shown in Fig. 2(b) the regression line was found to be:

$$y = 0.9053x + 0.7852. \tag{4}$$

From the regression line (plotted in red) shown in Fig. 2(b) it had a slope of $p = 0.9053$ which corresponded to an angle of 42° with respective to the horizontal line. In the case where an intercept of $d = 0.7852$ implied that the subject had a severe forward trembling pattern than the former case.

For comparisons, we also performed the experiments on the left-right trembling directions. It is obvious that the slopes found from the left-right trembling directions were smaller than those in Fig. 2. This indicated that the subject's trembling patterns showed a tendency towards the left-right direction.

These two illustrated cases indicated that regression lines can provide valuable quantitative information on the trembling directions and magnitudes. This is quite useful for physicians to measure the rehabilitation progress on Parkinson's disease subjects. Next we will extend the results from regression model to group similar patterns of tremor by using the algorithm of self-organizing maps (SOM).

Based on the three centroids from clustering the trembling signals we can find a regression line to identify the trembling direction of a subject. In our study we found that even from the same subject the measured data may not have consistency but did not show significant difference from time to time experiments. Therefore, a careful evaluation on the trembling patterns may help discover the overall tendency of a subject.

We used the nntool in Matlab toolbox to train the 4×4 SOM that contained 8 inputs, including 2-dimensional coordinates of three centroids, and slope and intercept of the regression line. We are interested in recognizing the hand trembling patterns in two perspectives: direction and severity. In our experiments, subjects' trembling directions were subdivided into left-right (LR), 45°, and forward-backward (FB). As for the severity of tremor it was grouped to mild (ML), middle (MD) and severe (SE). As a result, there are nine combinations of direction and severity for training the SOM model as shown in Table 1.

Eighteen trembling patterns chosen from different subjects were used to train the SOM model that gave the results as shown in Fig. 3. The numerals in the 4×4 SOM map indicated the number of patterns grouped into the node. Distribution of the training patterns on the SOM map was given in Table 2. The results clearly indicated that the forward-backward patterns were grouped into the upper left hand side nodes while the left-right patterns were categorized to the right hand side nodes. As for the 45° direction patterns they were clustered either to the upper part or to the lower part of the map. Besides that, upper part of the nodes contained mild trembling patterns and

Table 1. Trembling patterns for SOM model.

	LR	45°	FB
ML	LR-ML	45°-ML	FB-ML
MD	LR-MD	45°-MD	FB-MD
SE	LR-SE	45°-SE	FB-SE

Table 2. Distribution of training patterns.

FB-ML:1	FB-ML:1	45°-ML:2	LR-ML:2
FB-SE:2	FB-MD:2	n/a	LR-MD:1
n/a	n/a	n/a	LR-MD:1
45°-MD:2	45°-SE:1	45°-SE:1	LR-SE:2

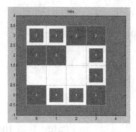

Fig. 3. Training results from 18 patterns by SOM.

intermediate nodes had the middle patterns while the sever patterns were distributed to the bottom part of the map. Based on the training results we can conclude that SOM has the capability to classify the trembling patterns measured from Parkinson's disease subjects.

4 Experimental Results and Analysis

To verify the effectiveness of the proposed system the experiment results were compared from the viewpoint of SOM map. After finding the three centroids from FCM model we can calculate the Euclidean distances D_{left} and D_{right} between the center centroid to the two outmost centroids, respectively. Sum of the two distances, $D_{sum} = D_{left} + D_{right}$, can provide a quantitative measure of trembling severity. As to measure the trembling directions, for the left-right tremor the angle is about 0° and for the forward-backward tremor the angle is about 90°. Slope obtained from the linear regression model provided us an angle of trembling pattern, θ. When the trembling angles were associated with the directions they can help physician to judge the severity of tremor.

Three participants were instructed to walk straight for 30 s. During the experiments it was hard to measure the standard trembling patterns from the subjects. For example, even for the left-right trembling patterns they showed deviations to the desired 0°. As for the severe degrees we are only interested in the relative magnitudes from the same direction. If we compared the magnitudes from different directions they may have larger differences. But these will not affect the identification of trembling patterns.

Table 3. Accuracy of test results.

10-fold cross-validation	Training data	Test data	Accuracy
Test 1	G1, G2	G3	88.9 %
Test 2	G1, G3	G2	100.0 %
Test 3	G2, G3	G1	77.8 %

Our database contained 27 experimental data that included 9 trembling patterns. Data were subdivided into 3 groups (each containing 9 patterns) that had 2 groups' data for training and the remaining group for test. We used 10-fold cross validation to verify the effectiveness of the SOM performance. The accuracy was shown in Table 3. The best classification accuracy was 100 % while the worst one was 77.8 % and the average accuracy was 88.9 %.

After reviewing the three misclassified patterns the errors were from the mild and middle trembling patterns. One left-right middle pattern was misclassified to the left-right mild pattern and two forward-backward mild patterns were grouped to the forward-backward middle pattern. Two forward-backward mild patterns that were erroneously grouped to the forward-backward middle pattern were caused by the intercept from the linear regression line. There was no clear boundary to define the severity from one level to the next. That may cause the SOM to be unable to differentiate the levels of severity. Still, the results indicated that the SOM classification was effective at differentiating hand trembling directions but required deep investigation on recognizing the trembling severity.

5 Conclusions

The designed cane can be easily attached to any commercial cane or umbrella and has many useful functions as pedometer, gait identification, fall detection and trembling measurement. We applied the self-organizing feature maps (SOM) to classify the trembling patterns. The SOM map provided users a visualization interface to recognize the trembling patterns which need intensive care on the side that the users may have high risk to fall over during walking. Experimental results indicated that the SOM classification was effective at differentiating hand trembling directions but required detailed investigation on recognizing the trembling severity.

Although the current designed sensor device can be easily attached to a cane or umbrella, the minimization of the device size so that it can be directly embedded inside the cane is worthy of investigation. Besides that, only 9 different trembling patterns were considered under three major directions and three different levels of severity in this study. More classifications on trembling patterns are under investigation to help PD subjects early identify the direction they have high risk of falling over.

Acknowledgments. This work was supported in part by the Ministry of Science and Technology, Taiwan under Grants MOST103-2221-E-027-122-MY2 and by a joint project between the National Taipei University of Technology and Mackay Memorial Hospital under Grants NTUT-MMH-105-04 and NTUT-MMH-104-03.

References

1. Rigas, G., Tzallas, A.T., Tsipouras, M.G., Bougia, P., Tripoliti, E.E., Baga, D., Fotiadis, D.I., Tsouli, S.G., Konitsiotis, S.: Assessment of tremor activity in the Parkinson's disease using a set of wearable sensors. IEEE Trans. Inf. Technol. Biomed. **16**, 478–487 (2012)
2. Yoneyama, M., Kurihara, Y., Watanabe, K., Mitoma, H.: Accelerometry-based gait analysis and its application to Parkinson's disease assessment—part 1: detection of stride event. IEEE Trans. Neural Syst. Rehabil. Eng. **22**, 613–622 (2014)
3. Yoneyama, M., Kurihara, Y., Watanabe, K., Mitoma, H.: Accelerometry-based gait analysis and its application to Parkinson's disease assessment—part 2: a new measure for quantifying walking behavior. IEEE Trans. Neural Syst. Rehabil. Eng. **21**, 999–1005 (2013)
4. Milosevic, M., McConville, K., Sejdic, E., Masani, K., Kyan, M., Popovic, M.: Visualization of trunk muscle synergies during sitting perturbations using self-organizing maps (SOM). IEEE Trans. Biomed. Eng. **59**, 2516–2523 (2012)
5. Zhang, Y., Pan, G., Jia, K., Lu, M., Wang, Y., Wu, Z.: Accelerometer-based gait recognition by sparse representation of signature points with clusters. IEEE Trans. Cybern. **45**, 1864–1875 (2015)
6. Hsu, Y.-L., Chung, P.-C., Wang, W.-H., Pai, M.-C., Wang, C.-Y., Lin, C.-W., Wu, H.-L., Wang, J.-S.: Gait and balance analysis for patients with Alzheimer's disease using an inertial-sensor-based wearable instrument. IEEE J. Biomed. Health Inf. **18**, 1822–1830 (2014)
7. Kurihara, Y., Watanabe, K., Yoneyama, M.: Estimation of walking exercise intensity using 3-D acceleration sensor. IEEE Trans. Syst. Man Cybern. Part C Appl. Rev. **42**, 495–500 (2012)
8. Mercado, J., Chu, G., Imperial, E., Monje, K., Pabustan, R., Silverio, A.: Smart cane: instrumentation of a quad cane with audio-feedback monitoring system for partial weight-bearing support. In: Proceedings of IEEE International Symposium on Bioelectronics and Bioinformatics, pp. 1–4 (2014)
9. Bhatlawande, S., Mahadevappa, M., Mukherjee, J., Biswas, M., Das, D., Gupta, S.: Design, development, and clinical evaluation of the electronic mobility cane for vision rehabilitation. IEEE Trans. Neural Syst. Rehabil. Eng. **22**, 1148–1159 (2014)

Research on Freshness Detection for Chinese Mitten Crab Based on Machine Olfaction

Peiyi Zhu$^{(\boxtimes)}$, Chensheng Chen, Benlian Xu, and Mingli Lu

School of Electrical and Automation Engineering,
Changshu Institute of Technology, Hushan Road, Changshu, Jiangsu, China
zhupy@cslg.edu.cn

Abstract. Aquatic products freshness detection is an important topic in the current issue of food quality and safety. In this paper, we presented an automatic device based on electronic nose for evaluation freshness of Chinese mitten crab. The crabs were stored at 4 °C for nine days. Electronic nose sensor responses of each sensor over the array were collected from the living crab samples in parallel with data from microbiological analysis for total volatile basic nitrogen (TVB-N). Qualitative interpretation of response data was based on sensory evaluation discriminating samples in three quality classes (fresh, semi-fresh, and spoiled). Principal component analysis (PCA), linear discriminant analysis (LDA), kernel principal component analysis (KPCA) and Laplacian Eigenmap (LE) were developed to classify crab samples in the respective quality class with response data. Experiment results indicated that LE outperform other methods and achieve the highest recognition accuracy for crabs with three quality classes.

Keywords: Electronic nose · Chinese mitten crab · Freshness · Laplacian Eigenmap

1 Introduction

Chinese mitten crabs contained abundant nutrient elements such as protein and multivitamin which were beneficial to human health. The consumption amount of crab had been increased significantly in recent years. In the condition of food security issues had become increasingly frequent occurrence, it was crucial to ensure the quality of crab for consumers. However, crabs deteriorate rapidly in state of dying and postmortem as a consequence of microbial breakdown mechanisms [1]. So freshness is an important factor to be considered to evaluate the quality attribute of crabs when consumed [2].

Among the traditional methods for meat freshness detection including sensory [3], chemical [4] or microbiological methods [5]. However, these reliable and accurate methods were not suitable for fast detection of freshness due to the fact that they were commonly laborious, destructive and unable to evaluate a large number of samples [6], and we didn't find any method which was proposed for living crab freshness detection in literature. Our research aimed at using automatic freshness detection device combined with seven metal semiconductor sensor array, then PCA, LDA, KPCA, LE algorithms were developed to classify whole living crab samples in the respective quality class with response data.

Y. Tan et al. (Eds.): ICSI 2016, Part II, LNCS 9713, pp. 223–230, 2016.
DOI: 10.1007/978-3-319-41009-8_24

2 Device and Materials

2.1 Automatic Freshness Detection Device Design

The automatic freshness detection device was proposed based on mechanism of bionic olfactory and static headspace. The main components of the device included acquisition platform, sensor array, data acquiring and signal conditioning, which is showed in Fig. 1.

For the purpose of whole living detection for crabs, the design of acquisition platform took gas consumption into consideration, that was, shorten the length of the pipeline and test chamber used for collecting gas. Selecting appropriate sensor array combined with characteristic odor volatilized by the living crabs could greatly improve the performance of the device. The specific models and target gas of sensors mainly included Figaro series and MQ series as shown in Table 1.

The distribution of sensor array was placed in same cross-section in consequence of their absorption. These seven sensors converted smell concentration information into measurable electrical signal which were collected by USB6341 data acquisition card through pickup circuit [7]. The sensor response recorded by a computer was sampled as

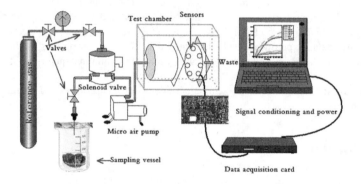

Fig. 1. Schematic diagram of the automatic freshness detection device design.

Table 1. Sensor models and its sensitive gas, detection ranges

Sensor models	Sensitive gas	Detection ranges
MQ3	Ethanol vapor	10–1000
MQ135	Ammonia, sulfide, benzene vapor	10–1000
TGS822	Alcohol, organic solvents	50–5000
TGS2600	Air quality	1–30
TGS2602	Volatile organic compounds	1–30
TGS2620	Organic solvents or other volatile gases	50–5000
QS-01	Peculiar smell	1–100

voltage (V) and its changes relied on odor concentration caused by freshness of crabs. Signal conditioning circuit including amplification and filter was set up for processing weak signal in living state. The MATLAB 2011(b) software was used for data acquisition and analysis.

2.2 Sample Preparation

The crabs weight about 105 ± 5 g were purchased from trading market of Yangcheng lake in Suzhou City, Jiangsu Province. The samples used in the experiment were initially as fresh as possible and placed in foam box equipped with ice. After that, they were immediately stored in refrigerator and kept at a constant temperature of 4 °C. Eight parallel samples which need warming two hours out from refrigerator were set up every day. Each parallel sample performed same operation until the ninth day.

The relevant parameters of experiment were set in connection with actual situation. During the time that sensor array were heating, crab samples back to room temperature were placed in beakers of five hundred milliliters and sealed for 40 min. The measurement interval of device was 1 s. The device sucked the headspace gases of samples through the sensor array at 600 ml/min for 100 s. Once measurement was completed, sensors would be washed by zero gas for 120 s prior to the next sample measurement, and the available samples were properly stored for later use.

3 Methodology

3.1 Date Pre-processing

It indicated that each sample data was matrix of 7 * 100 from the preset parameters. Furthermore, there were 8 * 7 * 100 data points daily due to 8 parallel samples. The response of each of the 9 days in an array was represented by seven sensors, which resulted in a matrix of 9 * 8 * 7 * 100. In order to decrease data dimensionality some signal analysis tools were used previous to classification process. In this paper data preprocessing including mean filter, baseline processing and removing abnormal data were essential. Flow of the pre-processing was shown in Fig. 2. The procedure of feature selection which directly impacted the accuracy and stability of subsequent established classification model sought most effective feature from the original feature space. How to make the selected feature neither create dimension disaster nor topically reflect the overall odor message was very important. The researchers usually chose mathematical characteristics that can be described quantitatively by computer, such as average value, median, integral, differential, etc. In order to reflect the concentration of odor information and obtain information related to time variable, we ultimately chosen steady-state and transient information combined with the entire response curve, that was, steady-state response and three coefficients acquired from instantaneous curve fitted by polynomial.

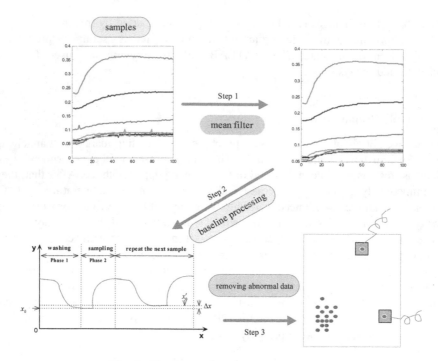

Fig. 2. Key steps of preprocessing procedure

3.2   Feature Extraction Based on LE

Although the nonlinear feature extraction method is the main interest of this paper, three other feature extraction methods were selected for comparison purposes. After the mean filter, baseline processing and removing abnormal data processes, a feature extraction stage was applied to data. Four techniques were applied in total including Principal component analysis (PCA), Linear discriminant analysis (LDA), Kernel principal component analysis (KPCA) and Laplacian Eigenmap (LE).

3.2.1   Principal Component Analysis

Principal component analysis (PCA) was a powerful linear unsupervised pattern recognition method which was usually successfully used in gas sensor applications. Assumed that $F_1, F_2, \ldots, F_m$ were m principal components. If $Var(F_i)$ was maximum, then more information would be included in F_i, namely first principal component. As original variables X was a matrix, its covariance matrix S was firstly calculated by

$$S = \frac{1}{n-1} \sum_{i,j}^{n} (X_i - \bar{X})(X_j - \bar{X}) \tag{1}$$

The ultimate m principal components were finally selected through cumulative variance contribution rate more than 85 %.

3.2.2   Linear Discriminant Analysis

Linear discriminant analysis basic principle was found best projection direction that made multiclass samples could be distinguished. In general, average of various classes of samples u_i, average of all samples u were given by

$$u_i = \frac{1}{n_i} \sum_{x \in classi} x \qquad (2)$$

$$u = \frac{1}{m} \sum_{i=1}^{m} x_i \qquad (3)$$

where, n_i and m were the number of samples of i category and all samples, then dispersion matrix S_b, S_w could be calculated as follows

$$S_b = \sum_{i=1}^{c} n_i (u_i - u)(u_i - u)^T \qquad (4)$$

$$S_w = \sum_{i=1}^{c} \sum_{x_k \in classsi} (u_i - x_k)(u_i - x_k)^T \qquad (5)$$

Finally, the problem of research the best projection vector was converted into solving eigenvalues of $S_w^{-1} S_b$ matrix.

3.2.3   Kernel Principal Component Analysis

Kernel principal component analysis (KPCA) was based on kernel principle function which projected input space into high dimensional feature space in that mapping data was analyzed by PCA. It was available that KPCA took advantage of nonlinear characteristics and enhance capability of processing nonlinear data. Then, the crucial issue lied in the selection of kernel function and its parameters. Gaussian kernel function and its standard deviation were finally determined depending on the classification performance.

3.2.4   Laplacian Eigenmap

Laplacian eigenmap (LE) was commonly used as a manifold learning algorithm for nonlinear dimensionality reduction. An undirected-entitled graph was utilized to describe a manifold and achieve low-dimensional representation through embedded graph. In other words, the local adjacency relationship of graph which was re-painted in the low-dimensional space were kept. The neighborhood graph in which vertices represented data and edge weights denoted relationships between data were constructed. We connected each sample point x_i with its nearest K points (namely neighborhood $\Gamma(i)$) using K-nearest neighbor method. Thermonuclear function was chosen to determine the size of weights between points. The weight matrix W_{ij} was represented by

$$W_{ij} = \begin{cases} \exp(-\|x_i - x_j\|^2/2 \times \sigma^2) & j \in \Gamma(i) \\ 0 & otherwise \end{cases} \qquad (6)$$

where, thermonuclear width parameter $\sigma = 1$, then eigenmap was built by calculating the generalized eigenvectors of Laplace operator L so that obtain a low dimensional embedding coordinates. Just as

$$Ly = \lambda Dy \qquad (7)$$

4 Experiences and Results

The four charts in Fig. 4 shown that values of the seven sensors changed with the variety and intensity of the volatiles emitted by the crab during storage. Figure 3(a) has shown response of the sensor array before exposure to the volatiles. It could be actually defined as baseline that was reference information relative to other samples. And Fig. 3 (b), (c), (d) has shown the variations of the sensor array for three different storage days i.e. day 1, day 5 and day 9.

All sensors shown a monotonous increase in their response towards the headspace of crabs, however, trend of each sensor were not the same. The phenomenon depended on different sensitivity of individual sensors. As can be seen from Fig. 3(b), sensor responses were intensive, changed little which attributed to odor concentration in primary stage. Figure 3(c), (d) reflected more diversity of sensor curves with the increase of storage time.

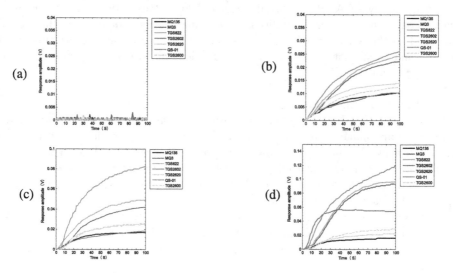

Fig. 3. Response of seven metal semiconductor sensor arrays toward the crab samples during storage: (a) blank sample, (b) day 1, (c) day 5, (d) day 9. (Color figure online)

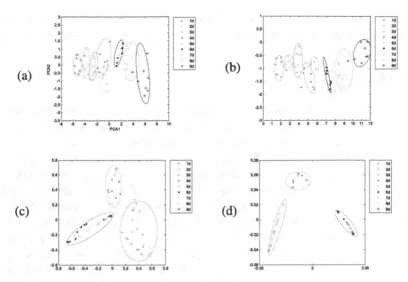

Fig. 4. (a) PCA analysis results, (b) LDA analysis results, (c) KPCA analysis results, (d) LE analysis results (Color figure online)

Next, classification results obtained for Chinese mitten crabs are presented using the database described in Fig. 3. Figure 4 shows the classification results using each one of the four feature extraction methods.

Figure 4 has shown the projections of the experimental results on a two-dimensional plane. The results of samples by PCA analysis, LDA analysis, KPCA analysis and LE analysis were respectively displayed in Fig. 4(a), (b), (c), (d). The first two components PC1 and PC2 captured 97.48 % of data variance from Fig. 4(a). Figure 4(a), (b) revealed that the data of adjacent two days overlapped each other and had no apparent classification. The phenomena mainly related to weak signals measured in whole living detection. This indicated that the method of traditional linear dimension reduction i.e. PCA and LDA were not available for solving classification of crab freshness. Although daily data remained unclassifiable and partial samples intertwined, three distinct categories were emerged from the two-dimensional visual interface of Fig. 4(c), (d). From the perspective of storage time, crab samples can be basically divided into fresh, medium and aged. These three groups that can't be obtained by traditional linear dimension reduction method respectively correspond to samples having undergone up to day 3, samples from days 4 to 5 and samples from days 6 to 9. These classifications were consistent with the measured results of physical and chemical indicator (that is TVB-N). Then we can considered that the classified information acquired by nonlinear dimensionality reduction algorithms were effective. Furthermore, it was available to found that the trend of data just like a manifold over time from Fig. 4(d). It explained that the low-dimensional manifold structure was recovered from the original high-dimensional space by LE. On the whole, nonlinear dimensionality reduction algorithms can better reflect significant classified information with respect to linear dimension reduction algorithms. What is more, introducing LE algorithm into the field of electronic nose and applying it in aquatic freshness pattern recognition would greatly retain nonlinear and linear features contained in raw signals.

5   Conclusions

This paper provided an electronic nose system which is able to automatically detect and evaluate the states of freshness of crabs samples. The automatic freshness detection device was proposed based on the whole living detection. According to the characteristic odor distributed by crab in the living state, we chose the corresponding sensitive sensors. With respect to complex multivariate statistical variables, data preprocessing including mean filter, baseline processing and removing abnormal data were essential. Then we selected steady-state response and three coefficients acquired from instantaneous curve as features. Linear dimension reduction i.e. PCA and LDA and nonlinear dimensionality reduction algorithms i.e. KPCA and LE were used for extracting effective features in favor of categories. We could see from the two-dimensional visual interface that samples of adjacent two days couldn't be completely separated through PCA and LDA. However, it was possible to distinguish among three stages of freshness correspond to fresh crabs (i.e., those stored a maximum of 3 days), medium crabs (i.e., stored 4–5 days) and aged crabs (i.e., stored 6 days or more) by KPCA and LE. The same pattern is occurred as TVB-N analysis, so our method is viable.

Acknowledgments. It is a project supported by the Project of Talent Peak of Six Industries of Jiangsu Province (2014-NY-021) and Qing Lan Project of Jiangsu Province, the Prospective Study in Changshu Institute of Technology (QZ1502).

References

1. Aleman, P.M., Kakuda, K., Uchiyama, H.: Partial freezing as a means of keeping freshness of fish. Bull. Tokai Reg. Fish. Res. Lab. **106**, 11–26 (1982)
2. Guohua, H., Lvye, W., Yanhong, M., et al.: Study of grass carp (ctenopharyngodon idellus) quality predictive model based on electronic nose. Sens. Actuators B **166**, 301–308 (2012)
3. Alimelli, A., Pennazza, G., Santonico, M., et al.: Fish freshness detection by a computer screen photoassisted based gas sensor array. Anal. Chim. Acta **582**(2), 320–328 (2007)
4. Bindu, J., Ginson, J., Kamalakanth, C.K., et al.: Physico-chemical changes in high pressure treated indian white prawn (fenneropenaeus indicus) during chill storage. Innovative Food Sci. Emerg. Technol. **17**, 37–42 (2013)
5. Özogul, Y., Özyurt, G., Özogul, F.: Freshness assessment of European eel (anguilla anguilla) by sensory, chemical and microbiological methods. Food Chem. **92**(4), 745–751 (2005)
6. Dai, Q., Cheng, J.-H., Sun, D.-W., et al.: Potential of visible/near-infrared hyperspectral imaging for rapid detection of freshness in unfrozen and frozen prawns. J. Food Eng. **149**, 97–104 (2015)
7. Jun, Z., XiaoYu, L., Wei, W., et al.: Determination freshwater fish freshness with gas sensor array. Comput. Sci. Inf. Eng. **5**, 221–224 (2009)

Image Classification and Encryption

Texture Feature Selection Using GA for Classification of Human Brain MRI Scans

M. Nouman Tajik[1](✉), Atiq ur Rehman[1], Waleed Khan[1],
and Baber Khan[2]

[1] Department of Electrical Engineering, Comsats Institute of Information
Technology (CIIT), Abbottabad, Pakistan
tajiknomi@gmail.com, khanwaleed247@gmail.com,
atiqjadoon@ciit.net.pk
[2] Department of Electronics Engineering,
International Islamic University (IIUI), Islamabad, Pakistan
baber.khan@iiu.edu.pk

Abstract. Intelligent Medical Image Analysis plays a vital role in identification of various pathological conditions. Magnetic Resonance Imaging (MRI) is a useful imaging technique that is widely used by physicians to investigate different pathologies. Increase in computing power has introduced Computer Aided Diagnosis (CAD) which can effectively work in an automated environment. Diagnosis or classification accuracy of such a CAD system is associated with the selection of features. This paper proposes an enhanced brain MRI classifier targeting two main objectives, the first is to achieve maximum classification accuracy and secondly to minimize the number of features for classification. Two different machine learning algorithms are enhanced with a feature selection pre-processing step. Feature selection is performed using Genetic Algorithm (GA) while classifiers used are Support Vector Machine (SVM) and K-Nearest Neighbor (KNN).

Keywords: Feature selection · Brain MRI · Genetic algorithm · Support vector machine · Classifier · Machine learning · Supervised learning

1 Introduction

Since MRI does not use any ionizing radiation, it is generally safe compared to CT (Computed Tomography) scan. Manual diagnosis of medical images is subjective, time consuming and costly. Looking at an image (visual perception) and interpreting what is seen is often prone to errors due to technician oversights [1]. At the same time due to various imaging constraints and tissue characteristics, automated classification of brain MRI into normal and abnormal studies is also quite difficult.

Classification is an automated process that intends to order every information/ data/instance in specific class, in light of the data portrayed by its features. However, without previous knowledge, useful features cannot be determined for classification. So initially it requires an introduction of large number of features for classification of a particular dataset. Introducing a large number of features may include irrelevant and

© Springer International Publishing Switzerland 2016
Y. Tan et al. (Eds.): ICSI 2016, Part II, LNCS 9713, pp. 233–244, 2016.
DOI: 10.1007/978-3-319-41009-8_25

redundant features which are not helpful for classification and this can even lessen the performance of a classifier due to large search space known as "the curse of dimensionality" [2]. This problem can be subsided by selecting just relevant features for grouping. By omitting irrelevant and unnecessary features, feature selection reduces the training time and minimizing the feature set, thus improving the performance of classifier [3, 4].

During the analysis of tissue in MRI by radiologists, image texture plays a pre-dominant role. In fact texture (in) homogeneity is one of the most common individual MRI features used for tumor diagnosis [5]. Studies have shown that texture information can improve accuracy of classification and produce comparable/preferable results to radiologists when used for machine classification of MRI tissue [6]. Different families of texture calculation methods are being used for MRI analysis and it has been shown that combination of texture features from different families can lead to better classification performance [7].

Feature selection has two principal goals; the first one is to boost the classification performance by minimizing the error rate and the second is to reduce feature set. These goals are paradoxical, and the ideal choice should be made in the vicinity of a tradeoff between them. Treating feature selection as a multi-objective problem can obtain a set of non-dominated feature subsets to meet different requirements in real-world applications. Although GA, multi-objective optimization, and feature selection have been individually investigated frequently, there are very few studies on multi-objective feature selection.

The general objective of this paper is to build up a multi objective approach which can achieve high classification accuracy and at the same reducing the feature set to minimum possible features. The goal is achieved by employing GA for feature selection along with two machine learning classifiers for testing the performance. Classifiers used are SVM and KNN, both of which are tested with different classifier parameters and results are compared based on sensitivity, specificity and accuracy. Feature set is composed of two different texture feature families' i.e Grey-Level Co-Occurrence Matrix (GLCM) and Discrete Wavelet Transform (DWT). Feature set contains total 63 features and among those 51 features are extracted using GLCM and 12 are DWT features.

The paper is presented as follows: Sect. 2: Preprocessing. Section 3: Feature extraction process. Section 4: Feature selection using GA. Section 5: Classification via supervised learning. Section 6: Experimental results Sect. 7: Conclusion.

2 Preprocessing

Brain MRI scans contain some redundant tissues and skull portion with an absence of hard intensity boundaries [8] that needs to be removed in order to achieve better performance. Therefore, to remove the unwanted portion of the scan, three threshold boundaries i.e. Lower(LB), Medium(MB) and Upper boundary (UB) are selected from image histogram as shown in Fig. 1.

The objective is to remove all pixels within LB and UB. Simple binarization technique cannot be used for this purpose because some intensity pixels of unwanted

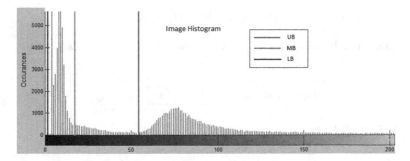

Fig. 1. MRI scan histogram (Color figure online)

skull matter will be left around the brain portion. In order to avoid loss of information in Region of interest (ROI) and to accurately remove the outer strip, the following sequence of steps are followed.

Step 1: (Dividing Image): MRI brain image is divided in two equal halves, shown in Fig. 2(b).

Step 2: (Stripping Outer portion): Using one part at a time, traversing each row of original image $F(x, y)$ from top-left and removing all pixels between MB and UB, A flag $f \in Z_2$ is used to make sure that pixels are removed only from the outer strip ($f = 1$) and not from the ROI ($f = 0$). Once all rows are traversed, the 2nd part is flipped and the same procedure to applied to it. This elimination of pixels makes sure that all intensity values between MB and UB become zero (from outer strip only). Once the outer strip is removed, two portions are merged together. Stripped mask $g(x, y)$ is formed from the original image $F(x, y)$ using (1.1).

$$g(x, y) = \begin{cases} \{(MB < F(x,y) < UB) \& AND \, (f = 1)\}0 \\ \quad\quad else \quad\quad\quad\quad F(x,y) \end{cases} \quad (1.1)$$

Step 3: (Binary Image): Using binary mask $h(x, y)$ from (1.2) a binary image is obtained to eliminate the background pixels, resulting image from step 2 and step 3 is shown in Fig. 2(c).

$$h(x, y) = \begin{cases} LB < g(x,y) < UB & 0 \\ F(x, y) \geq UB & 1 \end{cases} \quad (1.2)$$

Step 4: (Filling Holes): Step 3 result in the loss of pixels from brain area. In order to fill those holes, a matrix of ones $M(x, y)$ of dimension n*n is used where $\{n | n = 2r + 1, r \in \mathbb{N}^+\}$. This mask is applied on all pixels (except the boundary) of image in order to determine its surrounding region. If the area around the pixel is of the majority white, it fills the hole by putting value '1' in that position and value '0' vice versa. Pixel-by-pixel multiplication of $M(x, y)$ with binary image $h(x, y)$ extracts n*n values around single pixel p is represented in (1.3).

$$I(x, y)_{n*n} = M(x, y) * h(x, y) \tag{1.3}$$

Summing up values of $I(x, y)_{n*n}$ provides number of white pixels.

$$\text{Region identifier} = \sum_{k=1}^{n} \sum_{l=1}^{n} I(k, l)_{n*n} \tag{1.4}$$

Holes are filled using (1.5)

$$J(x, y) = \begin{cases} \text{Region identifier} \geq \frac{n*n}{2} & 1 \\ \text{else} & 0 \end{cases} \tag{1.5}$$

Thus, for every pixel p (except boundary) in binary image $h(x, y)$ (1.3), (1.4) and (1.5) are evaluated, resulting image is shown in Fig. 2(d).

Step 5: (Brain Matter Extraction): Once the holes are filled, final gray scale image $S(x, y)$ is obtained by taking product of Binary image $J(x, y)$ and original image $F(x, y)$ using (1.6), result is shown in Fig. 2(e, f).

$$S(x, y) = J(x, y) * F(x, y) \tag{1.6}$$

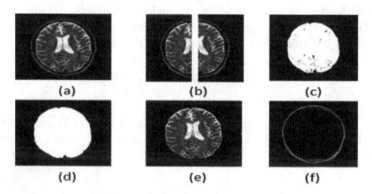

Fig. 2. Preprocessing steps (a) Original scan (b) Dividing image (c) Stripping and binarization (d) Holes Filling (e) Brain Matter (f) Stripped portion

3   Feature Extraction

3.1   GLCM (Gray-Level Co-occurrence Matrix)

GLCM is used to represent the statistical texture features over a spatial domain. It has been proven to be a very powerful tool for image segmentation [9]. Features produced using this method is known as Haralick features, after Haralick et al. [10]. Co-occurrence

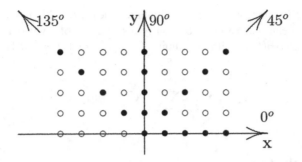

Fig. 3. GLCM angles

matrix is often formed by using two offsets i-e distance (d = 1, 2, 3...) and angle (θ = 0, 45, 90, and 135). For a given distance d, four angular GLCM are calculated as shown in Fig. 3.

The number of gray levels is reduced to avoid the computation cost because GLCM are sensitive to size of texture samples. Sixteen gray levels for GLCM are computed which are sufficient to differentiate brain MRI textures [11]. To avoid direction dependency, Haralik also suggested using the angular mean $M_T(d)$, variance $V_T^2(d)$ and range $R_T(d)$ of GLCM given in (2.1), (2.2) and (2.3) respectively.

$$M_T(d) = \frac{1}{N_\emptyset} \sum_\emptyset T(d, \emptyset) \tag{2.1}$$

$$R_T(d) = Max[T(d, \emptyset)] - Min[T(d, \emptyset)] \tag{2.2}$$

$$V_T^2(d) = \frac{1}{N_\emptyset} \sum_\emptyset [T(d, \emptyset) - M_T(d)]^2 \tag{2.3}$$

Where $N_\emptyset$ represents the number of angular measurements and $T(d, \emptyset)$ are scalar texture measures.

A total of 17 GLCM features proposed by Haralik are initially computed for d = 1 at four different θ values i.e. 0, 45, 90, and 135. $M_T(d)$, $V_T^2(d)$ and $R_T(d)$ are computed from the extracted features, so a feature vector of 51(17 × 3) GLCM features is obtained.

3.2   Wavelet Transforms (WT)

WT provides a multi-scale analysis of an image such as Information of Horizontal (HL), Vertical (LH), Diagonal (HH) and Approximation (LL). The advantage of DWT compared to Fourier transform is that it provides both frequency and temporal details [12].

Haar, Daubechies and Symlets are used to assess the relating transforms. Total of twelve features are extracted (Four features for each wavelet). Two out of four features are the mean and variance of energy distribution of level one 2D transform and other two features are mean and variance of energy distribution of level two 2D transform.

For a sub band $b(l, h)$ with the limits $1 < l < L$ and $1 < h < H$ the energy is calculated as:

$$Energy = \frac{1}{LH} \sum_{l=1}^{L} \sum_{h=1}^{H} |b(l, h)|^2 \tag{2.4}$$

4   Feature Selection

The main assumption when using feature selection is that there are a lot of redundant or irrelevant features which sometimes reduces the classification accuracy [13]. Features are evaluated against a fitness function, thus selecting the best rated features among the feature set.

A feature selection is an operator f_s which maps from m dimensional (input) space to n dimensional (output) space given in mapping.

$$f_s : R^{r \times m} \mapsto R^{r \times n} \tag{3.1}$$

Where $m \geq n$ and $m, n \in Z^+$; $R^{r \times m}$ is matrix containing original feature set having r instances; $R^{r \times n}$ is a reduced feature set containing r instances in subset selection.

4.1   Discrete Binary Genetic Algorithm

GA is a heuristic process of natural selection which is inspired from the procedures of evolution in nature. This algorithm uses the Darwin's theory "Survival of fittest" motivated by inheritance, mutation, selection, and crossover. In comparative terminology to human genetics, gene represent feature, chromosome are bit strings and allele is the feature value [14]. From algorithm perspective, population of individuals represented by chromosomes are the arrangement of binary strings in which each bit (gene) represents a specific feature within a Chromosome (bit strings). Chromosomes are evaluated using Objective function (fitness function) which ranks individual chrome by its numerical value (fitness) within a population.

Process of GA

Step 1 (Generation begins): A random population $(a_{11}a_{12} \ldots a_{nm})_2$ matrix p of size n x m is generated using population size n and number of features m shown in (3.2).

$$p = \begin{bmatrix} a_{11} & \cdots & a_{1m} \\ \vdots & \ddots & \vdots \\ a_{n1} & \cdots & a_{nm} \end{bmatrix} \tag{3.2}$$

Where $m, n \in Z_2$

Step 2 (Tournament): This phase selects the best-fit individuals for reproduction. Two chromosomes (parent chromes) with the highest fitness will take part in cross over.

Step 3 (Cross Over): Analogous to biological crossover, it is the exchange of bits within the selected parents to produce offspring. Number of bits b selected from parent p_n is computed using (3.3), where parameter: $0 < k < 1$ is a crossover probability.

$$b = kx\ p_n \tag{3.3}$$

Step 4 (Mutation): Mutation refers to the change (growth) in the genome of chromosome, flipping of bit strings (genes) of chromosome.

Step 5 (Fitness Evaluation): Analogous to "survival of fittest", chromosomes with a certain level of fitness will survive for next generation while the others whose fitness is less than the threshold value will be discarded.

5 Classification via Supervised Learning

Classification is the method of identification, discrimination of objects or patterns on the basis of their attributes. It is done using supervised learning. In this type of machine learning, machine classifies objects on the basis of previous knowledge. The system is trained using some attributes (features) along with their label, these attributes are used by classifier to guess the unknown objects.

Two classifiers are used here viz. KNN and SVM.

5.1 K Nearest Neighbor Classifier—KNN

It is a simple algorithm that stores all the available patterns/cases and classifies new patterns/cases based on distance function. In binary classification, there are only two classes $\{C_i, C_j\}$; a new unknown case U_n is classified as C_i if the majority of observation specified by parameter k is C_i and vice versa. The parameter k is user-adjustable parameter such that $\{k = 2r + 1, r \in Z^+\}$. The Euclidean distance d between two points a and b in the plane with coordinates (x, y) is given by (4.1)

$$d^2 = (x - a)^2 + (y - b)^2 \tag{4.1}$$

KNN faces genuine difficulties when pattern of distinctive classes overlap in vector space, and sometimes it shows ignorance when it arranges patterns on the premise of checking more number of neighbors which are far separated than the least number of neighbors which are more close together [15]. Three different values of k (1, 3 and 5) are tested for KNN classifier.

5.2 Support Vector Machine—SVM

A supervised classifier, previously, found best for soft tissues classification by Juntu et al. [16]. It is also found to be the second most efficient classification method in all 17 families available today [17]. A hyperplane is made by this classifier in high dimension space which is used as a boundary to classify patterns. A good SVM model is the one which has its hyperplane at quite a large distance from the input data. In this paper, Linear and Polynomial (degree 3) kernel is used for testing SVM accuracy.

6 Experimental Results

Project is carried out on Intel(R) (Core(TM) i5-4530 s CPU @ 2.30 GHz, with 4.00 GB of RAM).MATLAB 8.1.0 (R2013a) is used for simulation. A set of 60 images of size (256 × 256) with a format of Portable Network Graphics (png) are taken from Harvard Medical School website http://www.med.harvard.edu/AANLIB/ among which 20 are normal and remaining 40 are abnormal scans. The abnormal scans consist of three diseases viz. glioma, visual agnosia and meningioma. Samples from dataset are shown in Fig. 4.

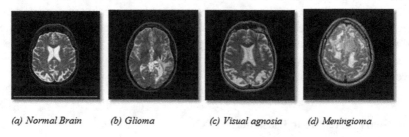

(a) Normal Brain   (b) Glioma   (c) Visual agnosia   (d) Meningioma

Fig. 4. Sample images from dataset

Total number of features extracted from both families is:

Number of features:
GLCM = 51
Wavelet = 12
Total = 63

Configuration used for GA is shown in Table 1. Validation is done using 10-fold cross Validation; the data set is divided into ten equal parts (60/10) where each part is known as fold. Nine folds (i.e. 54 images) are used for training and the remaining one fold (i.e. 6 images) is used for testing. The process is repeated 10 times taking one fold from dataset in each iteration to acquire mean accuracy.

Algorithm pseudocode

Input : *[Population size 'n' , Number of features 'm' , Generation_Limit , Offspring_Limit, Training data, Testing data]*
Ouput : *[Best Chromosome, Accuracy]*
Generation = 1; k = 1;
While Generation < Generation_Limit, do
    Step 1: *P(x,y) <--- Generate_Pop(n,m)*   // *Generate Binary Population Matrix P(x,y)*
    While offspring < Offspring_Limit, do
        Step 2A: *[s1,s2,s3,s4] <--- Rand_4_Chrome(P(x,y))*   // *Pick Random 4 solutions from P(x,y)*
        Step 2B: *[P1, P2] <--- Tornament(s1,s2,s3,s4)*   // *Pick Best 2 solutions (i.e. Parent chromes)*
        Step 3: *[P1', P2'] <--- Cross_Over(P1,P2, C_p)*   // *Cross the Parents Chromes with Cross probability C_p*
        Step 4: *[P1",P2"] <--- Mutation(P1', P2', M_p))*   // *Change the Offspring Chrome with specified probability*
    M_p
        [P(k,m),P(k+1,m)] <--- [P1",P2"]   // *Put the offspring chromes in Population Matrix P(x,y)*
        k = k+2   // *Incriment P(x,y) index for future offsprings*
        Offspring++   // *Incriment the offspring counter*
  Step 5: *Best_Chrome <--- Fitness_Evaluation(P(x,y))*   // *The best solution in Population Matrix for Generation N*
  Generation ++   // *Incriment the Generation counter*
  k = 1   // *Initialize k for every new Generation*
Step 6: *Mean_Accuracy <--- Ten_fold(Best Chrome)*   // *Evaluating Mean accuracy using ten-fold validation*

Classification accuracies of selected features using GA-SVM and GA-KNN with different classification parameters are shown in Table 2.

Table 1. Genetic algorithm configuration

Parameters	Value
Population size	500
Genome length	63
Population type	Binary string
Fitness function	KNN/SVM
Number of generations	40
Offspring	500
Cross over	Uniform Crossover
Mutation	Bit inversion
Cross over probability	0.5
Mutation probability	0.2
Selection scheme	Tournament of size 2

Table 2. Classification accuracy of GA-SVM and GA-KNN

Sr	No. of features	SVM kernel	Accuracy	Sensitivity	Specificity
1	3	Linear	98.34 %	95.23 %	100 %
2	12	Linear	96.67 %	100 %	95 %
3	2	Polynomial degree 3	96.67 %	100 %	95 %
4	15	Polynomial degree 3	95 %	90 %	97.5 %
Sr	No. of features	KNN value of 'K'	Accuracy	Sensitivity	Specificity
1	3	K = 1	93.34 %	90 %	95 %
2	2	K = 3	86.6 %	75 %	92.5 %
3	5	K = 5	96.67 %	100 %	95 %

A comparison between selected and isolated texture families in terms of classification accuracy is shown in Table 3. Optimum selected features for both classifiers are shown in Tables 4A and 4B.

For visualation purpose, we had used scatter plot to show how the data are correlated to their respective classes. The plot is made using three isolated features which are deduced from the GA-SVM. Samples are concentrated in the vicinity of its class-type which shows how effectively the data is classified by the SVM (linear), shown in Fig. 5.

Table 3. Isolated feature families vs. selected features

Texture family	KNN k = 1	KNN k = 3	KNN k = 5	SVM polynomial degree 3	SVM Linear
All features (GLCM + Wavelet)	71.67 %	70 %	70 %	96.67 %	96.67 %
GLCM features	86.67 %	76.67 %	75 %	92 %	96.67 %
Wavelet features	71.66 %	66.67 %	66.67 %	81.67 %	80 %
Reduced features	93.34 %	86.6 %	96.67 %	96.67 %	98.34 %

Table 4A. Optimum selected features for KNN

GA-KNN (k = 1)	GA-KNN (k = 3)	GA-KNN (k = 5)
(i) Difference entropy (mean) (ii) Mean Y(mean) (iii) Standard deviation X (range)	(i) Variance (range) (ii) Entropy (range)	(i) Cluster prominence (mean) (ii) Entropy (mean) (iii) Mean of energy distribution of 2D Haar transform (iv) Mean of energy distribution of 2D Daubechies 5 transform (v) Mean of energy distribution of 2D Symantic transform
Three/93.34 %	Two/86.6 %	Five/96.67 %

Table 4B. Optimum selected features for SVM

GA-SVM polynomial degree 3	GA-SVM linear
(i)Homogeneity(mean) (ii)Sum average (mean)	(i)Difference entropy(mean) (ii)Mean Y(mean) (iii)Deviation X(range)
Two/96.67 %	Three/98.34 %

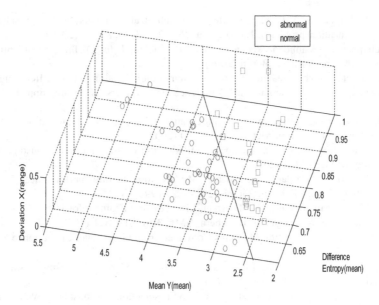

Fig. 5. Scatter plot of normal and abnormal scans using only three features (Color figure online)

7 Conclusion and Future Work

Combining two different texture families for classification has enhanced the accuracy of the classifier. Results reveal the superiority of combination of texture families over the isolated texture family for classification. Similarly feature selection process has also enhanced the classification results. It is clearly seen, that without feature selection the classifier performance is weak, when compared with classification after feature selection. GA-SVM has improved classification accuracy using least number of features. The proposed work can be extended to classify different abnormalities in brain, for example, Alzheimer's disease, visual agnosia, Glioma with tumor, Herpes encephalitis with a tour, bronchogenic carcinoma and Multiple sceloris with a tour. We encourage our readers to analyze the method and results in the proposed paper to help us set a better approach, thereby, reducing computational cost and further increase mean-accuracy for the classification of T2-Weighted Human Brain MRI scans.

References

1. Castellino, R.A.: Computer aided detection (CAD): an overview. Cancer Imaging **5**, 17–19 (2005)
2. Gheyas, I.A., Smith, L.S.: Feature subset selection in large dimensionality domains. Pattern Recognit. **43**(1), 5–13 (2010)
3. Dash, M., Liu, H.: Feature selection for classification. Intell. Data Anal. **1**(1–4), 131–156 (1997)

4. Unler, A., Murat, A.: A discrete particle swarm optimization method for feature selection in binary classification problems. Eur. J. Oper. Res. **206**(3), 528–539 (2010)
5. De Schepper, A., Vanhoenacker, F., Parizel, P., Gielen, J. (eds.): Imaging of Soft Tissue Tumors, 3rd edn. Springer, Berlin (2005)
6. Atiq ur, R. et al.: Hybrid feature selection and tumor identification in brain MRI using swarm intelligence. In: 11th International Conference on Frontiers of Information Technology (2013)
7. García, M.A., Puig, D.: Improving Texture Pattern Recognition by Integration of Multiple Texture Feature Extraction Methods, pp. 7–10 (2002)
8. Sidra et al.: Improved tissue segmentation algorithm using modified Gustafson-kessel Clustering for brain MRI (2014)
9. du Buf, J.M.H., Kardan, M., Spann, M.: Texture feature performance for image segmentation. Pattern Recogn. **23**(3/4), 291–309 (1990)
10. Haralick, R.M., Shanmugam, K., Dinstein, I.H.: Textural Features for Image Classification. IEEE Trans. Syst. Man Cybern. B Cybern. **SMC-3**(6), 610–621 (1973)
11. Fritz et al.: Statistical Texture Measures Computed from Gray Level Concurrence Matrices, 5 November 2008
12. Prasad, B.: Speech, Audio, Image and Biomedical Signal Processing Using Neural Networks. Springer, Berlin (2008). 356 p
13. Bruzzone, L., Persello, C.: A novel approach to the selection of robust and invariant features for classification of hyperspectral images. Department of Information Engineering and Computer Science, University of Trento (2010)
14. Sivanandam, S.N., Deepa, S.N.: Introduction to Genetic Algorithms. Springer, Heidelberg (2008)
15. Yao, M.: Research on learning evidence improvement for KNN based classification algorithm. Int. J. Database Theory Appl. **7**(1), 103–110 (2014)
16. Juntu, J., De Schepper, A.M., Van Dyck, P., VanDyck, D., Gielen, J., Parizel, P.M., Sijbers, J.: Classification of Soft Tissue Tumors By Machine Learning Algorithms (2011)
17. Fernández-Delgado, M., et al.: Do we need hundreds of classifiers to solve real world classification problems? J. Mach. Learn. Res. **15**, 3133–3181 (2014)

Spiking Neural Networks Trained with Particle Swarm Optimization for Motor Imagery Classification

Ruben Carino-Escobar[1,2(✉)], Jessica Cantillo-Negrete[1],
Roberto A. Vazquez[2], and Josefina Gutierrez-Martinez[1]

[1] Technological Research Subdirection, Instituto Nacional de Rehabilitacion,
Mexico City, Mexico
ricarino@inr.gob.mx
[2] Intelligent Systems Group, Facultad de Ingeniería, Universidad La Salle,
Mexico City, Mexico
ravem@lasallistas.org.mx

Abstract. Spiking neural networks (SNN) have been successfully applied in pattern classification problems. However, their performance for solving complex problems such as electroencephalography (EEG) classification has not been widely assessed. It is necessary to consider new approaches to select relevant information and for training SNN in order to improve their accuracy when applied to complex data classification. In this paper, we present a novel channel selection and classification method based on SNN trained with Particle Swarm Optimization (PSO) for the classification of EEG signals associated to motor imagery. The proposed method was able to correctly identify the most relevant channels for different motor imagery tasks. The SNN trained with PSO achieved good classification performances for a well-studied public database using a minimal number of EEG channels, showing advantages against other approaches, regarding both performance and system requirements.

Keywords: Pattern recognition · Supervised training · Electroencephalography

1 Introduction

Artificial neural networks (ANN) are mathematical abstractions of the behavior of real biological neurons. ANN are suitable for classification problems, and have been tested for numerous practical applications [1]. To date, three generations of ANN can be identified. The third generation of ANN, called spiking neural networks (SNN), incorporates a simulation time window into the mathematical abstraction of the neuron, which allows these models to achieve a higher degree of biological realism than previous generations of ANN [2]. When SNN are excited with an input stimulus, the output of each neuron emits a spike train generated within the simulation time window of the neuron model. The information contained within these spike trains can be coded in the number of spikes per spike train (rate-coding). In order to solve practical classification problems with SNN a training strategy must be selected. Different approaches for training SNN have been assessed; most of them use the time information of the

© Springer International Publishing Switzerland 2016
Y. Tan et al. (Eds.): ICSI 2016, Part II, LNCS 9713, pp. 245–252, 2016.
DOI: 10.1007/978-3-319-41009-8_26

output spikes trains for adjusting the synaptic weights of the neurons which encompass the SNN [3–5]. Other approaches have used evolutionary computation for training SNN, like the genetic algorithm (GA) [6], artificial bee colony algorithm (ABC) [7], and particle swarm optimization (PSO) [8].

PSO is a promising method for SNN training, since it is an optimization method for which extensive research has been conducted in order to improve its convergence and avoid stagnation [9]. PSO was introduced by Kennedy and Eberhart and is inspired in the behavior shown by bird flocks while searching for food. Each individual in the population, referred to as a particle, explores the problem's search space based on its position and velocity in each generation of the algorithm. Position and velocities of each particle are updated with respect to the best solution achieved so far by the current particle (local search), and the one achieved by all particles (global search) [10].

A practical application that can be benefitted from a SNN trained with PSO is pattern recognition in electroencephalography (EEG). EEG signals are acquired with electrodes placed in the brain scalp, and encompass electrical information of the brain cortex. Different patterns associated to clinical and physiological conditions like epilepsy, movement, motor imagery (MI), cognition and sleep are encoded within the spatial, time and frequency information of the EEG. Particularly, MI can be differentiated using EEG information. Classification of different types of MI, for example: left hand, right hand and feet MI, are used for physiotherapy purposes and as control signals for brain-computer interfaces [11]. Some recent studies have shown the effectiveness of applying PSO for MI classification. For example, Xu et al. applied PSO for finding the optimal frequency and time bands for classification of MI of index fingers [12]. A combination of PSO and GA based k-means clustering for classification of right and left hand MI was used by Tiwari et al. [13]. Wei and Wei used binary PSO for optimizing the frequency bands for CSP feature extraction followed by LDA classification of right foot and hand MI [14]. Atyabi et al. applied PSO for both dimensionality reduction and for selecting spatial and frequency features extracted from the EEG, in order to integrate a subject-independent MI classification scheme [15]. The goal of this paper is to propose and validate a SNN trained with PSO which can classify different classes of MI from EEG signals. Wavelet transform (WT) is used to extract time-frequency features from the EEG signals. In addition, an automatic channel selection method based on spiking neurons is suggested. The proposed methods are tested on a publicly available dataset, and compared with other approaches comprised of up-to-date feature extraction and classification algorithms.

2   Methods

2.1   MI Database from BCI Competition III

Dataset IVa from BCI competition III was employed for testing the proposed methodology, this dataset is publicly available from [16]. The database is comprised of right hand and right foot MI, and information of 5 subjects is used. Each subject performed a total of 280 trials, with each MI trial encompassed by a 3.5 s interval. 118 EEG channels were acquired at 100 Hz, and band-pass filtered from 0.05 Hz to 250 Hz.

2.2   Pre-processing and Feature Extraction

Each EEG channel was band-pass filtered between 8–30 Hz (30th order FIR filter) in order to preserve information of the frequency bands associated to MI. Time-Frequency (TF) information from each EEG channel was extracted by means of WT. Each channel was convoluted with complex Morlet wavelets, as explained by Tallon-Braudy et al. [17]. The wavelet family used was defined by a ratio of 6 (f/σ_f), with f ranging from 8 Hz to 30 Hz using a resolution of 0.5 Hz. Preprocessing and processing of the EEG signals were done using the free toolbox Fieldtrip for Matlab®. Applying WT, 11 sub-bands with a resolution of 2 Hz were extracted in the range of 8-30 Hz, and from the 0.5–2.5 s time interval after the MI tasks onset. Time-frequency power extracted with WT is then averaged so that the information for each channel is contained in a 250 element vector. With this methodology both time and frequency information is extracted for each EEG channel. Figure 1 illustrates the feature extraction procedure.

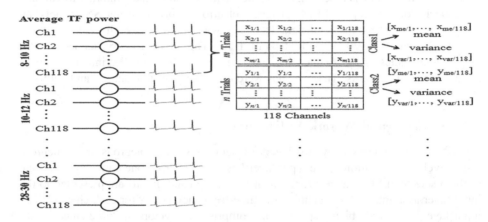

Fig. 1. Channel selection and feature extraction stage

2.3   Spiking Neuron Model

The spiking neuron model described by Izhikevich (IZ) was selected for the SNN. This model has a good biological realism as well as a low computational cost. IZ model is encompassed by two differential equations presented in [18]. Euler's method was used for solving the model and its parameters were set in order to reproduce the behavior of regular spiking neurons [18].

2.4   Proposed Algorithm for Automatic Selection of EEG Channels

The algorithm is based on feature vectors extracted from each EEG channel which encompass the simulation input current of the IZ models. A rate coding scheme is adopted in order to quantify the output of each IZ model. This procedure is repeated for all trials and EEG channels of each subject. This produces a two-dimensional matrix where the number of rows is computed in terms of the number of trials used for training, and the number of columns in terms of the number of channels. Figure 1 illustrates the proposed methodology for EEG channel selection. After the matrixes for each class are computed for each one of the 11 extracted frequency bands, the variances and the mean values per class for all channels are computed. Then a quality index for MI classification is computed for each channel and frequency band according to Eq. 1.

$$Minx_{/ch} = abs\left[x_{mean/ch} - y_{mean/ch}\right] / \left[mean\left(x_{variance/ch}, y_{variance/ch}\right)\right]. \tag{1}$$

where $Minx$ describes the quality for any given channel. A larger value of $Minx$ indicates that the channel has more discriminative information. x and y are the number of spikes counted for each class. After, each of the EEG channel's $Minx$ values for each of the 11 frequency sub-bands are computed, the 3 channels with the highest $Minx$ per frequency band are selected for classification. This makes a total of 33 channels, with a chance of some channels repeating across the frequency sub-bands. In order to validate the capacity of the proposed EEG channel selection algorithm using SN models, an analysis of determination coefficient (r^2) [19] was applied to each channel at the frequency range of 8–30 Hz with steps of 1 Hz. This gives a matrix named feature map, which indicates the frequencies and channels with more discriminative information between MI tasks. The r^2 values were computed for all channels and subjects.

2.5   Spiking Neural Network Architecture

The SNN is encompassed by two layers of neurons with 11 neurons and 1 neuron, respectively. Each neuron in the input layer has 3 inputs, each one corresponds to one of the 3 selected EEG channels for a given frequency band. Inputs are presented to the input layer as a time series containing the time-frequency power for each channel. The input layer outputs are binary spike trains comprised of vectors of 250 elements with values ranging from 0 to 1 (1 if a spike was elicited in the given simulation time step and 0 if otherwise). Each neuron of the input layer has a single output connection to the synapse of the output layer neuron. The output layer neuron's inputs are 11 spike trains. The output of this neuron is also a 250 element binary spike train, and the number of spikes per spike train is computed, which gives the output of the entire SNN. Input current for synaptic connections are computed based on Eq. 2.

$$I_{mn} = Input_{mn} * W_{mn}. \tag{2}$$

where I is the simulation current of the Izhikevich neuron model for the mth neuron located on the nth layer. $Input$ is a vector of 250 elements with the time-frequency power extracted for a given channel, for the case of the input layer. For the case of the output layer it is a vector of 250 binary elements. And W is the synaptic weight of the mth neuron of the nth layer which is element-wise multiplied by the $Input$. An illustration of the proposed SNN architecture is shown in Fig. 2.

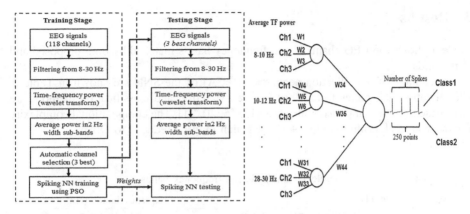

Fig. 2. Method for classification of MI tasks (left) and topology of the SNN (right).

2.6   PSO Implementation for Computing Synaptic Weights of the Proposed SNN

PSO was used for computing synaptic weights so that the SNN could perform MI classification for each subject. The selected PSO algorithm was proposed by Shi and Eberharth [9]. The fitness function is the classification error of the SNN. Equations 3 and 4 encompass the implemented PSO.

$$V_{i+1} = wv_i + c_1 r_1 (PBest - x_i) + c_2 r_2 (GBest - x_i). \tag{3}$$

$$x_{i+1} = x_i + (v_i). \tag{4}$$

where v is the velocity of the particle of the swarm in the ith generation. w is a descending inertia value set from 1 to 0. c_1 and c_2 are positive constants, set to 1 which coupled to r_1 and r_2, balance local and global search. x is a vector of the same size as the input search space, which denotes the position of each particle in the search space. *PBest* and *GBest* are the best position achieved by a given particle and by the swarm respectively. For the present application the search space is comprised of 44 dimensions, linked to the 44 synaptic weights that need to be adjusted in the SNN. The minimum and maximum position (x) and velocity (v) values were set in the -50000 to 50000 range. The value of w ranges from 1 to 0. For the training phase the PSO algorithm was executed with 50 particles during 200 generations.

2.7   Training and Testing of the Database with a SNN

Two classes of MI were classified using the data of five healthy subjects (aa, al, av, aw and ay) of BCI III competition database. The procedure for training and testing the data from each subject is shown in Fig. 2. For each subject's data, classification with the proposed SNN was repeated 10 times with the same training and testing samples as the ones specified by BCI competition III.

3 Results

Table 1 shows the EEG channels and frequency bands matched by both proposed and r^2 methods.

Percentage of classification accuracies (%CA) of the SNN are shown in Table 2. Classification accuracies from other feature extraction and classification methods are also presented for comparison purposes [13, 15, 20, 21]. The computational complexity for the whole algorithm is $O(n) + O(1)$, which may be reduced to $O(n)$.

Table 1. Channels and frequency bands

Subject	Channels	Frequency bands (Hz)
aa	CCP5 CP5 PCP5 CP4 CP3	12–16, 20–28
al	C3 C6 CCP5 CCP6 CCP3	10–30
av	CP5 CCP6 CP3 CP4	8–10, 16–20, 22–24, 26–28
aw	CCP5 CCP6 C4 PCP5 CP3 FC4 PCP3 CFC4	8–12, 16–30
ay	FT10 CP5	12–14, 20–24

Table 2. Performances of proposed methodology with respect to other approaches. CC = Correct Classification

Method	Feature extraction	Classification	% CC (Number of used channels)				
			aa	al	av	aw	ay
SNN + PSO	WT	SNN	74 (11)	84 (9)	66 (16)	68 (18)	62 (22)
Wei et al.	BPSO-CSP	LDA	88 (14)	95 (14)	74 (14)	85 (14)	92 (14)
Yijun et al.	CSP	LDA	95 (118)	100 (118)	80.6 (118)	100 (118)	98 (118)
Yu et al.	PCA	SVM	63 (118)	86 (118)	64 (118)	78 (118)	82 (118)
Atyabi et al.	PSO-DR	SVM, ELM, ANN	63 (12)	74 (12)	58 (12)	70 (12)	68 (12)

4 Discussion

The proposed algorithm for channel selection was able to correctly choose high discriminative EEG channels from different frequency bands. This algorithm does not involve the computation of a covariance matrix as r^2, and therefore can be a suitable alternative for high dimensionality spaces like high density EEG recordings. The proposed SNN achieved classification accuracies above the level of chance (50 %) for all tested subjects. This was achieved with the information of a low number of channels. This shows that the PSO algorithm is suitable for training of SNN applied to real classification problems. The maximum and minimum positions values set in PSO are important for training the SNN, since they set the weights for the SNN and allow the synapses of each neuron to behave as excitatory or inhibitory. The values set in this experiment allow a single input to be large or small enough to produce such kind of behaviors in the neuron models encompassed in the SNN. The channel selection coupled to the SNN, integrates an automatic classification system, capable of simplifying the dimensional search space. The results obtained with the proposed methods may seem low compared to those from other studies. However, for the proposed SNN no manual selection of EEG channels, spatial filters, algorithms for noise reduction like independent component analysis, or elimination of noisy trials were employed [13, 21]. Additionally, unlike the methodology proposed by Yijun et al., no time segments before the trial onset were used [20], and unlike the approach from Atyabi et al. only data from a single subject was needed for classification [15]. The computational complexity of the proposed algorithm is competitive when compared with other approaches, since the most computational demanding segment of it, which involves training or testing the SNN has a fixed computational cost. This is achieved with the channel selection section of the algorithm which generates an input of fixed size for the SNN section. However training the SNN with PSO is still a computational extensive task not suitable for online training. This issue can be addressed by parallelizing both the PSO and SNN implementations. The SNN approach allows to directly identify which EEG channels have the most discriminative information, which is not possible with the compared methods since both CSP and PCA employ a transformation of the original dimension space. This EEG channel information could be useful for clinical purposes such as evaluating brain plasticity and for adjusting the parameters of a BCI training scheme using neuromodulation.

5   Conclusion

The present work shows that SN models coupled to PSO may be used successfully for both channel selection and classification of time-frequency features of the EEG. The method is encompassed by a channel and frequency selection stage, followed by a classification stage, both based on SN models. The proposed method is not limited by statistical restrictions like other approaches, and is capable of achieving competitive results with a small number of automatically selected EEG channels. However training time will impede online training of a BCI with the proposed approach due to the computational intensive process of training the SNN with PSO. For future work a parallel implementation of the algorithm will be done and further offline as well as online tests will be performed for testing the proposed approach for BCI applications.

Acknowledgments. The authors thank CONACYT and Universidad La Salle for the economic support under grant SALUD-2015-2-262061 and NEC-03/15, respectively.

References

1. Dreyfus, G.: Neural Networks Methodology and Applications. Springer, Heidelberg (2005)
2. Gerstner, W., Kistler, W.M.: Spiking Neuron Models. Cambridge University Press, Cambridge (2002)

3. Gutig, R., Sompolinsky, H.: The tempotron: a neuron that learns spike timing-based decisions. Nat. Neurosci. **9**(3), 420–428 (2006)
4. Bohte, S.M., Kok, J.N., Poutré, H.L.: Error-backpropagation in temporally encoded networks of spiking neurons. Neurocomputing **48**, 17–37 (2002)
5. Song, S., Abott, L.F.: Cortical development and remapping through spike timing-dependent plasticity. Neuron **32**, 339–350 (2001)
6. Cachón, A., Vázquez, R.A.: Tuning the parameters of an integrate and fire neuron via a genetic algorithm for solving pattern recognition problems. Neurocomputing **148**, 187–197 (2015)
7. Vazquez, R.A., Garro, B.A.: Training spiking neural models using artificial bee colony. Comput. Intell. Neurosci. (2015). doi:10.1155/2015/947098
8. Garro, B.A., Vázquez, R.A.: Designing artificial neural networks using particle swarm optimization algorithms. Comput. Intell. Neurosci. (2015). doi:10.1155/2015/947098
9. Shi, Y., Eberharth, R.C.: A modified particle swarm optimizer. In: Proceedings of IEEE International Conference on Evolutionary Computation, pp. 69–73 (1998)
10. Clerc, M., Kennedy, J.: The particle swarm-explosion, stability and convergence in a multidimensional complex space. IEEE Trans. Evol. Comput. **6**(1), 58–78 (2002)
11. Alonso-Valerdi, L.M., Salido-Ruiz, R.A., Ramirez-Mendoza, R.A.: Motor imagery based brain-computer interfaces: an emerging technology to rehabilitate motor deficits. Neuropsychologia **79**(B), 354–363 (2015)
12. Xu, P., Liu, T., Zhang, R., Zhang, Y., Yao, D.: Using particle swarm to select frequency band and time interval for feature extraction of EEG based BCI. Biomed. Sig. Process. **10**, 289–295 (2014)
13. Tiwari, P., Ghosh, S., Sinha, R.K.: Classification of two class motor imagery tasks using hybrid GA-PSO based K-means clustering. Comput. Intell. Neurosci. (2015). doi:10.1155/2015/945729
14. Wei, Q., Wei, Z.: Binary particle swarm optimization for frequency band selection in motor imagery based brain-computer interfaces. Bio-Med. Mater. Eng. **26**, S1523–S1532 (2015)
15. Atyabi, A., Luerssen, M.H., Powers, D.: PSO-based dimension reduction of EEG recordings: implications for subject transfer in BCI. Neurocomputing **119**, 319–331 (2013)
16. BCI competition III. http://www.bbci.de/competition/iii/desc_IVa.html
17. Tallon-Baudry, C., Bertrand, O., Delpuech, C., Pernier, J.: Oscillatory gamma-band (30–70 Hz) activity induced by a visual search task in humans. J. Neurosci. **17**(2), 722–734 (1997)
18. Izhikevich, E.M.: Simple model of spiking neurons. IEEE Trans. Neural Netw. **14**, 1569–1572 (2003)
19. Wonnacott, T.H., Wonnacott, R.: Introductory Statistics. Wiley, New York (1977)
20. Blankertz, B., et al.: The BCI competition III: validating alternative approaches to actual BCI problems. IEEE Trans. Neural Syst. Rehabil. **14**(2), 153–159 (2006)
21. Yu, X., Chum, P., Sim, K.B.: Analysis of the effect of PCA for feature reduction in non-stationary EEG based motor imagery of BCI system. Optik-Int. J. Light Electron Opt. **123**(3), 1498–1502 (2014)

Methods and Algorithms of Image Recognition for Mineral Rocks in the Mining Industry

Olga E. Baklanova[1](✉) and Mikhail A. Baklanov[2](✉)

[1] D.Serikbayev East-Kazakhstan State Technical University,
Ust-Kamenogorsk, Kazakhstan
OEBaklanova@mail.ru
[2] Tomsk State University, Tomsk, Russia
baklanov.ma@gmail.com

Abstract. This paper describes development of methods and algorithms of image recognition for mineral rocks. Algorithms of the cluster and morphological analysis to determinate colors and shapes for composition of rocks are described. This approach is actual because of existence of objects with similar color-brightness characteristics, but different shapes or objects that have similar color-brightness characteristics. Preliminary determination of group membership allows reducing the computational complexity of classification. Sorting by group produces color of the object is determined at the stage of segmentation. Few examples of segmentation algorithms in the solving of mineral rock recognition problems are described and discussed.

Keywords: Pattern recognition · Segmentation of color images · Cluster analysis · Morphological analysis · Mineral rocks · Mining industry

1 Introduction

Knowledge of the modal composition of rock or ore is very important to solve the mineralogical and technological issues. Mineralogical and petrographic characteristics of rocks are determined by the macro- and microscopic examination of samples and thin sections [1].

According to the results of macro- and microscopic studies of rock up a summary petrographic characteristics and define the scope of the possible use of the rock [2]. Macroscopic examination of the rock is carried out visually with a magnifying glass or microscope with the following description of ores and rock cores [3].

Microscopic study of rock from thin sections includes a definition [4]:

- Mineralogical composition;
- Quantitative composition of rock-forming minerals;
- Morphology of minerals and their relationships;
- Structure;
- Diagnostic petrographic constants of minerals;

© Springer International Publishing Switzerland 2016
Y. Tan et al. (Eds.): ICSI 2016, Part II, LNCS 9713, pp. 253–262, 2016.
DOI: 10.1007/978-3-319-41009-8_27

- The presence of harmful impurities;
- Availability of secondary structures (newly-formed minerals, veins, and others.) with their quantification;
- Petrographic name of the rock.

In this paper it describes the algorithms for image recognition of mineral rocks, using the methods of cluster and morphological analysis.

2   Materials and Methods

2.1   Methods of Identification of Mineral Rocks Images

Often different minerals on the micrographs correspond to objects of different types of shapes and colors. This allows the identification of various minerals in the form and color of objects. In some cases, also take into account the polarization of certain minerals sample [5]. In this case, it is necessary to make several pictures, accompanying it by turning the sample.

Consider a sample of slag copper anode as an example shown on Fig. 1. Micrographs of this sample were kindly provided Eastern Research Institute of Mining and Metallurgy of Non-ferrous Metals (Kazakhstan, Ust-Kamenogorsk).

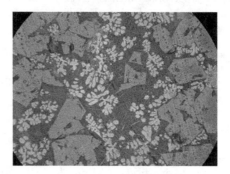

Fig. 1. Micrograph of a sample of slag copper anode, increasing in 500 times

According to experts on microscopy of minerals from Eastern Research Institute of Mining and Metallurgy of Non-ferrous Metals at this picture there is no minerals having dependent on the direction of the plane of polarization of light. In this picture you can detect metallic copper and the following minerals: cuprite Cu_2O, magnetite Fe_3O_4, Delafosse $CuFeO_2$, silicate glass.

Cuprite Cu_2O can be identified as follows: it is characterized by the shape of a round shape, color - it is light gray (sometimes with a slight bluish tint). Figure 2 shows the graphical representation of cuprite.

Fe_3O_4 magnetite on micrographs may also be detected by color and shape. Color of magnetite on micrographs is dark gray. Shape is angular, as expressed by

Fig. 2. Cuprite on micrographs (Color figure online)

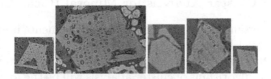

Fig. 3. Magnetite on micrographs (Color figure online)

Fig. 4. Delafossite on micrographs (Color figure online)

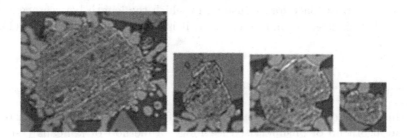

Fig. 5. Metallic copper on micrographs (colour figure online)

technologists, "octahedral". Figure 3 shows magnetite apart from other minerals picture.

Delafossite $CuFeO_2$ micrographs can allocate to the needle shape and gray (with a brownish tint) color. On Fig. 4 it can be seen delafossite on the micrographs.

Metallic copper on the micrographs can be found on the following criteria: color - yellow, shape - round, without flat faces. Figure 5 represents a micrograph metallic copper.

Silicate glass - is a dark gray mass fills the rest of the space that is left of the other minerals.

These data indicate that for real micrographs slag samples (and some other minerals) is possible to use automated qualitative assessment of the mineral composition.

After receiving the full image it is often needed to treat it, mainly to simplify further analysis.

2.2   Methods of Cluster Analysis for Mineral Rocks Images

Clustering - is the automatic partitioning of a set of elements into groups according to their similarity. Elements of the set can be anything, for example, data or characteristics vectors. Themselves groups are also called clusters [6].

In our case, using algorithms of cluster analysis will be the identification of ore minerals by color and texture characteristics of color-coded minerals identified in images taken in reflected light using a microscope [7].

In general, the K-means method segments the image on K different clusters (areas) located far away from each other based on certain criteria [8].

Segmentation method "K- means" is implemented through a two-step algorithm that minimizes the sum of distances "point-to-centroid" obtained by summing over all K clusters. Another words, the purpose of the algorithm is to minimize variability within clusters and maximize variability between clusters [9].

The purpose of cluster analysis - to implement such a partition of the n-dimensional feature space for k-clusters, in which the length between centroids of the resulting clusters would be greatest, it is shown in the expression (1).

$$d_{i,j} \to max, \tag{1}$$

where $d_{i,j}$ - distance between centroids for i-th and j-th cluster, $i = 0, \ldots, k$, $j = 0, \ldots, k$.

In this case, the most appropriate method of solving the problem of clustering is classic algorithm of unsupervised learning - a method of k-means (k-means method). Clustering incrementally in this case is as follows:

1. Specifies the number of clusters K, you want to find.
2. It is randomly selected K vectors ' from the set of vectors in selected space. These vectors are centroids of the clusters on the initial calculation stage.
3. Calculate the distance from each vector space used to each of the obtained centroids in step 2. It can be used metric (2)-(3) to determine the distance.

$$D_{(x,y)k} = \sqrt{\sum_{p=1}^{n} (P_{x,y}^p - P_k^p)^2}, \tag{2}$$

$$D_{(x,y)k} = \sum_{p=1}^{n} |(P_{x,y}^p - P_k^p)|, \tag{3}$$

where:
- (x, y) – coordinates of the observation,
- $k \in [1, K]$ – cluster index,
- n – dimensionality of the used feature space,
- $p \in [1, n]$ – index of the feature observations.

4. Determine the centroid of the cluster to which the distance from the observation is smallest. This cluster matched the observation.
5. Goes through all available vectors and then recalculate centroids for each resulting cluster according (4).

$$P'^n_{(x,y)k} = \frac{1}{S(k)} \sum_{s=1}^{S_k} (P^n_{(x,y)s}), \tag{4}$$

where:
- k - cluster index,
- $S(k)$ - number of observations related to the cluster index k,
- s - indexes of the observations,
- P'^n_k - new value n-th feature of centroid cluster k.

6. Steps 3–5 iterative process stops when the process centroids changes stops or centroids will be fluctuate around some stable values. If the step of centroids change reached a predetermined value also possible to stop iterations.

After completion of the segmentation algorithm described program may provide additional information such as:

- sum of distances "point-to-centroid";
- coordinates of centroid as well as some other data.

Algorithm K-method can converge to a local optimum, when the separation points move any point to another cluster increases the resultant sum of the distances. This problem can be solved only by a reasonable (successful) choice of initial points [10].

2.3   Methods of the Morphological Analysis of Mineral Shapes

Identification of the classification parameters is one of the primary task in pattern recognition [11].

It is offered the following description of the basic model of the object on the basis of morphological features (5–8):

$$M = \langle C, F, G \rangle, \tag{5}$$

$$C = \langle H, Sc, V \rangle, \tag{6}$$

$$F = \langle A \rangle, \tag{7}$$

$$G = \langle S, \beta \rangle, \tag{8}$$

where:

- C - cortege of metrics color of the object;
- F - cortege of morphological metrics of the object;
- G - geometrical metrics of the object;
- H - tone, Sc - saturation; V - value;
- A - number of allocated erosion circles;
- S - area of the object, β - the ratio of the long axis to the short one.

Proposed formalized description is focused on the entire spectrum of morphologically recognizable object parameters.

3   Algorithms of the Morphological Analysis of Mineral Shapes

It is necessary to select the most common forms in order to estimate the form of segments. This will determine the least expensive methods from a computational point of view. Specific minerals meet the objects defined by the color and structure [5].

3.1   Algorithm for Undivided Circles

The basis of the algorithm is the consistent application of morphological operation "erosion" [11] drawn up by the mask of image segments. As a result, a sequence of "erosion" of the image is decomposed into its constituent objects - circumference. It is defined at each stage in the process of "erosion" by the number of objects that are able to separate from each other. When the binary image will be nothing left process ends. The number of circles is defined as the greatest number of distinct objects that were recorded on successive stages of "erosion".

Counting the number of circles has the following algorithm:

1. It is provided binary mask object.
2. Applies morphological operation "erosion" for the binary mask.
3. Counts the number of separate circles. The resulting value is recorded in the computer memory.
4. If the number of segments in the image is greater than zero, then go to step 2.
5. Sequentially retrieves from computer memory and is determined the greatest among them.

Figure 2 shows an image of cuprite. You can see in the figure that the object is characterized adjoining circles.

3.2 Algorithm for Round Shape

The algorithm is based on determining the degree of sphericity. Quantify estimation the form of micro-objects is possible using the following metrics - sphericity coefficient β. Coefficient is focused on determining the ratio of the axes of classified objects. It is enough measurement and the least of the longest axis of the object to calculate the degree of sphericity.

Calculation of the coefficient will be based on (9)

$$\beta = \frac{a}{b}, \tag{9}$$

where:

- a – the size of the short axis (in arbitrary units);
- b – the size of the long axis (in arbitrary units).

Perfectly spherical object is characterized by the fact that the shortest and longest axis are equal, therefore, $\beta = 1$. In the case of elliptical form factor will decrease and tends to 0 as the object degenerates into a line. Figure 5 shows an image of metallic copper, which objects have a rounded shape, without flat edges.

3.3 Algorithm for Elongated Thin Shape

The algorithm is based on the definition of the area object to select objects of small area. Selecting objects of small area is required not only for labelling but also for implementing the screening, due to the fact that besides providing micrographs objects representing technological interest, and also a variety of artefacts small area.

Figure 4 shows an image of delafossite, which objects have a characteristic needle-like shape.

4 Results and Discussion

Nowadays developed automated image recognition system for assessing the qualitative composition of mineral rocks consists of 7 main subsystems [7]:

1. Research and getting micrograph rock.
2. Input and identification micrograph rock.
3. Pre-processing: improving the quality.
4. Definition of image reduction threshold [12].
5. Select the feature vector for cluster analysis.
6. Cluster analysis of color image to determine the mineralogical composition of rocks [10].
7. Morphological analysis of mineral shape to determine the mineralogical composition of rocks.

The search of an object is executed based on combined segmentation while accounting for object classification. For methodologies oriented at the stage of a preliminary search for an object on an image, which precedes classification, segmentation of the entire image using a several variable vector is not effective from the point of computational costs. We cab raise the efficiency if we use methodologies based on preliminary search for an object. If we have an object on an image, the segmentation is only carried out in the window of a preemptively fixed size that includes the found object, this process considerably improves the speed of automated analysis, by the means of exclusion of low-information parts of an image.

The result of the segmentation is shown in Fig. 6. Various minerals marked in different colors. In this case, the metallic copper is red, magnetite - blue cuprite - orange.

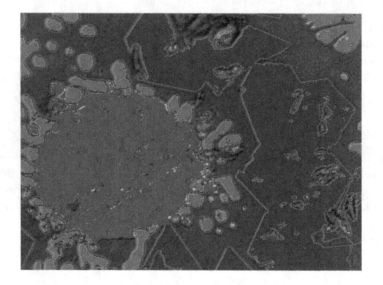

Fig. 6. The result of segmentation. (Colour figure online)

Each cluster includes a certain number of points. Given the ratio of the number of points allocated in each cluster with a number of common points can be displayed relative rates of minerals in rock samples [10]. Various minerals marked in different colors. In this case, the metallic copper is red, magnetite - blue cuprite - orange.

Considered sample has the following content of useful elements:

– Magnetite - 28.45 %;
– Metallic copper - 18.45 %;
– Cuprite - 7.92 %.

5   Conclusion

Petrography is the science that studies the material composition of the rocks. Historically minerals were determined by color and shape. It is possible to classify object's sample by the algorithm of selective classification. A selection here is preliminary separation of sample objects into two groups. This approach is used because of possible presence of objects with similar color-brightness characteristics, but different shapes or objects that have similar color-brightness characteristics. Preliminary determination to the groups allows to reduce the computational complexity of classification. The group depends on color of the object that was determined at the stage of segmentation. Results of studies demonstrate different color spaces by k-means clustering. It was supposed the technique of precomputing the values of the centroids. Methods and algorithms of computer vision of mineral rocks, in particular problems of the algorithm for automatic segmentation of color images of ores, using the methods of morphological and cluster analysis are considered. The program complex in the language C# Visual Studio 2015 was developed for check of results of research.

References

1. Harvey, B., Tracy, R.J.: Petrology: Igneous, Sedimentary, and Metamorphic, 2nd edn. W.H. Freeman, New York (1995)
2. Chris, P.: Rocks and minerals. In: Smithsonian Handbooks. Dorling Kindersley, New York (2002)
3. Clarke, A.R., Eberhardt, C.N.: Microscopy Techniques for Materials Science, 459 p. Woodhead Publishing, CRC Press, Cambridge (2002)
4. ISO 25706-83. Interstate standard. Magnifying glasses. Types, key parameters. General technical requirements. Date of Introduction 1984-01-01. [Electronic resource] (1984). http://docs.cntd.ru/document/gost-25706-83
5. Farndon, J.: The Practical Encyclopedia of Rocks and Minerals: How to Find, Identify, Collect and Maintain the World's Best Specimens, with Over 1000 Photographs and Artworks. Lorenz Books, London (2006)
6. Mandel, I.D.: Cluster analysis. Finance and Statistics, Moscow, 176 p. (1988)
7. Baklanova, O.E., Shvets, O., Uzdenbaev, Z.: Automation system development for micrograph recognition for mineral ore composition evaluation in mining industry. In: Iliadis, L., Maglogiannis, I., Papadopoulos, H. (eds.) Artificial Intelligence Applications and Innovations. IFIP AICT, vol. 436, pp. 604–613. Springer, Heidelberg (2014). doi:10.1007/978-3-662-44654-6
8. Odell, P.L., Duran, B.S.: Cluster Analysis: A Survey. Springer, Heidelberg (1974)
9. Huang, Z.: Extensions to the k-means algorithm for clustering large data sets with categorical values. Data Min. Knowl. Discovery 2, 283–304 (1998)
10. Baklanova, O.E., Shvets, O.Y.: Methods and algorithms of cluster analysis in the mining industry - solution of tasks for mineral rocks recognition. In: Proceedings of the 11th International Conference on Signal Processing and Multimedia Applications (SIGMAP 2014), pp. 165–171. SCITERPRESS Science and Technology Publications, Lda (2014) doi:10.5220/0005022901650171
11. Gonsalez, R.C., Woods, R.E.: Digital Image Processing, 3rd edn. Pearson Education, Upper Saddle River (2011). p. 976

12. Baklanova, O.E., Shvets, O.Y.: Development of methods and algorithms of reduction for image recognition to assess the quality of the mineral species in the mining industry. In: Chmielewski, L.J., Kozera, R., Shin, B.-S., Wojciechowski, K. (eds.) ICCVG 2014. LNCS, vol. 8671, pp. 75–83. Springer, Heidelberg (2014). doi:10.1007/978-3-319-11331-9

Image Encryption Technology Based on Chaotic Hash Function and DNA Splicing Model

Guoyu Lv, Changjun Zhou$^{(\boxtimes)}$, Hongye Niu, and Bin Wang

Key Laboratory of Advanced Design and Intelligent Computing,
Dalian University, Ministry of Education, Dalian 116622, China
zhou-chang231@163.com

Abstract. The security of communication and preservation for digital image has great significance in modern society. In this paper, DNA splicing model was combined with hash function to encrypt grayscale image. In the proposed algorithm, hash function was used to scramble the positions of pixel values from original image. DNA splicing model and XOR operation were used to diffuse pixel values. Additionally, the chaotic sequences which were generated by the chaotic map were used to encrypt grayscale image. Experiment results and simulation analysis have proved that the algorithm has many advantages. Its key space is huge. It has stronger key sensitivity. It is safe enough to resist all types of recognized attacks including exhaustive attacks, statistical attacks and differential attacks.

Keywords: Chaotic system · Hash function · Splicing model · Image encryption

1 Introduction

Digital image has become an important measure to transfer information for human beings, because of its strongly comprehensiveness and intuition. Besides that it is able to convey the image information vividly, so it is widely used by people in daily life. But, some criminals use technical means to steal information by attacking the digital image [1]. For ensure the security of the image, a lot of image encrypt algorithm have been proposed [2–5].

The chaos theory has gotten more and more attention and application in the design of secure communication system based on its sensitivity to initial conditions and parameters, its noise characteristics and so on [6]. It has formed the basic theory of chaotic systems gradually [7]. The application of chaos theory makes the security of image improved in the image encryption field. Hash function is an important part of modern cryptography encryption algorithm, which has let the encrypted data more secure because of his characteristics like compressibility and easy to calculate and the characteristics of its anti-conflict resistance [8]. DNA computing has attracted extensive attention and has been applied to the study of image encryption due to its intrinsic parallelism in recent years [11]. And in this paper, we proposed a new image

Y. Tan et al. (Eds.): ICSI 2016, Part II, LNCS 9713, pp. 263–270, 2016.
DOI: 10.1007/978-3-319-41009-8_28

encryption technology which combined with chaotic system, Hash function and DNA splicing model. Its aim is to get better encryption effect.

2   The Related Works

2.1   Chens Chaotic System

The topological structure of Chens chaotic system is more complex and its dynamic behavior makes it has a better application prospect in the fields of encryption [9], there is its dynamic equation:

$$
\begin{cases}
\dot{x} = a(y - x); \\
\dot{y} = -xz - dx + cy - q; \\
\dot{z} = xy - bz; \\
\dot{q} = x + k;
\end{cases}
\tag{1}
$$

The a, b, c, d, k, of the formula are the parameters of the system, when the $a = 36$, $b = 3, c = 28, d = 16$, and $-0.7 \leq k \leq 0.7$, Chens chaotic system is chaos mapping, it can produce the four chaotic sequence, so the k value was set to 0.4.

2.2   DNA Splicing Model

The DNA splicing model is presented by Head et al. [10], there are a lot of developments of models of DNA computing after that. The use of DNA splicing model in the process of encryption: intercepting chaotic sequences which is generated from the image matrix which is operated by chaotic function, and then the matrix was divided into four parts according to certain rules S1, S2, S3, S4, next the column of each part is a sub sequence T1, T2, T3, T4, after that T1 and T3, T2 and T4 were exchanged of each sub sequence if they meet the conditions, and then four parts are merged into a matrix.

2.3   XOR Operation of DNA Sequence

In this paper the image matrix was transformed into the form of DNA chain through DNA coding rules. The XOR operation is one of the most common ways that used for DNA sequences. XOR rule was used to scramble image, at the same time the XOR operation was used to decrypt image, due to the XOR operation is reflexive. According to the above mentioned encoding rules: C standing for 00, G standing for 11, A standing for 01, T standing for 10, this is the XOR rule of DNA sequence such as shown in Table 1.

Table 1. XOR operation rules for DNA sequence

XOR	G	C	T	A
G	C	G	A	T
C	G	C	T	A
T	A	T	C	G
A	T	A	G	C

3 Image Encryption Algorithm

3.1 Generation of the Key

The key is the encryption parameters of the cryptography and it is one of the important factors that can affect the encryption effect directly.
    Key generation steps:

First step: the initial key was given is "1234567890123456", and the sum of the pixel values of the image was calculated.
Second step: an add operation was done for the initial key. The pixel value of the image and the initial key were added up and then there was a new key.
Third step: the encrypted key was assigned into 4 parts averagely to 4 initial values of chaotic system; the 4 initial values became the initial conditions of Chens chaotic system.

3.2 Specific Progress of Encryption

There is the encryption process in more details:
    First step: original image whose size is (m, n) is transformed into binary matrix A (m, n). The four initial values x_0, y_0, z_0, q_0 that generated from the process of the key generation became initial value of Chens chaotic system.
    $x(x_1, x_2, x_3, \ldots, x_n)$, $y(y_1, y_2, y_3, \ldots y_n)$, $z(z_1, z_2, z_3, \ldots z_n)$, $q(q_1, q_2, q_3, \ldots q_n)$ are gotten, the parameters a, b, c, d, k are the given system parameters.
    Second step: the hash function H (key) = keymod3 model is used to handle image matrix and then the obtained data is saved into the hash table C, it is the key image. And the hash table D became the hash table when the chain address law is in conflict.
    Third step: the hash function is used to scramble the pixel value of the image; specific methods below is used to scramble the pixel value:
    The first number of the first line and the first column of the image matrix A is used to divide 3, and then there was their remainder which is stored in $C(i,j)$, the value of remainder can only be 0, 1, 2. And then there is the hash table data for C, whose i belongs to $1\ldots m$, and j belongs to $1\ldots n$, and the columns of the matrix $D(1, count1)$, $D(2, count2), D(3, count3)$ is used to store the remainder whose value is 0, 1, 2 of pixel values of $A(i,j)$, besides the initial value of $count1, count2, count3$ was 0, and then there is an operation of up.
    When $C(i,j) = 0$, if $D(1, count1) \neq 0$, and then $S(i,j) = D(1, count1)$, $Count1 = Count1 - 1$, S becomes permutation matrix.
    Fourth step: the Table 2 of DNA encoding table is used to encode the matrix which consist of the scrambled pixel value, and then the EA matrix is gotten.
    Fifth step: the value of diffusion of the image would be gotten from XOR operation which is used on the EA image pixel matrices. EA matrix is divided into four parts, and each part of it is bR to deal the EA matrix with the x and y sequences which are generated by Chens chaotic system by XOR operation:

Table 2. Correlation coefficients of two adjacent pixels of the proposed cipher and other schemes

	Plain image			Cipher image		
	Horiz.	Vert.	Diag.	Horiz.	Vert.	Diag.
Our method	0.8902	0.9156	0.9224	−0.0401	−0.0321	−0.0428
In Ref. [3]	0.9707	0.9733	0.9122	0.0012	0.0026	0.0021
In Ref. [4]	0.9831	0.9689	0.9671	0.014	0.0092	0.0051

$$\begin{cases} x(i) = floor((m/4) \times x(i)) & i = 1, 2, \ldots, m/4; \\ y(j) = floor((n/4) \times y(j)) & j = 1, 2, \ldots, n/4; \end{cases} \tag{2}$$

And, $i = 1, 2, \cdots, m/4, j = 1, 2, \cdots, n/4, x(i)$ and $y(j)$ represent for the pixel value of i and j.

The (x, y) and the following formula is used to diffuse bR matrix diffusion:

$$bR(i,j) = bR(i,j) \oplus bR(x(i), y(j)); \tag{3}$$

And, $i = 1, 2, \ldots, m/4, j = 1, 2, \ldots, n/4$, matrix DA is gotten by the operation..

Sixth step: DNA splicing model and chaotic system are used to diffuse pixel value. Firstly, the chaotic sequences was cut, $z(i) = x(i), q(j) = y(j)$; and then the pixel values of the matrix A are divided into four parts $S1, S2, S3, S4$, every column of each part was a sub sequence $T1, T2, T3, T4$, they are dealt with following formula:

$$\begin{cases} MA\{i\} \leftrightarrow MB\{i\}, & if \; x(i) + y(i) < 1 \\ no \; operation, & else \end{cases} \tag{4}$$

$$\begin{cases} MC\{i\} \leftrightarrow MD\{i\}, & if \; z(i) + q(i) < 1 \\ no \; operation, & else \end{cases} \tag{5}$$

And $i = 1, 2, 3, \ldots, n$, there are four new parts $U1, U2, U3, U4$. The four parts are mixed into the new matrix SA; there is the diffusion of the pixel values of image.

Seventh step: the encrypted image can be gotten by using DNA decoding to decode the SA.

The decrypt image process is the inverse process of encrypt image, so I will not elaborate further here.

4   Experimental Results and Analysis of Security

The software for the experiment of cryptography and decryption was MATLAB (R2012a), and the image "Lenna. bmp" was the image of experiment. We used the attacks which were official in the field of image encryption to attack the encrypted image, such as exhaustive attack, statistical attack and differential attack. The proposed algorithm of encryption was confirmed the safety of the attack (Fig. 1).

(a)   (b)

Fig. 1. Experimental results: (a) is the original picture, (b) is the encrypted picture.

4.1 Analysis of the Susceptibility of the Key

In this algorithm, the key space is huge, so image could be encrypted with different keys. The key was used in this paper was "123456789123456", and it had 128 bits so it proved key space was so large that it could resist the exhaustive attack.

The sensitivity of the key is very important to the security of the image. The chaotic system is extremely sensitive to the parameters of the system parameters and initial values. Therefore, the initial key is "123456789123456", and when the original key is changed to "123456799123456", the enemy still cannot get the original image from the encrypted image The experimental results of the key sensitivity shown in Fig. 2.

(a)   (b)

Fig. 2. Key sensitivity test: (a) is the decrypted picture by the right key, (b) is the decrypted picture by the wrong key.

4.2 Analysis of Statistical Attack

Gray histogram is one of the indexes of the evaluation of encryption algorithm's safety. In the evaluation of the function of gray level, the greater differences between encrypted image and the original image is, the similarity between encrypted image and the original image is worse, and the image is safer. When the gray histogram of image encryption is close to a smooth curve, indicating that the effect of resisting statistical attack is better. Figure 3 is the gray histogram of the original image and gray histogram of encrypted image.

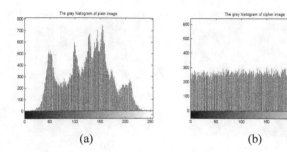

Fig. 3. Gray histogram analysis: (a) is the gray histogram of plain image, (b) is the gray histogram of cipher image.

4.3 The Correlation Between Neighbor Pixels

The correlation of the adjacent coefficients and its coefficients are used to reflect the position of pixel of image. In this paper the correlation of adjacent coefficients is used to analyze the location of the image pixel. The comparative analysis of correlation of original image and encrypted image from the horizontal, vertical and diagonal direction are done by using the following formula:

$$E(x) = \frac{1}{N}\sum_{i=1}^{N} x_i \tag{6}$$

$$\text{cov}(x,y) = \frac{1}{N}\sum_{i=1}^{N} (x_i - E(x))(y_i - E(y)) \tag{7}$$

$$D(x) = \frac{1}{N}\sum_{i=1}^{N} (x_i - E(x))^2 \tag{8}$$

$$r_{xy} = \frac{\text{cov}(x,y)}{\sqrt{D(x)} \times \sqrt{D(y)}} \tag{9}$$

x and y are the gray value of the neighbor pixels, $E(x)$ is mean value, $D(x)$ is variance, $\text{cov}(x,y)$ is covariance.

In the Table 2, the results from Refs. [3, 4] are used to compare with our entropies about "lenna.bmp". Our method has a better performance on the correlation between neighbor pixels.

There are about 1000 pairs of chosen adjacent pixels in the same position in the test of the pixel correlation. The distribution maps of the correlation of the original image and the encrypted image are shown in Fig. 4.

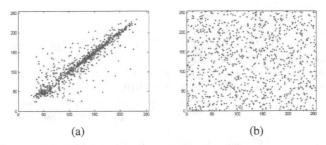

(a)   (b)

Fig. 4. Correlation coefficient analysis: (a) is the correlations of two horizontally adjacent pixels in the plain image, (b) is the correlations of two horizontally adjacent pixels in the cipher image.

4.4 Entropy of Information

Information entropy is a index that is used to a measure the expected value of the appearance of the random variable Following is the formula of information entropy:

$$H(X) = - \sum_{i=0}^{n} P(x_i) \log_2 P(x_i) \tag{9}$$

x_i is the i pixel of the gray image, $P(x_i)$ is the probability of the appearance of x_i, and n is the size of the image.

In the Table 3, the results from Refs. [3, 4] are used to compare with our entropies about "lenna.bmp", our method is better than the Ref. [4], but not as good as the Ref. [3]. So our method is a new image encryption algorithm, it can provide adequate protection for the image.

Table 3. The entropy analysis of the proposed cipher and other schemes

	Our method	In Ref. [3]	In Ref. [4]
Entropy	7.9967	7.9968	7.9939

5 Conclusion

In this paper, we proposed a new digital image encryption algorithm. Its basic idea is: firstly, the Chens chaotic system is used to scramble image. The XOR operation, hash function and DNA splicing model are used to diffuse the values of image pixel to obtain encrypted image. All the analysis of attacks the encrypted image have indicated that this algorithm has a good effect of encryption.

Acknowledgement. This work is supported by the National Natural Science Foundation of China (Nos. 61572093, 61402066, 61402067, 61370005), the Basic Research Program of the Key Lab in Liaoning Province Educational Department (Nos. LZ2014049, LZ2015004).

References

1. Wen, C.C., Wang, Q., Miao, X.N., Liu, X.H., Peng, Y.X.: Digital image encryption: a survey. J. Comput. Sci. **39**, 6–9 (2012)
2. Cheddad, A., Condell, J., Curran, K., Mckevitt, P.: A hash-based image encryption algorithm. J. Opt. Commun. **283**(6), 879–893 (2010)
3. Zhang, Q., Guo, L., Wei, X.P.: A novel image fusion encryption algorithm based on dna sequence operation and hyper-chaotic system. J. Light Electron Opt. **124**(18), 3596–3600 (2013)
4. Wang, X.Y., Wang, X.J., Zhao, J.F., Zhang, Z.F.: Chaotic encryption algorithm based on alternant of stream cipher and block cipher. J. Nonlinear Dyn. **63**, 587–597 (2011)
5. Xu, J., Li, S.P., Dong, Y.F., Wei, X.P.: Sticker DNA computer model – PartII: application. J. Chin. Sci. Bull. **49**(9), 863–871 (2004)
6. Fu, C., et al.: A chaos-based digital image encryption scheme with an improved diffusion strategy. J. Opt. Express **20**(3), 2363–2378 (2012)
7. Elnashaie, S., Abashar, M.E.: On the chaotic behaviour of forced fluidized bed catalytic reactors. J. Chaos Solitons Fractals **80**(3), 13–17 (2011)
8. Xu, J., Dong, Y.F., Wei, X.P.: Sticker DNA computer model – Part I: theory. J. Chin. Sci. Bull. **49**(8), 772–780 (2004)
9. Zhou, K., Wei, C.J., Cheng, Z., Huang, Y.F.: Study of DNA computing model. J. Comput. Eng. Appl. (2009)
10. Head, T., Rozenberg, G., Bladergroen, R.S., Breek, C.K., Lommerse, P.H., Spaink, H.P.: Computing with DNA by operating on plasmids. J. BioSyst. **57**(2), 87–93 (2000)
11. Zhou, C.J., Wei, X.P., Zhang, Q., Liu, R.: DNA sequence splicing with chaotic maps for image encryption. J. Comput. Theoret. Nanosci. **7**(10), 1904–1910 (2010)

Design of a Low-Latency Multiplication Algorithm for Finite Fields

Kee-Won Kim[1] and Seung-Hoon Kim[2(✉)]

[1] College of Convergence Technology, Dankook University, Yongin, Republic of Korea
nirkim@dankook.ac.kr
[2] Department of Applied Computer Engineering, Dankook University, Yongin, Republic of Korea
edina@dankook.ac.kr

Abstract. Finite field arithmetic operations have received much attention in error-control codes, cryptography, etc. Among the basic arithmetic operations over $GF(2^m)$, the multiplication is the most important, complex, and time consuming. In this paper, we present a low-latency finite field arithmetic algorithm for multiplication which is a core algorithm for division and exponentiation operations. In order to obtain a dedicated pipelined algorithm for finite field multiplication based on all-one polynomial, we split its computation into two parts, design the algorithm to simultaneously compute them, and reduce computation delay. Also, we show the detailed derivation process in order to obtain the proposed algorithm for the pipelined computation. Therefore, the proposed algorithm can lead to a hardware architecture which has a considerably low latency. The multiplier applying the proposed algorithm will be a highly modular architecture and be thus well suited for VLSI implementations.

Keywords: Finite fields · All-one polynomial · Multiplication · VLSI

1 Introduction

The finite field $GF(2^m)$ is increasingly important for many applications in cryptography and algebraic coding theory [1–3]. Among the finite field arithmetic operations, the multiplication is the most important. Other time-consuming finite field arithmetic operations such as exponentiation, division, and multiplicative inversion can be carried out by repeated multiplications. However, the multiplication is time-consuming and has a complicated circuitry. Thus, an algorithm of efficient multiplication with low complexity and latency, high throughput rate, and short computation delay is required.

The representation of the field elements plays an important role in the efficiency of the algorithms for the arithmetic operations. There are three main types of bases over finite fields, polynomial (standard) basis, normal basis, and dual basis. Each basis has its distinct advantage. Many architectures of finite field multiplication using various bases have been proposed [4–25].

© Springer International Publishing Switzerland 2016
Y. Tan et al. (Eds.): ICSI 2016, Part II, LNCS 9713, pp. 271–278, 2016.
DOI: 10.1007/978-3-319-41009-8_29

The architectures based on polynomial basis [4–8] have low complexities and their sizes can be easily extended to meet various applications due to their simplicity, regularity, and modularity. The major benefit of the normal basis [9–14] is that the squaring of an element is computed by a cyclic shift in the binary representation.

An efficient multiplication algorithm requires an irreducible polynomial for the construction of the field. For example, $GF(2^m)$ multipliers using some popular polynomials, such as all-one polynomials (AOPs) and trinomials [16–25] have a low circuit complexity.

Wang and Lin [26] proposed a systolic architecture for finite field multiplication, which consists of regularly connected identical cells and requires a latency of $3m$ clock cycles. However, it results from directly unrolling iterative algorithms, does not fully exploit inherent parallelism, and requires a large area and a large latency overhead to be fully pipelined. Lee et al. [22] proposed efficient systolic multipliers with low-latency and low-complexity, defined by all-one and equally spaced polynomials. Recently, Kim et al. [25] proposed a multiplier with lower complexity than Lee et al.'s bit-parallel multiplier [22].

In this paper, we induce an efficient multiplication algorithm for reduction of hardware complexity of typical architectures. The proposed algorithm enables multiplication to operate in pipelined computation so that two modules can be computed in parallel. Also, we design modules to have the low latency. Therefore, our algorithm can lead to a low-latency hardware architecture for the multiplication over $GF(2^m)$.

The remainder of this paper is organized as follows. In Sect. 2, we review a mathematical background and propose a low-latency multiplication algorithm over $GF(2^m)$. Finally, Sect. 3 gives our conclusions.

2   The Proposed Multiplication Algorithm over $GF(2^m)$

2.1   Mathematical Background

A polynomial of the form $G = g_m x^m + g_{m-1} x^{m-1} + \cdots + g_1 x + g_0$ over $GF(2)$ is called an all-one polynomial (AOP) of degree m if $g_i = 1$ for $i = 0, 1, 2, \ldots, m$ [16]. It has been shown that an AOP is irreducible if and only if $m+1$ is a prime and 2 is a generator of the field $GF(m+1)$ [17]. For $m \leq 100$, the values of m for which an AOP of degree m is irreducible are 2, 4, 10, 12, 18, 28, 36, 52, 58, 60, 66, 82, and 100 [22–25].

Suppose that $A = a_m x^m + a_{m-1} x^{m-1} + \cdots + a_1 x + a_0$ and $B = b_m x^m + b_{m-1} x^{m-1} + \cdots + b_1 x + b_0$ are two elements in the field $GF(2^m)$, where both A and B are represented with the extended basis $\{1, x, x^2, \cdots, x^m\}$. According to $x^{m+1} = 1$ and $x^{m+i} = x^{i-1}$, the result of the multiplication A and B over $GF(2^m)$ can be obtained by:

$$P = AB = \left(\sum_{j=0}^{m} a_j x^j \right) \left(\sum_{i=0}^{m} b_i x^i \right)$$

$$= \sum_{i=0}^{m} \sum_{j=0}^{m} a_j b_{\langle i-j \rangle} x^i = \sum_{i=0}^{m} \sum_{j=0}^{m} a_{\langle i-j \rangle} b_j x^i, \tag{1}$$

where $\langle k \rangle$ denotes the residues of k modulo $m+1$.

2.2 The Proposed Multiplication Algorithm

Assume that $A = a_m x^m + a_{m-1} x^{m-1} + \cdots + a_1 x + a_0$ and $B = b_m x^m + b_{m-1} x^{m-1} + \cdots + b_1 x + b_0$ are two elements in the field $GF(2^m)$, where x is a root of the irreducible AOP of degree m. Let k be an integer such that $0 \le k \le m$ and j be $\langle (i+k)/2 \rangle$. Then $j = \langle (i+k)/2 \rangle$ must be $0 \le j \le m$ for $0 \le i \le m$. Thus, we can substitute $j = \langle (i+k)/2 \rangle$ into the subscripts of $a_j b_{\langle i-j \rangle}$ in (1) to obtain:

$$P = AB = \sum_{i=0}^{m} \sum_{k=0}^{m} a_{\langle (i+k)/2 \rangle} b_{\langle (i-k)/2 \rangle} x^i. \tag{2}$$

Splitting the right side of (2) into two terms, we have

$$P = \sum_{i=0}^{m} \sum_{k=0}^{m/2} a_{\langle (i+k)/2 \rangle} b_{\langle (i-k)/2 \rangle} x^i + \sum_{i=0}^{m} \sum_{k=m/2+1}^{m} a_{\langle (i+k)/2 \rangle} b_{\langle (i-k)/2 \rangle} x^i. \tag{3}$$

In the second term on the right side of (3), k is in $m/2 + 1 \le k \le m$. Let j be $k - 1/2$. Because of $m/2 + 1 \equiv 1/2 \bmod (m+1)$, $j = k - 1/2$ must be $0 \le j \le m/2 - 1$. Thus, we can substitute $k = j + 1/2$ into the subscripts of $a_{\langle (i+k)/2 \rangle} b_{\langle (i-k)/2 \rangle}$ in (3) and then obtain:

$$P = \sum_{i=0}^{m} \sum_{j=0}^{m/2} a_{\langle (i+j)/2 \rangle} b_{\langle (i-j)/2 \rangle} x^i + \sum_{i=0}^{m} \sum_{j=0}^{m/2-1} a_{\langle (i+j)/2+1/4 \rangle} b_{\langle (i-j)/2-1/4 \rangle} x^i. \tag{4}$$

In (4), let $P = S + T$ where

$$S = \sum_{i=0}^{m} \sum_{j=0}^{m/2} a_{\langle (i+j)/2 \rangle} b_{\langle (i-j)/2 \rangle} x^i \tag{5}$$

and

$$T = \sum_{i=0}^{m} \sum_{j=0}^{m/2-1} a_{\langle (i+j)/2+1/4 \rangle} b_{\langle (i-j)/2-1/4 \rangle} x^i. \tag{6}$$

From (5), let

$$C^{(j)} = \sum_{i=0}^{m} a_{\langle (i+j)/2 \rangle} b_{\langle (i-j)/2 \rangle} x^i. \tag{7}$$

Then

$$S = \sum_{j=0}^{m/2} \sum_{i=0}^{m} a_{\langle (i+j)/2 \rangle} b_{\langle (i-j)/2 \rangle} x^i = \sum_{j=0}^{m/2} C^{(j)}. \tag{8}$$

The recurrence equation of S from (8) can be formulated as follows:

$$S^{(j)} = S^{(j-1)} + C^{(j-1)}, \tag{9}$$

where $S^{(-1)} = 0$, $C^{(-1)} = 0$, and $0 \leq j \leq m/2 + 1$.

From (7) and (9), we can obtain the coefficients of $S^{(j)}$ and $C^{(j)}$ as follows:

$$s_i^{(j)} = s_i^{(j-1)} + c_i^{(j-1)}, \tag{10}$$

$$c_i^{(j)} = a_{\langle (i+j)/2 \rangle} b_{\langle (i-j)/2 \rangle}, \tag{11}$$

where $s_i^{(-1)} = c_i^{(-1)} = 0$, $0 \leq i \leq m$, and $0 \leq j \leq m/2 + 1$.

From (6), let

$$D^{(j)} = \sum_{i=0}^{m} a_{\langle (i+j)/2+1/4 \rangle} b_{\langle (i-j)/2-1/4 \rangle} x^i. \tag{12}$$

Then,

$$T = \sum_{j=0}^{m/2-1} \sum_{i=0}^{m} a_{\langle (i+j)/2+1/4 \rangle} b_{\langle (i-j)/2-1/4 \rangle} x^i = \sum_{j=0}^{m/2-1} D^{(j)}. \tag{13}$$

The recurrence equations of T from (13) can be formulated as follows:

$$T^{(j)} = T^{(j-1)} + D^{(j-1)}, \tag{14}$$

where $T^{(-1)} = 0$, $D^{(-1)} = 0$, and $0 \leq j \leq m/2$.

From (12) and (14), we can obtain the coefficients of $T^{(j)}$ and $D^{(j)}$ as follows:

$$t_i^{(j)} = t_i^{(j-1)} + d_i^{(j-1)}, \tag{15}$$

$$d_i^{(j)} = a_{\langle (i+j)/2+1/4 \rangle} b_{\langle (i-j)/2-1/4 \rangle}, \tag{16}$$

where $t_i^{(-1)} = d_i^{(-1)} = 0$, $0 \leq i \leq m$, and $0 \leq j \leq m/2$.

After computing $S^{(m/2+1)}$ and $T^{(m/2)}$, the product of A and B is obtained by computing $P = S^{(m/2+1)} + T^{(m/2)}$. To clarify the understanding of the proposed method we explain our method through an Example 1.

Example 1: Let two elements in $GF(2^4)$ be given by $A = a_4 x^4 + a_3 x^3 + a_2 x^2 + a_1 x + a_0$ and $B = b_4 x^4 + b_3 x^3 + b_2 x^2 + b_1 x + b_0$. Assume that $P = p_4 x^4 + p_3 x^3 + p_2 x^2 + p_1 x + p_0$ is denoted by the product of A and B over $GF(2^4)$. From (5) and (6), the product of A and B over $GF(2^4)$ is $P = S + T$, where

$$S = \sum_{i=0}^{4} \sum_{j=0}^{2} a_{\langle (i+j)/2 \rangle} b_{\langle (i-j)/2 \rangle} x^i \tag{17}$$

and

$$T = \sum_{i=0}^{4} \sum_{j=0}^{1} a_{\langle(i+j)/2+1/4\rangle} b_{\langle(i-j)/2-1/4\rangle} x^i. \tag{18}$$

By above equations, we can derive Algorithm 1 for the low latency multiplication for finite fields. The input values of Algorithm 1 are two elements over $GF(2^m)$, A and B. According the above equations, CAL_S and CAL_T functions in Algorithm 1 can be executed simultaneously since there is no data dependency between computing S and T.

According to (7) and (9), Table 1 presents the calculation of $C^{(j)}$ and $S^{(j)}$ ($0 \le i \le 4$ and $0 \le j \le 3$). Also, Table 2 presents the calculation of $D^{(j)}$ and $T^{(j)}$ ($0 \le i \le 4$ and $0 \le j \le 2$) from (12) and (14). After computing $S^{(3)}$ and $T^{(2)}$, the product of A and B is obtained by computing $P = S^{(3)} + T^{(2)}$. The proposed multiplication algorithm is regular and simple and is well-suited to implementing systolic architecture by fully exploiting inherent parallelism of the input datum. Based on the proposed Algorithm 1, the hardware architecture can be efficiently composed. As can be seen in Algorithm 1, CAL_S and CAL_T functions can be executed simultaneously and can be induced very similarly. So, CAL_S and CAL_T functions are also induced as an identical type. It means two functions can be computed with the same structure. Thus the architecture based on the proposed algorithm can reduce almost a half of hardware complexity compared to the original architecture. Also, the basic computation time in one loop of each function has been reduced from $T_{AND} + T_{XOR}$ to T_{XOR}, where T_{AND} and T_{XOR} denote AND and XOR delay time, respectively.

Table 1. Calculation of $C^{(j)}$ and $S^{(j)}$ over $GF(2^4)$

j	$C^{(j)}$					$S^{(j)}$
	$c_0^{(j)}$	$c_1^{(j)}$	$c_2^{(j)}$	$c_3^{(j)}$	$c_4^{(j)}$	
0	a_0b_0	a_3b_3	a_1b_1	a_4b_4	a_2b_2	$S^{(0)} = S^{(-1)} + C^{(-1)}$
1	a_3b_2	a_1b_0	a_4b_3	a_2b_1	a_0b_4	$S^{(1)} = S^{(0)} + C^{(0)}$
2	a_1b_4	a_4b_2	a_2b_0	a_0b_3	a_3b_1	$S^{(2)} = S^{(1)} + C^{(1)}$
3	-	-	-	-	-	$S^{(3)} = S^{(2)} + C^{(2)}$

Table 2. Calculation of $D^{(j)}$ and $T^{(j)}$ for $GF(2^4)$

j	$D^{(j)}$					$T^{(j)}$
	$d_0^{(j)}$	$d_1^{(j)}$	$d_2^{(j)}$	$d_3^{(j)}$	$d_4^{(j)}$	
0	a_4b_1	a_2b_4	a_0b_2	a_3b_0	a_1b_3	$T^{(0)} = T^{(-1)} + D^{(-1)}$
1	a_2b_3	a_0b_1	a_3b_4	a_1b_2	a_4b_0	$T^{(1)} = T^{(0)} + D^{(0)}$
2	-	-	-	-	-	$T^{(2)} = T^{(1)} + D^{(1)}$

Algorithm 1. The proposed multiplication algorithm over $GF(2^m)$

Input : A, B

Output : $P = AB$

1 in parallel do:

2      $S \leftarrow CAL\_S(A, B)$

3      $T \leftarrow CAL\_T(A, B)$

4 end do

5 $P \leftarrow S + T$

6 return P

7

8 function $CAL\_S(A, B)$

9     $C^{(-1)} \leftarrow 0,\ S^{(-1)} \leftarrow 0$

10   for $j = 0$ to $m/2 + 1$ do

11   for $i = 0$ to m do

12   in parallel do:

13         $c_i^{(j)} \leftarrow a_{\langle(i+j)/2\rangle} b_{\langle(i-j)/2\rangle}$

14         $s_i^{(j)} \leftarrow s_i^{(j-1)} + c_i^{(j-1)}$

15   end do

16   end for

17   end for

18   return $S^{(m/2+1)}$

19 end function

20

21 function $CAL\_T(A, B)$

22     $D^{(-1)} \leftarrow 0,\ T^{(-1)} \leftarrow 0$

23   for $j = 0$ to $m/2$ do

24   for $i = 0$ to m do

25   in parallel do:

26         $d_i^{(j)} \leftarrow a_{\langle(i+j)/2+1/4\rangle} b_{\langle(i-j)/2-1/4\rangle}$

27         $t_i^{(j)} \leftarrow t_i^{(j-1)} + d_i^{(j-1)}$

28   end do

29   end for

30   end for

31   return $T^{(m/2)}$

32 end function

3 Conclusion

In this paper, we have proposed a parallel pipelined algorithm for multiplication over finite fields based on all-one polynomial. We have induced an efficient algorithm which is highly suitable for the design of pipelined structures. The proposed algorithm enables finite field multiplication to operate in pipelined computation so that two modules can be computed in parallel. Also, we design modules in our algorithm to have the low latency. Moreover, our algorithm enables the computation to share the hardware architecture so that we expect that it reduce not only time complexity but also hardware complexity compared to the recent study. Therefore, we expect that our algorithm can be efficiently used for various applications including crypto coprocessor design, which demand high-speed computation, for security purposes.

Acknowledgments. The authors would like to thank the anonymous referees for their valuable comments. This research was supported by Basic Science Research Program through the National Research Foundation of Korea(NRF) funded by the Ministry of Education (NRF-2015R1D1A1A01059739).

References

1. Blahut, R.E.: Theory and Practice of Error Control Codes. Addison-Wesley, Reading (1983)
2. Lidl, R., Niederreiter, H.: Introduction to Finite Fields and Their Applications. Cambridge University Press, New York (1994)
3. Diffie, W., Hellman, M.: New directions in cryptography. IEEE Trans. Inf. Theory **22**(6), 644–654 (1976)
4. Kim, K.W., Kim, S.H.: A low latency semi-systolic multiplier over $GF(2^m)$. IEICE Electron. Express **10**(13), 20130354 (2013)
5. Kim, K.W., Lee, W.J.: An efficient parallel systolic array for AB^2 over $GF(2^m)$. IEICE Electron. Express **10**(20), 20130585 (2013)
6. Choi, S.H., Lee, K.J.: Low complexity semi systolic multiplication architecture over $GF(2^m)$. IEICE Electron. Express **11**(20), 20140713 (2014)
7. Choi, S.H., Lee, K.J.: Efficient systolic modular multiplier/squarer for fast exponentiation over $GF(2^m)$. IEICE Electron. Express **12**(11), 20150222 (2015)
8. Kim, K.W., Jeon, J.C.: A semi-systolic montgomery multiplier over $GF(2^m)$. IEICE Electron. Express **12**(21), 20150769 (2015)
9. Wang, C.C., Truong, T.K., Shao, H.M., Deutsch, L.J., Omura, J.K., Reed, I.S.: VLSI architectures for computing multiplications and inverses in $GF(2^m)$. IEEE Trans. Comput. **c–34**(8), 709–717 (1985)
10. Reyhani-Masoleh, A., Hasan, M.A.: A new construction of Massey-Omura parallel multiplier over $GF(2^m)$. IEEE Trans. Comput. **51**(5), 511–520 (2002)
11. Reyhani-Masoleh, A., Hasan, M.A.: Fast normal basis multiplication using general purpose processors. IEEE Trans. Comput. **52**(11), 1379–1390 (2003)
12. Fan, H., Dai, Y.: Key function of normal basis multipliers in $GF(2^n)$. Electron. Lett. **38**(23), 1431–1432 (2002)
13. Sunar, B., Koc, C.K.: An efficient optimal normal basis Type II multiplier. IEEE Trans. Comput. **50**(1), 83–87 (2001)
14. Takagi, N., Yoshiki, J., Takagi, K.: A fast algorithm for multiplicative inversion in $GF(2^m)$ using normal basis. IEEE Trans. Comput. **50**(5), 394–398 (2001)
15. Lee, C.Y., Chiou, C.W., Lin, J.M.: Low-complexity bit-parallel dual basis multipliers using the modified Booth's algorithm. Comput. Electron. Eng. **31**(7), 444–459 (2005)
16. Itoh, T., Tsujii, S.: Structure of parallel multipliers for a class of fields $GF(2^m)$. Inf. Comput. **83**(1), 21–40 (1989)
17. Hasan, M.A., Wang, M.Z., Bhargava, V.K.: Modular construction of low complexity parallel multipliers for a class of finite fields $GF(2^m)$. IEEE Trans. Comput. **41**(8), 962–971 (1992)

18. Paar, C.: A new architecture for a parallel finite field multiplier with low complexity based on composite fields. IEEE Trans. Comput. **45**(7), 856–861 (1996)

19. Koc, C.K., Sunar, B.: Low-complexity bit-parallel canonical and normal basis multipliers for a class of finite fields. IEEE Trans. Comput. **47**(3), 353–356 (1998)

20. Sunar, B., Koc, C.K.: Mastrovito multiplier for all trinomials. IEEE Trans. Comput. **48**(5), 522–527 (1999)

21. Diab, M., Poli, A.: New bit-serial systolic multiplier for $GF(2^m)$ using irreducible trinomials. Electron. Lett. **27**(20), 1183–1184 (1991)

22. Lee, C.Y., Lu, E.H., Lee, J.Y.: Bit-parallel systolic multipliers for $GF(2^m)$ fields defined by all-one and equally-spaced polynomials. IEEE Trans. Comput. **50**(5), 385–393 (2001)

23. Lee, C.Y., Lu, E.H., Sun, L.F.: Low-complexity bit-parallel systolic architecture for computing $AB^2 + C$ in a class of finite field $GF(2^m)$. IEEE Trans. Circ. Syst. II **48**(5), 519–523 (2001)

24. Lee, C.Y., Horng, J.S., Jou, I.C., Lu, E.H.: Low-complexity bit-parallel systolic montgomery multipliers for special classes of $GF(2^m)$. IEEE Trans. Comput. **54**(9), 1061–1070 (2005)

25. Kim, T.W., Lee, W.J., Kim, K.W.: Bit-parallel systolic multiplication architecture with low complexity and latency in $GF(2^m)$ using irreducible AOP. J. Korean Inst. Inf. Technol. **11**(3), 133–139 (2013)

26. Wang, C.L., Lin, J.L.: Systolic array implementation of multipliers for finite fields $GF(2^m)$. IEEE Trans. Circ. Syst. II **38**(7), 796–800 (1991)

Data Mining

A Directional Recognition Algorithm of Semantic Relation for Literature-Based Discovery

Xiaoyong Liu[✉], Hui Fu, and Chaoyong Jiang

Department of Computer Science, Guangdong Polytechnic Normal University,
Guangzhou 510665, Guangdong, China
lxyong420@126.com

Abstract. Literature-Based Discovery (LBD), a kind of knowledge discovery algorithm, is proposed by Don R. Swanson, which can assist the researchers to recognize implicit knowledge connection and further accelerate the generation of new knowledge. However, most of algorithms in the field of LBD mainly start from the co-occurrence of terms to find connections between terms, and barely consider the semantic relation actually existing between pairs of terms. In this paper, a kind of directional recognition algorithm of semantic relation is put forward to recognize the directionality of semantic relation existing between pairs of terms. This algorithm will automatically judge the direction of semantic relation based on WordNet and JWNL. The numerical experiment results have indicated that the algorithm proposed in this paper can well recognize the directionality of the semantic relation.

Keywords: Data mining · Natural language processing · Semantic relation · WordNet · JWNL

1 Introduction

Don R. Swanson was firstly proposed Literature-Based Discovery (LBD), a kind of knowledge discovery algorithm in 1986. Through development for about 30 years, many scholars have participated into the research on this algorithm, which has greatly improved and enhanced LBD. The algorithm can release the researchers from the limitations of the familiar and narrow research field, and we can avoid islanding phenomenon by virtue of this algorithm and efficiently support interdisciplinary intersecting innovation. The generalized discovery procedure of LBD proposed by Swanson can be briefly described as below. Find the intermediate word set B which co-occurs with A after starting from document set containing term A. Then further search document set b by starting from B, and recognize term set C which co-occurs with B from b. Later, starting from A and C to carry out the third search, if A and C occurs simultaneously in the same document, delete C, and if not, it can conclude there is recessive connection between A and C. And the obtained A-B-C is a kind of recessive connection knowledge. Based on research of Swanson, some scholars have successively raised some improved LBD algorithms. Yetisgen-Yildiz and Pratt have

© Springer International Publishing Switzerland 2016
Y. Tan et al. (Eds.): ICSI 2016, Part II, LNCS 9713, pp. 281–288, 2016.
DOI: 10.1007/978-3-319-41009-8_30

proposed an open LBD system, - LitLinker [1]. The algorithm starts from the initial concept-Starting Term (e.g. migraine-), and finds the word that directly links with the initial concept, - linking terms. Then it will carry out information retrieval through linking terms and conducts another text mining on the obtained document to recognize each word that links with the linking term, - target terms. Weeber has developed a Literby system [2]. This algorithm realizes drug discovery by integrating the drug discovery process of Vos and LBD process of Seanson. This system is the extension of DAD system and still adopts the Two-Step knowledge discovery algorithm. In other words, it firstly generates hypothesis by using open LBD technology, and then verifies the obtained hypothesis based on close LBD technology. LRD proposed by Kostoff contains two improvements. Firstly, manpower is used in the system to carry out semantic filters according to the semantic type of term, so as to eliminate terms irrelevant to the subject of search; secondly, the documents under search are divided into two types, core document and extended document. The former refers to documents which directly links with the subject of search and the latter stands for documents obtained in the secondary search after reducing the search terms into new search terms based on semantics. Ingrid Petric and Tanja Urbancic et al. have taken Bisociation phenomenon [3] proposed by Koestler A in The Act of Creation as guidance, and considered the rare terms in document set are more important for the generation of potential discovery. Under guidance of the thought, they have raised RaJolink algorithm [4]. RaJolink algorithm contains two steps, generation of hypothesis and verification of hypothesis. The former period contains two steps for looking for rare term, and hypothesis forms through rare term. Supphachai Thaicharoen, Tom Altman and Katheleen Gardiner et al. [5] have proposed to apply WARMR algorithm based on Inductive Logic Programming (ILP) data mining method into research on LBD, to find the intermediate word set b shared by document set A and C. The algorithm is a kind of close LDB technology. Two cases in medical science are studied, namely, Raynaud's disease and fish oils and Down syndrome and cell polarity. Hu et al. has put forward a new LBD algorithm based on semantics, Bio-SARS [6, 7], but he has not explored the semantic relation between pairs of term, but limit the semantic type of term in the two steps of LBD to reduce the eventually generated recessive correlation knowledge.

In the current research in connection with LBD, generally only the correlated pair of term between A—B—C can be recognized, but instructions of relation between the three is not enough. And researchers can hardly recognize the type of relation between A and C, and is also short of judgment of the directionality of this kind of relation. Hence, an algorithm able to recognize the direction of semantic relation between pairs of term is designed in this paper, which can be applied to recognize the semantic relation of certain semantic relation existing between pairs of term.

2   Algorithm of Judging Direction of Semantic Relation – LSJ

The main problem to be solved in this paper is to recognize the directionality of semantic relation between pairs of terms in LBD algorithm, and the major method of that is to recognize the direction of relation based on WordNet and JWNL. Realize the judgment of relation through many look-ups of words by using WordNet forest of

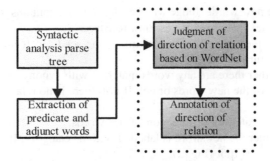

Fig. 1. Thought of judging direction of relationship

synonyms. The main thought of recognition algorithm for semantic relation proposed in this paper is as indicated in Fig. 1. The syntactic analysis parse tree set is the set of sentences processed through syntactic analysis, and the corresponding predicate and adjunct words (e.g. adverb) are extracted from the syntactic analysis parse tree.

2.1   Definition of Relation Directionality

The meaning of directionality of relation can be explained based on examples as below:

Relation in forward direction stands for the fact that mutual promotion exists between the pairs of term (e.g. increase, improve, etc.);

Relation in reverse direction stands for the fact that mutual weakening effect exists between the pairs of term (e.g. decrease, reduce, inhibit).

The undirected relation indicates neither mutual promotion nor weakening effect exists between the pairs of term (e.g. affect).

For instance, if A→B indicates increase and B→C presents decrease, it can be known the relation between A and C is a kind reverse relation. That is to say, the appearance of C can be reduced by increasing A.

If A→B presents decrease and B→C means "influence", it can be known C can be influenced by decreasing A. Of course, further expert knowledge or deep text mining is needed to recognize how it influences in detail.

Since there are many words representing relation of direction, and many words may have different meanings in different fields, the assistance of expert knowledge is required during judgment to avoid confusion and omission of the meaning of the word.

2.2   Introduction of Algorithm

The specific implementation steps of LSJ, judging algorithm of directionality of relation proposed in this paper are as below:

Step 1: Define the seed set that indicates the relation of direction (Provide seed set by virtue of expert knowledge in the field);

Step 2: Input the new words. Judge whether the seed set contains this word or not. If not, turn to the next step, and if yes, turn to Step5;

Step 3: Judge in the category of synonyms of the seed set. If not, turn to the next step, and if yes, turn to Step 5;

Step 4: See whether there are any words matched with synonyms of the seed words in the synonyms of the new words or not. If not, there is no relation, and if yes, turn to Step 5;

Step 5: Label the category of direction of the words;

Step 6: See whether the new word is labelled with its category. If not, continue Step 2–Step 5, and if yes, quit (Fig. 2).

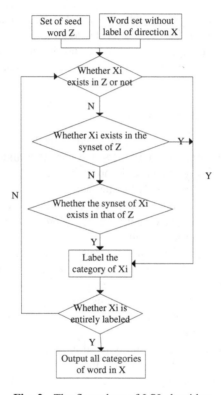

Fig. 2. The flow chart of LSJ algorithm

3   Numerical Experiments

In this section, two experiments are used to verify the algorithm proposed in this paper. At first, the tools used in the algorithm are introduced, and then two experiments are carried out for test, and result analysis is conducted.

3.1   Introduction of Tools

3.1.1   Architecture of WordNet

In this paper, WordNet [8] acts as the key basis for judging architecture of relation. WordNet is actually an English dictionary, but due to the semantic information it includes, it's different from the common dictionaries. WordNet divides the entries according to the meaning, and each entry set with the same meaning is called a synset. WordNet provides short and synoptical definition for each synset, and includes the semantic relation between different synsets.

In WordNet, the noun, verb, adjective and adverb are respectively organized a network of synonym. Each synset stands for a basic semantic concept, and the sets are mutually connected by various connections (a synonym will appear in the synset of each of its meaning). The main part of noun network is in superior-inferior relation, which occupies almost 80 % of the relation. At the top layer of the hierarchy, 11 most abstract concepts exist, called unique beginners, such as entity ("specific existence with or without life"), psychological feature ("psychological characteristics of life organism"), etc.

3.1.2   Introduction of JWNL

JWNL [9] (Java WordNet Library) is an API designed especially for visiting dictionary of WordNet relation. In addition to the provision of data access, it also can conduct relation finding and semantic processing. It uses Java to edit, and is suitable for all editions following WordNet2.0. The subsequent editions of JWNL1.3 all support three ways to visit WordNet, normal file distribution, database and in-memory map.

3.2   Results and Analysis of Numerical Experiment Based on LSJ

LSJ algorithm is coded by JAVA, to apply WordNet3.0 and JWNL1.4.1. At first, define the seed set which includes some words that are mainly used to define the seed words to recognize forward and reverse relation. In practice, expert knowledge in the field can be used to set the words in the seed set. The seed set defined in the experiment in this section is as shown in Table 1.

Table 1. Seed set of terms of relation in pair

Relation type	Seed word
Forward relation	boost, heighten, raise, increase
Reverse relation	reduce, cut_down, drop, let_down
Undirected relation	affect, influence, cause

In order to indicate the effect and efficiency of verification of LSJ algorithm, two test sets are used in this section to test the algorithm. For the first test set, individual words are extracted from the synset of the seed sets for test and verification, and the second test set mainly comes from the synset of the seed set and that of the test set.

3.2.1 Test of Once Extended Word Set of Seed Set

In the algorithm test in this section, we firstly recognize the synonyms of relevant words in the seed set through WordNet. Generally, since a word has several synonyms (For example, the term "boost" can be a noun or verb, and it has dozens of synonyms in WordNet), only 20 are randomly picked from the synset of the seed set to form the test set for verification of algorithm. The data sets used for test of three kinds of relation of direction are as shown in Tables 2, 3 and 4.

Table 2. Test set of synonyms in forward relation

Seed words	Synset
boost	Hike up; lift; bring up; elevate; encourage; promote; advance; cost; increase; encouragement
heighten	Deepen; sharpen
increase	Gain; growth; increment; step-up
raise	Get up; grow; rear; evoke; upgrade

Table 3. Test set of synonyms in reverse relation

Seed words	Synset
reduce	Trim; shrink; boil down; dilute; abridge
cut_down	Slash; cut back; cut out; push down; fell
drop	Knock off; deteriorate; degenerate; devolve; dismiss; flatten; discharge; dip
let_down	Take down; lower

Table 4. Table of synonyms in undirected relation

Seed words	Synset
affect	Impact; bear upon; involve; regard; impress
influence	Charm; mold
cause	Get; make; have

It can be found from the computational results of the three test sets above that LSJ can reach a recognition rate of 100 % for the directionality, which shows that for the synonyms in the seed set, LSJ algorithm has very high recognition rate.

3.2.2 Test of Twice Extended Word Set of the Seed Set

The experiments in the second group is to verify the accuracy of judgment of directionality of relation conducted by LSJ algorithm based on the synset of seed set. The test set in this section is formed by the words extracted from the synset of seed set, and the relevant words are as indicated in Table 5.

When selecting the test set, the synonyms of the seed set must be eliminated, or otherwise, the test results will have no meaning of reference.

Table 5. Test set of the second extended word set of the see set

Relation type	Seed words
Forward relation	plus, arise, rise, benefit
Reverse relation	strike down, fall, depress, lour, subside
Undirected relation	imply, shock, suffer

It can be found through experimental verification that LSJ can also reach a recognition rate of 100 % on the test set obtained through second extension of seed set, which indicates the application of the algorithm in the second extended is also feasible.

3.2.3 Discussion

The test in this section is mainly used for verifying the validity of LSJ algorithm based on WordNet on the judgment of directionality of relation. Although the test results on the two data sets have indicated the algorithm is relatively effective in judging directionality of words through one extension and twice extensions of the seed set, a problem also exists. That is, the accuracy of the computational results of algorithm is directly related to the selection of seed set, and if the selection of seed set is not suitable, the judgment of directionality of certain words will probably be wrong. Hence, in order to solve this problem, we should extend the scale of seed set according to expert's advice, to make the seed set reach a better coverage rate, so as to guide the accurate judgment of algorithm on the directionality of relation.

4 Conclusion

In this paper, a kind of algorithm of judging directionality (LSJ) is proposed to recognize the directionality of semantic relation between pairs of term. At first, the directionality of relation is defined in this paper, and it's indicated some words can reflect the directional relation between itself and the related pair of word (e.g. increase, decrease, etc.); then LSJ algorithm is raised to carry out judgment of directionality of relation; subsequently, WordNet required by the operation of algorithm and the tool for visit, JWNL (Java WordNet Library), are introduced. At last, two test sets are applied to verify the performance of LSJ algorithm, and the issues in the algorithm and solutions are pointed out.

Acknowledgments. This work has been supported by grants from Program for Excellent Youth Scholars in Universities of Guangdong Province (Yq2013108). The authors are partly supported by the Key Grant Project from Guangdong provincial party committee propaganda department, China (LLYJ1311), Guangdong Natural Science Foundation (no. 2015A030313664, no. 2015A030310340), Special funds for science and technology development of Guangdong Province in 2016 (collaborative innovation and construction of platform: Big data analysis platform with accurate and real-time information services for scientific and technological innovation talent), Guangzhou science and technology project (no. 201510020013), Special Program for Applied Research on Super Computation of the NSFC-Guangdong Joint Fund (the second phase), and Guangdong Provincial Application-oriented Technical Research and Development Special fund project (no. 2015B010131017).

References

1. Yetisgen-Yildiz, M., Pratt, W.: Using statistical and knowledge-based approaches for literature-based discovery. J. Biomed. Inform. **39**, 600–611 (2006)
2. Weeber, M.: Drug discovery as an example of literature-based discovery. In: Džeroski, S., Todorovski, L. (eds.) Computational Discovery 2007. LNCS (LNAI), vol. 4660, pp. 290–306. Springer, Heidelberg (2007)
3. Koestler, A.: The Act of Creation. MacMillan Company, New York (1964)
4. Urbančič, T., Petrič, I., Cestnik, B.: RaJoLink: a method for finding seeds of future discoveries in nowadays literature. In: Rauch, J., Raś, Z.W., Berka, P., Elomaa, T. (eds.) ISMIS 2009. LNCS, vol. 5722, pp. 129–138. Springer, Heidelberg (2009)
5. Thaicharoen, S., Altman, T., Gardiner, K., et al.: Discovering relational knowledge from two disjoint sets of literatures using inductive logic programming. In: 2009 IEEE Symposium on Computational Intelligence and Data Mining, pp. 283–290. IEEE Press, New York (2009)
6. Hu, X., Zhang, X., Yoo, I., Zhang, Y.-Q.: A semantic approach for mining hidden links from complementary and non-interactive biomedical literature. In: 6th SIAM International Conference on Data Mining, Bethesda, MD, USA, pp. 200–209 (2006)
7. Hu, X., Zhang, X., Yoo, I., Wang, X., Feng, J.: Mining hidden connections among biomedical concepts from disjoint biomedical literature sets through semantic-based association rule. Int. J. Intell. Syst. **25**(2), 207–223 (2010)
8. About Wordnet. http://wordnet.princeton.edu/
9. JWNL User's Manual. http://sourceforge.net/apps/mediawiki/jwordnet/index.php?title=User%27s_Manual#Source

Research on Pattern Representation and Reliability in Semi-Supervised Entity Relation Extraction

Feiyue Ye and Nan Tang[(✉)]

Shanghai University, Shanghai, China
{yefy, tangnan}@shu.edu.cn

Abstract. This paper proposes a bootstrapping-based method to extract multiple entity relations. Compared with previous entity relation extraction methods, this method analyzes the syntax and semantics of sentences based on traditional context pattern representation. In this way, the features of keyword with the nearest syntactic dependency, phrase structure distance and semantics are extracted so as to form new semantic patterns. To reduce the noise caused by pattern extension, patterns and instances are adopted to verify their reliability mutually. In addition, by combining the information entropy of patterns, accurate and significant instances are selected. Experimental results show that this method effectively improves the quality of patterns and obtains favorable extraction results.

Keywords: Entity relation · Semi-supervised · Semantic pattern · Reliability

1 Introduction

Relation extraction task was first put forward in the last message understanding conference in 1998. In recent years, with the continuous development of relation extraction technology, researchers have proposed a number of ways to achieve relation extraction. Relation extraction methods are mainly based on rules and machine learning. Machine learning based method is divided into supervised machine learning, semi-supervised machine learning and unsupervised machine learning according to the need of manual labeling training corpus. Supervised machine learning needs to construct training data with some form of expression, and then it trains with machine learning algorithm like Classification of classifiers such as support vector machine (SVM) and winnow. According to the expression of the training data, it can be divided into two categories: feature vector based method [1] and kernel function method [2, 3]. Unsupervised relation extraction mainly uses clustering method, and it regards the same class in the same relation.

Rule based method requires domain experts to construct a large-scale knowledge base while supervised method requires a large number of annotated corpora, which are required to pay a lot of manpower and material resources. In order to achieve a better balance between system performance and artificial intervention, more and more researchers have been paying attention to the semi-supervised learning method.

© Springer International Publishing Switzerland 2016
Y. Tan et al. (Eds.): ICSI 2016, Part II, LNCS 9713, pp. 289–297, 2016.
DOI: 10.1007/978-3-319-41009-8_31

2 Related Work

The Bootstrapping method is a widely used semi-supervised learning method, which sometimes refers to self-training. It is first used to carry out the relation extraction between entities in 1998 when Brin proposed DIPPRE. As merely small quantities of useful data are needed by the bootstrapping method, it can effectively extend these data through continuous learning until the data with a required scale are acquired.

In terms of pattern representation, relation patterns are learned based on the context of entities with pattern representation being <left, tag1, middle, tag2, right> [4–6]. The feature-based extraction for relation description patterns selects certain meaningful linguistic features from specified context windows by analyzing the existing basic linguistic features to form patterns so as to improve the applicability of patterns to linguistic phenomena [1]. However, owing to the complexity and diversity of grammatical structures, it is difficult to extract effective grammatical features.

From the perspective of improving the quality of patterns, Snowball [4] select instances with high reliability in each iteration and put them into positive instance sets so that the generated patterns can be used in the next iteration. However, this method significantly depends on information redundancy. The Espresso method proposed by Pantel and Pennacchiotti [7] adopts the web expansion strategy which mines a large number of patterns from the whole Web by obtaining redundant information. Whereas, the patterns extracted from the Web using this method are possibly not usable in the small data set.

The contributions of this paper are as follows: (1) On the basis of description patterns of context relation, this paper presents a new method of pattern representation by combining syntactic and semantic features, which is more efficiently to show the semantic information contained in relation instances. (2) To further improve the quality of patterns, the authors put forward to use patterns and instances to mutually validate their reliability and to select accurate and significant instances by combining the information entropy of patterns.

3 Pattern Representation and Reliability

This paper extracts various entity relations through bootstrapping-based semi-supervised learning. This method is mainly composed of the following four steps.

(1) The sentences containing given relation seeds are found from corpora to undergo syntactic and semantic analysis. In this way, corresponding characteristics are obtained to construct initial relation description patterns P;
(2) All the relation patterns generated in the second step in the same category are clustered to generate more generalized patterns. In this way, though there are few generalized patterns, they are more applicable;
(3) The generalized patterns obtained in the third step are matched in the initial corpus so as to extract candidate instances T which belong to target relations;
(4) After the reliability of candidate relation instances are calculated, instances with reliability being larger than threshold value α are selected and then to be used in the next iteration. If there is no new instance generated, the iteration is stopped.

3.1   Relation Description Patterns

As specific information occurs in contexts in various forms, merely depending on contextual words cannot effectively distinguish relations. Therefore, by combining syntactic and semantic features, this paper obtains patterns which can distinguish different relations so as to improve the expression ability and applicability of patterns.

Feature Extraction. *The Feature of Keyword with the Nearest Syntactic Dependency.* In previous pattern representations, the entities in sentences are served as cores [5, 6]. However, after analyzing a large amount of corpora, it is found that the patterns of specific relations generally take keywords as cores. For instance, the keyword "is-a" is contained in hyponymy relations in general, while "locate" is possibly included in position relations. By further analyzing the syntax of sentences, it is found that the entities in sentences inevitably occur in dependency structures, which certainly reflects the relational features between corresponding entities. Figure 1 is an example of syntactic analysis. The entity pair (上海, 长江三角洲) is connected with the phrase structure through keyword "位于". In addition, after conducting lots of experiments on the results of syntactic analysis [8], it is found that keywords play an important role in obtaining entity boundaries and connecting entity relations.

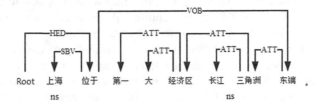

Fig. 1. Syntactic analysis of dependency structure

Therefore, this paper applies the keywords in the nearest common dependency structure of two entities as the feature of the pattern. To be specific, the dependent relation sequences $<E_1, w_{11}, w_{12}, \cdots, w_{1m}, Root>$ and $<E_2, w_{21}, w_{22}, \cdots, w_{2n}, Root>$ of two entities E_1 and E_2 are found separately. Then, the word w firstly matched in these two sequences is acted as a keyword.

The Feature of Phrase Structure Distance. Traditionally, distance is calculated based on the distance between two words in word sequences [6, 9]. As this method fails to take the long-distance matching or modified relation of words in sentences into consideration, it lacks grammatical and semantic expression abilities. Hereinto, Nguyen et al. [10] extracted semantic relations by using different tree kernel functions. The results suggest that the extraction based on traditional phrase structure trees shows a better performance. Inspired by this conclusion, this paper performs syntactic analysis to sentences so as to obtain phrase structure trees. Then, according to the positions of word nodes on these trees, the relative distance of words is calculated. The specific formula for calculating this distance is illustrated in formula (1).

$$D(w_1, w_2) = dist(w_1, w_2)/depth(w_1, w_2). \tag{1}$$

Where $dist(w_1, w_2)$ is the path length between words w_1 and w_2, and $depth(w_1, w_2)$ represents the depth of the common father node of the two words.

Semantic Feature. The words in the context of relation instance provide more adequate semantic information for relation, so the entity relation extraction should take the context of words' semantic features into account. This paper uses HowNet to calculate the semantic similarity between two words. HowNet is a common knowledge base, it describes concepts represented by words while concept represented by sememes. This paper uses the literature [11] mentioned method based on HowNet to calculate semantic similarity $Sim(w_1, w_2)$ between word w_1 and word w_2, it can make full use of the rich semantic information of each concept in HowNet.

Pattern Representation and Relevant Definitions. According to the above selected features and the context pattern representation, the new pattern is designed as:

$$P = \ <V_l, E_1, V_m, E_2, V_r, K> . \tag{2}$$

Where V_l, V_m and V_r represent the context vectors of two entities separately. The weighted word in a vector V is $w = \{w_d, w_t\}$, in which w_d and w_t denote the word and the weight of the word. E_1 and E_2 are two entities which contain themselves and their category mark tag t. Besides, K refers to the selected keyword.

It is found in previous research that, the expression ability of words in the context for relation instances is associated with the relative location between the words and the relation instances [12, 13]. Considering this, the weights of words are computed based on the information gain of word locations. Combining the new pattern representation and taking the semantic distance from words to entities and keywords into account, the weights of words are calculated using formula (3).

$$Weight(w, V, P) = \sum_{c' \in C} G(D(w, c')) / \sum_{w' \in V} \sum_{c' \in C} G(D(w', c')). \tag{3}$$

Where C represents the set of entities and keywords of pattern P, and $D(w, c')$ is the proposed function, which used to calculate the semantic distance between words. $G()$ is described in [12] in detail, which is a function to calculate an information gain of the relative distance.

3.2   Extraction and Generalization of Raw Patterns

By finding sentences containing the given relation seeds in the corpus, the original relation extraction pattern is established according to the pattern representation proposed in Sect. 3.1. However, by doing so, numerous original patterns are generated. In addition, each of the original patterns shows low abstraction and cannot be directly used to extract the relation instances of target relations. So, the original patterns need to be clustered to improve the generalization of patterns on the premise of ensuring the high selectivity of patterns, so as to increase the efficiency of relation extraction. The similarity between two weighted words w_1 and w_2 is defined as $WSim(w_1, w_2)$.

$$WSim(w_1, w_2) = Sim(w_1.w_d, w_2.w_d) * (\sum_{i=1}^{2} w_i.w_t/2). \tag{4}$$

The similarity between two patterns P_1 and P_2 is $PSim(P_1, P_2)$ and expressed by formula (5).

$$PSim(P_1, P_2) = \begin{cases} V_lSim * \omega_l + V_mSim * \omega_m + V_rSim * \omega_r & other \\ 0 & P_1.E_1.t \neq P_2.E_1.t \ or \ P_1.E_2.t \neq P_2.E_2.t \end{cases}. \tag{5}$$

Where ω_l, ω_m and ω_r denotes the weights of different contexts of the pattern, and $\omega_l + \omega_m + \omega_r = 1$. $VSim$ represents the similarity of the vectors corresponding to patterns P_1 and P_2, and is calculated using the greedy algorithm. A pair of weighted words with highest similarity is selected in the two vectors and their similarity is computed using formula (4). Similarly, pairs of weighted words with highest similarity are selected until the two vectors are empty. Finally, the similarities of all the weighted word pairs are added together and the obtained result is served as the similarity of the two vectors. The patterns with similarity exceeding the threshold δ are generalized after being clustered. The generalization of patterns is similar to the computation of similarity of vectors, that is, two words with highest similarity are chosen in each time and merely the one most frequently occurs in all the patterns is reserved.

3.3   Reliability of Patterns and Instances

Owing to the complexity and diversity of Chinese information, it is inevitable to extract patterns with low quality. Therefore, the patterns need to be filtered further to reduce noise. While performing bootstrapping-based relation extraction, reliable patterns are obtained according to relation instances; correct instances are extracted based on patterns in turn. So, the reliability of pattern and instance are closely related to each other. In the research, based on the positive and negative matching ratios used in previous study [4], patterns and instances are employed to verify their reliability respectively. The pattern P to be evaluated is calculated using the following formula.

$$Conf(P) = \sum_{i=1}^{|T^+|} Conf(T_i^+)/(\sum_{i=1}^{|T^+|} Conf(T_i^+) + \sum_{i=1}^{|T^-|} conf(T_i^-)). \tag{6}$$

Where T^+ and T^- are the positively and negatively matched relation instances of relation description patterns respectively. If T is the original seed, then $Conf(T) = 1$.

Instances in a same category, even a same instance, probably present various relations. Taking an instance (America, Obama) for an example, it may show "Employ" relation and "BornIn" relation. Therefore, the extracted patterns are possibly related to various relations. To avoid the occurrence of such uncertainty, the reliability of instances is calculated by combining information entropy. According to the definition of information entropy, the less the entropy is, the higher the certainty and thereby the higher the classifying reliability of instances. The information entropy of pattern P is redefined by:

$$H(P) = (\log n + \sum_{i=1}^{n} p_i \log p_i) / \log n. \tag{7}$$

Where n represents the number of relation classifications and p_i is the average possibility that the current instance is classified in i category.

After computing the reliability of all the patterns, the instance generated in the iteration can be evaluated. As each instance is not just extracted by one pattern, the reliability of instance T extracted using these patterns are computed using formula (8).

$$Conf(T) = 1 - \prod_{i=1}^{|P|} (1 - Conf(P_i) \cdot H(P_i)). \tag{8}$$

Then, the relation instances with reliability exceeding the threshold α are selected and included in the seed set to carry out the next iteration.

4 Experimental Results

The tagged corpus of People's Daily published in 1998 is applied as the experimental data. The syntax and phrase structure of the corpus are analyzed using LTP-Cloud. Moreover, Located (位置关系), BornIn (出生地关系) and Employ (雇佣关系) relations are utilized as extracted entity relations. As the extraction of entity relations is also regarded as a classification, Precision, Recall, and F value in information retrieval are employed as the final evaluation indexes, shown as below.

$$\text{Precision} = N_1/N_2. \tag{9}$$

$$\text{Recall} = N_1/N_3. \tag{10}$$

$$F = 2 \times \text{Precision} \times \text{Recall}/(\text{Precision} + \text{Recall}). \tag{11}$$

Where N_1 and N_3 are the numbers of the accurately extracted instances and corresponding instances in the corpus, and N_2 is the total number of the extracted instances.

To compare the results of relation extraction of the pattern representation proposed in the study and the traditional context pattern representation, comparative experiments are performed. Method 1 is how the pattern represented in literature [6], which vectoring the context of entities and the weight of words in vector quantity is decided by the distance of word sequences. Method 2 is the adopted one in this paper. The experimental results are displayed in Table 1.

Table 1. Comparative experimental results statistics

Relation	Method 1			Method 2		
	Precision	Recall	F	Precision	Recall	F
BornIn	71.73 %	74.65 %	73.16 %	75.82 %	77.95 %	76.87 %
Located	75.89 %	81.74 %	78.71 %	77.61 %	84.48 %	80.90 %
Employ	61.50 %	68.08 %	64.62 %	68.87 %	74.63 %	71.63 %

The above results manifest that, while adding more syntactic and semantic features, Method 2 is apparently superior to Method 1 in the entity relation extraction. It is believed that the high-quality patterns are contributed to the high accuracy, and the new pattern is more efficiently to show the semantic information contained in relation instances. The method proposed in the research is effective, especially for Employ relation, the F value of which is improved by 7.01 % compared with traditional method. By analyzing incorrect instances, the authors find that incorrect keywords are supposed to be extracted in syntactic analysis due to the nonstandard expression of sentences. This phenomenon reduces the quality of patterns and thereby the accuracy.

Based on Methods 2 and 3 is conducted with the application of the reliability of patterns and instances in the research, and the results are illustrated in Fig. 2.

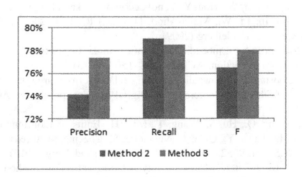

Fig. 2. Comparative experimental results (Color figure online)

It can be seen from the above results that except for the recall ratio which is declined by 0.51 %, the accuracy of Method 3 is increased by 3.28 %, indicating that the selective method for patterns and instances put forward in the research is effective. Compared with the approach based on the positive and negative matching ratios method, the proposed method can filter uncertain patterns and instances in iteration. While these patterns probably contain correct instance relations, the Recall is likely to be reduced by using the method. By using the proposed method, the patterns can be selected more accurately, thus improving the accuracy significantly.

5   Conclusions

In the research, the sentences first undergo syntactic and semantic analysis to extract features of keyword with the nearest syntactic dependency, phrase structure distance, and semantics, which are employed to construct new patterns. On this basis, the bootstrapping-based method is utilized to extract multiple relations. To reduce the noise contained in the data of bootstrapping-based relation extraction, patterns and instances are applied to verify their reliability mutually. Meanwhile, accurate and significant instances are selected by combining the information entropy of patterns. In this way, the relation extraction model is enhanced with higher extraction performance.

In the future, relation instances are expected to be mined further to improve the extraction accuracy and obtain more semantic information. Besides, the coreference resolution which is not discussed in the research will be studied intensively.

References

1. Jiang, J., Zhai, C.X.: A systematic exploration of the feature space for relation extraction. In: HLT-NAACL 2007, pp. 113–120. Association for Computational Linguistics, Rochester (2007)
2. Zhang, J., Ouyang, Y., Li, W., Hou, Y.: A novel composite kernel approach to chinese entity relation extraction. In: Li, W., Mollá-Aliod, D. (eds.) ICCPOL 2009. LNCS, vol. 5459, pp. 236–247. Springer, Heidelberg (2009)
3. Plank, B., Moschitti, A.: Embedding semantic similarity in tree kernels for domain adaptation of relation extraction. ACL 1, 1498–1507 (2013)
4. Agichtein, E., Gravano, L.: Snowball: extracting relations from large plain-text collections. In: Proceedings of 5th ACM Conference on Digital libraries, pp. 85–94. ACM, New York (2000)
5. Li, J., Cai, Y., Wang, Q., Hu, S., Wang, T., Min, H.: Entity relation mining in large-scale data. In: Liu, A., Ishikawa, Y., Qian, T., Nutanong, S., Cheema, M.A. (eds.) DASFAA 2015 Workshops. LNCS, vol. 9052, pp. 109–121. Springer, Heidelberg (2015)
6. He, T., Xu, C., Li, J., et al.: Named entity relation extraction method based on seed self-expansion. Comput. Eng. 32(21), 183–184 (2006)
7. Pantel, P., Pennacchiotti, M.: Espresso: Leveraging generic patterns for automatically harvesting semantic relations. In: Proceedings of 21st International Conference on Computational Linguistics and The 44th Annual Meeting of the Association for Computational Linguistics, pp. 113–120. Association for Computational Linguistics, Stroudsburg (2006)
8. Guo, X.Y., He, T., et al.: Chinese named entity relation extraction based on syntactic and semantic features. J. Chin. Inf. Process. 28(6), 183–189 (2014)
9. Li, H., Wu, X., Li, Z., et al.: A relation extraction method of Chinese named entities based on location and semantic features. Appl. Intell. 38(1), 1–15 (2013)
10. Nguyen, T.V.T., Moschitti, A., Riccardi, G.: Convolution kernels on constituent, dependency and sequential structures for relation extraction. In: Proceedings of 2009 Conference on Empirical Methods in Natural Language Processing, pp. 1378–1387. Association for Computational Linguistics, Stroudsburg (2009)

11. Liu, C., Li, S.J.: Word similarity computing based on HowNet. Comput. Linguist. Chin. **7** (2), 59–76 (2002)

12. Lu, S., Bai, S., et al.: An unsupervised approach to word sense disambiguation based on sense-words in vector space model. J. Softw. **13**(6), 1082–1089 (2002)

13. Chen, C., He, L., Lin, X.: REV: extracting entity relations from world wide web. In: Proceedings of 6th International Conference on Ubiquitous Information Management and Communication, pp. 1–5. ACM, New York (2012)

Pushing Decision Points Backward to the Latest Possible Positions with a Workflow Log

Su-Tzu Hsieh[1], Ping-Yu Hsu[2(✉)], Ming Shien Cheng[3], and Hui-Ting Huang[2]

[1] Department of Information Management and Technology,
Xiamen University Tan Kah Kee College, Technology Developing Area,
Zhangzhou City, Fujian Province, China
sthsieh@xujc.com, helenhsieh@mgt.ncu.edu.tw
[2] Department of Business Administration, National Central University,
No. 300 Jhongda Road, Jhongli City, Tao Yuan County, Taiwan
pyhsu@mgt.ncu.edu.tw
[3] Department of Industrial Engineering and Management,
Ming Chi University of Technology, No. 84, Gongzhuan Rd.,
Taishan Dist., New Taipei City 24301, Taiwan
mscheng@mail.mcut.edu.tw

Abstract. Currently, enterprises face more competitive and complex environments than ever before. Companies would suffer devastating setbacks if business changes were ignored. Enterprises could leverage competitive advantages if business changes were automatically adopted in processes by stimulating workflow redesign. Business changes imply new business processes with decision points. The effects of business changes on business operations are firmly recorded by system logs. Pushing decision points to the latest possible position may be a feasible method while facing business changes. Business changes also imply redesigning business processes. This study proposes an algorithm for pushing decision points backward, which enables decision points in business processes to be identified automatically. This study also proposes a new workflow graphic based on the algorithms for pushing decision points backward. The new workflow graphic from the results of this study can contribute to redesigning businesses.

Keywords: Pushing decision points backward · Business process mining · Workflow mining · System log mining

1 Introduction

Currently, enterprises face hypercompetitive and highly complex environments [1]. To survive in these challenging environments, several enterprises build various products to fulfill elusive customer needs [2] Customers can change or even cancel orders immediately prior to shipping. The change of customer needs can disrupt company operation and lead to high inventory costs. Hence, pushing decision points backward to

© Springer International Publishing Switzerland 2016
Y. Tan et al. (Eds.): ICSI 2016, Part II, LNCS 9713, pp. 298–305, 2016.
DOI: 10.1007/978-3-319-41009-8_32

reduce business risks is crucial. Pushing decision points backward can increase the flexibility of business processes and reduce the effects of customer order changes. This research provides an algorithm for pushing decision points to the latest moment by pulling generic operation tasks ahead. Pushing decision points to the latest positions can increase the flexibility of business processes.

Previous related studies have primarily focused on calculating cost savings, product improvement, and lead time reduction [6, 7] hiegh efficiency in resources and time can be achieved if decision points are successfully pushed backward. However, no studies have covered the subject of how to push decision points backward in a business workflow.

Designing a new workflow is complex and time consuming [3] Fortunately, business workflow analysis from an information system log is possible. With the prevalence of enterprise systems, such as enterprise resource planning, customer relationship management, and business process management, logs exist in various companies. Hence, several researchers have devoted themselves to workflow research, with the main focus on workflow reconstruction from logs [3, 4, 11]. Few researchers have focused on workflow improvement [4]. The present study is one of the first studies to propose an algorithm for pushing decision points backward.

This research focuses on an algorithm that can effectively push decision points backward according to workflow logs, which contain sequences of task execution and the input and output resources of each task. By comparing the required input and output resources between any two consecutive tasks, pushing the decision point to the latest position is plausible.

The remainder of the research is organized into five sections. The literature on workflow mining and decision point movement is discussed in Sect. 2. Section 3 delineates the data structure used in the workflow log algorithm and proposes conditions for pushing decision backward to the latest possible points. Section 4 explains the algorithm. The conclusion and discussion of future work are provided in Sect. 5.

2 Related Works

Identifying decision points and pushing them to the latest position are crucial in reducing risk and cost savings; however, only limited related studies have focused on the subject. Grigori et al. proposed predefined metrics for predicting workflow instance behaviors [8]. Castellanos et al. used a predefined metric to predict and control instantly the improvement of business operations [9]. Two studies employed output data of instances of workflow to predict and control consequent values of the metric. However, previous researchers did not discuss algorithms for postponing decision points.

An algorithm for moving decision points was proposed by Subramaniam. He pulled decision points ahead through a process workflow with a new workflow graphic [10]. The results by Subramaniam revealed that decision adjustments reduce the operation execution time and redundancy among tasks. However, decision points should be pushed backward rather than pulled ahead because fewer tasks gain more business flexibility after decision points, as evidenced by previous studies. Pushing-backward strategies entail pushing customization operations backward till the customer order is confirmed [5].

This research proposes pushing decision points backward by mining the relationships between decision points and their descendant tasks from the execution log of the workflow to gain the flexibility of the business workflows and enable enterprises to quickly respond to customer requirements.

3 Data Structure

The purpose of this research is to push decision points backward on a log-based workflow as much as possible. Before the algorithm can be presented, the notation of the workflow, log, and decision points must be formally defined.

Definition 1 (Workflow trace, wt). Workflow log Lp indicates workflow traces of workflow process P.

1. Workflow trace wt is defined as a string of <input(i), i, output(i)> where i is an instance of a task, input(i) is a set of data fed to task i, and output(i) is a set of data generated by task i.
2. Workflow log Lp is defined as a set of workflow traces.

Definition 2 (Work flow process graph, Gp). The workflow graph Gp of a work flow process model P is composed of (A, c, m, t, s, e, Path), where

1. A is a set of activity aggregations, $A = c \cup m \cup t \cup s \cup e$.
2. a is an activity. An activity denotes the least independent work step, which can refer to a task node and choice node of a decision point and a merge point in the process. $a \in A$
3. c is a choice node in the workflow process. $c \in A$.
4. m is a merge node in the workflow process. $m \in A$
5. t is a task between c and m. $t \in A$
6. s is a starting activity, which is the beginning of the workflow process. $s \in A$.
7. e is the final activity, which is the end of the workflow process. $e \in A$

Definition 3 (Workflow data, WD). Workflow data are defined as the aggregated set of related data for executing process instances, workflow data WD = {a. v}. Every single instance of a task is recorded as input (a) and denoted in the input data table as output (a)

$\subseteq a \in A$ Input (a) = {$av_{in}|v_{in}$ is the data item that a consumes}, where v_{in} workflow data

$\subseteq a \in A$ Output (a) = {$av_{out}|v_{out}$ is the data item that a consumes}, where v_{out} workflow data

This section focuses on developing formal construction rules for a well-formed workflow, of which branches start and end with a corresponding pair of choice and merge nodes. Let T be a set of task nodes.

Definition 4 (Path, Path (c_i, m_i)). Multiple tasks are connected to each other between decision point c and merge point m into a path. A path in the workflow is defined as follows:

1. Path (c_i, m_i) denotes a set of possible path aggregations from a decision node and a merge node of task i. Path(c_i, m_i) = {Pathj(c_i, m_i)}.
2. Pathj(c_i, m_i) is the jth path between a decision node and a merge node of task i.
3. c denotes a choice node, which represents *XOR* of repel. Only one exclusive or alternative control flow can be chosen from c.
4. m denotes a merge point; an m node merges two repelling alternative controls flowing from a choice node into one control flow.

Definition 5 (Module operation process workflow, module (ti, ti + 1)). Multiple activities are boundaries connected to each other in a module operation process flow.
   A module operation flow in a workflow process is defined as follows:

1. t denotes task nodes between ci and mi. A task is assumed to be naturally sorted by a consequence of job order sequences of operational steps.

Figure 1 illustrates an example of pushing a decision point from a blurred industry case. A Taiwan-based liquid-crystal-display television (TV) manufacturer operates both in Taiwan and in mainland China. To simplify business scenarios, this research assumes two hypothetical customers, denoted as CTR_V and CTR_H. Each customer has its own sales and distribution channels in the United States and Europe. Figure 1 outlines the business process, starting from taking orders and ending at the point of shipping the products. The first task in the workflow is a decision point of product choice. At this task node, the product type is fixed with a customer order: a TV broadcasting system either in the United States or in Europe. Any change after (ORD_TAKE) could result in excess inventory problems, because products are not interchangeable between CTR_V and CTR_H, with each requesting a fixed TV

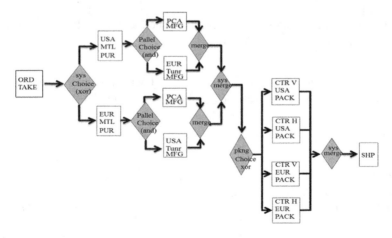

Fig. 1. Business process prior to pushing a decision point backward

broadcasting system. Figure 1 indicates the inflexibility of the decision node starting from the first task of customer taking the order.

Pushing decision points backward is defined as pushing the differentiation of decision points until the latest possible point [3]. Decision points can be pushed backward if an input of descendant tasks is reported as not being influenced by an output of decision points, and pushing backward exerts no effect on other descendant operations after the pushing backward movement. This research argues that decision points can be pushed if they fulfill the following three conditions.

Condition 1: If any existing first task z in any path between ci and mi is not affected by ci, then decision point ci can be pushed to a position after task z. By contrast, none of the subsequent tasks in the path is moved if the first task is influenced by ci. Newly pushed backward decision points can be denoted as the final critical decision point LP (ci) to distinguish them from original ci. The purpose of this research is to push decision points backward to the latest possible position point LP(ci) automatically.

1. Output (c_i) is the output data aggregation of data items of Viout.
2. $\exists z \in$ path (c_i, m_i), such that input $(z) \cap$ output $(c_i) = \phi$.

Condition 2: To ensure completeness of decision points, the new path volume after decision points are pushed backward is the same as the original path volume. None of the paths is causal eliminated by the movements of pushing backward that limits the effect of pushing backward to a minimum. For example, if there are two subsequent paths following the decision points, after the decision points are pushed backward, these two subsequent paths should still be remained.

1. New path (c_i, m_i): new path aggregation set after pushing decision points backward
2. $|$New path $(c_i, m_i)| = |$path $(c_i, m_i)|$

Condition 3:

1. NFP (z) is the aggregated set of first tasks in a new module of pushed backward decision points.
2. LMP(y) is the aggregated set of final tasks in the module (ti, ti + 1) sequence.
3. Output (LMP(y)) $\cap$ input (NFP (z)) $= \phi$.

The final task in the new pulling ahead module LMP(y) is not affected by the first tasks in the new module of pushed backward decision points NFP (z).

$$\text{input}(z) \cap \text{output}(c_i) = \phi.$$

According to Definition 3 (module operation process workflow, module (*USA_MTL_PUR, EUR_MTL_PUR*), the two tasks USA_MTL_PUR and EUR_-MTL_PUR can be bounded into one task module of module (*USA_MTL_PUR, EUR_MTL_PUR*).

The first subsequent-aggregated module follows decision point ci is NFP(z) = {*USA_MTL_PUR, EUR_MTL_PUR*}. The input of ORD_TAKE is not affected by the output of EUR_MTL_PUR and USA_MTL_PUR. That is, input (*ORD_TAKE*) ∩ output (*EUR_MTL_PUR*), and output (*USA_MTL_PUR*) = ϕ.

If input (*CRT_V_USA_PACK*) ∩ output (*ORD_TAKE*) ≠ ϕ, then the task CRT_V_USA_PACK is affected by decision point ci (*ORD_TAKE*). Consequently, decision point c_i cannot be pushed backward, and neither the task CRT_V_USA_- PACK nor the subsequent task SHP can be moved ahead of the decision point. The push backward program stops at this point, and a new graphic is obtained as illustrated in Fig. 2.

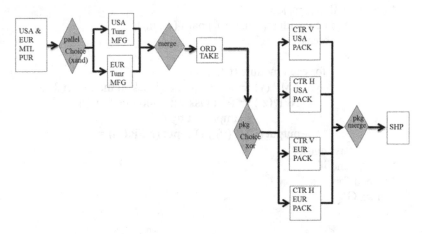

Fig. 2. Business process after a decision point is pushed backward

4 Algorithm for Pushing Decision Points Backward

The proposed algorithm for pushing decision points backward is executed through the following steps (Fig. 3):

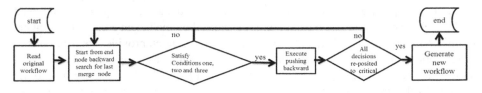

Fig. 3. Flowchart of the proposed algorithm

```
G' ← G
for each ci in Gp do
    for each ti between ci and mi  do
        z ← first then next sequenced task of path (ci , mi )
        if z (the first task in path (ci , mi ))
            such that ( input (z) ∩ output ( Ci ) = ⸖ )
        else break
        end if
        module (ti,ti+1)← z
    end for ti
    LP(ci ) ← discover ( ci , mi )
endfor ci

for each path (c_i , m_i )do
        if ( | new path (c_i , m_i)| ) ≠ | path (c_i , m_i)|
            break
        else
            for each module (t_i,t_{i+1}) do
                LMP (y) ← last sequenced task in module (t_i,t_{i+1})
                NFP (z ) ← first task after module (t_i,t_{i+1})
                            impacted by Ci
                output (LMP(y) ) ∩ input (NFP (z)) = ⸖
            endfor
        end if
    end for
Return G'
```

5   Conclusions, Limitations, and Future Works

By using the pushing-backward rules and algorithm that this research provides, decision points can be automatically pushed backward to the latest possible position if the input of descendant tasks is not influenced by the output of decision points. Thus, customization operations are pushed to the latest position. Companies can gain more flexibility and free time with customization tasks by pushing decision points to the latest possible position. Because generic tasks are pulled ahead, enterprises are relieved of the high inventory and high operation costs caused by inefficient early fixed decision points. With pushing decision points backwards, this research provides contributions to organizations in terms of lower inventory, higher manufacturing-operational efficiency, and higher business flexibility. By pushing decision points backward according to the proposed rules and algorithm, companies can build their core operation competencies in confronting business changes.

This research proposes rules and an algorithm for postponing decisions, which can automatically push decision points backward to the latest possible positions. These postponing rules and algorithms enable enterprises to push tasks involving differentiation until the latest possible position for business operations including manufacturing processes, supply chains, and distribution chains. They empower enterprises to perform mass customization effectively and to build competitive advantages.

This research focused on an *XOR* decision choice node only. The authors intend to research an *AND* decision node for fork processes in a future study. Additionally, top managers sometimes require more time-allowance to make good decisions. The next step of this research is implementing time factors to discover a decision time-allowance when the latest possible decision points exist with the earliest possible start points. Furthermore, this research is limited to the decision points that have been identified. Identifying a decision choice would be worthwhile.

References

1. Kuehnle, H.: Post mass production paradigm (PMPP) trajectories. J. Manuf. Technol. Manag. **18**(8), 1022–1037 (2007)
2. Lee, H.L., Billington, C.: Designing products and processes for postponement. In: Dasu, S., Eastman, C. (eds.) Management of Design: Engineering and Management Perspectives, pp. 105–122. Kluwer Academic Publishers, Boston (1994)
3. Wen, L., Wang, J., van der Aalst, W.M.P., Huang, B., Sunan, J.: A novel approach for process mining based on event types. J. Intell. Inf. Syst. **32**(2), 163–190 (2009)
4. Gaaloul, W., Gaaloul, K., Bhiri, S., Haller, A., Hauswirth, M.: Log-based transactional workflow mining. Distrib. Parallel Databases **25**(3), 193–240 (2009)
5. Bucklin, L.P.: Postponement, speculation and the structure of distribution channels. J. Mark. Res. **2**, 26–31 (1965)
6. Ma, S., Wang, W., Liu, L.: Commonality and postponement in multistage assembly systems. Eur. J. Oper. Res. **42**(3), 523–538 (2002)
7. Su, J.C.P., Chang, Y.-L., Ferguson, M.: Evaluation of postponement structures to accommodate mass customization. J. Oper. Manag. **23**(3–4), 305–318 (2005)
8. Grigori, D., Casati, F., Dayal, U., Shan, M.C.: Improving business process quality through exception understanding, prediction and prevention. In: Proceedings of the 27th International Conference on Very Large Data Bases, pp. 159–168. Morgan Kaufmann Publishers Inc., San Francisco, CA, USA (2001)
9. Castellanos, M., Casati, F., Shan, M.-C., Dayal, U.: iBOM: a platform for intelligent business operation management. In: ICDE Tokyo, Japan, pp. 1084–1095 (2005)
10. Subramaniama, S., Kalogerakia, V., Gunopulosa, D., Casatib, F., Castellanosc, M., Dayalc, U., Sayalc, M.: Improving process models by discovering decision points. Inf. Syst. **32**, 1037–1055 (2007)
11. Zhou, H., Lin, C., Deng, Y.P.: Process ming based on statistic order relationship of events. In: Advanced Material Research, 2nd International Conference on Opto-Electronics Engineering and Material Research, vol. 760–762, pp. 1959–1966, October 2013

A DPSO-Based Load Balancing Virtual Network Embedding Algorithm with Particle Initialization Strategy

Cong Wang[✉], Yuxuan Liu, Ying Yuan, Guorui Li,
and Qiaohong Wang

School of Computer and Communication Engineering,
Northeastern University at Qinhuangdao, Qinhuangdao 066004, China
{congw,yingyuan,lgr,wqh}@neuq.edu.cn,
liuyuxuan_neuq@sina.com

Abstract. Virtual network embedding (VNE) is helpful for effective sharing the underlying physical substrate network resources. In the existing particle swarm optimization-based VNE algorithms, the particle position initialization phase will produce a lot of resources fragment and prevent the underlying substrate network from receiving more resource requests. In order to improve the utilization rate and load balancing of the substrate network as well as accelerate the mapping efficiency of VNE algorithm, this paper presents a particle swarm optimization based load balancing VNE algorithm. A particle position initialization strategy is introduced by reducing the candidate substrate nodes to reject the inaccuracy caused by the random selection in position initialization. Simulation results show that the proposed algorithm can effectively increase the acceptance rate of virtual network requests and improve the long-term average revenue to cost ratio of the substrate network.

Keywords: Network virtualization · Virtual resources allocation · Virtual network embedding · Discrete particle swarm

1 Introduction

With the rapid development of network technology, the "rigid" problem in the Internet architecture has become increasingly prominent, which has become a hindrance to the development of the Internet applications. Network virtualization [1] is an emerging Internet technology and it makes it possible to coexist with multiple heterogeneous virtual networks without changing the underlying physical network architecture, and gradually become an important factor in the future Internet architecture to solve.

Optimal mapping of virtual network requests (VNRs) on the top of the substrate network has been called virtual network embedding (VNE) [2]. The key problem in VNE is to embed VN as far as reasonably possible so as to achieve high resource utilization and revenue of substrate network [3].

In this paper, we propose a discrete particle swarm optimization (DPSO)-based VNE algorithm referred as PIAS-VNE-PSO to solve VNE problem with path splitting. A particle position initialization strategy is introduced by reducing the candidate

© Springer International Publishing Switzerland 2016
Y. Tan et al. (Eds.): ICSI 2016, Part II, LNCS 9713, pp. 306–313, 2016.
DOI: 10.1007/978-3-319-41009-8_33

substrate nodes to reject the mortality caused by the random selection in position initialization.

The rest of this paper is organized as follows. Section 1 provides an overview of the related works. Section 2 formalizes the network model and the VNE problem itself. In Sect. 3, we present the details of the DPSO-based embedding algorithm. The simulation results of our algorithm is demonstrated and discussed in Sect. 4. Section 5 concludes this paper.

2 VNE Problem Formalization

In this section, we will first define the network model, then introduce VNE problem description and the system model.

2.1 Substrate Network

We model the substrate network as a weighted undirected graph $G_s = (N_s, L_s)$, where N_s and the L_s present aggregate of physical nodes and links respectively. The attribute of each physical node $n_s \in N_s$ can be CPU capacity, RAM, storage, etc., this paper only consider the CPU capacity, and let $c(n_s)$ denotes CPU resources that the node n_s have. Each physical link's attribute is the bandwidth resource, let $b(l_s)$ denotes bandwidth resources that the link l_s have. Let P_s denotes a collection of all acyclic paths on the substrate network.

2.2 Virtual Network Request

We model the virtual network request as a weighted undirected graph $G_v = (N_v, L_v)$, where N_v and the L_v present aggregate of the virtual nodes and links respectively. For each virtual node $n_v \in N_v$, the CPU request can be expressed as $c'(n_v)$. For each virtual link $l_v \in L_v$, the bandwidth request can be expressed as $b'(l_v)$.

2.3 System Model

On the premise of that the substrate network can support path splitting i.e. a virtual link can be mapped into more than one physical path between two physical nodes the corresponding two virtual nodes are mapped. Thus, we establish an integer linear programming model taking the cost of the physical network as a goal:

$$\text{Min}(\alpha_c \sum_{n_v \in N_v} c(n_v) + \beta_c \sum_{p \in \rho_v} hops(p) * b_s(p)) \tag{1}$$

where ρ_v represents all the paths assigned to virtual links, $hops(p)$ represents the hops on path p, while $b_s(p)$ represents the consumed bandwidth on path p.

Any embedding solution must satisfy nodes and links resources constraints:

$$\forall u \in N_v, \forall i \in N_s, x_i^u \times c'(u) \leq R_N(i) \tag{2}$$

$$\forall l_{ij} \in L_s, \forall l_{uv} \in L_v, f_{ij}^{uv} \times b'(l_{uv}) \leq R_L(P) \tag{3}$$

where x_i^u is a binary variable, if virtual node u is mapped onto physical node i, its value is 1; otherwise is 0. $R_N(i)$ is available CPU resources of node i, $R_L(P)$ is available bandwidth of physical path p.

For each virtual network request, one substrate node can only accommodate one virtual node. So the variable x_i^u and f_{ij}^{uv} should satisfy the following constraints:

$$\forall i \in N_s, \sum_{u \in N_v} x_i^u \leq 1 \tag{4}$$

$$\forall u \in N_v, \sum_{i \in N_s} x_i^u = 1 \tag{5}$$

$$\forall i \in N_s, \forall u \in N_v, x_i^u \in \{0, 1\} \tag{6}$$

$$\forall l_{ij} \in L_s, \forall l_{uv} \in L_v, f_{ij}^{uv} \in \begin{cases} [0, 1] & l_{ij} \text{ are mapped using path splitting;} \\ \{0, 1\} & l_{ij} \text{ is mapped onto a single path;} \end{cases} \tag{7}$$

The ratio of allocated CPU resource to the idle CPU resources of a physical node is define as:

$$\partial_i = \frac{c'(n_v)}{R_N(n_s)} \tag{8}$$

In order to ensure load balancing of substrate nodes, the following conditions need to be satisfied:

$$\forall i \in N_s \quad 0 < \partial_i = \frac{c'(n_v)}{R_N(n_s)} \leq \gamma \tag{9}$$

where γ is the preset constant used to adjust the load balancing, it meets: $0 < \gamma \leq 1$.

3   PIAS-VNE-PSO VNE Algorithm

3.1   DPSO for VNE

The particle swarm optimization algorithm [4] is based on the simulation of birds and fish, has the characteristic of both evolutionary computation and swarm intelligence It is a new global optimization evolutionary algorithm. Firstly, PSO algorithm generates initial population, then random initializes a group of particles in the searching space, and each particle corresponds to a feasible solution to the optimization problem which is determined by the objective function. Each particle will move in the search space, the direction and distance are determined by a speed. The PSO algorithm is widely used in the field of pattern recognition, multi-objective optimization, because this algorithm is simple conceptually and easy done. The parameters needed to be adjusted are less, and it can solve the complex optimization tasks effectively.

However, the original particle swarm optimization algorithm is used to solve the continuous problem and it is not suitable for the solution of the VNE problem. Therefore, a discrete particle swarm optimization algorithm (DPSO) is used in this paper. The parameters of the particles and related operations are redefined as follows:

Definition 1. The position of particle $X_i = [x_i^1, x_i^2, \ldots, x_i^N]$, $x_i^j \geq 0, j \in (1, N)$ is a feasible solutions of VNE. N represents the number of virtual nodes in a virtual network request, x_i^j is positive integer, represents the label of the substrate physical node to the j virtual node is mapped.

Definition 2. The speed $V_i = [v_i^1, v_i^2, \ldots v_i^N,]$, indicates the evolution direction of the current particle. v_i^j is a binary number, if its value is 0, it shows that we should select another physical node to map the corresponding virtual nodes, otherwise don't make any adjustments.

Definition 3. Subtraction $X_i \ominus X_j$, expresses the difference between two positions. If the value is same in the same dimension, then the corresponding dimension value in the result is 1, otherwise is 0.

Definition 4. Velocity addition $P_1 V_i \oplus P_2 V_j$, expresses using the probability of P1 and P2 respectively to maintain the speed of V_i and V_j.

Definition 5. The movement of particles $X_i \otimes V_i$, expresses to adjust the particle according to the speed V_i, if the speed is 0 in a dimension, then select another node to map the virtual nodes, otherwise don't make any adjustments.

According to the redefinition of the operations, a new formula for the velocity and position of the particle is obtained:

$$V_{i+1} = p_1 V_i \oplus p_2 (X_p^i \ominus X_i) \oplus p_3 (X_g \ominus X_i) \tag{10}$$

$$X_{i+1} = X_i \otimes V_{i+1} \tag{11}$$

where p_1 is the inertia weight, p_2 and p_3 represent cognitive weight and social weight respectively. There are all constant, and meet $p_1 + p_2 + p_3 = 1$, $(0 \leq p_1 \leq p_2 \leq p_3 < 1)$.

3.2   Particle Position Initialization Strategy

In the existing VNE algorithms based on PSO, position initialization procedure sorts the physical node and virtual node in descending order according to resources, then finds the minimum node resource request of the virtual network, and only considers the physical nodes that are greater than the minimum virtual node resource requests. In the process of particle initialization, the physical nodes are randomly assigned to each virtual node. In fact, although removing the physical nodes which are less than the minimum resource request from the physical nodes, it has little effect on the random distribution of subsequent nodes. To a large extent, they will still distribute to the virtual node resource that is less than its minimum requested. Thus, the initialization of the particle has improvement space, and it can accelerate the subsequent iterative process.

We build candidate nodes for each virtual node for a further accurate solution space. First, we remove those nodes and links whose idle resources don't meet the requirement of minimum resource requests. Then create a list of candidate nodes, from which we allocate positions for virtual nodes until all nodes have been allocated. Here we add a factor in formula (9) to adjust load balance. Particle initial allocation strategy is described as follows.

Algorithm 1: Particle initialization.

1.   Sort physical nodes and virtual nodes according to the resources in descending order, and initial a population of particles which denotes by χ;
2.   Create particle list *Plist*;
3.   Create list *Vnlist*, join the virtual node in turn;
4.   Create a list of candidate nodes for each virtual node;
5.   It must satisfy that all the candidate nodes' free resources are greater than or equal to the minimum resource request of the virtual node, and satisfy the load balancing constraints defined by formula (9);
6.   **While (χ >0) do**
7.   get a virtual node from *Vnlist* in proper order;
8.   If the candidate node's list of the virtual nodes getting from 5 is not empty, then random assign a physical node for it;
9.   Remove the allocated node from the list of all candidate nodes;
10.   Add particles into *Plist*;
11.   **End while**.

The particle position initialization strategy and the random initialization are similar in using the descending order to sort the physical and virtual nodes according to resources. The difference is that our particle position initialization strategy assigns a list of candidate nodes for each virtual node and remove the physical node which is less than this virtual node's request. Thus the accuracy of the node assignment has been greatly improved and reduce the uncertainty of random assignment. At the same time, the assigning sequence is in descending order of virtual nodes, which can improve the accuracy of distribution.

3.3   DPSO-Based VNE Algorithm

In our intact VNE algorithm, we use the particle position initialization strategy mentioned above to generate particles. Then we map virtual links to the underlying substrate network through multi-shortest paths mechanism. If single physical path doesn't meet the requirement of virtual bandwidth request, we map it by adding suboptimal shortest paths, meanwhile setting hops no greater than a given constant to control the length for saving bandwidth. Formula (1) is used as the fitness function to calculate the individual best *Pbest* and global best *gbest*, and using formula (10) and formula (11) to update the velocity and position vectors of particle. The details of our proposed PIAS-VNE-PSO VNE algorithm is as follows:

Algorithm 2: PIAS-VNE-PSO VNE algorithm.

1. Executing the particle position initialization strategy in Algorithm 1;
2. **while** (not exceed the max iteration counts or *gbest* changed in five rounds) **do**
3.   **For each** particle in *plist* **do**;
4.   Creating a queue *Vlqueue* for virtual links, *VlqueueCount=Vlqueue.size()*;
5.   **while** (*VlqueueCount* ≥ 1) **do**
6.   *Vlqueue* dequeues; Assign the bandwidth request to l_v , hops=0, *pathcount*=0, l_s =0, *available*=false;
7.   **while** (*hops<ε || pathcount<MAX_PATHCOUNT*) **do**
8.   search the substrate path according to Floyd-Warshall [5] shortest path algorithm and get link bandwidth l_s'. Let $l_s = l_s + l_s'$ and remove the links from underlying substrate netwok.
9.   if ($l_s \geq l_v$), *available*=true;
10.   **End while**
11.   if (*available*=false), reinitialize this particle.
12.   using formula (1) as fitness function to calculate *Pbest* and *gbest*;
13.   Updating velocity and position of current particle according to formula (10) and formula (11).
14.   **End while**
15.   **End for**
16.   if *gbest* is exist, output the corresponding position as an feasible solution;
17. **End while**

Algorithm 2 uses the particle position initialization strategy to form the initial particles, then maps links according to the virtual nodes mapping solution indicated by the positions. If the current position is infeasible, the position will be reinitialized. The stop criterion of the algorithm is the iterative rounds exceed a given maximum number of iterations or the global optimal position *gbest* doesn't change during latest five rounds.

4   Experimental Results and Analysis

In order to evaluate the algorithm proposed in this paper, we compared our proposed algorithm with two existing algorithms: D-ViNE-SP [6], VNE-R-PSO [7] to observe their performance. The substrate network is set to having 100 nodes and 500 links. The physical node CPU capacity and physical link bandwidth are uniform distribution from 50 to 100. For every virtual network request, the number of virtual nodes uniform distribute from 2 to 20. Both virtual network node CPU request and virtual bandwidth request are uniform distribute from 3 to 50. He maximum number of iterations is 20.

p_1, p_2, p_3 in formula (10) are set to 0.1, 0.2 and 0.7, and γ in formula (7) is set to 0.8, 0.9 and 1, which is used to adjust load balancing.

Figure 1 shows the variation of Virtual network request acceptance ratio changing with time of all three algorithms, where the value of load factor γ of our proposed PIAS-VNE-PSO algorithm is set to be 1, 0.8, 0.9. As seen in the results, virtual network request acceptance ratio of all algorithms declines rapidly originally, and tends to be stable later. Our proposed PIAS-VNE-PSO algorithm performs better than other

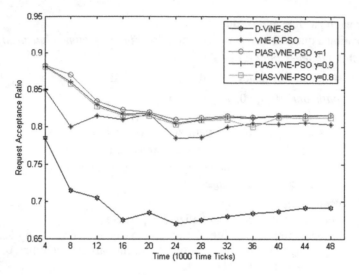

Fig. 1. The Virtual network request acceptance ratio under different γ.

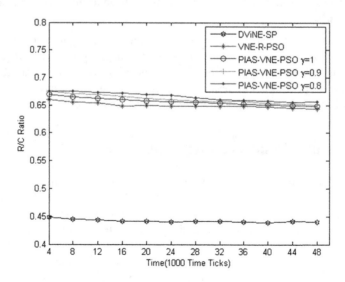

Fig. 2. The long-term of R/C Ratio of the substrate network.

algorithms, while the smaller of load factor, the lower of virtual network request acceptance ratio it obtains.

Figure 2 is the comparison of long-term revenue to cost (R/C) ratio between PIAS-VNE-PSO algorithm and the other two existing algorithms. From the results, we can see that the long-term R/C ratio of our proposed PIAS-VNE-PSO algorithm is a bit higher than the other two algorithms. The reason is that the particle position initialization strategy can make the searching in DPSO more efficient. With the time going, bigger γ incurs better R/C ratio. This is because that in order to reach load balance, the allocable resources of the substrate networks are reduced.

5 Conclusion

This paper proposes a particle swarm optimization based load balancing VNE algorithm with particle position initialization strategy, which can improve the success searching probability of particles and significantly reduce the time cost of iterations. Simulation results show that our proposed VNE algorithm can achieve better virtual network requests acceptance rate and better long-term revenue to cost ratio.

Acknowledgment. This work was supported in part by the following funding agencies of China: National Natural Science Foundation of China (61300195, 61402094), Natural Science Foundation of Hebei Province (F2014501078), The Research Fund for the Doctoral Program of Higher Education of China (Grant no. 20120042120009), and the Science and Technology Support Program of Northeastern University at Qinhuangdao (Grant no. XNK201401).

References

1. Chowdhury, N.M., Boutaba, R.: Network virtualization: state of the art and research challenges. Commun. Mag. IEEE **47**(7), 20–26 (2009)
2. Fischer, A., Botero, J.F., Till Beck, M., De Meer, H., Hesselbach, X.: Virtual network embedding: a survey. Commun. Surv. Tutorials IEEE **15**(4), 1888–1906 (2013)
3. Fischer, A., Beck, M.T., De Meer, H.: An approach to energy-efficient virtual network embeddings. In: 2013 IFIP/IEEE International Symposium on Integrated Network Management (IM 2013), pp. 1142–1147. IEEE, May 2013
4. Kennedy, J., Eberhart, R.C.: A discrete binary version of the particle swarm algorithm. In: 1997 IEEE International Conference on Systems, Man, and Cybernetics, Computational Cybernetics and Simulation, vol. 5, pp. 4104–4108. IEEE, October 1997
5. Djojo, M.A., Karyono, K.: Computational load analysis of Dijkstra, A*, and Floyd-Warshall algorithms in mesh network. In: 2013 IEEE International Conference on Robotics, Biomimetics, and Intelligent Computational Systems (ROBIONETICS), pp. 104–108. IEEE, November 2013
6. Lischka, J., Karl, H.: A virtual network mapping algorithm based on subgraph isomorphism detection. In: Proceedings of the 1st ACM Workshop on Virtualized Infrastructure Systems and Architectures, pp. 81–88. ACM, August 2009
7. Cheng, X., Zhang, Z.B., Su, S., Yang, F.C.: Virtual network embedding based on particle swarm optimization. Dianzi Xuebao (Acta Electronica Sinica) **39**(10), 2240–2244 (2011)

Sensor Networks and Social Networks

MISTER: An Approximate Minimum Steiner Tree Based Routing Scheme in Wireless Sensor Networks

Guorui Li[1($\boxtimes$)], Ying Wang[2], Cong Wang[1], and Biao Luo[1]

[1] School of Computer and Communication Engineering, Northeastern
University at Qinhuangdao, Qinhuangdao 066004, Hebei, China
lgr@neuq.edu.cn, congwl981@gmail.com,
luobiao@ise.neu.edu.cn
[2] Department of Information Engineering, Qinhuangdao Institute of Technology,
Qinhuangdao 066100, Hebei, China
wyqhd@hotmail.com

Abstract. In this paper, we propose the MISTER (approximate Minimum Steiner TreE based Routing) scheme which can construct the routing paths among sinks in multi-sink wireless sensor networks or cluster heads in clustered wireless sensor networks. By extracting a hierarchically well-separated tree based on the proposed scheme, we can build a bounded approximate minimum Steiner tree among multiple sinks or cluster heads. The experiments show that the MISTER scheme can reduce the average hop counts among multiple sinks or cluster heads. Therefore, we can save the energy of wireless sensor nodes and prolong the lifetime of wireless sensor networks.

Keywords: Wireless sensor networks · Minimum Steiner tree · Routing · Distributed · Hieratically well-separated tree

1 Introduction

As one of the most emerging technologies in the future, Wireless Sensor Networks (WSNs) are autonomous wireless networks which consist of dozens or even thousands of tiny, inexpensive and battery-powered wireless sensor nodes [1]. Those nodes are usually equipped with several different types of sensors, such as temperature, humid, light and air pressure sensors, to perceive the surrounding environment and report the sensed data to one or multiple sinks. The typical applications of wireless sensor networks include industrial process control, environment monitoring, military surveillance and smart city, etc. [2].

However, the energy resources in typical wireless sensor nodes are usually limited and un-renewable. When synchronizing data or exchanging information among sinks in multi-sink WSN or cluster heads in clustered WSN, it is very important to just select necessary intermediate forwarding nodes along the routing paths between source and destination sinks or cluster heads. Thus, the network delay and energy consumption of the whole sensor network can be reduced to a relatively low level.

© Springer International Publishing Switzerland 2016
Y. Tan et al. (Eds.): ICSI 2016, Part II, LNCS 9713, pp. 317–323, 2016.
DOI: 10.1007/978-3-319-41009-8_34

In this paper, we proposed the MISTER scheme to build the aforementioned intermediate forwarding paths among multiple sinks or cluster heads. By creating a hierarchically well-separated tree progressively, we can construct a bounded approximate minimum Steiner tree among multiple sinks or cluster heads. Furthermore, the routing paths can be extracted from the generated hierarchically well-separated tree by finding the common ancestor of the source and destination nodes.

The rest of this paper is organized as follows. Section 2 introduces the related works in the minimum Steiner tree problem. Section 3 describes the MISTER scheme in detail. Section 4 presents the experimental results. Finally, Sect. 5 concludes this paper.

2 Related Works

In a multi-sink WSN or clustered WSN, finding the minimum communication paths among sinks or cluster heads can be abstracted as a minimum Steiner tree problem. Minimum Steiner Tree (MST) problem is a classical combinatorial optimization problem. Given a weighted graph $G = (V, E)$ and a vertex subset $R \subseteq V$, a minimum Steiner tree is a tree connecting all the vertexes in R with minimum weights [3].

It is a NP problem to find a minimum Steiner tree in a graph. However, we can extract a hierarchically well-separated tree that approximates the original graph by a logarithmic distortion [4]. Then, we can easily build the minimum Steiner tree based on the extracted hierarchically well-separated tree.

An α-Hierarchically well-Separated Tree (α-HST) is a weighted tree satisfies (1) the weights from each node to its children are the same, (2) all root-to-leaf paths have the same hop distance, and (3) the edge weights from the root to leaf decrease by a factor of α. Figure 1 shows an example of 2-HST. The leaf nodes in the HST are actual vertexes in the original graph and other nodes are dummy nodes which can be mapped to one or many vertex(es) in the original graph.

Fakcharoenphol et al. proposed a centralized algorithm to extract a random 2-HST with logarithmic approximation to the original graph [5]. That is, for any $u, v \in G$, the distance of u, v in HST $d_T(u, v)$ and that in original graph $d(u, v)$ satisfies $d_T(u, v) \geq d(u, v)$ and $E[d_T(u, v)] \leq O(\log n)d(u, v)$. However, it is a centralized algorithm with

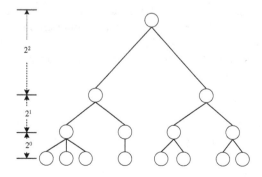

Fig. 1. An example of the 2-HST.

top-down building fashion. Gao et al. proposed a distributed random HST based scheme to extract the same hierarchically well-separated tree with bottom-up building fashion [4, 6]. Compared with the centralized algorithm, the distributed scheme is more appropriate to wireless sensor networks. We will compare our proposed scheme with the random HST based scheme in this paper. For convenience, we call sink nodes in multi-sink WSN and cluster heads in clustered WSN as skeleton nodes.

3 The MISTER Scheme

In the random HST based scheme [4], the ID of sensor nodes and skeleton nodes are chosen randomly. Thus, the structure of the generated random hierarchically well-separated trees are stochastic and various. Most of the time, the communication path among skeleton nodes should circumvent around peripheral nodes, which is not energy efficient.

In the MISTER scheme, we assign the ID of sensor nodes and skeleton nodes incrementally from the center of the network deployment area to the periphery. The structure of the generated hierarchically well-separated tree will be centripetal. In other words, the tree will root in the center of the network deployment area and extend outward. Thus, the communication path among skeleton nodes will be centripetal and no circumvention paths around peripheral nodes are required. Therefore, the network delay and energy consumption can be reduced.

The pseudocode of the MISTER scheme is presented in Table 1. After extracting the HST, the communication path between source and destination skeleton nodes can be built by finding the common ancestor of them in the HST.

Table 1. The MISTER scheme.

INPUT: A wireless sensor network with n-m sensor nodes and m skeleton nodes
Assign the ID of sensor nodes and skeleton nodes incrementally from center to the periphery
Choose a random number β uniformly from [0.5, 1]
Candidate set $S=\{n$-m sensor nodes and m skeleton nodes$\}$
WHILE $|S| \neq 1$
Each node $u \in S$ floods the network up to distance $2^{i+1}\beta$
Each node $u \in S$ selects the node $v \in B(u, 2^{i+1}\beta)$ with the lowest ID
Each node $u \in S$ nominates node v as its ith level ancestor
Remove non-nominated nodes in S
END
OUTPUT: A hierarchically well-separated tree

In Fig. 2, we present an instance of the running process of the MISTER scheme. There are 7 sensor nodes and 3 skeleton nodes in the network. Figure 2(a) shows the initial status of the MISTER scheme. The circle nodes and square nodes represent sensor nodes and skeleton nodes, respectively. The nodes in the candidate set S are marked with color.

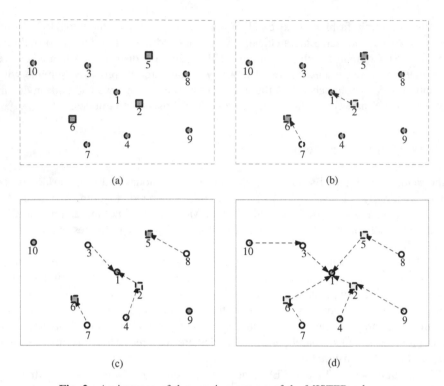

Fig. 2. An instance of the running process of the MISTER scheme.

Initially, all sensor nodes and skeleton nodes are in the candidate set S. In Fig. 2(b), node 2 and 7 nominate node 1 and 6 as their ancestors, respectively. After that, node 2 and 7 are removed from the candidate set S. In Fig. 2(c), node 4, 3 and 8 nominate node 2, 1 and 5 as their ancestors. Then, node 4, 3 and 8 are excluded from the candidate set S. Finally, node 6, 10, 5 and 9 nominate node 1, 3, 1, and 2 as their ancestors, respectively. Then, only node 1 remains in the candidate set S. Hence, a hierarchically well-separated tree is built progressively by the MISTER scheme in three iterations. The final hierarchically well-separated tree is shown in Fig. 2(d).

4   Experiments

We evaluate the performance of the proposed MISTER scheme in this section. The simulation was carried out based on the Wislab WSN simulator on a DELL desktop computer with dual core CPU and 4G memory. We deployed $40 \sim 100$ wireless sensor nodes and $4 \sim 8$ skeleton nodes in a 400×400 m^2 surveillance area. The experiment results were averaged after 100 trials.

We show two instances of the hierarchically well-separated tree built by the random HST based scheme and our proposed MISTER scheme in Figs. 3 and 4, respectively. The black dots and green dots represent sensor nodes and skeleton nodes, respectively.

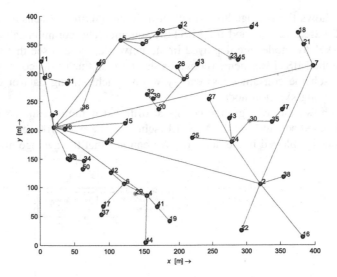

Fig. 3. An instance of the random HST based scheme.

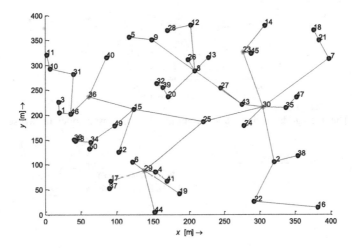

Fig. 4. An instance of the MISTER scheme.

Meanwhile, the solid lines represent the routing paths among skeleton nodes and the dotted lines represent that among sensor nodes. It should be pointed out that we keep the display IDs of all nodes in these two figures same for the sake of easy comparison.

We can see that the hierarchically well-separated tree in Fig. 3 is unbalanced. In other words, the root of the tree deviates from the center of the network. On the contrary, the hierarchically well-separated tree in Fig. 4 is balanced and roots at the center of the deployment area. Therefore, the hop counts among skeleton nodes in the HST generated by the MINSTER scheme are much shorter than that of the random HST based scheme.

Figure 5 shows the average hop counts at different number of sensor nodes in the random HST based scheme and the MISTER scheme. The communication radius is 50 m and 4 skeleton nodes are deployed in the network. We can see that the average hop counts in the MISTER scheme is obvious shorter than that of the competitor. Thus, the MISTER scheme consumes less energy when synchronizing data or exchanging information among skeleton nodes.

Figure 6 shows the average hop counts at different communication radius in the random HST based scheme and the MISTER scheme. There are 100 sensor nodes and 4 skeleton nodes deployed in the network. We can see that the average hop counts in

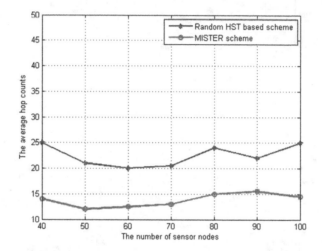

Fig. 5. The average hop counts at different number of sensor nodes in two schemes.

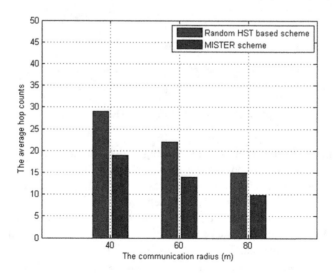

Fig. 6. The average hop counts at different communication radius in two schemes.

the MISTER scheme is obvious shorter than that of the random HST based scheme. Meanwhile, the average hop counts of these two schemes decrease with the increase of the communication radius.

5 Conclusion

We proposed the MISTER (approximate MInimum Steiner TreE based Routing) scheme in this paper. By iteratively nominating the ancestors in a gradually expanded circle, a balanced hierarchically well-separated tree can be built. The MISTER scheme consumes less energy than the existed scheme when synchronizing data or exchanging information among sinks in multi-sink WSN or cluster heads in clustered WSN.

Acknowledgments. The work in this paper has been supported by the National Natural Science Foundation of China (Grant nos. 61402094, 61300195), the Research Fund for the Doctoral Program of Higher Education of China (Grant no. 20120042120009), the Natural Science Foundation of Hebei Province (Grant nos. F2014501078, F2016501076), and the Science and Technology Support Program of Northeastern University at Qinhuangdao (Grant no. XNK 201401).

References

1. Kulkarni, R.V., Forster, A., Venayagamoorthy, G.K.: Computational intelligence in wireless sensor networks: a survey. IEEE Commun. Surv. Tutor. **13**(1), 68–96 (2011)
2. Rashid, B., Rehmani, M.H.: Applications of wireless sensor networks for urban areas: a survey. J. Netw. Comput. Appl. **58**, 1–28 (2015)
3. Huang, F.K., Richards, D.S.: Steiner tree problems. Networks **22**(1), 55–89 (1992)
4. Gao, J., Guibas, L., Milosavljevic, N., Zhou, D.: Distributed resource management and matching in sensor networks. In: 8th ACM/IEEE International Conference on Information Processing in Sensor Networks, pp. 97–108. IEEE Press, New York (2009)
5. Fakcharoenphol, J., Rao, S., Talwar, K.: A tight bound on approximating arbitrary metrics by tree metrics. J. Comput. Syst. Sci. **69**(3), 485–497 (2004)
6. Zhou, D., Gao, J.: Maintaining approximate minimum steiner tree and k-center for mobile agents in a sensor network. In: 29th IEEE Conference on Computer Communications, pp. 1–5. IEEE Press, New York (2010)

An Improved Node Localization Method for Wireless Sensor Network Based on PSO and Evaluation of Environment Variables

Qingjian Ni[✉]

School of Computer Science and Engineering, Southeast University,
Nanjing, China
nqj@seu.edu.cn

Abstract. Node localization is an essential problem for some engineering applications in wireless sensor network (WSN). The problem characteristics of node localization in WSN are analyzed firstly, and then a framework and a strategy for the settings of important parameters are given to solve such problems with particle swarm optimization (PSO). Furthermore, the environment variable of WSN is evaluated and an improved node localization method of wireless sensor network based on PSO is proposed. In the proposed node localization method, the environment variable of WSN is evaluated during the process of ranging and the fitness function is designed according to the environment variable, and then the PSO algorithm is adopted to solve the node localization problem. Compared with the traditional methods, the experiments results show that the proposed method has good performance. Besides, we also discussed the relationship among the number of nodes, the number of failure nodes, the positioning error and the mean distance-measuring error.

Keywords: Particle swarm optimization (PSO) · Wireless sensor network (WSN) · Node localization · Environment variable

1 Introduction

Wireless sensor network (WSN) is a network consisting of a series of sensor nodes, which can transmit information through wireless signal. In WSN, several base stations on the ground are used as data processing center to help process data. Sensor nodes of WSN can collect information in the region to be monitored according to specific requirements. However, the performance of each sensor node is limited in computing capacity, energy and transmission bandwidth.

Node localization in WSN is a very critical issue, and it directly affects the efficiency of the entire network. According to different node localization methods, node localization in WSN can be divided into two categories: range based localization and non ranging based localization. Range based localization locates nodes by measuring the distance between nodes, while non ranging based localization locates nodes by network connectivity. In WSN, the nodes whose positions are known are called anchor nodes, and the others are called ordinary

© Springer International Publishing Switzerland 2016
Y. Tan et al. (Eds.): ICSI 2016, Part II, LNCS 9713, pp. 324–332, 2016.
DOI: 10.1007/978-3-319-41009-8_35

nodes. The location of an ordinary node can be determined by measuring the distance between the ordinary node and the anchor nodes.

The node localization problem in WSN can be described as an optimization problem [1]. Evolutionary computation is a method which can be applied to solving complex optimization problem [1,2]. And PSO algorithm is widely used to solve all kinds of complex optimization problems because it is easy to implement and has fast convergence rate. PSO algorithm has been introduced to solve the node localization problem in WSN [3,7]. However, most research used the basic PSO algorithm when selecting PSO variants. The setting of fitness function needs further discussion. This paper estimates the environment variables during distance measuring in WSN, and designs a reasonable fitness function with the environment variables. Furthermore, we propose a novel node localization method based on environment variables, and use PSO algorithm with dynamic inertia weight to solve the problem. Experiment results show that the proposed strategy has good positioning accuracy and high practicability.

The remainder of this paper is organized as follows. Section 2 detailed describes the node localization problem in WSN and the basic concept of PSO. The proposed node localization method based on PSO is introduced in Sect. 3. And the experimental settings and results are analyzed in Sect. 4. Section 5 gives a conclusion of this paper and points out the further work.

2 Node Localization Problem in WSN and PSO Fundamentals

Node localization in WSN is a NP-hard problem [3]. Therefore, in practical engineering applications we can only find approximate optimal solution to this problem. The positioning error is an important measure of the algorithm performance.

In the node localization problem in this paper, ordinary nodes and anchor nodes are randomly deployed in a two-dimensional plane. The anchor nodes can determine the coordinates themselves, and the coordinates of the ordinary nodes need to be solven in this paper. These ordinary nodes are called target nodes.

A simple localization method is based on distance measurement. Target nodes can determine their own coordinates by measuring the distance between target nodes and anchor nodes.

However, if a node can only receive the signal from less than three anchor nodes, then the node is considered as a failure node. If the resulting location of a node is out of the predetermined region, the node can also be considered as a failure node.

Node localization in WSN is mainly based on received signal strength indicator (RSSI), time difference of arrival (TOA), angle of arrival (AOA) and so on [4]. These ranging technologies have ranging errors. The ranging error comes from many random factors, such as the wireless signal model is not accurate or there is an obstacle between the nodes. There will be errors in the positioning, and the error can be represented by a random variable. According to the central limit theorem, it can be assumed that the error is Gaussian random variable. In this paper, the error is considered as an important factor influencing the final positioning error.

PSO is one of the evolutionary computation methods, which is inspired by birds, fish and other group behavioral mechanisms. Since PSO is easy to implement and has relatively fast convergence rate, it is widely used in solving optimization problems in practical engineering applications. The basic process of PSO with inertia weight is as follows.

1. The individual particles are randomly distributed in the solution space;
2. For each particle, update the speed and position;
3. Update the best position of each particle and the optimal position of its neighborhood particles;
4. Repeat step 2 to 4 until the termination condition is satisfied.

The update equations of particle's velocity and position are as Eqs. (1) and (2).

$$
\begin{aligned}
V_i(t) = \omega \times V_i(t-1) + \phi_1 \times rand_1 \times (P_i - X_i(t-1)) \\
+ \phi_2 \times rand_2 \times (P_g - X_i(t-1))).
\end{aligned}
\tag{1}
$$

$$
X_i(t) = V_i(t) + X_i(t-1).
\tag{2}
$$

where ω is the inertia weight. The inertia weight value can affect search capability of PSO algorithm. When ω is large, PSO algorithm tends to explore the entire solution space, and is not easy to fall into local optimum, but not easy to converge. Conversely, when ω is small, it is easy to fall into local optimum but easy to converge. Therefore, in order to achieve a balance between search ability and convergence speed, we can use dynamic inertia weight. Initially, the weight is large, and as the iteration increase, the weight becomes smaller. As is proposed in [5], we can calculate the inertia weight according to Eq. (3).

$$
\omega = (\omega_1 - \omega_2) \times \frac{MaxIter - CurIter}{MaxIter} + \omega_2.
\tag{3}
$$

where ω_1 is 0.9, and ω_2 is 0.4. $MaxIter$ Is the maximum number of iterations, $CurIter$ is the current number of iteration.

There are many excellent PSO variants, which have good performance in solving optimization problems in specific engineering fields [6]. In this paper, PSO with inertia weight is used to solve node localization problems in wireless sensor networks.

3   Proposed Node Localization Method Based on PSO

3.1   Related Work

There are some initial results for the node localization problem using PSO. From the above discussion, we know that it requires at least three adjacent anchor nodes to determine the position of the target node. When there are enough adjacent anchor nodes and long transmission radius of node signal, node localization method based on PSO algorithm can achieve good results.

In literature [8], all nodes are distributed in a plane, maximum likelihood estimation method is used to solve the node localization problem. The variance of the distance-measuring error is used as the fitness function of this problem, and then such localization problem could be considered as finding the minimum value of the fitness function. Literature [8] uses PSO algorithm to solve the problem, and compares the results with simulated annealing algorithm. Literature [9] uses Gauss-Newton iterative method and PSO algorithm for node localization problems. The experiment deploys nodes in a real environment and applies localization algorithm to microcontrollers, using pedometers to estimate distances. The experimental results show in certain circumstances PSO algorithm is better than Gauss-Newton iterative method on positioning accuracy. Literature [10] also uses maximum likelihood estimation method, and changes the localization problem into finding the optimal value of a nonlinear function. All nodes are deployed by a small unmanned vehicle, which is equipped with a pedometer and a compass. So during the deployment process, the node can get the relative position of other nodes.

The focus of above studies are usually not the model of node localization problems, and most of them used the basic PSO algorithm. This paper will introduce environment variables, and use PSO with dynamic inertia weight to solve the problem.

3.2   Node Localization Model Using Environment Variable Estimation

In this paper, all nodes in WSN are distributed in a plane area, the distance measurement of nodes will generate errors by the environment. According to the central limit theorem, the error is a random number consistent with the Gaussian distribution. Assuming this error α is consistent with the Gaussian distribution, and the standard deviation is σ , which is determined by the actual environment.

$$\alpha \sim Gaussian(0, \sigma^2) \,. \tag{4}$$

Let d represent the actual distance, and d' represent measuring distance. The positioning process is as follows.

1. Anchor nodes send their coordinate information to the other anchor nodes, and uses ranging technique to measure the distance between the anchor node itself and another anchor node nearby. Assuming the coordinates of anchor node A are (x_1, y_1), and the coordinates of anchor node B are (x_2, y_2), the measuring distance between nodes is d'. Then a sample of Gaussian distribution is obtained, as shown in Eq. (5).

$$g = \sqrt{(x_1 - x_2)^2 + (y_1 - y_2)^2} - d' \,. \tag{5}$$

All anchor nodes can obtain a sample of Gaussian distribution $g_1, g_2, g_3, \ldots, g_m$ by the above method, and then find the estimate value σ' of the standard deviation α by Eq. (6).

$$\sigma' = \sqrt{\frac{\sum_{i=1}^{N} (g_i - E(g))^2}{N-1}}. \tag{6}$$

2. For an ordinary node, the coordinates are (x, y). Measure the distance between the node itself and the anchor node nearby. Assuming there are m anchor nodes near the ordinary node, the coordinates are $(x_i, y_i), i = 1, 2, 3, \ldots, m$, then the measurement result is a collection $d_1', d_2', d_3', \ldots, d_m'$. According to Eq. (7), we can obtain another sample of Gaussian distribution $h_1', h_2', h_3', \ldots, h_m'$, and then use this sample to calculate the standard deviation by Eq. (6). The absolute value of the difference between the standard deviation ε calculated in this step and σ' calculated in step 1 can be used as the fitness function of PSO algorithm.

$$h_i = \sqrt{(x - x_i)^2 + (y - y_i)^2} - d_i'. \tag{7}$$

3. For every ordinary node, the method in step 2 is used until all the nodes have been processed.

$$f(x, y) = \frac{1}{m} \sum_{i=1}^{m} (\sqrt{(x - x_i)^2 + (y - y_i)^2} - d_i')^2. \tag{8}$$

In addition, only using the above function as the fitness function of PSO algorithm may not obtain accurate node coordinates. In literature [8], Eq. (8) is used as the fitness function of PSO.

In this paper, weighted average of the function in literature [8] and the function in step 2 are used as the fitness function.

4   Experimental Results and Analysis

4.1   Experimental Setting

In this paper, m ordinary nodes are randomly deployed in a 20×20 plane, and n anchor nodes are also randomly deployed in the plane. These anchor nodes can determine their coordinates, while the ordinary nodes do not know their own positions. The coordinates of ordinary nodes are the locations of the target nodes to solve in this paper.

In a 20×20 plane, coordinates are represented by (x, y). All the ordinary nodes and anchor nodes are randomly deployed in the plane, and we need to locate the ordinary nodes according to the method described in Sect. 3.2. In the experiment, if the number of neighbors of a ordinary node is less than 3, then this node can not be located and will be regarded as a failure node. If the positioning result of an ordinary node is outside the 20×20 plane, this node is also a failure node. Node positioning error is defined as the Euclidean distance between the actual position and the result position. The mean positioning error of all the nodes which are not failed is used as the positioning error of the experiment. All the experiments were repeated 30 times, and the average is used as the final result.

4.2   Results and Analysis

The main parameters affecting the experiment results are the mean distance-measuring error σ, the number of anchor nodes and the maximum signal receiving distances between nodes. The number of anchor nodes and the maximum signal receiving distance between nodes will have impact on the number of neighbors of the ordinary nodes.

Figures 1, 2 and 3 show the relationship among the number of anchor nodes, the number of failure nodes, the positioning error and the mean distance-measuring error.

As can be seen in Fig. 1, when the transmission range is too small and the anchor nodes are few, most nodes failed in the localization. One reason is that there are no sufficient anchor nodes in the neighborhood to achieve the position of nodes. Another reason is that the positioning error is large, and the final position of the node is outside the region specified in the experiment.

In Fig. 2, the abscissa is the number of anchor nodes, and the ordinate is the positioning error. As can be seen from Fig. 2, when the number of anchor nodes and the transmission range increase, the positioning error decreased significantly. This shows that the node localization performance will be better when the number of anchor nodes in the neighborhood of the ordinary nodes decreases.

From Fig. 3, it can be seen that when the number of anchor nodes is small, the distance-measuring error has a greater impact on the final positioning result. When using more anchor nodes, the final positioning results were less affected by the distance-measuring error.

Figures 4 and 5 show the comparison results between the proposed method and traditional PSO methods when transmission is 5. As is shown in Fig. 4,

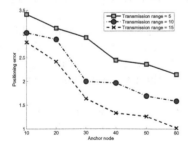

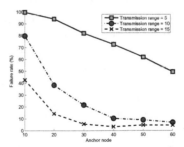

Fig. 1. The relationship between the number of anchor nodes and the number of failure nodes

Fig. 2. The relationship between the number of anchor nodes and the positioning error

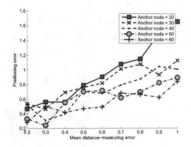

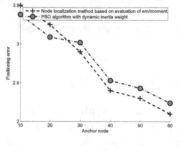

Fig. 3. The relationship between the positioning error and the mean distance-measuring error

Fig. 4. The relationship between the proposed method and traditional PSO method

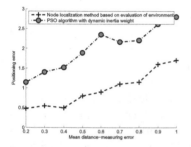

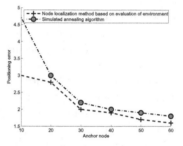

Fig. 5. The relationship between the proposed method and traditional PSO method

Fig. 6. The relationship between the proposed method and simulated annealing algorithm

when anchor nodes are few, traditional localization based on PSO algorithm has smaller positioning error. When the number of anchor node is larger, the proposed method in this paper has better performance. In Fig. 5, in different mean distance-measuring errors, the proposed method in this paper has better performance than the traditional localization method.

Figure 6 shows the comparison results between the proposed method in this paper and the node localization method based on simulated annealing algorithm in [8] when the transmission range is 10. Under the same circumstances, the positioning error of the proposed method is smaller than the positioning error of simulated annealing algorithm.

According to the above analysis, it can be seen that the proposed node localization method for wireless sensor network based on evaluation of environment variables has higher positioning accuracy.

5   Conclusions

In this paper, according to the characteristics of node localization in WSN, environment variable is evaluated in the ranging process, and then the fitness function is designed combining with environment variable and maximum likelihood estimate method, ultimately, we proposed a new node localization strategy based on PSO. Specifically, in the simulation experiments, we used the PSO with variable inertia weight to solve the problem of node localization. Based on the experimental results, we analyzed the relationship between the number of anchor nodes, the number of failure nodes, and the positioning error. We also compared the proposed method with the traditional method based on classical PSO and simulated annealing algorithm. Experimental results show that the proposed node localization method can get better positioning results. In the future, we plan to further improve the proposed model for node localization problem. We also try to use several newer PSO variants to solve such problems, compare and design better PSO variants suitable for solving such problems.

Acknowledgements. This paper is supported by the Fundamental Research Funds for the Central Universities.

References

1. Kulkarni, R.V., Venayagamoorthy, G.K.: Particle swarm optimization in wireless-sensor networks: a brief survey. IEEE Trans. Syst. Man Cybern. Part C Appl. Rev. **41**(2), 262–267 (2011)
2. Del Valle, Y., Venayagamoorthy, G.K., Mohagheghi, S., Hernandez, J.C., Harley, R.G.: Particle swarm optimization: basic concepts, variants and applications in power systems. IEEE Trans. Evol. Comput. **12**(2), 171–195 (2008)
3. Low, K.S., Nguyen, H., Guo, H.: Optimization of sensor node locations in a wireless sensor network. In: Fourth International Conference on Natural Computation, vol. 5, pp. 286–290. IEEE (2008)
4. Aspnes, J., Eren, T., Goldenberg, D.K., Morse, A.S., Whiteley, W., Yang, Y.R., Anderson, B.D., Belhumeur, P.N.: A theory of network localization. IEEE Trans. Mob. Comput. **5**(12), 1663–1678 (2006)
5. Shi, Y., Eberhart, R.C.: Parameter selection in particle swarm optimization. In: Porto, V.W., Waagen, D. (eds.) EP 1998. LNCS, vol. 1447, pp. 591–600. Springer, Heidelberg (1998)
6. Bonyadi, M.R., Michalewicz, Z.: A fast particle swarm optimization algorithm for the multidimensional knapsack problem. In: IEEE Congress on Evolutionary Computation (CEC), pp. 1–8. IEEE (2012)
7. Cao, C., Ni, Q., Yin, X.: Comparison of particle swarm optimization algorithms in wireless sensor network node localization. In: IEEE International Conference on Systems, Man and Cybernetics (IEEE SMC), pp. 252–257. IEEE (2014)
8. Gopakumar, A., Jacob, L.: Localization in wireless sensor networks using particle swarm optimization. In: IET International Conference on Wireless, Mobile and Multimedia Networks, pp. 227–230. IET (2008)

9. Guo, H., Low, K.S., Nguyen, H.A.: Optimizing the localization of a wireless sensor network in real time based on a low-cost microcontroller. IEEE Trans. Ind. Electron. **58**(3), 741–749 (2011)
10. Low, K.S., Nguyen, H., Guo, H.: A particle swarm optimization approach for the localization of a wireless sensor network. In: IEEE International Symposium on Industrial Electronics, pp. 1820–1825. IEEE (2008)

Efficient Routing in a Sensor Network Using Collaborative Ants

Md. Shaifur Rahman[1,2], Mahmuda Naznin[1(✉)], and Toufique Ahamed[1]

[1] Department of Computer Science and Engineering, Bangladesh University
of Engineering and Technology, Dhaka, Bangladesh
shaifur@cs.stonybrook.edu, {mahmudanaznin,tofiqueahmed}@cse.buet.ac.bd
[2] Department of Computer Science, State University of New York at Stony Brook,
Stony Brook, USA

Abstract. In a Wireless Sensor Network (WSN), due to energy constraints and remote deployment in harsh environment, centralized routing is difficult. In a WSN, if more sensors are active better paths are available. But it causes more energy consumption. Conventional shortest path routing causes repeated use of some nodes which causes power failure of those nodes and the routing holes pop up. In our research, we propose an efficient collaborative routing on the improvisation of ant colony meta-heuristics. We construct the best possible routing by building load balanced virtual circuits dynamically. We consider on-demand load condition to the network and span virtual circuits between source-destination pairs using collaborative ants. To validate our method we do experiments, and we also compare our method to a relevant agent based routing technique for WSNs. We find that, our method works better.

Keywords: Routing hole · Evolutionary algorithm · ACO · Forward and backward ant

1 Introduction

In a Sensor Network, sensors detect an object of interest, its sensed or processed data starts to flow towards the recipients or the sink nodes. In a typical application scenario, there may be multiple sensing operations running simultaneously feeding multiple recipients of the same or different sensing events [12]. As a result, it may happen that the same event is detected by multiple sensors and same nodes are participating in multiple routing paths. The nodes who actively participate in routing spend more energy and they may deplete faster and a *routing hole* or *disconnected closed area* may pop up in the network (see Fig. 1). Another challenge is involved with the routing hole that, whenever a hole is formed, the nodes in the vicinity of the hole try to fill up the gap and they become out of power quickly. These holes in WSN have the tendency to get bigger and deteriorating the routing scenarios faster [12].

In Fig. 1(a), Node 1 wants to send continuous stream of data to Node 22 in some intervals. Considering the delay and energy loss in route finding on each

© Springer International Publishing Switzerland 2016
Y. Tan et al. (Eds.): ICSI 2016, Part II, LNCS 9713, pp. 333–340, 2016.
DOI: 10.1007/978-3-319-41009-8_36

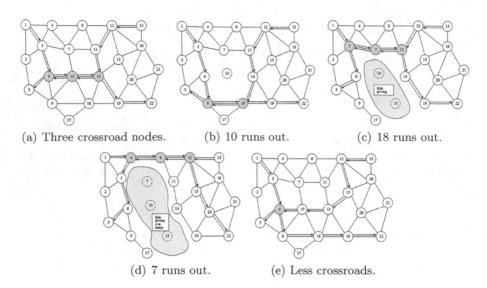

(a) Three crossroad nodes.   (b) 10 runs out.   (c) 18 runs out.

(d) 7 runs out.   (e) Less crossroads.

Fig. 1. Energy depletion and hole formation scenario.

packet basis, they can establish a *Virtual Circuit* between them and the route is <1, 3, 6, 10, 14, 19, 22>. Similarly, Node 13 with another service also wants to send packets to Node 5, and it finds a good route which is <13, 12, 11, 14, 10, 6, 5>. There are 3 nodes that fall on both of the routes. Nodes 6, 10 and 14 those are shown in Fig. 1(a). These nodes are active for data forwarding or routing, and may die out quickly. If the nodes die out, nearby nodes will be used for routing those packets and the hole will get bigger. Therefore, it is important to have load balancing, dynamic routing strategy to prevent hole formation and another is to maintain the reliability through multiple routes among the source-destination pairs. We feel that cooperative agents can help to find better routing path dynamically in an uncertain network. Ant Colony Optimization (ACO) motivate us [3] to formulate a routing algorithm that would ensure joint load distribution when multiple event horizons and multiple recipients of data-flows crop up in the WSN. It should also minimize the number of nodes on the routing path to reduce *intra* and *inter node delay*. We use ACO because, (a) ACO follows distributed method, (b) ACO can enforce network-wide exploration, and (c) ACO has sufficient parameters for fine-tuning. We organize our paper as follows. In Sect. 2, we discuss the related research, in Sect. 3, we formulate the problem and propose our solution. Section 4 provides the experimental results and finally, we give the conclusion in Sect. 5.

2   Related Work

In directed diffusion [2], sensors measure events and create gradients of informa-tion in their respective neighbourhoods. However, it suffers from energy loss due to blind flooding. Agent based Rumor Routing [4] is a nice variation of directed diffusion where the key idea is to route the queries to the nodes those have observed a particular event rather than flooding. However, the lack of optimal route and reliability are two drawbacks of this routing. Niannian Ding et al. [6] were the first to successfully incorporate the notion of ACO in WSN. But their effort concentrated on routing towards the central base station unlike in a decentralized WSN. ACO based technique [1,6,9,10], is used in WSN. The authors devised Virtual Circuit based routing technique also. But these lack the dynamic behaviour of WSNs and VCs are fixed based on geographic locations. T. Camilo et al. in [5] proposes a simple ACO integration in their routing protocol with significant overhead per packet. It is still challenging problem for WSN.

3 Problem Formulation

In Fig. 1(a), the nodes like Nodes 6, 9 and 14 be on multiple routes are called *crossroad* nodes. There might be other discrete routing activities for these *crossroad* nodes apart from those packet streams they are sitting on. Hence, their energy reservation deplete fast. If a node n sits on δ route of continuous packet streams, each generating packet at the rate of r_p, energy depletion rate r_ϵ can be related to other parameters as follows- $r_\epsilon \propto \delta(n) * r_p$ [12]. ϵ is the current energy level of the node. Where $\delta(n) =$ The number of entries in the *Virtual Circuit (VC) Table* of Node n. Since, the stream of such packets on such routes would be transmitted for quite a long interval, prior set-up of a VC is recommended. We consider current δ of a node for selecting as a routing node. Of the three crossroad nodes of Fig. 1(a), Node 10 goes low on energy and it schedules a sleep cycle (for recharge or replacement) after sending notification to its neighbours. This notification cascades up to the receiver nodes 5 and 22, who set up different routes to nodes 13 and 1 respectively excluding node 10. Since no other routing information has not been taken into account for the new route setup, it ends up with 2 more crossroad nodes- 9 and 18, as shown in Fig. 1(b). As shown in Fig. 1(c), just like Node 10, Node 18 gives up after sometime, and a big hole pops up in the network. Again, source destination pairs reschedule their respective routes and Nodes 3, 7 and 11 become new crossroad nodes. Figure 1(d) shows that, crossroad node 7 becomes the new toll and the hole consisting of nodes 7, 10 and 18 becomes even bigger threatening connectivity among source-receiver pairs. In our protocol, we feel that if we do not take only the hop count and remaining energy levels in the selection of router nodes but also the current scenario of other routes, it may buck down the number of crossroad nodes among different packet streams running simultaneously. This motivation, as shown in Fig. 1(e), renders VC which is quite different from Fig. 1(a). We find that, the number of crossroad nodes is reduced. We now define the relevant data structures and parameters for our model as follows.

In *Virtual Circuit Table (VCTable)*, any node participating in the data routing must have built such a table where *NeighbourID* is the neighbour receiving the data packet from *SourceID* and forwards it to the node with *DestinationID*.

Just like a VC in [7], the packet header contains only $<source, destination>$ information. *Ant Agent Log Table (AntLog)* is the log table for all incoming Ants. When a Forward Ant agent (yet to find the destination) is received, an entry is made to the *AntLog* table. All entries except *To* is extracted from the agent header. The *To* entry is filled after computation of heuristics. The agent, after necessary header modification is forwarded to *To* neighbour. However, if the receiving node is the destination itself, no entry is made to this table. Rather, the *Forward Ant* is converted to a *Backward Ant* and the packet is sent back to the sender. After receiving from a *Backward Ant* agent (one that has reached the destination and now backtracking to source) the entry to backtrack i.e. *From* is retrieved after matching *Source ID, Destination ID, Iteration No.* and *Ant No.*. All ants of the same iteration are numbered from 1 through some n_k. *ETL* is *Entry To Live* field, after an interval of *ETL*, the corresponding entry is deleted assuming that Ant agent has been lost due to some fault or timed out. *Ant Colony System Table (ACSTable)* is the main control table. There can be multiple instances or routes in a routing node. Each of tables is keyed on (Source ID, Destination ID). When the source starts to send out Ant agent for the first time, it does not know Destination ID. Hence, it uses *attrib* field of the header to indicate the value of interest. The setting of the value must be meaningful to all other nodes. When a forward agent without *Destination ID* reaches a routing node, *ACSTable* is keyed on *(Source ID, attrib)*. When the corresponding agent backtracks, the table is renamed by the key on *(Source ID, Destination ID)*. Hence, an *ACSTable* consists of *<Source ID, Destination ID>* pair. The *ACSTable* handles the key retrievals. The number of rows of the table primarily equals to the number of neighbours. However, when cycles or dead-ends are detected from a neighbour node, it is dropped from the *ACSTable*.

Finding Route: We now define agent's parameter. *Hop Count* is increased by each node until the destination is reached. *nInfoTimestamp* is used for determining the freshness of ϵ, δ and ϕ. Each Ant agent carries this information when traversing from one node to another node. *PTL* is a time interval after which all VC table information built up from the current Ant process is ruined (to enforce another route finding current Ant process is discarded.) *sID* is Source Identifier, *antNo* is the number of ant agents in the same iteration, *iterationNo* is the current iteration number of Ant Process Spanning from Source *sID*, S_ϕ is a source geographic location information, D_ϕ is destination geographic location information, r_τ is Pheromone Intensity Decay Interval, d_τ is the amount by which to reduce pheromone, β is the heuristics prioritization index in terms of Hop Count, ϵ is the current energy level of the remitting node, δ is the current Virtual Circuit Table's size of the remitting node, ϕ is geographic location information of remitting node, *nInfo.Timestamp* is the time stamp to verify staleness of the above mentioned metrics, *DER* is Dead End Reached Flag which turned on when cycle or dead-end detected, *attrib* is the primary Interest of sinks or data-centric convergence *dID* is the destination ID, retrieved only if such sensor destination matching attribute value ever found, the *Entry To Live* is the longevity of cached

Agent in intermediate nodes, PTL is process to live, after this duration nodes on the routes discard route data enforcing another route finding ACS.

Here, we define heuristic parameters. τ is the measure of experience of the previous agents, and it is reduced after every interval of r_τ by an amount d_τ. η_F and ρ_F are the heuristic for the *forward route* computation from the source to destination. Similarly, η_B and ρ_B are the heuristics for the *backward route* computation from the destination to source. Computation is as follows.

$$\tau = \tau + \frac{1}{HC} - d_{\tau(after\ each\ \tau_r)}, \quad \eta_F = \frac{\epsilon}{(\delta+1)(Euclidian(D_\phi,\phi)+1)},$$

$$\rho_F = \frac{\tau(i)\eta_F(i)^\beta}{\sum_{i \in rows\ of\ ACSTable} \tau(i)\eta_F(i)^\beta}, \quad \eta_B = \frac{\epsilon}{(\delta+1)(Euclidian(S_\phi,\phi)+1)},$$

$$\rho_B = \frac{\tau(i)\eta_B(i)^\beta}{\sum_{i \in rows\ of\ ACSTable} \tau(i)\eta_B(i)^\beta}$$

Here, ϵ = the current energy level, δ = the current number of entries in VCTable, ϕ = geographic position of the neighbour that sends the agent; and S_ϕ and D_ϕ are the geographic positions of the source and the destination respectively. This information is saved in the agent header.

- A Forward Ant is forwarded to a neighbour randomly selected according to: $\forall_{row\ i\ of\ acsTable}\{NeighborID(i), \rho_F(i)\}$.
- A Backward Ant is forwarded to a random neighbour generated according to: $\forall_{row\ i\ of\ acsTable}\{NeighborID(i), \rho_B(i)\}$.

In [3,8], there are chronological developments of ACO for application to render approximate solutions for computationally-intensive problems domains. For multi-objective optimization scenario, [4] gives a framework in term of Pareto Ant Colony Optimization (PACO). However, due to the limited energy of sensor nodes, these cannot be applied since these methods have computational overload.

In Algorithm 1, we define our routing method. When a new Ant agent packet arrives, the event handler determines whether it is a *Forward Ant* or *Backward Ant*, and whether it should run either of Forward Ant Process or Backward Ant Process. If an entry in *AntLog* times out, it is discarded by extraction. When no subsequent Ant agent is sent for a long time and a time-out occurs, the *makeSummary* function enforces an inference of the routes, and priorities of them to integrate into *Virtual Circuit* table from corresponding *ACSTable* based on the perspective probabilities. Periodic decrement of pheromone happens and pheromone level is updated.

Forward Ants and Backward Ants: The step handles all incoming Forward Ant agents. At first dead ends or cycles are detected. Cycle exists when agent already registered in *AntLog* table comes as forward agent, but not backward agent. Dead end exists when a node has only a single neighbour. Whatever the case, any thrown up Ant agent always carries location (if any), energy, and VC count information from previous node to current node in its path. Subsequent nodes getting the agent checks time stamp carried to decide whether that information was stale or not. After these processing, the node finally prepares the heuristics metric values. Then tables are updated accordingly- Ant agent information goes to *AntLog* table and heuristics computed goes to *ACSTable*.

Algorithm 1. ACO Based Routing

```
 1: if Event = NEW_PACKET_ARRIVAL(p) then
 2:     if packetType(p) = ANT_AGENT then
 3:         h ← read Ant Agent Header of packet p
 4:         if h.forwardAnt = TRUE then
 5:             processForwardAnt(h)
 6:         end if
 7:         if h.forwardAnt = FALSE then
 8:             processBackwardAnt(h)
 9:         end if
10:     end if
11: else if Event = ENTRY_TIME_OUT(i) then
12:     extract < AntLog > (Index = i)
13: else if Event = PROCESS_TIME_OUT(tID) then
14:     VCTable(s, d) ← makeSummary(acsTable((tID)) >)
15:     delete < acsTable((tID) >)
16: else if Event = PHEROMONE_UPDATE(tID) then
17:     for each i ∈< acsTable(tID)(NeighborID) do
18:         τ(i) ← max(0, τ(i) − d_τ)
19:     end for
20: end if
```

Before sending the ant to a random neighbour according to the computed probability distribution, its header is enriched with the processing nodes own information regarding energy level, location etc. The prime metric *hop count (HC)* is incremented in every node except the detection of dead end and cycles and until the agent reaches destination and turns into a Backward Ant. Backward Ants are processed differently from the way forward ones are processed. First of all, backward heuristics are calculated. This information makes a reverse near optimal path from destination to source, which may not be the same as the forward path (source to destination). Before sending the agent to the previously immediate remitting neighbour, information is popped from *AntLog* table. If the processing node is the source itself, it increases the total agent count sent in the current iteration and initiate next iteration process.

As iteration number goes up, weight given to the tendency of exploration, is decreased gradually by decreasing β per iteration. This process also justifies whether terminating condition has been met; if not, it initializes the header parameters and starts to send out Ant agents of the next iteration. One important thing is that both the source and the destination sterilize the energy level, $h.\epsilon$ (=1) and VC table count $h.\delta$ (=0) parameters. Based on these parameters a node is selected as the best router node.

4   Experimental Results

In this section, we report the experimental results for validating our method. For our experiment, we use parameters in [11]. We use a discrete event simulation model where a central event queue continuously dispatches registered events and dispatched events may span further events. Uniform distribution U(0,1) is used to generate random numbers. We assume that the medium is lossless. The measured values are the average of several runs with the same source-destination pair parameter settings. For performance measure, we use *Set-up Delay* which is the count of packets needed to establish a Virtual Circuit and packet delivery cost which is *hop count per packet*. In our method, if hop count per packet is less, that means the less crossroad nodes are required. The average of runs is computed.

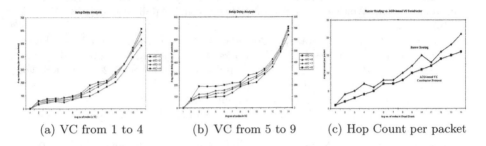

(a) VC from 1 to 4   (b) VC from 5 to 9   (c) Hop Count per packet

Fig. 2. The effect on Set-up Delay (a), (b) and (c) hop count.

In Fig. 2(a) and (b), set-up delay of Virtual Circuits (VCs) are plotted against the average number of nodes in the routes. Inner curves represent lower number of the concurrent VCs. Outer curves represent the higher number of VCs being present in the WSN. The measure of set-up cost in term of packets sent by source gives a clear picture of rationality of setting cost of a VC. For example, for a scenario where the average number of nodes present in the VC is around 12 (high) and the number of concurrent VCs is 8, setting up another VC needs to send only 20 packets. In Fig. 2(c), we compare our algorithm in term of the average number of hops required for a packet to pass through the routes. We find that, our method works better by providing less hop counts per packet. It means that less crossroad nodes are involved.

5   Conclusion

In our paper, we study the challenging routing hole problem in a WSN which is a multiple objective optimization problem. Routing is degraded due to the formation of routing holes. We consider on-demand load condition of the network and construct virtual circuits between source-destination pairs of the network nodes. We compute the best possible load balanced routes dynamically by using co-operative decisions of distributed ant agents. The multiple distributed forward and backward ant agents facilitate quick finding of better routes. To validate our method, we compare our method to a relevant agent based routing technique for WSN. We find that, our method works better by providing the less hop counts which indicates that in our method, less node are involved for QoS routing.

References

1. Amiri, E., Keshavaraz, H., Alizadeh, M., Zamani, M., Khodadadi, T.: Energy efficient routing in Wireless Sensor Networks based on fuzzy ant colony optimization. Int. J. Distrib. Sens. Netw. **2014**, 1–17 (2014). Hindawi
2. Intanagonwiwat, C., Govindan, R., Balakrisnan, D.: Directed diffusion: a scalable and robust communication paradigm for sensor networks. In: Proceedings of ACM Mobicom, MA, USA, pp. 56–57 (2000)
3. Bonabeau, E., Dorigo, M.: Swarm Intelligence: From Natural to Artificial Systems. Oxford University Press, New York (1999)
4. Braginsky, D., Estrin, D.: Rumour routing algorithm for sensor networks. In: Proceedings of the Workshop on Sensor Networks and Applications (WSNA), GA, USA (2002)
5. Camilo, T., Carreto, C., Silva, J.S., Boavida, F.: An energy-efficient ant-based routing algorithm for Wireless Sensor Networks. In: Dorigo, M., Gambardella, L.M., Birattari, M., Martinoli, A., Poli, R., Stützle, T. (eds.) ANTS 2006. LNCS, vol. 4150, pp. 49–59. Springer, Heidelberg (2006)
6. Ding, P.N., Liu, X.: Data gathering communication in wireless sensor networks using ant colony optimization. In: Proceedings of IEEE International Conference on Robotics and Biomimetics (ROBIO) (2004)
7. Doerner, K., Gutjahr, W.-J., Hartl, R.F., Strauss, C., Stummer, C.: Pareto ant colony optimization: a meta-heuristic approach to multiobjective portfolio selection. Ann. Oper. Res. **131**, 79–99 (2004). Springer-Verlag
8. Dorigo, M., Stutzle, T.: Ant Colony Optimization. MIT Press, Cambridge (2004)
9. Okedem, S., Karaboga, D.: Routing in Wireless Sensor Networks using an Ant Colony Optimization (ACO) router chip. Sensors **9**(1), 909–921 (2009). MDPI, Switzerland
10. Pei, Z., Deng, Z., Yang, B., Cheng, X.: Application-oriented Wireless Sensor Network communication protocols and hardware platforms: a survey. In: Proceedings of IEEE International Conference on Industrial Technology, pp. 1–6 (2008)
11. Saleem, M., Khayam, S.A., Farooq, M.: A Formal performance modelling framework for bio-inspired ad-hoc routing protocols. In: Proceedings of the Annual Conference on Genetic and Evolutionary Computation (GECCO), GA, USA, pp. 103–110 (2008)
12. Stojmenovic, I.: The Handbook of Sensor Networks: Algorithms and Architectures. Wiley, Hoboken (2005)
13. Zhu, X.: Pheromone based energy aware directed diffusion algorithm for Wireless Sensor Network. In: Huang, D.-S., Heutte, L., Loog, M. (eds.) ICIC 2007. LNCS, vol. 4681, pp. 283–291. Springer, Heidelberg (2007)

Community-Based Link Prediction
in Social Networks

Rong Kuang[✉], Qun Liu, and Hong Yu

Chongqing Key Laboratory of Computational Intelligence,
Chongqing University of Posts and Telecommunications,
Chongqing 400065, People's Republic of China
kuangrongcom@163.com

Abstract. Link prediction has attracted wide attention in the related fields of social networks which has been widely used in many domains, such as, identifying spurious interactions, extracting missing information, evaluating evolving mechanism of complex networks. But all of the previous works do not considering the influence of the neighbors and just applying in small networks. In this paper, a new similarity algorithm is proposed, which is motivated by the herd phenomenon taking place on network. Moreover, it is found that many links are assigned low scores while it has a longer path. Therefore, if such links the longer path has not been taken into account, which can improve the efficiency of time further, especially in large-scale networks. Extensive experiments were conducted on five real-world social networks, compared with the representative node similarity-based methods, our proposed model can provide more accurate predictions.

Keywords: Link prediction · Community structure · Common neighbor · Herd phenomenon

1 Introduction

The graphs are always used to describe the complex networks in which nodes stand for individuals or organizations and edges represent the interaction among them. The purpose of link prediction is to estimate the possibility of unknown links according to the known links [1]. And the link prediction has been applied in many relation networks such as underground relationships between terrorists [2], prediction of being actor [3], recommendation of friends for new members [4], and so on.

Finding the similar nodes based on similarity calculation are basic methods in link prediction which have low space complexity and low time consumption. In these methods, for each pair of nodes, x, y, the $s(x, y)$ is used to define the similarity between x and y [5]. The core of these methods is to find a good criterion of similarity which plays an effective role on providing appropriate result in link prediction.

This paper is organized as follows. A short description of studies that had been done is presented in Sect. 2. In Sect. 3, the problem of link prediction and typical evaluation methods are described. A novel method is put forward in Sect. 4. Section 5 contains the results analysis of experiments by comparing our method with other

© Springer International Publishing Switzerland 2016
Y. Tan et al. (Eds.): ICSI 2016, Part II, LNCS 9713, pp. 341–348, 2016.
DOI: 10.1007/978-3-319-41009-8_37

previous works. Finally in Sect. 6, we summarize the features of the proposed model and the prospect for the future.

2 Related Work

The mainly methods used in link prediction can be classified into two categories: topology-based methods and learning-based methods. Topology-based methods only use the local node features which have low space complexity and low time consumption. Zhang and Wu [6] put forward that 3 hops common neighbor give valuable contributions to the connection likelihood. Yin et al. [7] proposed the node link strength algorithm (SA), which considered both common neighbors and link strength between each common neighbor node. The accurate community structure [8] can also improve the accuracy of prediction. Valverde-Rebaza and Lopes [9] combined topology with community information by considering users' interests and behaviors. The learning-based methods often have better performance than topology-based algorithms, many studies [10–12] show that using attributes of nodes and links (such as users' ages, interests, characteristics and friends) can significantly improve the link prediction performance. Menon and Elkan [13] treated link prediction as matrix completion problem and extend matrix factorization method to solve the link prediction problem. Ozcan and Sule [14] utilizes the network temporal information along with modeling the combination of topological metrics and link occurrences information, this method can predict new link information and repeat occurrences of existing links.

Our approach is based on similarity by considering the information of communities. Meantime we introduce a new method which based on herd phenomenon to show the influence of neighbors. The result shows that our model has good performance.

3 Problem Description and Evaluation Method

3.1 Problem Description

Given an undirected simple network G (V, E), where V represents the set of vertices, E represents the set of edges. For each pair of nodes, $x, y \in V$, every algorithm referred to in this paper assigns a score s_{xy}. This score can be viewed as a measure of similarity between nodes x and y, higher score implies higher likelihood that two nodes are connected, and vice versa.

Empirically, the known link set E would normally be randomly divided into a training set of E_p and a test set of E_t before the experiment, which the training set contains 90 % links and the remaining 10 % of links are constituted in the test set.

3.2 The Algorithm Based on Node Similarity

We will apply four representative traditional algorithms which based on node similarity to five data sets, and compared the accuracy with our method later.

Common Neighbors (*CN*) [15]: it means the possibility of a link between two nodes is equal to the number of their common neighbors. Let $\Gamma(x)$ stand for the neighbor of x, then the score of *CN* can be calculated as follows:

$$S_{xy}^{CN} = |\Gamma(x) \cap \Gamma(y)|. \tag{1}$$

Adamic-Adar Index (*AA*) [16]: at first, it is used to calculate the similarity of two pages. Both common neighbor nodes and the degrees of common neighbor nodes are taken into account.

$$S_{xy}^{AA} = \sum_{z \in \Gamma(x) \cap \Gamma(y)} \frac{1}{\log \Gamma(z)}. \tag{2}$$

Resource Allocation Index (*RA*) [17]: namely, the problem whether the node x and node y generate a link in the network is converted to analyze the process of x transferring resources to y, and the calculation of the similarity of x and y is converted to calculate the amount of resources that y has received from x.

$$S_{xy}^{RA} = \sum_{z \in \Gamma(x) \cap \Gamma(y)} \frac{1}{\Gamma(z)}. \tag{3}$$

Cohesive Common Neighbors (*CCN*) [6]: the three hops common neighbor can also give valuable contributions to the connection likelihood.

$$S_{xy}^{CCN} = \sum_{z \in \Gamma(x) \cap \Gamma(y)} \frac{1}{\Gamma(z)} + \sum_{\substack{m \in \Gamma(x), n \in \Gamma(y) \\ m \in \Gamma(n)}} \frac{1}{\Gamma(m) * \Gamma(n)}. \tag{4}$$

3.3 Evaluation Metrics

The area under the receiver operating characteristic curve (*AUC*) [18] and precision [19] are widely used to quantify the accuracy of link prediction. Providing the rank of all non-observed links, the *AUC* value can stand for the probability that a randomly chosen missing link is given a higher score than a randomly chosen nonexistent link.

$$AUC = \frac{n' + 0.5n''}{n}. \tag{5}$$

Different from *AUC*, *Precision* can be defined as the ratio of the number of relevant links to selected links. In our experiments, all the missing and nonexistent links are ranked in decreasing order firstly. The formula of *Precision* is as following:

$$Precision = m/L. \tag{6}$$

In which, we concentrate on the *top-L* (here $L = 100$) links in recovery, and m represents the numbers of actual links in the testing set E_p.

4 Community-Based Evolution Algorithm

4.1 Problem Presentation and Evolution Model

Social networks are highly structured, and now the scale of the network is very huge. Compared with traditional algorithm model, it is difficult to apply to large datasets. Inspired by birds of a feather flock together, and Abir et al. [8] proved that constructing communities can improve the accuracy of prediction, and it can also greatly advance time efficiency, that is the reason we introduce community.

We assume that the scale of a community will not change along with the evolution. The evolution model of the communities is defined as follows:

Community Detection. We introduce density-based link clustering algorithm [20] to divide the network into communities. For convenience, we denote the community set as C in the following text.

Evolution.

(a) Starting from a node p in community C_i (one of community in C).
(b) Selecting nodes around p which $L_{pq} < AVG_L$ (L_{pq} represents the distance between node p and node q, AVG_L represents the average distance of the network.) and $D_q \geq D_p$ (D_x represents the degree of node x). Q denotes the set of the selected node.
(c) Calculating the similarity between p and Q_i by Eq. (7).
(d) Inserting the link to network whose score are relatively high.
(e) Repeat steps (a) to (d) until every node in C_i is considered, as well as every community in C.

4.2 *SN* (Similar to the Neighbors) Algorithm

In this paper, the difference with previous algorithms is that we start from a single node p, and then we just find possible related nodes around p, rather than compare to all other nodes. In addition, the relationship of the node p's neighbor with node q can also make important influence on the result when we analyze whether node p has a connection with node q. Therefore we need consider not only the relationship of p and q, but also the potential influence between neighbors of p and q. Finally, we proposed a method based on the similarity with neighbors which we call it Similar to the Neighbors (*SN*). The definition of the influence of neighbors can be expressed as follow:

$$S_{xy}^{SN} = (1 - \delta) \sum_{z \in \Gamma(x) \cap \Gamma(y)} \frac{1}{\Gamma(z)} + \delta \max_{z \in \Gamma(x)} \sum_{t \in \Gamma(z) \cap \Gamma(y)} \frac{1}{\Gamma(t)}. \qquad (7)$$

5 Experiments and Results Analysis

5.1 Data Analysis

We collect five real-world social network datasets, which are different from previous. All these networks contain large amounts of data. (1) *Facebook* [21]: Facebook is a social networking service website. The data set contains 4039 users and 88234 friendship links. (2) *Twitter* [21]: Twitter is an American social networking and micro-blog services, and it is one of the ten most visited Internet sites over the world. And it includes 81304 users and 1768149 connections. (3) *YouTube* [22]: YouTube is the world's largest video site, which contains 1134890 nodes and 2987624 edges. (4) *Sina web*: Sina web provides micro-blog serve, users can use one sentence to express what he saw, heard and thought or sent a picture to share with friends. The data set we obtained contains 60955 nodes and 311056 edges. (5) *DBLP* [22]: DBLP integrates English literature in computer field. It contains 317080 authors and 1049866 connected. All figures are typical of social networks and their topological features are shown in Table 1.

Table 1. The basic topological features of five example networks.

Networks	$|V|$	$|E|$	$<k>$	$<d>$	C
Facebook	4039	88234	43.691	3.693	0.606
DBLP	317080	1049866	11.256	4.416	0.632
YouTube	1134890	2987624	5.265	4.069	0.081
Sina web	60955	311305	3.802	4.251	0.083
Twitter	81306	1768149	43.494	3.583	0.565

$|E|$ and $|V|$ are the total numbers of links and nodes respectively. $<k>$ is the shortest average degree of the example networks. $<d>$ is the average shortest distance between node pairs. C stands for the clustering coefficient of every network.

5.2 Results

In order to get the proportion of parameters in Eq. (7). Figure 1 shows the performance of SN measured by AUC as a function of δ. According to Fig. 1, AUC changes with δ. We choose δ when AUC obtain the highest, and then compare with other methods later.

From the Tables 2 and 3, the simulation results of our algorithm and other previous algorithms are shown, where the black and bold items represent the highest accuracies. We simplify the computational complexity by introducing community information.

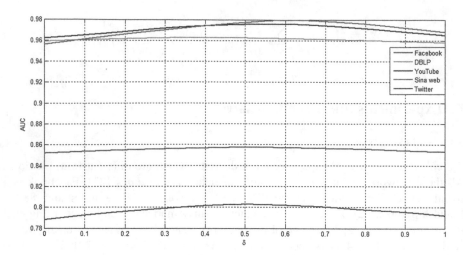

Fig. 1. The performance of SN measured by AUC as a function of δ in Eq. (7) (Colour figure online)

Table 2. The prediction accuracy measured by *AUC* on five real-world networks. Each number is obtained by averaging over 10 implementations with independently random partitions of testing set and training set. The abbreviations, *AA*, *CN*, *RA* and *CCN*, stand for Adamic-Adar Index, Common neighbors, Resource Allocation Index and Cohesive Common Neighbors, respectively. The parameter for *SN*, ε and δ are divergence within different data sets, the parameter values get from Fig. 1.

Measures	CN	AA	RA	CCN	SN
Facebook	**0.861**	0.849	0.852	0.856	0.858
DBLP	0.950	**0.963**	0.959	0.961	**0.963**
YouTube	0.956	0.974	0.962	0.975	**0.976**
Sina web	0.953	0.965	0.956	0.978	**0.980**
Twitter	0.729	0.748	0.788	**0.790**	0.789

Table 3. The prediction accuracy measured by the precision metric on five real-world networks.

Measures	CN	AA	RA	CCN	SN
Facebook	0.149	**0.367**	0.326	0.315	0.356
DBLP	0.132	0.526	0.229	**0.531**	0.416
YouTube	0.207	0.349	0.326	0.478	**0.504**
Sina web	0.413	0.469	0.545	0.538	**0.547**
Twitter	0.139	**0.212**	0.181	0.203	0.199

As can be seen form Table 2, our algorithm outperforms other algorithms in *YouTube* and *Sina web*, and the accuracies of *CCN* are close to our method in these two networks, our model also performs well in *DBLP*. As can be seen from Table 2, all method performance are poor in *Facebook* and *Twitter*, just because the method based on common neighbors will lost the meaning of original resource allocation in rich-club [23] networks.

As the result Table 3 we can see, the algorithm we purposed, can give the better forecast results than other algorithms in the first type of network. From the above discussion, we know that fewer common neighbors of two nodes will cause about lower scores of these both nodes, though they have connection among them. Therefore, the accuracy of the forecasts will cut down when the pair of nodes, which get lower score, are ignored. In our model, motivated by the herd phenomenon which means any node likes to follow the leader neighbors as in the reality, we consider the influence of the neighbors when analyzing whether there is a relationship between two nodes. Thus, it can improve the score among many pair of nodes. We can also see from Table 3, *AA* get better score than other methods in the second type of network.

6 Conclusion

In this paper, we propose a new algorithm to predict the missing links in the network, and also compare it with some other typical link prediction algorithms based on nodes similarities. The difference between our approach and other previous work is that we introduced the community message which motivated by birds of a feather flock together phenomenon, and it can greatly improve time efficiency, and ensure the accuracy.

Numerical results on the five real-world data sets of social network indicate that: (1) There is no such algorithm which can be applied to all kinds of networks due to the different characteristic of networks. (2) Many social networks can be roughly divided into three categories, we called Hub network (clustering coefficient is very small and have low average degree), Uniform network (high clustering coefficient and large degree) and Trend center network (contains small average degree of node and have high clustering coefficient). The algorithms based on common neighbors have better performance in Hub network, but all existing algorithms get poor scores in Uniform network called rich-club network.

In this paper, we found that the social network can be divided into three types roughly by its average degree and clustering coefficient. In the future, we will committed to improve the prediction accuracy of each type of network through using different methods respectively.

Acknowledgments. This work is partly funded by the National Nature Science Foundation of China (61075019) and the Natural Science Foundation of Chongqing (CSTC2014jcyjA40047) and the Municipal Education Commission research project of Chongqing (KJ1400403).

References

1. Lü, L.Y., Zhou, T.: Link prediction in complex networks: a survey. Phys. A: Stat. Mech. Appl. **390**, 1150–1170 (2011)
2. Clauset, A., Moore, C., Newman, M.E.J.: Hierarchical structure and the prediction of missing links in networks. Nature **453**, 98–101 (2008)

3. O'Madadhain, J., Hutchins, J., Smyth, P.: Prediction and ranking algorithms for event-based network data. ACM SIGKDD **7**(2), 23–30 (2005)
4. Han, X., Wang, L.Y., Chen C., Farahbakhsh, R.: Link prediction for new users in social networks. In: ICC, pp. 1250–1255 (2015)
5. Lü, L.Y., Pan, L.M., Zhou, T., Zhang, Y.C., Stanley, H.E.: Toward link predictability of complex networks. PANS **112**, 2325–2330 (2015)
6. Zhang, W.Y., Wu, B.: Accurate and fast link prediction in complex networks. In: Natural Computation (ICNC), pp. 653–657 (2014)
7. Yin, G., Yin W.S., Dong, Y.X.: A new link prediction algorithm: node link strength algorithm. In: SCAC, pp. 5–9 (2014)
8. Abir, D., Nilloy, G., Soumen, C.: Discriminative link prediction using local links, node features and community structure. In: Data Mining (ICDM), pp. 1009–1018 (2013)
9. Valverde-Rebaza, J., Lopes, A.: Exploiting behaviors of communities of twitter users for link prediction. Soc. Netw. Anal. Min. **3**, 1063–1074 (2013)
10. Li, X., Chen, H.: Recommendation as link prediction in bipartite graphs: a graph kernel-based machine learning approach. Decis. Support Syst. **54**, 880–890 (2013)
11. Scellato, S., Noulas, A., Mascolo, C.: Exploiting place features in link prediction on location-based social networks. In: Proceedings of the 17th ACM SIGKDD International Conference on Knowledge Discovery and Data Mining, San Diego, pp. 1046–1054 (2011)
12. Chen, Z., Zhang, W.: A marginalized denoising method for link prediction in relational data. In: Proceedings of the 2014 SIAM International Conference on Data Mining, Philadelphia, pp. 298–306 (2014)
13. Menon, K., Elkan, C.: Link prediction via matrix factorization. In: Proceedings of the 2011 European Conference on Machine Learning and Knowledge Discovery in Databases, Athens, pp. 437–452 (2011)
14. Ozcan, A., Sule, O.G.: Multivariate temporal link prediction in evolving social networks. In: 2015 IEEE/ACIS 14th International Conference on Computer and Information Science (ICIS), pp. 185–190 (2015)
15. Zhu, M., Zhou, Y.: Density-based link clustering algorithm for overlapping community detection. J. Comput. Res. Dev. 2520–2530 (2013)
16. Newman, M.E.J.: Clustering and preferential attachment in growing networks. Phys. Rev. E **64**, 025102 (2001)
17. Adamic, L.A., Adar, E.: Friends and neighbors on the web. Soc. Netw. **25**, 211–230 (2003)
18. Zhou, T., Lü, L., Zhang, Y.C.: Predicting missing links via local information. Eur. Phys. J. B **71**, 623–630 (2009)
19. Hanley, J.A., McNeil, B.J.: A method of comparing the areas under receiver operating characteristic curves derived from the same cases. Radiology **148**, 839–843 (1983)
20. Zhu, M., Meng, F.R., Zhou, Y.: Density-based link clustering algorithm for overlapping community detection. 2520–2530 (2013)
21. Herlocker, J.L., Konstan, J.A., Terveen, L.G., Riedl, J.T.: Evaluating collaborative filtering recommender systems. ACM TOIS **22**, 5–53 (2004)
22. McAuley, J., Leskovec, J.: Learning to discover social circles in ego networks. Adv. Neural Inf. Process. Syst. **25**, 548–556 (2012)
23. Zhou, S., Mondragon, R.J.: The rich-club phenomenon in the internet topology. IEEE Common. Lett. **8**, 180–182 (2004)

Comparative Statistical Analysis of Large-Scale Calling and SMS Network

Jian Li, Wenjun Wang, Pengfei Jiao$^{(\boxtimes)}$, and Haodong Lyu

Tianjin University, Tianjin 300072, China
pjiao@tju.edu.cn

Abstract. Mobile phone call and SMS are one the most popular communication means in modern society. The interactions between individuals result in a complex community structure that embody the social evolution. The real time call and SMS records of 36 million mobile phone users provide us with a valuable proxy to understand the change of communication behaviors embedded in social networks. Mobile phone users call each other and send SMS forming two paralleled directed social networks. We perform a detailed analysis on these two weighted networks and their derivative networks by examining their degree, weight, strength distribution, clustering coefficients and topological overlapa, as well as the correlations among these quantities. We focus on comparing the statistical properties of these networks and try to discover and interpret the discrepancy between calling and SMS networks. The finings shows that these networks have many structural features in common and exhibit idiosyncratic features when compared with each other. These findings offer insight into the pattern differences between the two large networks.

Keywords: Mobile phone network · Human dynamic · Complex network

1 Introduction

Mobile phones are wildly used in modern society. People mainly used phone to make calls and send SMS before the birth of smartphone. With the boom of mobile Internet, instant message applications such as WeChat change people's communication patterns. It was released in October 2015 by the Ministry of Industry and Information Technology of China that there were 1.302 billion mobile phone users and a quarter of them were 4G network users. The massive data brought by billions of users provide us with opportunities to understand human's communication patterns and the evolution of social networks.

Gonzalez et al. studied mobile phone records of more than 6 million anonymous mobile phone users over a period of 6 months in an European country and found that the density function followed a shifted power law with an exponential cutoff. Though the individual's communication patterns might not be homogeneous, the communication pattern of human beings are largely predictable at the aggregate level [1].

© Springer International Publishing Switzerland 2016
Y. Tan et al. (Eds.): ICSI 2016, Part II, LNCS 9713, pp. 349–357, 2016.
DOI: 10.1007/978-3-319-41009-8_38

Individual communication patterns are too frequently changing to be explored. However, the records of mobile phone provide us with opportunity to study human communication dynamics via complex networks. We know phone call and SMS are two wildly used communication methods in our daily life, and the duration analysis on two consecutive calls [2,3] and short message correspondences has achieved significant gains that the distribution of inter-communication durations have a fat tail and human interactions exhibit non-Poissonian characteristics. However, the communication networks' evolution and comparative analysis between calling network and SMS network have long been ignored.

In this paper we present a comparative analysis on calling and SMS networks constructed from a data set of 36 million anonymous users' mobile phone records over one month. We investigate these weighted, large scale, peer to peer social interaction networks, with emphasis on the eight derivative networks of calling and SMS networks. The eight networks include directed network, mutual network, statistically validated directed network, statistically validated mutual network and their corresponding giant components. We carry out a systematic analysis of basic and advanced network characteristics, and our work mainly focus on the qualitative analysis and the comparison between the calling and SMS networks in the idiosyncratic characteristics.

2   Data and Network Construction

The networks are constructed from the detailed mobile phone records of more than 36 million mobile phone users. We split the data set into call set and SMS set. In each part we treat all users in our data set as nodes and a directed edge is drawn from a call maker to a call receiver. Both data sets have an original network, a mutual network and corresponding statistically validated network. The statistically validated network is performed by comparing the number of connection between each pair of originator and receiver with a null hypothesis of random matching between the originator and receiver. The method is a variant of the concept proposed in Ref. [4] which evolves to different versions in different systems [5–7]. The statistically validated method is attaine by considering the number of calls (SMS) originated by the originator and the number of call (SMS) received by the receiver. Follow that, the compatibility is checked between the number of connections between them an the null hypothesis that these calls an SMS are attained by setting reveivers randomly. The test allows us to assign a p-value to every tested originator and receiver pair and the p-values are then compared with the 1 % of a statistical threshold. If the p-values are less than the threshold, we assume the connection has a social original. More detail information can be found in Ref. [4]

It is generally held that the frequency of SMS usage is much higher than calling, but our statistical results of mobile phone records show a contrary phenomenon. Due to the rapid development of instant messaging applications and SMS has been on a decline reported by the MIIT, we believe the result achieved

Table 1. Sizes of the calling and SMS networks and their giant components. N_{node} and N_{edge} are respectively the number of nodes and edges of a calling network. N_{Comp}, $N_{GC,node}$ and $N_{GC,edge}$ are respectively the number of components, the number of nodes and edges of the giant component of network.

CALL	N_{node}	N_{edge}	N_{Comp}	$N_{GC,node}$	$N_{GC,edge}$
DCN	35,637,916	166,315,663	402,940	34,567,395	165,614,958
SVDCN	18,514,998	77,367,187	418,599	17,407,810	46,083,991
MCN	5,302,334	25,937,081	164,756	4,917,763	51,401,150
SVMCN	4,190,455	8,983,869	157,462	3,822,224	17,520,235
SMS	N_{node}	N_{edge}	N_{Comp}	$N_{GC,node}$	$N_{GC,edge}$
DMN	16,261,236	104,211,414	211,992	15,752,994	52,010,649
SVDMN	9,820,402	35,026,653	254,541	9,179,469	17,199,919
MMN	322,117	403,552	9,110	11,889	25,099
SVMMN	138,247	77,443	63,893	60	124

in our study are reasonable. All the detail information about the network are summarized in Table 1.

3   Network Characteristics

3.1   Degree and Degree-Degree Distribution

As a basic network characteristic, degree distributions of networks are shown in Fig. 1. Since DCN and DMN and their statistical validated networks are directed, we respectively show their in-degree and out-degree distributions in Fig. 1(a) and (b). We find that all the eight probability distributions are well fitted by an exponentially truncated power law [10]: $P(k) = ak^{-\gamma_k}e^{-k/k_c}$

The distributions are all skewed with a fat tail, which indicates that most users communicate with only a few individuals. There exists noticeable differences between in-degree and out-degree distribution in panels (a) and (b). The out-degree distributions have a fatter tail and narrower distribution range than in-degree distributions. This indicates that most users' initiative usage frequency of mobile phone is lower than passive usage frequency.

In general, communication interaction via SMS between individuals should be more frequently than interactions through phone calls, because message reply is more convenience than dialing phone number and SMS is easily formed dialogic conversation. In panels (c) and (d), the degree distributions of MMN and SVMMN are quite sparse and the distribution range are also very narrow. It appears plausible that the mutual network is dominated by trusted interaction, that's to say people only reply to whom they know well. So we can safely infer that users mainly use phone call for interaction rather than SMS. An interesting phenomenon in (a) and (c) is that the MCN's right-end tail is much narrower

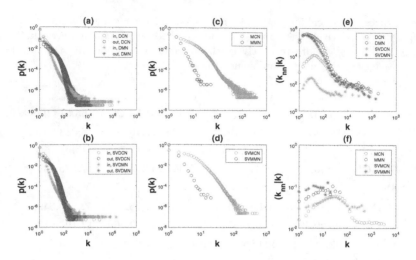

Fig. 1. Degree and Degree-degree distribution. (a) Distributions of in-degree and out-degree of DCN and DMN. (b) Distributions of in-degree and out-degree of SVDCN and SVDMN. (c) Degree distributions of MCN and MMN. (d) Degree distributions of SVMCN and SVMMN. (e) Average nearest neighbor degree $\langle k_{nn}|k\rangle$ as a function of degree k for DCN, DMN, SVDCN and SVDMN. (f) Average nearest neighbor degree $\langle k_{nn}|k\rangle$ as a function of degree k for MCN, MMN, SVMCN and SVMMN.

than DCN's, because the calling criterion for the construction of MCN is capable of filtering out most unusual calls associated with sale calling and hot lines.

Social networks are expected to be assortative: people with many friends tend to connect with other people who also have many friends. Nodes in the networks are not independent if the degrees of adjacent nodes have degree-degree correlation. In a practical way, we define the average nearest neighbor degree of a node $k_{nn,i} = (1/k_i)\sum_{j\in N(v_i)} k_j$, where N_{vi} denotes the neighbor nodes of node v_i and edge direction is ignored when dealing with directed networks. By averaging this over all nodes in the network of a given degree k, we can calculate the average degree of nearest neighbors with degree k denoted by $\langle k_{nn}|k\rangle$, which corresponds to $\sum_{k'} k'P(k'|k)$ [11]. It's said that the network exhibits assortative mixing if $\langle k_{nn}|k\rangle$ monotonously increasing and disassortative mixing if it monotonously decreasing as a function of k [12].

We present the results of degree-degree distributions in Fig. 1. In order to clearly demonstrate the degree correlation between adjacent nodes, we bin the experimental results. In panel (e) and (f), all distributions exhibit a similar feature that the curves keep increasing before reaching the peak and then decrease gradually. This interesting phenomenon shows that the network is assortative mixing if the k value is below a certain threshold and disassortative mixing if the k value is the threshold.

In panel (e), we find in SMS's directed networks that nodes with k less than 10 only interact with large degree nodes. However, we can't find similar feature in SMS's mutual networks. This can partly be explained by the fact that nodes with a few edges often only connect with the nodes with large degree which usually corresponds to short messages robots or short message advertisers.

3.2   Edge Weight Distribution

For calling and SMS networks, we define the edge weight w_{ij}^N as the number of calls or messages occurred between user i and user j and the edge direction in the networks is ignored. Here we focus on the distribution comparison between SMS networks and calling networks. Figure 2 shows the distribution of the main networks from the two data sets. From the Fig. 2 we can see that all the distribution curves of SMS have a shallower slope and a wider distribution range. This indicates that the number of interactions between users by SMS are higher than phone call. It can be partly explained by the convenience of SMS and its relatively cheaper price. In Fig. 2(a) and (b), the distributions of DCN and SVDCN exhibit an obvious Kink at $w^N \approx 80$. It's not clear why directed calling networks have kicks but mutual networks and SMS networks do not. We can use a bi-power-law distribution to fit the data: $p(w) \sim w^{-\alpha_1}, 1 < w < 80; p(w) \sim w^{-\alpha_2}, w > 80$

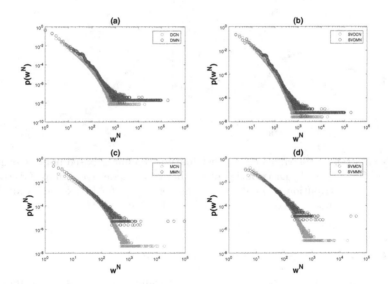

Fig. 2. Edge weight distribution. (a) Distribution of number-based edge weight w^N for DCN and DMN. (b) Distribution of number-based edge weight w^N for SVDCN and SVDMN. (c) Distribution of number-based edge weight w^N for MCN and MMN. (d) Distribution of number-based edge weight w^N for SVMCN and SVMMN.

3.3   Node Strength Distribution

For each node in the networks, we define the node strength S_i^N based on the number of calls or messages, and $s_i^N = \sum_{j \in N_i} w_{ij}^N$ as the total number of calls or messages user made where w_i^N is the calls or SMS between i and j. As the edges in directed networks have directions, we will further distinguish the in-degree and out-degree node strength in our experiments. We present the node strength distributions in Fig. 3, and the distributions for all networks in our experiments can be fitted by an exponentially truncated power-law function $p(s) \sim -\gamma_s exp(-s/s_c)$. In panel (a) and (b) we find that theere is no difference in in-degree and out-degree distributions among the four directed networks. But there is a remarkable difference between the distributions of the networks and the corresponding statistical validated networks. This seems to indicate that the statistical validated method is able to filter out the users who rarely contact with others.

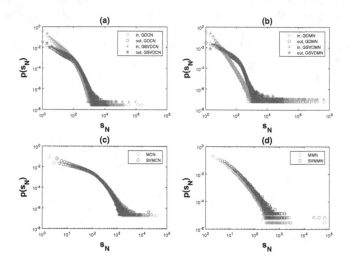

Fig. 3. Node strength distribution. (a) Distribution of number based node strength for GDCN and GSVDCN. (b) Distribution of number based node strength for GDMN and GSVDMN. (c) Distribution of number based node strength for MCN and SVMCN. (d) Distribution of number based node strength for MMN and SVMMN.

3.4   Clustering Coefficient and Topological Overlap

In order to quantify the local cohesiveness around the nodes, we import the concept of clustering coefficient in our study, and use it as an important method to metric the calling and SMS networks. The clustering coefficient of node i is defined as $C_i = 2t_i/[k_i(k_i-1)]$, where t_i denotes the number of triangles around node i [8] and the edge direction is ignored in directed networks. Panels (a) and (b) in Fig. 4 show the dependence of $\langle C|k \rangle$ on k for the eight networks. We find that the curves in the panels are fitted by the function $\langle C|k \rangle \sim k^{-1}$ which is commonly found in many empirical networks [9]. Every user has a quite small clustering coefficient value in the networks and high-degree users have a relativity low clustering coefficient. This indicates that triangle relationships are quite frequent in local structure of the networks and users with few contacts may only connect with people in the same social circle. And this can partly be explained by the fact that people have been accustomed to using mobile phone for communication in their real world social networks.

In panel (a) we find that even though the clustering coefficient variation tendency of calling networks and SMS networks are similar, there still exist some noteworthy discriminations between the four curves. The curves of DCN and DMN are above their corresponding statistical validated networks. This observation reflects the fact that statistical validated approach, while minimizing the presence of links not related to an underlying social relationship, may also remove some edges with real social relationships. The calling networks' clustering coefficient curves are also above the SMS networks' and this reflects the fact that user's phone call friend circle is bigger than SMS's. Phone call play a more important role than SMS in people's daily social contact.

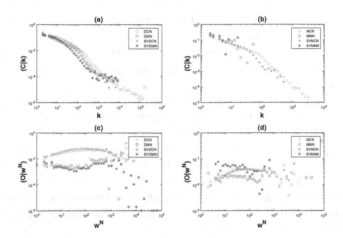

Fig. 4. Clustering coefficient and Topological overlap. (a) Average clustering coefficient $\langle C|k \rangle$ as a function of k for DCN, DMN and their SV networks. (b) Average clustering coefficient $\langle C|k \rangle$ as a function of k for MCN, MMN and their SV networks. (c) Average overlap $\langle O|w^N \rangle$ as a function of number-based edge weight w^N for DCN, DMN and their SV networks. (d) Average overlap $\langle O|w^N \rangle$ as a function of number-based edge weight w^N for MCN, MMN and their SV networks.

Finally we study the properties of links and their neighbor nodes. We quantify the topological overlap of the neighborhood of two connected nodes i and j by the relative overlap of their common neighbors. It's defined as

$$O_{ij} = \frac{n_{ij}}{(k_i - 1) + (k_j - 1) - n_{ij}} \tag{1}$$

where k_i and k_j are the degrees of the two nodes and n_{ij} is the number of the common neighbors of node i and j. When we calculate the overlap for the directed networks the edge directions is ignored. We present the result in Fig. 4 (a) and (b) and find that the two curves for DMN and SVDCN are quite similar, while the curve for the DCN is lower than SVDCN. This contrasts with the phenomenon in Ref. [7]. This indicates a significant fraction of abnormal nodes have been removed by the statistical validated method and the nodes in the network contact more closely. In addition all curves of calling exhibit an increasing trend and seem to decrease after $w^N \approx 100$, but the curves of SMS fluctuate more furiously. This remarkable difference reflects the SMS communication network is quite different from the phone call network.

4 Discussion

Due to the dramatic development of science and technology we have the opportunity to study large-scale complex networks and allow us perceive our social from a totally new perspective. In this paper we constructed networks from mobile phone records of 36 million users and used the cumulative number of calls and SMS as a measure of the strength of the social tie. We construct the social communication network by taking users as nodes and the connections between users as edges. We can study human communication patterns and human dynamics at the societal level.

From the original networks, we construct directed networks, mutual networks and their statistical validated networks. Since statistical validation is used to remove the edges that have no social origin, we take these statistical validated networks as control to be compared with their original networks. In order to understand the networks, we investigate the networks from the distributions of the degree, the edge weight, the node strength and etc. By comparative analysis of these network attributes, we found that these networks share many common properties and also exhibit idiosyncratic characteristics in both qualitative and quantitative ways.

Our work mainly focus on the comparison between calling networks and SMS networks. As we know that traditional mobile communication means have been challenged by mobile Internet. We all experience the powerful influence but we can't quantity the influence brought by mobile Internet. So through comparison between calling networks and SMS networks, we can find the discrepancy between calling and SMS and quantify the influence of burgeoning mobile Internet.

Our work find that usage of SMS is performance poor than calling no matter from quality of user or number of user. Users prefer to use phone call rather than SMS for mutual social communications. The communities constructed by SMS are sparse and unstable. And these phenomenon might be used to formation explaination and deletion of social ties indicate the presence of different elementary mechanisms governing social dynamics under different culture and social norm.

The result of our work can be used as a basis for social communication pattern evolution models of our society. In particular the basic mechanisms of the remarkable discrepancy between SMS and calling need a detailed investigation.

The lessons learnt from our work are not limited to understanding human social networks, but may be used in other domains as well. Finally, we believe our work provides a meaningful method for the study on the evolution of social communication patterns.

Acknowledgement. This work was supported by the major research plan of the National Natural Science Foundation (91224009, 51438009), Technology Commission (13ZCZDGX01099), and the Ocean Public Welfare Scientific under Grant No. 201305033

References

1. Yan, X.Y., Han, X.P., Wang, B.H., et al.: Diversity of individual mobility patterns and emergence of aggregated scaling laws. Sci. Rep. **3**(9), 454–454 (2013)
2. Gonzlez, M.C., Hidalgo, C.A., Barabsi, A.L.: Understanding individual human mobility patterns. Nature **453**, 779–782 (2008)
3. Candia, J., Gonzlez, M.C., Wang, P., Schoenharl, T., et al.: Uncovering individual and collective human dynamics from mobile phone records. J. Phys. A Math. Theor. **41**(22), 1441–1446 (2007)
4. Scebran, M., Palladini, A., Maggio, S., et al.: Statistically validated networks in bipartite complex systems. Plos One **6**(3), e17994 (2011)
5. Hatzopoulos, V., Iori, G., Mantegna, R.N.: Quantifying preferential trading in the e-MID interbank market. SSRN Electron. J. 15(4), 693–710(18) (2013)
6. Tumminello, M., Lillo, F., Piilo, J., et al.: Identification of clusters of investors from their real trading activity in a financial market. New J. Phys. (2011)
7. Li, M.X., Palchykov, V., Jiang, Z.Q., et al.: Statistically validated mobile communication networks: evolution of motifs in European and Chinese data. New J. Phys. (2014)
8. Watts, D.J., Strogatz, S.H.: Collective dynamics of small-world networks. Nature **393**, 440–442 (1998)
9. Szab, G., Alava, M., Kertsz, J.: Clustering in complex networks. In: Ben-Naim, E., Frauenfelder, H., Toroczkai, Z. (eds.) Complex Networks. LNP, vol. 650, pp. 139–162. Springer, Heidelberg (2004)
10. Clauset, A., Newman, M.E.J.: Power-law distributions in empirical data. SIAM Rev. **51**(4), 661–703 (2007)
11. Bril', A.I., Kabashnikov, V.P., Popov, V.M.: Dynamical and correlation properties of the Internet. Phys. Rev. Lett. **87**(25), 527–537 (2001)
12. Newman, M.E.J.: Assortative mixing in networks. Phys. Rev. Lett. **89**, 208701 (2002)

Neural Networks

Distributed Perception Algorithm

Anthony Brabazon[✉] and Wei Cui

Natural Computing Research and Applications Group, School of Business,
University College Dublin, Dublin, Ireland
anthony.brabazon@ucd.ie, will.weicui@gmail.com

Abstract. In this paper we describe the *Distributed Perception Algorithm* (DPA) which is partly inspired by the schooling behaviour of 'golden shiner' fish (*Notemigonus crysoleucas*). These fish display a preference for shaded habitat and recent experimental work has shown that the fish use both individual and distributed perception in navigating their environment. We assess the contribution of each element of the DPA and also benchmark its results against those of canonical PSO.

1 Introduction

The last few decades have seen significant and growing interest in *biomimicry* or 'learning from the natural world', with many disciplines turning to natural phenomena for inspiration as to how to solve particular problems. Examples include the inspiration for engineering designs based on structures and materials found in nature. Another strand of 'learning from nature' concerns the development of computational algorithms whose design is inspired by underlying natural processes which implicitly embed computation [5]. Mechanisms of collective intelligence and their application as practical problem-solving tools, has attracted considerable research attention leading to the development of several families of swarm-inspired algorithms including, ant-colony optimisation [6–8], particle swarm optimisation [9,11,12], bacterial foraging [14,15], honey bee algorithms [16] and a developing literature on fish school algorithms. A critical aspect of all of these algorithms is that powerful, emergent, problem-solving occurs as a result of the sharing of information among a population of agents in which individuals only possess local information.

A number of previous studies have previously employed a fish school metaphor to develop algorithms for optimisation and clustering. Two of the better-known approaches are Fish School Search (FSS) [1,2] and the Artificial Fish Swarm Algorithm (AFSA) [13]. A practical issue that arises in attempting to develop an algorithm based on the behaviour of fish schools is that we have surprising little hard data on the behavioural mechanisms which underlie their activity. At the level of the individual, agents respond to their own sensory inputs and to their physiological/cognitive states [10]. It is not trivial to disentangle the relative influence of each of these. At group-level, it is often difficult to experimentally observe the mechanics of the movement of animal groups or fish schools. Much previous work developing fish school algorithms has relied on

© Springer International Publishing Switzerland 2016
Y. Tan et al. (Eds.): ICSI 2016, Part II, LNCS 9713, pp. 361–369, 2016.
DOI: 10.1007/978-3-319-41009-8_39

high-level observations of fish behaviour rather than on granular empirical data on these behaviours.

In this paper, extending initial work in [4], we examine the *distributed perception algorithm* (DPA). This algorithm draws inspiration from certain behaviours of the fish species 'golden shiners'. This is a fresh water fish which is native to North America, typically growing to about 4 to 5 inches in length. The species is strongly gregarious. Members of the species form shoals of up to about 250 individuals [17]. A recent study [3] investigated one aspect of the behaviour of golden shiners, namely their marked preference for shaded habitat. In order to investigate the mechanism underlying their observed collective response to light gradients, fish were tracked individually to obtain trajectories. The study examined the degree to which the motion of individuals is explained by individual perception (this would produce movement in the steepest direction of light gradient as seen by the individual fish) and social influences based on distributed perception (this would produce movement based on the position and movement of conspecifics). The study indicated that both mechanisms are used and that the relative importance of each is context dependent. For example, when the magnitude of the social vector was high (all conspecifics moving in similar direction) the social influence was dominant.

2   Distributed Perception Algorithm

In each iteration of the proposed DPA, a fish is displaced from its previous position through the application of a velocity vector:

$$p_{i,t} = p_{i,t-1} + v_{i,t} \tag{1}$$

where $p_{i,t}$ is the position of the ith fish at current iteration, $p_{i,t-1}$ is the position of the ith fish at previous iteration, and $v_{i,t}$ is its velocity.

The velocity update is a composite of three elements, prior period velocity, an individual perception mechanism, and social influence via the distributed perception of conspecifics. The update is:

$$v_{i,t} = v_{i,t-1} + DP_{i,t} + IP_{i,t} \,. \tag{2}$$

While the form of the velocity update bears a passing resemblance to the standard PSO velocity update, in that both have three terms, it should be noted that the operationalisation of the individual perception and distributed perception mechanisms is completely different to the memory-based concepts of p_{best} and g_{best} in PSO. The next subsection explains the operation of the two perception mechanisms.

2.1   Prior Period Velocity

The inclusion of a prior period velocity can be considered as a proxy for momentum or inertia. The inclusion of this term is motivated by empirical evidence from the movement ecology literature which indicates that organisms do not follow uncorrelated random walks but rather move with a 'directional persistence' [?].

2.2   Distributed Perception Influence

The distributed perception influence for the ith fish is determined by the following:

$$DP_i = \frac{\sum_{j=1}^{N_i^{DP}} (p_j - p_i)}{N_i^{DP}}, \qquad j \neq i \tag{3}$$

where p_i is the position of the ith fish, and the sum is calculated over all neighbours within an assumed range of interaction of the ith fish r_{DP}, that is $0 <| p_j - p_i | \leq r_{DP}$, where p_j is the position of the jth neighbouring fish, and N_i^{DP} is number of neighbours in the assumed range of interaction of the ith fish. If there are no neighbours in its assumed range of interaction, this term becomes zero.

2.3   Individual Perception Influence

Individual perception is implemented as follows. At each update, each fish assesses the local 'light' gradient surrounding it, by drawing N_i^{IP} samples within an assumed 'visibility' region of radius r_{IP}. While a real-world fish will have a specific angle of vision depending on its own body structure, we adopt a random sampling in a hypersphere around the fish on grounds of generality. The individual perception influence for the ith fish is determined by the function as below:

$$IP_i = \frac{\sum_{j=1}^{N_i^{IP}} (s_j - p_i) * fit_j}{\sum_{j=1}^{N_i^{IP}} fit_j}, \qquad j \neq i \tag{4}$$

where p_i is the position of the ith fish, r_{IP} is the radius of the assumed range within which the ith fish can sense environmental information, N_i^{IP} is the number of samples which the ith fish generates, s_j is the position of the jth sample ($0 <| s_j - p_i | \leq r_{IP}$), and fit_j is the fitness value of the jth sample.

3   Results

In this section we describe the experiments undertaken and present the results from these experiments. Four standard benchmark problems (Table 1) were used to test the developed algorithms. The aim in all the experiments is to find the vector of values which minimise the value of the test functions.

In our experiments we assess the performance of the DPA on the four benchmark problems and also investigate the importance of the three components in the DPA algorithm, namely momentum, DP and IP using the DeJong function. The aim is to examine whether these components play a significant role in determining the DPA's performance. Three algorithms are developed which switch off in turn the momentum, DP and IP influences, denoted as DPAa1, DPAa2 and

Table 1. Optimisation problems

DeJong	$F(x) = \sum_{i=1}^{n} x_i^2$	$[-5.12\ 5.12]^n$	0.0^n
Griewank	$F(x) = 1 + \sum_{i=1}^{n} \frac{x_i^2}{4000} - \prod_{i=1}^{n} \cos(\frac{x_i}{\sqrt{i}})$	$[-600\ 600]^n$	0.0^n
Rastrigin	$F(x) = 10n + \sum_{i=1}^{n}[x_i^2 - 10\cos(2\pi x_i)]$	$[-5.12\ 5.12]^n$	0.0^n
Rosenbrock	$F(x) = \sum_{i=1}^{n-1}[100(x_{i+1} - x_i^2)^2 + (1 - x_i)^2]$	$[-30\ 30]^n$	1.0^n

DPAa3 respectively. The performance of the three variants are compared with that of the standard DPA algorithm which has all three components (denoted as DPAa).

The second set of experiments examines the sensitivity of the DPA to changes in two of its parameters, namely the radius of perception (r_{DP}, r_{IP}) and the number of samples used in the simulated individual perception component (s). The chosen values of these parameters are shown in Table 2. From a biological point of view it is plausible to assume that fish have a bigger radius for DP than IP, namely $r_{DP} > r_{IP}$. The value chosen for the two radii is problem specific, as it is influenced by the choice of the number of fish (N), the radius (size) of the search space (R) and the dimensionality of the this space (D). In the DAPa algorithm, the values of r_{DP} and r_{IP} were chosen after initial experimentation as $\frac{R}{1.5 \sqrt[D]{N}}$ and $\frac{R}{1.8 \sqrt[D]{N}}$ so that in most cases each fish has neighbouring fish within the radius r_{DP}. In order to undertake some sensitivity analysis, two variants of the DPAa algorithm are developed. In the DPAb algorithm, the values of r_{DP} and r_{IP} are set to be twice as large as those in the DPAa algorithm. In the DPAc algorithm, the value of s is increased to 10 (as against 5 in the DPAa algorithm).

Finally, the results from the DPA are compared against those of canonical Particle Swarm Optimisation (PSO). In order to allow a reasonably fair comparison, we control for the number of fitness function evaluations between algorithms.

3.1   Hypotheses and Parameter Settings

In all the experiments, we undertake thirty runs of each algorithm and average the results obtained over these runs. In order to assess the relative performance of each algorithm variant we examine the statistical significance of differences in performance at a conservative 99 % level using a t-test.

The first set of hypotheses concern the testing of the importance of each component of the DPA. The null hypothesis is that the algorithm with a component turned off performs better than the canonical DPA (DPAa). Therefore three hypotheses are tested as follows.

- H_{a1}: the DPAa1 algorithm outperforms the DPAa algorithm;
- H_{a2}: the DPAa2 algorithm outperforms the DPAa algorithm;
- H_{a3}: the DPAa3 algorithm outperforms the DPAa algorithm;

Table 2. Parameter setting of algorithms

Algorithm	Radius of DP (r_{DP})	Radius of IP (r_{IP})	Number of samples in IP (s)	Velocity updating equation
DPAa	$\frac{R}{1.5\sqrt[D]{N}}$	$\frac{R}{1.8\sqrt[D]{N}}$	5	$v_{i,t} = v_{i,t-1} + DP_{i,t} + IP_{i,t}$
DPAa1	$\frac{R}{1.5\sqrt[D]{N}}$	$\frac{R}{1.8\sqrt[D]{N}}$	5	$v_{i,t} = 0 + DP_{i,t} + IP_{i,t}$
DPAa2	$\frac{R}{1.5\sqrt[D]{N}}$	$\frac{R}{1.8\sqrt[D]{N}}$	5	$v_{i,t} = v_{i,t-1} + 0 + IP_{i,t}$
DPAa3	$\frac{R}{1.5\sqrt[D]{N}}$	$\frac{R}{1.8\sqrt[D]{N}}$	5	$v_{i,t} = v_{i,t-1} + DP_{i,t} + 0$
DPAb	$\frac{R}{3\sqrt[D]{N}}$	$\frac{R}{3.6\sqrt[D]{N}}$	5	$v_{i,t} = v_{i,t-1} + DP_{i,t} + IP_{i,t}$
DPAc	$\frac{R}{1.5\sqrt[D]{N}}$	$\frac{R}{1.8\sqrt[D]{N}}$	10	$v_{i,t} = v_{i,t-1} + DP_{i,t} + IP_{i,t}$

Note: R is the radius of the search space, D is the dimension of the test problem and N is the number of fish.

The next set of hypotheses concern the analysis of the two variants with different parameter settings (DPAb and DPAc) of the canonical algorithm (DPAa).

- H_{ba}: the DPAb algorithm outperforms the DPAa algorithm;
- H_{ca}: the DPAc algorithm outperforms the DPAa algorithm;

The final set of hypotheses concern the analysis of the performance of the three versions of the canonical algorithm with PSO.

- H_{a0}: the PSO algorithm outperforms the DPAa algorithm;
- H_{b0}: the PSO algorithm outperforms the DPAb algorithm;
- H_{c0}: the PSO algorithm outperforms the DPAc algorithm.

In all experiments, 30 fish (in DPAs), or in the case of PSO 30 particles, are used.

3.2 Analysis of Components in DPA

The developed algorithms, DPAa, DPAa1, DPAa2 and DPAa3, were tested on the DeJong problem with 20, 40, and 60 dimensions respectively. Table 3 shows the best fitness value obtained from all 30 runs ('Best'), the average of the best fitness ('Mean') and its standard deviation over all 30 runs. The results show that the standard DPA algorithm (DPAa) significantly outperforms the other three algorithms, which indicates that none of the three components, momentum, DP and IP, are sufficient on their own to produce a good search process. It is also observed that DPAa3 (which has individual perception 'switched off') performs the worst. This is not surprising as the IP component is fitness-guided. We also carry out a statistical significance test of the differences on performance between the DPAa algorithm and the other three algorithms, namely DPAa1, DPAa2 and DPAa3 and the p-values are shown in Table 3. The results indicate that the DPAa algorithm outperforms the DPAa1, DPAa2 and DPAa3 algorithms on all tested problems at a significance level of 0.99.

Table 3. Results of component analysis

		DeJong (20D)	DeJong (40D)	DeJong (60D)
DPAa	Best	13.8490	39.8309	82.9444
	Mean	16.7810	53.3594	99.3693
	Std Dev	1.6621	4.4814	5.4045
DPAa1	Best	101.3639	146.8877	173.8090
	Mean	108.8814	154.1140	190.1412
	Std Dev	3.3446	4.5209	7.5146
	H_{a1}	0.00	0.00	0.00
DPAa2	Best	76.9869	211.8659	384.0815
	Mean	88.1540	261.8146	459.8905
	Std Dev	6.2977	17.8869	21.6424
	H_{a2}	0.00	0.00	0.00
DPAa3	Best	104.0716	346.6357	522.4203
	Mean	181.4443	409.8456	636.0020
	Std Dev	29.0740	38.0923	58.6390
	H_{a3}	0.00	0.00	0.00

3.3   Parameter Sensitivity Analysis

The results of the DPAa, DPAb and DPAc algorithms are shown in Table 4.
As can be seen, the DPAa and DPAc algorithms perform significantly better
than the DPAb algorithm, which indicates that the radii of DP and IP are a
critical factor in determining the performance of DPA algorithm. A larger radius
means that the fish can sample more broadly in the IP component and can be
influenced by more fish in the DP component of the algorithm. Comparing the
DPAa and DPAc variants of the algorithm (these focus on the sensitivity of the
results to the number of samples in the IP mechanism), the results show that the
DPAc tends to do slightly better than DPAa but this difference is not generally
statistically significant. This indicates that the results obtained are not crucially
dependent on the number of samples used in the IP mechanism.

3.4   Comparisons with Canonical PSO

The comparisons between the DPA algorithms and the PSO algorithm are shown
in Table 4. Two of the DPA algorithm variants, DPAa and DPAc, are seen to
outperform the PSO algorithm across all benchmarks. The standard deviations
of all the DPA algorithms are smaller than for the PSO algorithm.

Table 4. Results of algorithm comparison

		DeJong (60D)	Griewank (60D)	Rastrigin (60D)	Rosenbrock (60D)
PSO	Best	229.6262	705.1046	706.8452	1.8195E+8
	Mean	313.7544	1059.4572	809.0102	4.2095E+8
	Std Dev	41.3335	205.1225	57.0805	1.0482E+8
DPAa	Best	82.9444	314.6526	499.8868	0.4054E+8
	Mean	99.3693	340.4509	535.9554	0.5268E+8
	Std Dev	5.4045	14.8251	15.9696	0.0591E+8
	H_{a0}	0.00	0.00	0.00	
DPAb	Best	454.0024	1459.6416	872.9574	8.2966E+8
	Mean	488.4168	1670.7541	947.8529	9.1680E+8
	Std Dev	17.5184	71.9513	24.9544	0.3978E+8
	H_{b0}	1.00	1.00	1.00	
	H_{ba}	0.00	0.00	0.00	
DPAc	Best	66.9831	251.4181	492.2972	0.2669E+8
	Mean	88.2610	315.1278	530.1059	0.4270E+8
	Std Dev	8.2310	24.7452	13.4010	0.0689E+8
	H_{c0}	0.00	0.00	0.00	
	H_{ca}	1.00	1.00	1.00	

4   Conclusions

In this paper we describe the distributed perception algorithm which is inspired by schooling behaviour of 'golden shiner' fish. We assess the utility of the algorithm on a series of test problems and undertake an analysis of the algorithm by examining the importance of its component elements for the search process. The results obtained are benchmarked against those from particle swarm optimisation (PSO). The results indicate that the algorithm is competitive against canonical PSO and support a claim that algorithms employing fish-school behaviour mechanisms can be a useful addition to an optimisation toolkit. The current study indicates several interesting areas for follow up research. Obviously the results from any study only extend to the problems examined and future work is required to examine the utility of the algorithm. It is also noted that other fish school algorithms have been developed using search mechanisms inspired by various fish behaviours and as this area of research matures, it would be useful to integrate these into a broader, general, framework.

Another interesting avenue would be to investigate alternative ways of modelling the distributed perception (DP) mechanism. In this study, following [3], we assume that this sensory mechanism has a fixed metric range, in other words, a fish 'interacts' with all its neighbours within a defined radius. Alternative assumptions as to the nature of interaction range can be made including [18], a fixed number of nearest neighbours (topological range) or a shell of near neighbours (Voronoi range). In [18], a novel approach is adopted whereby each fish is assigned a 'visual field' and only neighbours within this field impact on the social information processed by that fish. An interesting finding of this work is that there is lower redundancy of information (transitivity) in visually-defined networks than in metric or topological networks. This could be a useful characteristic in the context of designing an optimisation algorithm, particularly for application in a dynamic environment.

Acknowledgement. This publication has emanated from research conducted with the financial support of Science Foundation Ireland under Grant Number 08/SRC/FM1389.

References

1. Bastos Filho, C., de Lima Neto, F., Lins, A., Nascimento, A., Lima, M.: A novel search algorithm based on fish school behavior. In: Proceedings of IEEE International Conference on Systems, Man and Cybernetics (SMC), pp. 2646–2651. IEEE Press (2008)
2. Bastos Filho, C., de Lima Neto, F., Sousa, M., Pontes, M., Madeiro, S.: On the influence of the swimming operators in the fish school search algorithm. In: Proceedings of IEEE International Conference on Systems, Man and Cybernetics (SMC), pp. 5012–5017. IEEE Press (2009)
3. Berdahl, A., Torney, C., Ioannou, C., Faria, J., Couzin, I.: Emergent sensing of complex environments by mobile animal groups. Science **339**, 574–576 (2013)
4. Brabazon, A., Cui, W., O'Neill, M.: Information propagation in a social network: the case of the fish algorithm. In: Krol, D., Fay, D., Gabrys, B. (eds.) Propagation Phenomena in Real World Networks, vol. 85, pp. 27–51. Springer, Switzerland (2015)
5. Brabazon, A., O'Neill, M., McGarraghy, S.: Natural Computing Algorithms. Springer, Heidelberg (2015)
6. Dorigo, M.: Optimization, Learning and Natural Algorithms, Ph.D. Thesis, Politecnico di Milano (1992)
7. Dorigo, M., Maniezzo, V., Colorni, A.: Ant system: optimization by a colony of cooperating agents. IEEE Trans. Syst. Man Cybern. Part B Cybern. **26**, 29–41 (1996)
8. Dorigo, M., Stützle, T.: Ant Colony Optimization. MIT Press, Cambridge (2004)
9. Engelbrecht, A.: Fundamentals of Computational Swarm Intelligence. Wiley, Chichester (2005)
10. Grunbaum, D., Viscido, S., Parrish, J.: Extracting interative control algorithms from group dynamics of schooling fish. In: Kumar, V., Leonard, N., Morse, A.S. (eds.) Cooperative Control. LNCIS, vol. 309, pp. 103–117. Springer, Heidelberg (2004)
11. Kennedy, J., Eberhart, R.: Particle swarm optimization. In: Proceedings of the IEEE International Conference on Neural Networks, pp. 1942–1948. IEEE Press (1995)

12. Kennedy, J., Eberhart, R., Shi, Y.: Swarm Intelligence. Morgan Kaufman, San Mateo (2001)
13. Li, X., Shao, Z., Qian, J.: An optimizing method based on autonomous animats: fish swarm algorithm. Syst. Eng. Theor. Pract. **22**, 32–38 (2002)
14. Passino, K.: Distributed optimization and control using only a germ of intelligence. In: Proceedings of the 2000 IEEE International Symposium on Intelligent Control, pp. 5–13. IEEE Press (2000)
15. Passino, K.: Biomimicry of bacterial foraging for distributed optimization and control. IEEE Control Syst. Mag. **22**, 52–67 (2002)
16. Pham, D., Ghanbarzadeh, A., Koc, E., Otri, S., Rahim, S., Zaidi, M.: The Bees Algorithm - a novel tool for complex optimisation problems. In: Proceedings of International Production Machines and Systems (IPROMS), pp. 454–459 (2006)
17. Reebs, S.: Can a minority of informed leaders determine the foraging movements of a fish shoal? Anim. Behav. **59**, 403–409 (2000)
18. Strandburg-Peshkin, A., Twomey, C., Bode, N., Kao, A., Katz, Y., Ioannou, C., Rosenthal, S., Torney, C., Wu, H., Levin, S., Couzin, I.: Visual sensory networks and effective information transfer in animal groups. Curr. Biol. **23**(17), R709–R711 (2013)

Predicting Virtual Machine's Power via a RBF Neural Network

Hao Xu[1], Xingquan Zuo[1(✉)], Chuanyi Liu[2], and Xinchao Zhao[3]

[1] School of Computer Science, Beijing University of Posts
and Telecommunications, Beijing, China
772000433@qq.com, zuoxq@bupt.edu.cn
[2] School of Software Engineering, Beijing University of Posts
and Telecommunications, Beijing, China
[3] School of Science, Beijing University of Posts and Telecommunications,
Beijing, China

Abstract. Data centers are growing rapidly in recent years. Data centers consume a huge amount of power, therefore how to save power is a key issue. Accurately predicting the power of virtual machine (VM) is significant to schedule VMs in different physical machines (PMs) to save power. Current researches rarely consider the impact of workload on this prediction. This paper studies the power prediction of VM under the multi-VM environment, with consideration of the impact of PMs' workload. A RBF neural network approach is proposed to predict the VM's power. Experiments show that the proposed approach is effective for VM's power prediction and can achieve average error less than 2 %, which is smaller than those of comparative models.

Keywords: Virtual machine · Neural network · Power prediction

1 Introduction

Cloud computing receives more attention over the years. Most of the data centers gradually adopt cloud computing to provide services. There are a large number of computing and storage devices in cloud computing, which consume a large amount of energy. As reported in [1], the energy consumption of data centers in USA is up to 61 million kilowatts-hour (MKWh) in 2006, with a total cost of $4.5 billion. By 2011, the energy consumption is increased to 100 billion *MKWh* (with cost of $7.4 billion). Energy cost occupies a large percentage of the total cost of cloud providers.

Cloud computing services can be divided into three models [2], namely Infrastructure as a Service (IaaS), Platform as a Service (PaaS), and Software as a Service (SaaS), among which IaaS is the most basic and widely used one [3]. For IaaS, when VMs are scheduled on PMs, the power consumption of a VM on a PM needs to be considered to save the energy cost. Therefore, accurately predicting the power of a VM on a PM is significant to reduce energy cost and increase cloud providers' profit.

There have been a few studies on prediction of VMs' power. Most of those researches ignore the impact of the workload of PM on VMs' energy. However, as stated in [4], VM's power has a significantly nonlinear relationship (especially for

© Springer International Publishing Switzerland 2016
Y. Tan et al. (Eds.): ICSI 2016, Part II, LNCS 9713, pp. 370–381, 2016.
DOI: 10.1007/978-3-319-41009-8_40

multi-core VM) with CPU utilization of PM. Ignoring the workload influence may decrease the predict precision. There is only one published work [5] considering this issue. In [5], multiple piecewise linear regression models are used to handle the non-linear relationship between the workload of PM and VM's energy. However, this method is complex since it involves multiple linear regression sub-models.

In this paper, an artificial neural network (ANN) model is proposed to predict the power of a VM, with consideration of workload impact. There have been no studies on such model to predict the power of a VM to our best knowledge.

2   Related Work

There are a number of literatures on VM's power prediction. *Kansal et al.* [6] presented a solution approach named *Joulemeter* to infer VM's power through resource usage in run-time. Literatures [7–9] adopted linear models to predict VM's power. Wen *et al.* [7] used the parameters metered by PMC, Chen *et al.* [8] used the usage of CPU and memory as input parameters, and Smith *et al.* [9] used the usage of CPU and memory, transfer of disk, and network as input parameters. Krishnan *et al.* [10] explored the feasibility and challenges in developing methods for black-box on-line monitoring of the power usage of a VM, and proposed a linear model to monitor system's power.

Bohra *et al.* [11] divided the workload of VM into CPU-intensive and I/O-intensive and presented linear power model for each of them. In order to effectively manage and schedule VMs, Quesnel *et al.* [12] proposed a linear model to estimate the total power consumption of a single VM, taking into account its static (e.g. memory) and dynamic (e.g. CPU) power of resources. In [13], a two dimensional lookup table (LUT) is constructed for each VM. The LUT includes CPU utilization, last level cache (LLC) miss rate, and the power value. VM's power is obtained by retrieving the LUT using the given CPU utilization and LLC miss rate. By dividing the CPU utilization into bins, *Dhiman et al.* [4] proposed a Gaussian mixture model in each bin, and then used Gaussian mixture vector quantization algorithm to find the nearest vector to predict the power. *Yang et al.* [14] adopted support vector regression to predict the power of VM.

All above models and methods have not considered the workload impact of PM on VM's power, which would result in the decrease of prediction accuracy when a PM contains multiple VMs. There exists only one literature [5] considering the workload impact. The CPU utilization is divided into several segments and a linear regression model is established for each segment. This model is complex as it contains multiple sub-models.

3   Power Prediction of Virtual Machine

In cloud environment, the power of a PM, denoted by P_{PM}, can be divided into two parts: (1) $P_{baseline}$, the power of idle PM, and (2) P_{active}, the power of VMs running on the PM. Thus, the power of a PM is

$$P_{PM} = P_{baseline} + P_{active}. \tag{1}$$

Generally, $P_{baseline}$ is a fixed value for a PM.

The framework of VM's power prediction is illustrated in Fig. 1. The framework adopts Client/Server (C/S) mode. VMs in the PM are regarded as Clients, and the power prediction model in the Server is considered as Server. The Server sends a request to a VM, and then the VM collects its' parameters and send them to the Server as the inputs of the model. The model outputs the prediction value of the VM's power. Assume that there are m VMs in the PM, then the PM's power is

$$P_{PM} = P_{baseline} + \sum_{i=1}^{m} P_{VM}^{i}. \tag{2}$$

where P_{VM}^{i} is the power of ith VM.

When scheduling VMs on PMs, we need to predict the power of a VM in different PMs so as to schedule it in the manner of minimizing energy consumption. If there is only one VM in a PM, it is easy to measure the VM's power as the VM's power equals P_{active}. However, when a PM contains more than one VM, the power of a VM needs to be predicted because we cannot measure the power of a single VM directly.

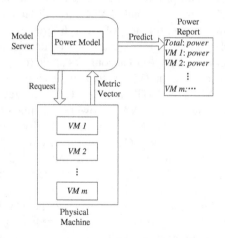

Fig. 1. Framework of VM's power prediction.

4 Proposed Prediction Method

A VM's power has a nonlinear relationship with the PM's workload. We have done experiments to demonstrate this, as shown in Fig. 2. The workload is expressed by the number of identical VMs on the PM. Figure 2(a) shows that with the increase of workload, the PM's power does not linearly increase. Figure 2(b) shows that the power of each VM is decreased with the increase of workload. Figure 2 reveals that the PM's workload has a significant impact on the power of a VM.

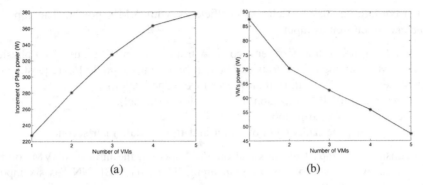

Fig. 2. (a) The nonlinear relationship of PM's workload and its power; (b) the impact of PM's workload on the power of a VM.

ANN is a nonlinear model and suitable to handle the nonlinear relationship. In this paper, we adopt ANN to predict the VM's power with consideration of PM's workload. There are many types of NNs, among which Radial Basis Function (RBF) NN has simple model structure and is easy to train. Hence, we use RBF NN to predict VM's power. The structure of RBF NN is shown in Fig. 3. RBFs in hidden layer are chosen as Gaussian functions.

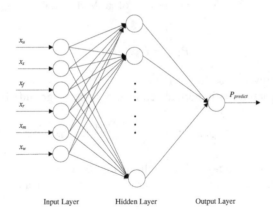

Fig. 3. Structure of RBF NN model for VM's power prediction.

4.1 Inputs of the RBF NN

To predict the power of a VM on a PM, we must choose parameters closely related to its power as the model inputs. To identify those parameters, all parameters of a VM are sorted according to their correlation coefficients with VM's power. The top five parameters are chosen as inputs:

1. x_r: Total number of transfers per second. A transfer is an I/O request to a physical device. Multiple logical requests can be combined into a single I/O request.
2. x_u: Percentage of CPU utilization in user level (application).
3. x_s: Percentage of CPU utilization in system level (kernel).
4. x_m: Percentage of used memory.
5. x_f: Number of page faults (sum of major and minor faults) per second.

To consider the impact of workload on VM's power, the number of VMs on the PM, denoted by x_w, is also chosen as an input. Thus, the RBF NN has six inputs, denoted by an input vector $x_{input} = \{x_u, x_s, x_f, x_r, x_m, x_w\}$.

4.2   Parameter Measurement and Normalization

Typically, a PM is virtualized into several VMs with the same configuration and the same application runs in each VM. To train the RBF NN, we need to get the training data; however, the power of a single VM cannot be directly measured when a PM contains multiple VMs. We use the following method to measure the power of a single VM. The same application runs simultaneously in each VM; in this case, each VM consumes the same power because each VM has the same configuration and executes the same application. Therefore, the power of a VM equals P_{active} divided by the number of VMs.

$$P_{VM} = P_{active}/m. \tag{3}$$

where P_{VM} is the power of a VM.

Each of input parameters is normalized as the same magnitude by

$$y = (y_{max} - y_{min})\frac{x - x_{min}}{x_{max} - x_{min}} + y_{min}. \tag{4}$$

where x and y are the original and normalized inputs, respectively; x_{max} (x_{min}) is the maximum (minimum) value of x; y_{max} and y_{min} are set to be 1 and -1, respectively.

4.3   Training of RBF NN

The training process of RBF NN is described as follows.

Input: Input training data contains n data, each of which is a measured parameter vector, denoted by $\{x_u^i, x_s^i, x_f^i, x_r^i, x_m^i, x_w^i\}$, where $i \in \{1, 2, \cdots n\}$. The training outputs are expressed by a vector, D, containing n power values.

Output: Trained RBF NN model.

Step 1. Initialize RBF NN. Calculate Euclidean distance among all input vectors, obtaining matrix C^1.

Step 2. Calculate the output of the hidden layer using C^1 as the inputs of the standard Gaussian function, and get the output matrix O^1. Define matrix R^1 by

$$R_{ij}^1 = (O_{ij}^1)^2 \quad i = 1, 2 \cdots, n; \ j = 1, 2 \cdots, n. \tag{5}$$

Define vector h^1

$$h_j^1 = \sum_{i=1}^{n} R_{ij}^1 \quad j = 1, 2 \cdots, n. \tag{6}$$

Calculate vector h^2

$$h_i^2 = (D_i)^2 \quad i = 1, 2 \cdots, n. \tag{7}$$

Calculate error vector:

$$e = (((O^1)^T D)^T)^2 \big/ ((h^2)^T h^1). \tag{8}$$

Step 3. Find the largest error e^i among the vector e. Then, add a hidden node with center $\{x_u^i, x_s^i, x_f^i, x_r^i, x_m^i, x_w^i\}$.

Step 4. Calculate Euclidean distance among all input vectors and the centers among all hidden nodes, obtaining C^2 and O^2. Then, calculate the weight matrix between hidden layer and output layer.

$$W = (O^2)^+ D. \tag{9}$$

where $(O^2)^+$ means generalized inverse of matrix O^2.

Step 5. Calculate the mean squared error (MSE) between network output vector U and actual output D.

$$MSE = \frac{1}{n} \sum_{i=1}^{n} (D_i - U_i)^2. \tag{10}$$

If MSE is less than a given threshold, the training process stops; otherwise, recalculate e using O^2 and repeat steps 3–5 until MSE is less than the threshold.

5   Experimental Results

The proposed RBF NN prediction model is applied to a real-world VM power prediction and compared against three other prediction models.

5.1   Experiment Settings

The experiment is implemented in a Lenovo physical server with six 2-core Xeon X5650 CPU (total 24 logical cores), 32 GB RAM and 1 TB hard disk. The server is regarded as the PM. The operation system is CentOS 6.5. Kernel virtual machine (KVM) is a Linux kernel module that can provide a full virtualization solution for Linux. We use KVM to create five VMs at most in the PM. The five VMs have the same configuration, and each of them has one Xeon X5650 CPU, 4 GB RAM and 70 GB hard disk.

Compiling Linux kernel is used as the test application program [15]. In each VM, the Linux Kernel is compiled from version 3.11.10 to 3.11.14. Each VM has 4 logical cores, thus that we can use at most 4 threads to run the compiling program. If only one thread is used, then the CPU utilization of a VM may vary between 10 % and 25 %. To make the utilization cover a wide range, we adopt multi-threading technology to run the compiling program. We use 1, 2, 3, and 4 threads to run the program, respectively, thus that the CPU utilization may vary in the ranges of 10 %–25 %, 40 %–55 %, 60 %–75 %, and 80 %–100 %.

An external hardware, AC power meter, is used to record the PM power. *Sysstat* [16] is a popular toolkit of Linux operating system. The analysis tool *Sar* [17] in *Sysstat* is used to collect the five parameters of each VM. The parameters of VM and PM's power are collected every 10 s to obtain the data set.

5.2   Comparative Models and Performance Metric

The RBF NN model (RBFNN) is compared against the following three models:

(1) Classical multivariate linear regression model (CMLR) [7–9]. It is a common model for VM's power prediction. The model is trained when PM contains only one VM (in this case, the power of the VM can be measured directly). CMLR typically uses CPU utilization, memory utilization and I/O amount as inputs. We also adopt those inputs for the model.

(2) Multivariate linear regression model (MLR), which is similar to the one in [5]. In this model, a sub-model is used for each case of workload. Its inputs are the same as those of the CMLR.

(3) BP NN (BPNN). This approach is constructed by replacing the RBFNN in our prediction approach with BPNN. Its inputs are the same as those of the RBFNN.

Note that the model is to predict the power of a single VM. The predicted PM's power, denoted by $P_{predicted}$, is calculated by

$$P_{predicted} = \sum_{i=1}^{m} P_{VM}^i + P_{baseline}. \qquad (11)$$

where P_{VM}^i is the predicted power of the ith VM. Prediction error is computed by

$$\varepsilon = \frac{\left| P_{measured} - P_{predicted} \right|}{P_{measured}}. \qquad (12)$$

where $P_{measured}$ is the measured PM's power. $P_{baseline}$ is 140 W for this PM. Average prediction error (AE) and maximum prediction error (ME) are used to evaluate model performance.

5.3   Experiments and Results

Let the PM contains 1, 2, 3, 4, and 5 VMs, respectively. When the PM contains only one VM, the VM's power is the PM's active power and can be directly measured. If the PM contains more than one VM, the method in Sect. 4.3 is used to measure the power of a single VM. For each case of workload (the number of VMs in the PM), 240 pairs of input vectors and output power values are measured. Thus, there are total 1200 data vectors.

CMLR is trained when the PM contains only one VM. In this case, 190 data are used as training data and other 50 data as test data. For each of other workload cases. 50 data are randomly chosen from the 240 data as test data. For MLR, a specific sub-model is established for each workload case, where 50 data are randomly chosen as test data and other 190 data as training data. For RBFNN, we randomly chose 50 data in each workload case as test data and other data as training data. Thus, there are $50 \times 5 = 250$ test data and $190 \times 5 = 950$ training data. For BPNN, the training and test data are chosen as the same as the RBFNN's.

In the case of the PM containing one VM, the comparison of prediction results is presented in Fig. 4 and Table 1. We can see that all models can predict well. When the PM contains two active VMs, Fig. 5 and Table 2 give the comparison of prediction results. We can see that the prediction error of CMLR is large, indicating that ignoring workload impact will decrease the prediction accuracy. MLR uses a sub-model to handle each workload case and can achieve better prediction results than CMLR. Figure 6 and Table 3 show the comparison results when the PM has three VMs. Figure 6 shows that the prediction error of CMLR is very large, which indicates that CMLR is not suitable for the multi-VM environment. Figures 7 and 8 and Tables 4 and 5 show the prediction results when the PM involves Tables 4 and 5 VMs.

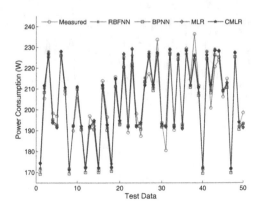

Fig. 4. Comparison of prediction results when PM contains one VM.

Table 1. Prediction error when PM contains one VM.

Models	RBFNN	BPNN	MLR	CMLR
AE	**1.36 %**	1.63 %	1.63 %	1.63 %
ME	6.50 %	6.55 %	6.26 %	6.26 %

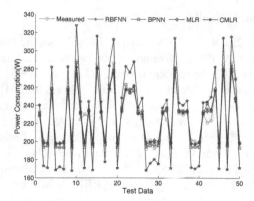

Fig. 5. Comparison of prediction results when PM contains two VMs.

Table 2. Prediction error when PM contains two VMs.

Model	RBFNN	BPNN	MLR	CMLR
AE	**1.84 %**	2.43 %	2.12 %	10.58 %
ME	13.4 %	15.8 %	13.8 %	25.7 %

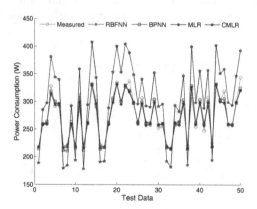

Fig. 6. Comparison of prediction results when PM contains three VMs.

Table 3. Prediction error when PM contains three VMs.

Model	RBFNN	BPNN	MLR	CMLR
AE	**1.54 %**	1.91 %	1.72 %	15.6 %
ME	6.88 %	6.86 %	6.77 %	29.6 %

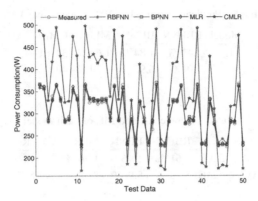

Fig. 7. Comparison of prediction results when PM contains four VMs.

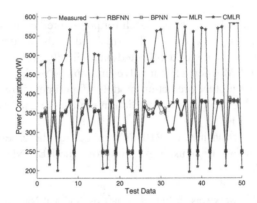

Fig. 8. Comparison of prediction results when PM contains five VMs.

Table 4. Prediction error when PM contains four VMs.

Model	RBFNN	BPNN	MLR	CMLR
AE	1.94 %	1.97 %	**1.70 %**	23.7 %
ME	7.24 %	8.33 %	7.68 %	36.7 %

Table 6 summarizes average AE and ME in all experiments. CMLR has the largest prediction error among all approaches and RBFNN achieves the smallest average AE. RBFNN improves the prediction precision by 0.15–15.49 % compared to other three approaches. The prediction error of RBFNN is slightly better than that of MLR. Both of them consider the workload impact. MLR uses several linear sub-models to handle the nonlinear relationship between workload and power, while RBFNN uses only one RBF NN to model such nonlinear.

Table 5. Prediction error when PM contains five VMs.

Model	RBFNN	BPNN	MLR	CMLR
AE	**1.42 %**	1.84 %	1.64 %	34.4 %
ME	7.01 %	6.98 %	7.39 %	40.5 %

Table 6. Average AE and ME of each model.

Model	RBFNN	BPNN	MLR	CMLR
Average AE	**1.61 %**	1.96 %	1.76 %	17.1 %
Average ME	8.20 %	8.90 %	8.38 %	27.75 %

6 Conclusion

In this paper, A RBF neural network approach for power prediction of VM is proposed. Firstly, parameters of VM are chosen and normalized as the model inputs. Then, the model is trained using the real-world collected data. Finally, the trained model is used to predict the power of VM when a PM involves multiple VMs. The proposed approach is compared against three other prediction models. Experimental results show that the RBFNN model can achieve an average prediction error less than 2 % and outperforms those comparative models.

Future research direction include RBF NN model for the case of PM containing more VMs. Another relevant extension is to improve the RBF NN to achieve smaller prediction error.

Acknowledgments. This work was supported in part by National Natural Science Foundation of China (61374204; 61375066).

References

1. U.S. Environmental Protection Agency ENERGY STAR Program: Server and Data Center Energy Efficiency Public Law 109-431 (2007)
2. NIST: The NIST definition of cloud computing. National Institute of Standards and Technology Special Publication 800-145, pp. 1–7 (2011)
3. Zuo, X., Zhang, G., Tan, W.: Self-adaptive learning PSO based deadline constrained task scheduling for hybrid IaaS cloud. IEEE Trans. Autom. Sci. Eng. **11**(2), 564–573 (2014)
4. Dhiman, G., Mihic, K., Rosing, T.: A system for online power prediction in virtualized environments using Gaussian mixture models. In: ACM/IEEE Design Automation Conference, Anaheim, pp. 807–812 (2010)
5. Li, Y.F., Wang, Y., Yin, B., Guan, L.: An online power metering model for cloud environment. In: IEEE International Symposium on Network Computing and Application, Cambridge, pp. 175–180 (2012)

6. Kansal, A., Zhao, F., Liu, J., Kothari, N.: Virtual machine power metering and provisioning. In: ACM Symposium on Cloud Computing, New York, pp. 39–50 (2010)

7. Wen, C.J., Long, X., Yang, Y., Ni, F., Mu, Y.F.: System power model and virtual machine power metering for cloud computing pricing. In: International Conference on Intelligent System Design and Engineering Applications, Hong Kong, pp. 1379–1382 (2013)

8. Chen, Q., Grosso, P., Veldt, K.V.D., Laat, C.D., Hofman, R., Bal, H.: Profiling energy consumption of VMs for green cloud computing. In: IEEE International Conference on Dependable, Autonomic and Secure Computing, Sydney, pp. 768–775 (2011)

9. Smith, J.W., Khajeh-Hosseini, A., Ward, J.S., Sommervile, I.: Cloud monitor: profiling power usage. In: IEEE Conference on Cloud Computing, Hawaii, pp. 947–948 (2012)

10. Krishnan, B., Amur, H., Gavrilovska, A., Schwan, K.: VM power metering: feasibility and challenges. ACM SIGMETRICS Perform. Eval. Rev. **38**(3), 56–60 (2010)

11. Bohra, A.E.H., Chaudhary, V.: VMeter: power modeling for virtualized clouds. In: IEEE International Symposium on Parallel and Distributed Processing, pp. 1–8. IEEE Press, Atlanta (2010)

12. Quesnel, F., Mehta, H.K., Menaud, J.M.: Estimating the power consumption of an idle virtual machine. In: Green Computing and Communications, Beijing, pp. 268–275 (2013)

13. Jiang, Z.X., Lu, C.Y., Cai, Y.S., Jiang, Z.Y., Ma, C.Y.: VPower: metering power consumption of VM. In: IEEE International Conference on Software Engineering and Service Science, Beijing, pp. 483–486 (2013)

14. Yang, H.L., Zhao, Q., Luan, Z.Z., Qian, D.P.: iMeter: an integrated VM power model based on performance profiling. Future Gener. Comput. Syst. **36**, 267–286 (2014)

15. Linux Kernel. https://www.kernel.org

16. *Sysstat*. https://github.com/sysstat/sysstat

17. *Sar*. http://linux.die.netman/1/sar

The Energy Saving Technology
of a Photovoltaic System's Control on the Basis
of the Fuzzy Selective Neuronet

Ekaterina A. Engel[1(✉)] and Igor V. Kovalev[2]

[1] Katanov State University of Khakassia, Shetinkina. 61,
655017 Abakan, Russian Federation
ekaterina.en@gmail.com
[2] Siberian State Aerospace University, Krasnoyarsky Rabochy Av. 31,
660014 Krasnoyarsk, Russian Federation
kovalev.fsu@mail.ru

Abstract. This paper presents the energy saving technology of a photovoltaic system's control. Based on the photovoltaic system's state, the fuzzy selective neural net creates an effective control signal under random perturbations. The architecture of the selective neural net was evolved using a neuro-evolutionary approach. The validity and advantages of the proposed energy saving technology of a photovoltaic system's control are demonstrated using numerical simulations. The simulation results show that the proposed technology achieves real-time control speed and competitive performance, as compared to a classical control scheme with a PID controller.

Keywords: Fuzzy neural net · Neuro-evolutionary approach · Random perturbations · Photovoltaic system

1 Introduction

Solar energy is being recognized as one of the main alternative clean energy sources due to its essentially non-polluting and inexhaustible nature. We consider a non-linear tracking problem [1] from photovoltaic applications, and this forms the motivation for the development of the energy saving technology of a photovoltaic system's control base on fuzzy selective neural net [2] as presented in this paper. Photovoltaic systems are non-linear and commonly suffer from restrictions imposed by the uncertainty of the environment as a result of sudden variations in the solar irradiation level. Therefore, unstable dynamics must be confronted when designing control systems to quickly track any reference command variations – in order to provide stability, disturbance attenuation, and reference signal tracking. Within the research literature, a whole array of differing control strategies has been proposed to deal with the tracking problem in photovoltaic applications. One of the most common control strategies involves the Proportional-Integral-Derivative (PID) controller, due to its simplicity and applicability [1]. But PID controllers for non-linear photovoltaic systems have slow response times to changing reference commands, take considerable time to settle down from

© Springer International Publishing Switzerland 2016
Y. Tan et al. (Eds.): ICSI 2016, Part II, LNCS 9713, pp. 382–388, 2016.
DOI: 10.1007/978-3-319-41009-8_41

oscillating around the target reference state, must often be designed by hand, and require extensive analysis of the system and dynamics. This process is generally difficult because it is hard to anticipate all operating conditions. The controller must coordinate the non-linear photovoltaic system properly, generating robust behavior to negotiate different terrain effectively while maintaining stability. Moreover, the non-linear photovoltaic (PV) system should be robust to different environmental conditions, in order to reliably generate maximum power. Therefore, automatic design methods utilizing intelligent techniques such as evolution, neural networks, and fuzzy logic are promising alternatives [1].

For real-life photovoltaic applications, the PV system's behavior can change; the system's parameters can exhibit random variations; the solar irradiation can fluctuate. Thus, neural-network-based solutions have been proposed to overcome these difficulties. But the neural network needs to become more adaptive. Adaptive behavior can be enabled by modifying the network to have fuzzy units which respond to the changing behavior of the photovoltaic system. Therefore, photovoltaic systems are better modeled as hybrid system since their behaviors result from the interaction between continuous and discrete dynamics. Several methods have been developed to overcome the aforementioned difficulties. The most important approaches are those that combine the basic model, which are differential equations, with the intelligent models, which are neural networks and fuzzy logic. This paper presents the energy saving technology of a photovoltaic system's control based on a fuzzy selective neural net.

2   The Energy Saving Technology of a Photovoltaic System's Control Based on a Fuzzy Selective Neural Net

Classically, neuro-adaptive control has been addressed using nonlinear systems of the form $\dot{x} = f(x) + g(x) u$, where the neural nets are used to approximate $f(x)$ and $g(x)$ [1]. Adaptive control methods such as linearization have been shown to be very effective for the control of a broad class of systems [1]. In contrast, in this paper, the function approximation capabilities of a fuzzy selective neural net [2] are exploited to approximate a nonlinear control law. This paper considers the development of an effective control method that remains easy to implement. The proposed fuzzy selective neural net is capable of handling uncertainties in both the system's parameters and in the environment.

In this paper, a photovoltaic system integrating segmented energy storage was considered as an example of a non-linear dynamical system. As formed, the fuzzy selective neural net energy-saving technology of the photovoltaic system's control creates the effective control signal, and identifies the system's state under random perturbations (Fig. 1). To make the energy saving technology of a photovoltaic system's control become adaptive, it needs to have some idea of how the actual PV system's behavior differs from its expected behavior, so that the fuzzy selective neural net can recalibrate its behavior intelligently during run time, and try to eliminate the constant tracking error. Thus, the input signal of the fuzzy selective neural net will be non-zero, and it will give useful feedback for inducing the technology to adapt to the dynamically changing PV system's conditions. This control approach does provide a

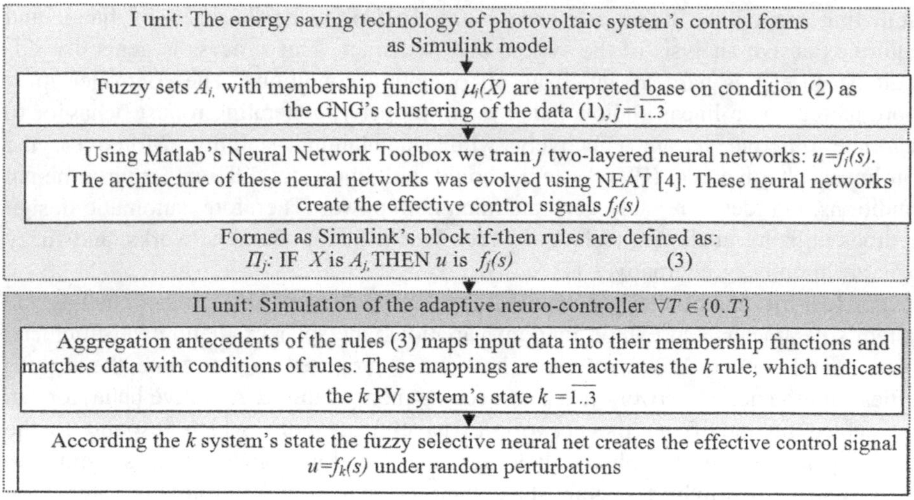

I unit: The energy saving technology of photovoltaic system's control forms as Simulink model
Fuzzy sets A_j, with membership function $\mu_j(X)$ are interpreted base on condition (2) as the GNG's clustering of the data (1), $j=1..3$
Using Matlab's Neural Network Toolbox we train j two-layered neural networks: $u=f_j(s)$. The architecture of these neural networks was evolved using NEAT [4]. These neural networks create the effective control signals $f_j(s)$
Formed as Simulink's block if-then rules are defined as: Π_j: IF X is A_j THEN u is $f_j(s)$ (3)
II unit: Simulation of the adaptive neuro-controller $\forall T \in \{0..T\}$
Aggregation antecedents of the rules (3) maps input data into their membership functions and matches data with conditions of rules. These mappings are then activates the k rule, which indicates the k PV system's state $k = \overline{1..3}$
According the k system's state the fuzzy selective neural net creates the effective control signal $u=f_k(s)$ under random perturbations

Fig. 1. The energy saving technology of the photovoltaic system's control based on the fuzzy selective neural net.

more intelligent method of implementing the control signal. The fuzzy selective neural net is trained based on the data

$$Z^i = (Ir^i,\ V^i,\ P^i,\ dI/dV^i,\ \Delta V^i),\qquad(1)$$

where $i \in \{1,...,\ 10^8\}$, I and V represent the current and voltage respectively, ΔV – the voltage increment, dI and dV represent (respectively) the current error and voltage error before and after the increment, Ir represent the solar irradiance, P – PV the system's power; $s = (Ir^i,\ V^i,\ P^i,\ dI/dV^i)$ – input signal of fuzzy selective neural net; $u = \Delta V$ – control signal and output signal of the fuzzy selective neural net. Data (1) have a training set of $9 * 10^7$ examples, and a test set of 10^7 examples.

First, in order to obtain the dynamics of the system, the fuzzy sets A_j ($j \in \{1, 2, 3\}$) with membership function $\mu_j(X)$ are formed using clustering of the GNG's – n^s (Growing Neural Gas [3]) of the data (1) is based on the condition (2):

$$\eta = \frac{1}{n^s}\left(\sum_{u=1}^{n^s}\left[\min_{\substack{i=1\\i\neq u}}^{n^s}\ [\mu_{A_u^s}(x),\ \mu_{A_i^s}(x)]\right]\right).\qquad(2)$$

Second, for each j an identifier is constructed by a two-layer feed forward neural network. The architecture of these neural networks was evolved using the NeuroEvolution of Augmenting Topologies method (NEAT) [4]. Third, an energy saving technology of the photovoltaic system's control is carefully designed to correctly tackle the control task under uncertainty of the photovoltaic system and of the environment.

2.1 Simulation of Photovoltaic System's Control

The Matlab Simulink model of the photovoltaic system integrating segmented energy storage contains: PV generator providing a maximum of 0.1 MW at 1 kW/m^2 solar irradiance; 5-kHz boost converter; 1980-Hz three-phase three-level Voltage Source Converter; 100-kVA 260 V/25 kV three-phase coupling transformer; 10-kvar capacitor bank; utility grid (25-kV).

This model uses a maximum power point tracking (MPPT) system with a duty cycle that generates the required voltage to extract maximum power. Many MPPT algorithms have been proposed such as voltage feedback, perturbation and observation, linear approximation, incremental conductance, actual measurement, fuzzy control and so on [5, 6]. To illustrate the benefits of the newly proposed energy saving technology of the photovoltaic system's control, the numerical examples from the previous Sect. 2 is revisited.

2.2 Simulation and Results

All the simulations for this study are implemented in MATLAB, Simulink. The true benefits of the proposed the energy saving technology of photovoltaic system's control are best demonstrated through a simulation study. Figure 2 shows the solar irradiance during the simulation time.

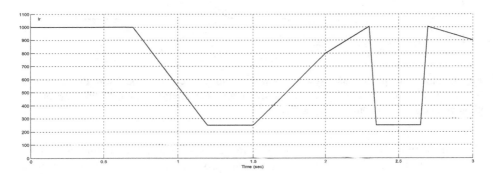

Fig. 2. Plot of solar irradiance.

In this comparison study, the performance of the energy saving technology of the photovoltaic system's control is compared against the standard model with the PID controller (base on perturbation & observation or incremental conductance algorithm), under the same conditions. Figures 3 and 4 show the simulation results. According to Fig. 3, the response time using the proposed method and the incremental conductance algorithm are not better than the one using the perturbation & observation algorithm in the first 0.5 s. In this comparison study, the performance of the energy saving technology of the photovoltaic system's control is compared against the standard model with the PID controller (base on perturbation & observation or incremental conductance

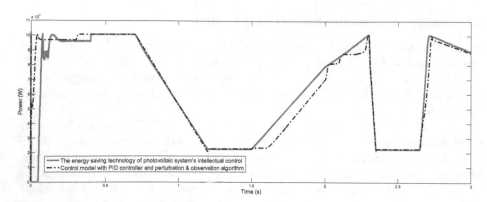

Fig. 3. Plot of the photovoltaic system's power provided by control model with PID controller based on perturbation & observation algorithm and the energy saving technology of the photovoltaic system's control respectively.

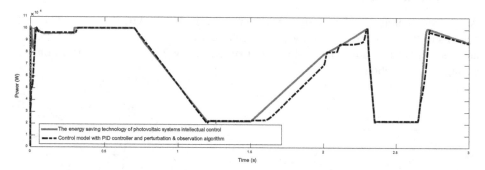

Fig. 4. Plot of the photovoltaic system's power provided by control model with PID controller based on perturbation & observation algorithm and the energy saving technology of the photovoltaic system's control respectively.

algorithm), under the same conditions. Figures 3 and 4 show the simulation results. According to Fig. 3, the response time using the proposed method and the incremental conductance algorithm are not better than the one using the perturbation & observation algorithm in the first 0.5 s. Overshoot is observed. This means that the neural network $f_1(s)$ which creates the control signal within the transient mode is the overshoot.

In order to overcome this overshoot, we retrain the neural network $f_1(s)$ based on the data (1) provided in the first 0.5 s by the control model with the PID controller, based on the perturbation & observation algorithm.

The energy saving technology of the photovoltaic system's control is more robust and provides more power (Figs. 4 and 5) in comparison with the control model with the PID controller (based on perturbation & observation, or the incremental conductance algorithm). The tolerance of a fuzzy selective neural net to noise is explained by two factors: the ability of the model to have similar responses for patterns contaminated with different intensities of noise, and the resilience to noise of the low similarity of the

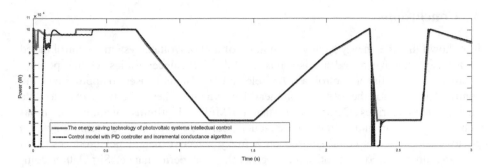

Fig. 5. Plot of the photovoltaic system's power provided by the control model with PID controller based on the incremental conductance algorithm and the energy saving technology of the photovoltaic system's control respectively.

responses for patterns of the different system's state defined by sudden variations in the solar irradiation levels.

It can be seen that the PV system's power is provided by the control model with the PID controller based on the incremental conductance algorithm's drop to zero. From time = 2.3 s to 2.4 s the PID controller takes on a huge numerical value of the control signal (some value of control signal $>10^{17}$) as a result of sudden variations in the solar irradiation levels (Fig. 6), while the energy saving technology of the photovoltaic system's control is provided by the MPP (Fig. 5).

The use of the fuzzy selective neural net provides a more suitable approach to the MPPT problem, with the pointing accuracy. Extensive simulation studies on the Simulink model have been carried out on different initial conditions, different disturbance profiles, and variation in photovoltaic system and solar irradiation level parameters. The results show that consistent performance has been achieved for the proposed energy saving technology of a photovoltaic system's control based on a fuzzy selective neural net with good stability and robustness as compared with the standard model with a PID controller (based on perturbation & observation, or the incremental conductance algorithm).

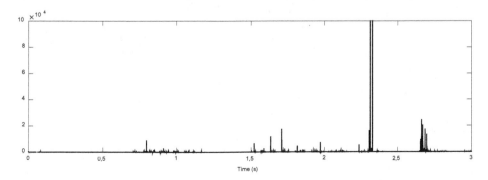

Fig. 6. Plot of the control signal provided by the PID controller base using the incremental conductance algorithm.

3 Conclusions

It is shown that the energy saving technology of a photovoltaic system's control based on a fuzzy selective neural net is robust to PV system uncertainties. Unlike popular approaches to nonlinear control, a fuzzy selective neural net is used to approximate the control law and not the system nonlinearities, which makes it suitable over a wide range of nonlinearities. Compared to standard MPPT algorithms, including perturbation & observation and incremental conductance, the energy saving technology of a photovoltaic system's control based on a fuzzy selective neural net produces good response time, low overshoot, and, in general, good performance. Simulation comparison results for a photovoltaic system demonstrate the effectiveness of the energy saving technology of a photovoltaic system's control as compared with the standard model with a PID controller (based on perturbation & observation, or incremental conductance algorithm). It is our contention that the proposed fuzzy selective neural net architecture can have generic control applications to other kinds of systems, and produce a competitive alternative method to neural networks and PID controllers.

Acknowledgments. The authors wish also to thank Daniel Foty and the reviewers for valuable comments.

References

1. Makarov, I.M., Lokhin, V.M., Manko, S.V., Romanov, M.P., Sitnikov, M.S.: Stability of intelligent systems of automatic control. Inf. Technol. **2** (2013)
2. Engel, E.A.: Solving control problems, decision making and information processing by fuzzy selective neural network. Inf. Technol. **2**, 68–73 (2012)
3. Sledge, I.J.: Growing neural gas for temporal clustering. In: IEEE (2008)
4. Stanley, K.O., Miikkulainen, R.: Competitive coevolution through evolutionary complexification. J. Artif. Intell. Res. **21**, 63–100 (2004)
5. Tavares, C.A.P., Leite, K.T.F., Suemitsu, W.I., Bellar, M.D.: Performance evaluation of photovoltaic solar system with different MPPT methods. In: 35th Annual Conference of IEEE on Industrial Electronics, IECON 2009, pp. 719–724 (2009)
6. Hua, C., Lin, J., Shen, C.: Implementation of DSP-controlled photovoltaic system with peak power tracking. IEEE Trans. Ind. Electron. **45**(1), 99–107 (1998)

Swarm Intelligence in Management Decision Making and Operations Research

An Augmented Artificial Bee Colony with Hybrid Learning

Guozheng Hu[1], Xianghua Chu[1(✉)], Ben Niu[1], Li Li[1], Yao Liu[1], and Dechang Lin[2]

[1] College of Management, Shenzhen University, Shenzhen, China
x.chu@szu.edu.cn, 11ii318@163.com
[2] Medical Business School, Guangdong Pharmaceutical University, Guangzhou, China

Abstract. Artificial bee colony as a recently proposed algorithm, suffers from low convergence speed when solving global optimization problems. This may due to the learning mechanism where each bee learns from the randomly selected exemplars. To address the issue, an augmented artificial bee colony algorithm, hybrid learning ABC (HLABC), is presented in this study. In HLABC, different learning strategies are adopted for the employed bee phase and the onlooker bee phase. The updating mechanism for food source position is enhanced by employing the guiding information from the global best food source. Eight benchmark functions with various properties are used to test the proposed algorithm, and the result is compared with that of original ABC, particle swarm optimization (PSO) and bacterial foraging optimization (BFO). Experimental results indicate that the designed strategy significantly improve the performance of ABC for global optimization in terms of solution accuracy and convergence speed.

Keywords: Artificial bee colony · Global optimization · Hybrid learning

1 Introduction

Swarm intelligence algorithm is originated from natural collective behaviors [1]. Simple individual behaves with self-adaption and perception to neighborhoods. Emerging intelligent behavior of decentralized and self-organized in group level lies on the interactions among these simple individuals, and between individuals and environment [2]. Drawn the inspirations from swarm intelligence, many modern meta-heuristics inheriting the relevant features such as self-regulation and self-adaption were proposed. For example, ant colony optimization (ACO) algorithm is simulated by the process of ants find shortest paths between their nest and food sources [3–5]; Particle swarm optimization (PSO) algorithm mimics birds keeping same formation by adjusting the relative velocity among them [6–8]; Bacterial Foraging Optimization (BFO) algorithm was proposed according to three types of basic rules of bacteria foraging behaviors [9, 10]. These nature-inspired meta-heuristic show a significant performance ability in dealing with combinatorial optimization problems and have been widely applied to different fields [11, 12]. Therefore, swarm intelligence

© Springer International Publishing Switzerland 2016
Y. Tan et al. (Eds.): ICSI 2016, Part II, LNCS 9713, pp. 391–399, 2016.
DOI: 10.1007/978-3-319-41009-8_42

algorithm is becoming an increasingly attractive research area for researchers across various disciplines including computer scientists, engineers, and operational researchers, just to name a few.

Artificial bee colony (ABC), a simple yet powerful population-based algorithm introduced by Karaboga D [13], is inspired by cooperative behaviors of honey bees. Initially, ABC algorithm was developed for numerical optimization problems. Later on, it was expanded for discrete and combinatorial optimization problems [14]. Many researchers have proved that the ABC has superiority in terms of convergence accuracy compared with other methods [15–17]. Many studies have been carried out on ABC which can be grouped into three directions. (1) Comparisons and modifications; These studies focus on evaluate the performance of ABC and to compare it with well-known algorithms such as GA, PSO and ACO [18, 19] Yet, some researchers have motivations to modify formulation and parameters to extend this algorithms to other areas such as integer programming [20], intrusion detection systems [21] and binary optimization [22]. (2) Hybrid approaches; ABC algorithm was combined with other methods component in order to enhance its power. Various ABC variants were proposed with new learning strategies for suiting different scenarios [23, 24]. (3) Applications; Since the standard and improved ABC algorithms have been effectively used for simulations, thus it has been applied in many fields, such as training neural network [25], engineering design [26], target identification [27] etc.

Based on our previous study, ABC may show a better exploration capability than many other algorithms, e.g. particle swarm optimization and bacterial foraging optimization. While promising, it is worth noticing that the convergence and exploitation capability of ABC is still less than satisfactory [28]. For example, the neighbor tracking strategy in ABC only focuses on randomly selected bees which may also lead to immature convergence over evolution [28]. This may be due to the fact that the same learning equation is adopted in both employed bee phase and onlooker bee phase. Since exploration and exploitation have different search interests over the optimization, the same learning mechanism may not be able to meet the needs of both of them at the same time. To address the issue, a hybrid learning artificial bee colony (HLABC) is proposed in this study. In HLABC, different learning strategies are adopted in the employed bee phase and the onlooker bee phase, respectively. The global best information of food source is introduced to enhance the convergence speed for global optimization problems.

This paper is organized as follows: a brief introduction of ABC in the Sect. 2, the Sect. 3 is a demonstrating of the improvement ABC, followed by the experiment improved ABC in Sect. 4, finally, drawn the conclusions in Sect. 5.

2   Introduction to ABC

In ABC, the actual position of the food source corresponds to a possible solution of optimization problems, and the richness of nectar represents the quality of the related solution. Usually, the number of employed bees is equal to that of the onlooker bees. The employed bee amount equals to the number of food means that the position of employed bee represents a feasible solution. Each employed bee in population randomly searches for a food. The search process in ABC are modeled as three phases as follows.

In general, as show in Eq. (1), ps randomly generated initial population contains D-dimensional vector, $X_i = \{x_{i,1}, x_{i,2}, \ldots, x_{i,d}\}$ represents the i_{th} population of a food source.

$$x_{i,d} = LB_d + (UB_d - LB_d) \times r \tag{1}$$

where $i = 1, 2, \ldots, ps$, $d = 1, 2, \ldots, D$, r is a random value between [0, 1], ps refers to the number of the employed bees, D denotes the problem dimension, LB means the lower bound, and UB is the upper bound.

Onlooker bees go to the region of food source explored by employed bees at X_i based on probability p_i defined as:

$$P_i = f_i / \sum_{i=1}^{Ps} f_i \tag{2}$$

where f_i refers to the fitness value of the i_{th} food source, which demonstrates the quality of the food source.

Information is shared by the employed bees after returning to the hive, the onlooker bees find its food source in the region of X_i by using the following equation

$$v_{i,d} = x_{i,d} + (x_{i,d} - x_{k,d}) \times r. \tag{3}$$

where v_{id} refers to the new potential food source, r is a random number between [0, 1], k denotes another random number different from i. In the original ABC, only one dimension is updated randomly. If the new food source I better than the old one food source X_i, then updated food source, otherwise the X_i unreplaced. Each onlooker bees chooses its food source according to the probability that given above. Hence a good food source should get an effective accommodated. Every bee will search for better food sources after a certain number of cycles (limit), if it can't find a better fitness food source, than the specific bee becomes the scout bee.

The framework of ABC algorithm is given below (Table 1):

Table 1. Framework of ABC

ABC algorithm
Initialization
DO while stop criteria not met
Employed bees onto the food sources and evaluate their fitness
Place the onlookers depending upon the fitness obtained by employed bees
Send the scouts for exploring new food sources
Update and memorize food sources and global best information
Terminate and output results

3 Proposed Strategy

As discussed in the previous sections, the algorithm structure of ABC into three stages: (1) the exploration stage based on the initialization and scout phase, (2) the exploration and exploitation stage based on the employed phase, (3) the deep exploitation stages based on the onlooker phase. It is worthy noticing that the same learning equation is adopted for both of the employed and onlooker phases, the production of new food sources is randomly selected a food source position and modified on the existing one as describe in Eq. (3). However, the probability of selecting a food source to learn is equally despite of the fitness value. Since exploration and exploitation have different search interests during the optimization, the same learning mechanism may not be able to meet the needs of both of them at the same time. More specifically, for current learning equation, the guiding exemplar is randomly selected from the neighbors which is good for global exploration, whilst it is less convergent for local exploitation. This may lead to low exploitation accuracy. Therefore, we developed an augment ABC by conducting different learning strategies in the employed bee phase and the onlooker bee phase, which is named as hybrid learning artificial bee colony.

In the proposed technique, the position of food sources updated using the following function:

$$v_{i,d} = x_{i,d} + (x_{i,d} - x_{k,d}) \times r + (x_{i,d} - x_{best}) \times r \tag{4}$$

where x_{best} is the current best food source found by the population. In Eq. (4), the individual not only randomly learns from its neighbor, but also move towards to the previous best food source found so far. This accelerates the convergence of the individuals which benefits the exploitation capability. Thus, in HLABC, the employed bee phase updates the bees using Eq. (3), while Eq. (4) is adopted for the onlooker bee phase. The comprehensive search capability of the proposed algorithm is justified in the next section.

4 Experiment

4.1 Benchmark Functions

To comprehensively test the performance of HLABC, we chose the functions with different features, such as unimodal, multimodal, shifted, rotated [29]. The purpose of shifting benchmark functions is to shift the optimal solution from its original position. If a function of multiple variables can be separated, it can be written as sum of one simple variable and can be solved by using D 1-dimensional searches. Hence, a function would not be easily solved by 1-dimensional search once it is rotated [30]. Therefore, the four types of benchmark functions are employed to justify the performance of the proposed algorithm.

The expressions of the basic test functions adopted in the experiment are listed below:

(1) Schwefel P2.22

$$f(z) = \max_{i=1,\dots,D} |z_i| \tag{5}$$

(2) Schwefel

$$f(z) = 418.982887273 \times D - \sum_{i=1}^{D} z_i \sin(\sqrt{|z_i|}) \tag{6}$$

(3) Non-continuousrastrigin

$$f(z) = \sum_{i=1}^{D} (y_i^2 - 10\cos(2\pi y_i) + 10)$$
$$\text{if } |z_i| < 0.5, y_i = z_i; \text{ else } y_i = round(2z_i)/2 \tag{7}$$

(4) Ackley

$$f(z) = -20\exp\left(-0.2\sqrt{\frac{1}{D}\sum_{i=1}^{D} z_i^2}\right) + 20 + e - \exp\left(\frac{1}{D}\sum_{i=1}^{D}\cos(2\pi z_i)\right) + f_{bias} \tag{8}$$

(5) Rosenbrock

$$f(z) = \sum_{i=1}^{D} \left(100(z_i^2 - z_{i+1})^2 + (z_i - 1)^2\right) \tag{9}$$

On the basis of the basic functions, the characteristics of the test problems are presented in Table 2.

4.2 Settings for Experiment and Algorithms

The presented algorithm is compared with original artificial bee colony (ABC), particle swarm optimization (PSO), and bacterial foraging optimization algorithm (BFO). For fair comparison, the population size is set to be 50, and the number of the dimension is set as 30 for all functions. The maximum number of function evaluations (FE) is set to be 1.5×10^5. In addition, the algorithms are independently run 50 times on each function. The mean and the standard deviation (SD) of the function values best are obtained for evaluation. Other control parameter of the algorithm and the schemes are presented below:

The BFO has three mechanisms in foraging optimization solutions: chemotaxis, reproduction and dispersion and elimination. The number of chemotactic steps, run unit step, reproduction and dispersion are set as 150, 4, 10 and 2, respectively. The first half of population survive and a surviving bacterium splits into two identical ones. Bacterium were chosen to be dispersed and moved to another position according to a preset

Table 2. Benchmark functions

Basic function	#	Multimodal	Shifted	Rotated	#	Multimodal	Shifted	Rotated
Schwefel P2.22	f_1	N	Y	N	–	–	–	–
Schwefel	f_2	Y	N	N	f_3	Y	N	Y
Non-continuousrastrigin	f_4	Y	Y	N	f_5	Y	Y	Y
Ackley	f_6	Y	Y	N	f_7	Y	Y	Y
Rosenbrock	f_8	N	Y	N	–	–	–	–

probability set as 0.25. In PSO, the additional inertia weight linearly decreases from 0.9 to 0.7 with the iterations. The learning factors are set to be 2. For ABC, the percentage of the employed bees and the onlooker bees are equally 50 % of the population size. The number of limits is set to 500. As there is no additional parameter introduced, HLABC algorithm has the same parameter settings with the original ABC algorithm.

4.3   Experimental Results and Discussions

Table 3 presents the result of the 50 runs of the four algorithms on the eight test functions.

From the table, it is observed that the performance of HLABC is better than that of ABC, PSO and BFO on most of the benchmark problems in terms of mean results. This demonstrates effectiveness of the proposed technique. Compared with original ABC, HLABC has a significant improvement in convergence on the complex multimodal problems. In other words, onlooker bees do the local search around the current optimal solution which after employed bees dimension variation, and the new hybrid learning mechanism introduced to onlooker bees enhances the local search capability. Also, the results indicates the capability of HLABC in getting out of the local optima while searching. The modified ABC algorithm obtains significant improvement on solution accuracy and convergence rate.

In order to show the convergence speed of the algorithms, Figs. 1, 2, 3 and 4 show the convergence characteristics of the algorithms on four representative problems:

Table 3. Results obtained by BFO, PSO, ABC and HLABC algorithms

f	BFO		PSO		ABC		HLABC	
	Mean	SD	Mean	SD	Mean	SD	Mean	SD
f_1	8.395E + 01	5.097E + 00	1.582E + 00	9.085E−01	1.141E + 00	6.274E−01	8.327E−01	4.440E−01
f_2	8.407E + 03	2.706E + 02	1.232E + 03	3.612E + 02	1.699E−08	9.963E−13	1.699E−08	9.963E−13
f_3	7.801E + 03	5.119E + 02	3.153E + 03	4.631E + 02	2.091E + 03	2.935E + 02	1.715E + 03	2.973E + 02
f_4	2.170E + 02	1.757E + 01	1.422E + 01	2.937E + 00	0.000E + 00	0.000E + 00	0.000E + 00	0.000E + 00
f_5	2.026E + 02	2.637E + 01	6.800E + 01	1.944E + 01	7.001E + 01	1.564E + 01	6.680E + 01	8.787E + 00
f_6	2.003E + 01	5.017E−02	1.631E−12	1.070E−12	4.050E−14	1.946E−15	3.837E−14	3.892E−15
f_7	1.992E + 01	2.397E−01	1.596E + 00	9.787E−01	1.794E + 00	2.099E−01	1.543E + 00	2.407E−01
f_8	3.843E + 10	5.924E + 09	6.828E + 01	8.782E + 01	9.015E + 00	1.578E + 01	1.217E + 01	1.487E + 01

unrotated unimodal function (f_1), rotated multimodal function (f_3), unrotated multi-modal function (f_6), and rotated multimodal function (f_7). It can be observed that HLABC converges faster than the other algorithms on the three representative functions since the early evolution. The curves indicate that the proposed strategy significantly enhances ABC's convergence speed on the testing functions. It is indicated that the proposed strategy significantly enhances ABC's convergence accuracy rate on the testing functions.

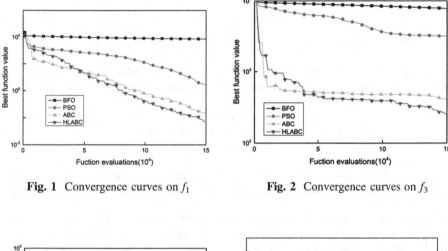

Fig. 1 Convergence curves on f_1 **Fig. 2** Convergence curves on f_3

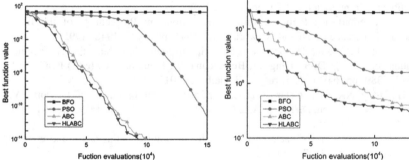

Fig. 3 Convergence curves on f_6 **Fig. 4** Convergence curves on f_7

5 Conclusion

This paper presents a HLABC algorithm for global optimization problems. In HLABC, a novel hybrid learning strategy is proposed for exploration and exploitation separately. The bees learns from it randomly selected neighbor in employed bee phase, while the bees moves towards to the neighbors and the global best food source over onlooker bee phase. The proposed method is compared with BFO, PSO and ABC. Experimental results demonstrate that HLABC outperforms the comparison algorithms in terms of solution accuracy and convergence speed. Also, it is seen from the experiment that the exploitation capability is improved on most of the testing problems. In the future work, we will apply the proposed method to more complex and large-scale experiments to justify the performance of HLABC. Applying the proposed algorithm for addressing more real-world problems is also our interests.

Acknowledgments. This work was supported by the national natural science foundation of china (71501132, 71571120 and 71371127).

References

1. Banks, A., Vincent, J., Anyakoha, C.: A review of particle swarm optimization. Part II: hybridisation, combinatorial, multicriteria and constrained optimization, and indicative applications. Nat. Comput. **7**, 109–124 (2008)
2. Banks, A., Vincent, J., Anyakoha, C.: A review of particle swarm optimization. Part I: background and development. Nat. Comput. **6**, 467–484 (2007)
3. Dorigo, M., Blum, C.: Ant colony optimization theory: a survey. Theor. Comput. Sci. **344**, 243–278 (2005)
4. Dorigo, M., Maniezzo, V., Colorni, A.: Ant system: optimization by a colony of cooperating agents. IEEE Trans. Syst. Man Cybern. Part B Cybern. Publ. IEEE Syst. Man Cybern. Soc. **26**, 29–41 (1996)
5. Mezura-Montes, E., DamiáN-Araoz, M., Cetina-DomiNgez, O.: Smart flight and dynamic tolerances in the artificial bee colony for constrained optimization. In: 2010 IEEE Congress on Evolutionary Computation (CEC), pp. 1–8 (2010)
6. Kennedy, J., Eberhart, R.: Particle swarm optimization. In: Proceedings of the IEEE International Conference on Neural Networks, vol. 1944, pp. 1942–1948 (1995)
7. Niu, B., Wang, J., Wang, H., Tan, L.: Bacterial-inspired algorithms for engineering optimization. In: Huang, D.-S., Jiang, C., Bevilacqua, V., Figueroa, J.C. (eds.) ICIC 2012. LNCS, vol. 7389, pp. 649–656. Springer, Heidelberg (2012)
8. Chen, S.H., Hong, X.U.: Parameter selection in particle swarm optimization. In: Porto, V. W., Saravanan, N., Waagen, D., Eiben, A.E. (eds.) EP 1998. LNCS, vol. 1447, pp. 591–600. Springer, Heidelberg (1998)
9. Passino, K.M.: Biomimicry of bacterial foraging for distributed optimization and control. IEEE Control Syst. **22**, 52–67 (2002)
10. Yang, C., Ji, J., Liu, J., Liu, J., Yin, B.: Structural learning of Bayesian networks by bacterial foraging optimization. Int. J. Approximate Reasoning **69**, 147–167 (2016)
11. Mallipeddi, R., Suganthan, P.N.: Problem Definitions and Evaluation Criteria for the CEC 2010 Competition on Constrained Real-Parameter Optimization. Nanyang Technological University, Singapore (2010)
12. Zamuda, A., Brest, J.: Population reduction differential evolution with multiple mutation strategies in real world industry challenges. In: Rutkowski, L., Korytkowski, M., Scherer, R., Tadeusiewicz, R., Zadeh, L.A., Zurada, J.M. (eds.) EC 2012 and SIDE 2012. LNCS, vol. 7269, pp. 154–161. Springer, Heidelberg (2012)
13. Karaboga, D., Basturk, B.: A powerful and efficient algorithm for numerical function optimization: artificial bee colony (ABC) algorithm. J. Glob. Optim. **39**, 459–471 (2007)
14. Szeto, W.Y., Wu, Y., Ho, S.C.: An artificial bee colony algorithm for the capacitated vehicle routing problem. Eur. J. Oper. Res. **215**, 126–135 (2011)

15. Balasubramani, K., Marcus, K.: A comprehensive review of artificial bee colony algorithm. Int. J. Comput. Technol. **5**, 15–28 (2013)
16. Karaboga, D., Basturk, B.: On the performance of artificial bee colony (ABC) algorithm. Appl. Soft Comput. **8**, 687–697 (2008)
17. Karaboga, D., Akay, B.: A comparative study of artificial bee colony algorithm. Appl. Math. Comput. **214**, 108–132 (2009)
18. Hai, S., Yasuda, T., Ohkura, K.: A self adaptive hybrid enhanced artificial bee colony algorithm for continuous optimization problems. Bio Syst. **132–133**, 43–53 (2015)
19. Banitalebi, A., Aziz, M.I.A., Bahar, A., Aziz, Z.A.: Enhanced compact artificial bee colony. Inf. Sci. **298**, 491–511 (2015)
20. Akay, B., Karaboga, D.: Solving integer programming problems by using artificial bee colony algorithm. In: Cucchiara, R., Serra, R. (eds.) AI*IA 2009. LNCS, vol. 5883, pp. 355–364. Springer, Heidelberg (2009)
21. Wang, J., Li, T., Ren, R.: A real time IDSs based on artificial bee colony-support vector machine algorithm. In: 2010 Third International Workshop on Advanced Computational Intelligence (IWACI), pp. 91–96 (2010)
22. Kashan, M.H., Nahavandi, N., Kashan, A.H.: DisABC: a new artificial bee colony algorithm for binary optimization. Appl. Soft Comput. **12**, 342–352 (2012)
23. Karaboga, D., Ozturk, C.: A novel clustering approach: artificial bee colony (ABC) algorithm. Appl. Soft Comput. **11**, 652–657 (2011)
24. Taherdangkoo, M., Bagheri, M.H., Yazdi, M., Andriole, K.P.: An effective method for segmentation of MR brain images using the ant colony optimization algorithm. J. Digit. Imag. **26**, 1116–1123 (2013)
25. Karaboga, D., Ozturk, C.: Neural networks training by artificial bee colony algorithm on pattern classification. Neural Netw. World **19**, 279–292 (2009)
26. Marzband, M., Azarinejadian, F., Savaghebi, M., Guerrero, J.M.: An optimal energy management system for islanded microgrids based on multiperiod artificial bee colony combined with Markov Chain. IEEE Syst. J. **100**, 1–11 (2015)
27. Bhandari, A.K., Kumar, A., Singh, G.K.: Modified artificial bee colony based computationally efficient multilevel thresholding for satellite image segmentation using Kapur's, Otsu and Tsallis functions. Expert Syst. Appl. **42**, 1573–1601 (2015)
28. Dervis Karaboga, B.G., Ozturk, C., Karaboga, N.: A comprehensive survey: artificial bee colony (ABC) algorithm and applications. Artif. Intell. Rev. **42**, 37 (2014)
29. Chu, X., Hu, M., Wu, T., Weir, J.D., Lu, Q.: AHPS 2: an optimizer using adaptive heterogeneous particle swarms. Inf. Sci. **280**, 26–52 (2014)
30. Liang, J.J., Qin, A.K., Suganthan, P.N., Baskar, S.: Comprehensive learning particle swarm optimizer for global optimization of multimodal functions. IEEE Trans. Evol. Comput. **10**, 281–295 (2006)

A Multiobjective Bacterial Optimization Method Based on Comprehensive Learning Strategy for Environmental/Economic Power Dispatch

Lijing Tan[1], Hong Wang[2], Fangfang Zhang[3], and Yuanyue Feng[3](✉)

[1] Department of Business Management, Shenzhen Institute of Information Technology, Shenzhen 518172, China
[2] Department of Mechanical Engineering, Hong Kong Polytechnic University, Hong Kong, China
[3] College of Management, Shenzhen University, Shenzhen 518060, China
yuanyuef@szu.edu.cn

Abstract. This article extends the bacterial foraging optimization (BFO) for addressing the multi-objective environmental/economic power dispatch (EED) problem. This new approach, abbreviated as MCLBFO, is proposed based on the comprehensive learning strategy to improve the search capability of BFO for the optimal solution. Besides, the fitness survival mechanism based on a health sorting technique is employed and embedded in reproduction mechanism to enhance the quality of the bacteria swarm. The diversity of the solutions is achieved by the combination of two typical techniques, i.e. non-dominance sorting and crowded distance. Experimental tests on the standard IEEE 30-bus, 6-generator test system demonstrate that the novel algorithm, MCLBFO, is superior to other well developed methods such as MOEA/D, SMS-EMOA and FCPSO. The results of the comparison indicate that MCLBFO is outstanding in handling optimization problems with multiple conflicting objectives.

Keywords: Multi-objective problems · Bacterial foraging optimization · Comprehensive learning strategy · Environmental/economic power dispatch

1 Introduction

Traditional electric power systems, known as Economic Power Dispatch (ED), are operated to minimize the cost in supply and transmission related processes while meeting all generation and systems constraints [1, 2]. However, the traditional model is proved to be not the best as the increasing public appeal to environmental protection [3]. Therefore, when it is related to environment issues, this single objective model, which only considers the cost, does not satisfy the public any more.

Actually, amendments to some rules of the clean air act have carried out to limit the emission of SO_2 and NO_x in 1990. Soon afterwards strategies used to reduce emissions were proposed and discussed accordingly [4]. Thus, the traditional problem of ED is

© Springer International Publishing Switzerland 2016
Y. Tan et al. (Eds.): ICSI 2016, Part II, LNCS 9713, pp. 400–407, 2016.
DOI: 10.1007/978-3-319-41009-8_43

transformed to the problem of Environmental/Economic Power Dispatch (EED) considering the conflicting objectives to minimize the generation cost and pollution emissions. Population based algorithms are demonstrated to have the advantages to address the EED by handling all the conflicting objectives simultaneously, and providing the results consisting of a set of satisfactory solutions (Pareto-optimal front). Since now, population based algorithms have been wildly used in solving EED, such as multi-objective particle swarm optimization (MOPSO) algorithms [5, 6], multi-objective evolutionary algorithms (MOEAs) [7, 8], Bees Algorithms [9, 10].

By getting inspiration from the foraging behavior of E. coli bacteria, bacterial foraging optimization was proposed by Passion in 2002 [11] and it has been improved for handling multi-objective problems [12–15]. However, most of them have only been tested by benchmark functions which are comparatively simple multi-objective issues with uncomplicated structure in objectives. Some of them are also applied for EED problem and solve it as multi-objective optimization problems [16–18], while the algorithms adopted for comparison seem to be outdated and numerous improvements have been proposed by other researchers.

In this article, an extension of the original Multi-objective Bacterial Foraging Optimization (MBFO) [14], referred as Multi-objective Comprehensive Learning Bacterial Foraging Optimization (MCLBFO), is presented to address the environmental/economic power dispatch (EED) problem. This newly proposed method considers the comprehensive learning between bacteria, the health sorting approach for reproduction, as well as the external achieves to handle the non-dominance choice for the improvement of the solution diversity.

The rest of the article is structured as follows: Sect. 2 provides a brief description of the original BFO, and the proposed MCLBFO are presented in this Sect. 3. Comparative studies and pertinent discussions are elaborated in Sect. 4, Finally, Sect. 5 provides the conclusions of this article.

2 Bacterial Foraging Optimization

Passion proposed Bacterial Foraging Optimization (BFO) in 2002 by getting inspiration from the foraging behavior of E. coli [11]. Generally, the foraging behavior of bacteria mainly includes three steps: chemotaxis, reproduction, and elimination and dispersal. A brief description of these three steps is illustrated as follows.

(1) Chemotaxis: Define $\theta^i(j, k, l)$ as the position of the ith bacterium at the jth chemotactic, the kth reproductive and the lth elimination and dispersal. The moving direction of each bacterium is adjusted as:

$$\theta^i(j+1, k, l) = \theta^i(j, k, l) + c(i)\frac{\Delta(i)}{\sqrt{\Delta^T(i)\Delta(i)}} \tag{1}$$

where: $C(i)$ indicates the unit moving length, and $\Delta(i)$ represents the direction angle of the ith step. The typical behaviors, e.g. running, tumbling during the bacterial foraging process are also incorporated into the chemotactic step.

(2) Reproduction: Reproduction as a basic life behavior of all kinds of species is also the key point for life preservation. The search efficiency of the bacterial colony increases along with the generation times. Define J_t (i, j, k, l) as the fitness value of the tth $(t = 1, 2)$ function of the ith bacterium at the jth chemotaxis, the kth reproduction and the lth dispersal. The health status of the ith bacterium can be formulated as:

$$J_t^i health = \sum_{j=1}^{N_c} J_t(i, j, k, l) \tag{2}$$

According to their health status, all bacteria are arranged in ascending order. The better ranking half which have better health status can be kept and the rest half should be eliminated. That is:

$$\theta^{i+Sr}(j, k, l) = \theta^i(j, k, l) \tag{3}$$

The healthy $S_r = S/2$ bacteria are now supposed to reproduce to keep the total population number unchanged while improving the quality of the population.

(3) Elimination and Dispersal: The bacteria individuals have the high chance to be impacted by the dynamic environment surrounding them. The one of main advantages of this strategy is to improve the diversity of the bacteria swarm. It could be formulated as:

$$\theta^i = x_{min} + (x_{max} - x_{min}) * rand \tag{4}$$

where x_{max} and x_{min} are the upper and the lower limitation of the initial position, and *rand* is a random value ranging from 0 to 1. The principle of elimination and dispersal in this paper refers to the literature published by Passino [11].

3 Multiobjective Comprehensive Learning Bacterial Foraging Optimization

In this section, MCLBFO is presented for addressing the multi-objective optimization problems. Two strategies, i.e. non-dominated sorting and crowding distance [19–21], are employed in MCLBFO. These two methods are regarded as the superiority tactics in addressing multi-objective problems. Except these two strategies, additional two mechanisms are modified for multi-objective optimization, i.e. comprehensive learning mechanism and health evaluation.

3.1 Comprehensive Learning Mechanism

The original comprehensive learning mechanism is proposed by Liang et al. [22] to enhance the capability of PSO. However, it is less applicable for multi-objective problems for Pareto optimal sets. To enhance the search accuracy, the comprehensive learning mechanism is modified for multi-objective optimization. In this learning mechanism, the moving direction of the ith bacterium in the dth dimension is updated as follows:

$$\theta_d(i,j+1,k,l) = \theta_d(i,j,k,l) + c(i)\frac{\Delta(i)}{\sqrt{\Delta^T(i)\Delta(i)}} + r*(pbest_{id} - \theta_d(i,j,k,l)) + \tag{5}$$

$$(1-r)*(rep_d - \theta_d(i,j,k,l))$$

$$pbest_{id} = \alpha * pbest_{compet} + (1-\alpha)*pbest_{id} \tag{6}$$

$$pbest_{compet} = b*pbest_n + (1-b)pbest_m \tag{7}$$

where n and m are two random generated individuals from population. $n \in \{1,\ldots,S\}$, $m \in \{1,\ldots,S\}$, and $n \neq m$. Three parameters r, a and b are randomely selected from $\{0, 1\}$, whilst are two constants given previously. $Pbest_{id}$ indicating the personal best of the dth dimension is obtained from either current position of individual $Pbest_{id}$ or from the best position of two randomly selected individual $Pbest_{compet}$. To obtain the satisfied solutions approaching to the Pareto optimal front, the position of individuals are also learned from external non-dominated archives Rep.

3.2   Health Assessment Based on Population Segmentation

In BFO, the bacteria with higher performance are used for reproduction according to their health indexes. For each bacterium in the colony, the ability to search for nutrients varies and is marked by the health indexes $J_t^i health$ ($t = 1, 2$) which are influenced by objective function values. In the multi-objective problems, the health condition of bacteria are different in definition. The particles approaching to the true Pareto front would be regarded as the better performance. Therefore, the reproduction of bacteria is formulated as follows:

Pseudo-code 1: *Reproduction based on health assessment*

For each bacterium ($i = 1,2,\ldots, n$)

$$J_{1\,health}^i = \sum_{j=1}^{Nc} J_1(i,j,k,l), \quad J_{2\,health}^i = \sum_{j=1}^{Nc} J_2(i,j,k,l), \quad J_{health}^i = J_{1\,health}^i + J_{2\,health}^i$$

end

Sort $\{J_{1\,health}^i\}$ in increasing order, and select the first 20% population, i.e. Pop_1

Sort $\{J_{2\,health}^i\}$ in increasing order, and select the first 20% population, i.e. Pop_2

Sort $\{J_{health}^i\}$ in increasing order, and select the first 60% population, i.e. Pop_3

Then, the new population is updated as: Pop=[Pop_1, Pop_2, Pop_3]

3.3　MCLBFO Algorithm

Based on the strategies described above, the Pseudo-code of MCLBFO for multi-objective problems are presented as follows.

Pseudo-code 2: *Multiobjective Comprehensive Learning Bacterial Foraging Optimization (MCLBFO)*
Begin
Initialization: Parameters' value, bacterial position, etc.
For k =1: Upper limitation of elimination process N_e
For j =1: Upper limitation of reproduction process N_r
For i = 1: Upper limitation of chemotaxis process N_c
For Each bacterium
Do Chemotaxis step using Eq. 5;
End for
Updating the archives: Non-Dominate sorting and crowded distance
End chemotaxis process
Do: Reproduction based on health assessment (refer to Pseudo-code 1)
End reproduction process
Do Elimination step using Eq. 4
End elimination process
Output: solutions in archives, i.e. Pareto optimal font

4　Application

To investigate its effectiveness, the proposed algorithm (MCLBFO) has been applied in the standard IEEE 30-bus 6-generator test system [23]. The objectives and constrains of the EED problem is described as a nonlinear combinatory multiple objective problem with multiply constraints in [24]. The total demand of the system is 2.834 *p.u.*

4.1　Parameter Settings and Multiobjective Methods

Transmission loss coefficient B_{ij}, upper and lower real power output of each generator, and values of fuel cost and emission coefficients are defined as the same in [25]. The numerical results of the novel algorithm are compared with six multi-objective approaches: BB-MOPSO [25], NSGAII [20], FCPSO [24], MOEA/D [21], MBFO [14] and SMS-EMOA [19].

Values of parameter used in the new technique are as follows: $p = 2$, $S = 100$, $N_c = 100$, $N_s = 5$, $N_{re} = 4$, $N_{ed} = 2$, $P_{ed} = 0.2$, $C = 0.1$. The parameter P_r in MCLBFO is a constant, and equals 0.5. Furthermore, the learning probability P_c is set as 0.1, and the algorithm is conducted for 20 runs to obtain the statistical results. Algorithms are initiated according to the above settings. The optimization process is implemented through the MATLAB.

4.2   Experimental Results and Analysis

For comparison purposes, the system is chosen as a typical case considering the transmission losses. This case considers several constraints, such as the generation capacity, the transmission losses and the power balance constraints. Two objectives, minimizing the fuel cost and minimizing emissions, are optimized simultaneously. The best values of fuel cost and emission optimized using the proposed algorithm as well as six other reported approaches are shown as Tables 1 and 2, respectively.

Table 1. Best results out of 20 trials for fuel cost

	BB-MOPSO	NSGAII	FCPSO	MOEA/D	SMS-EMOA	MBFO	MCLBFO
P_{G1}	0.1229	0.1151	0.1130	0.1186	0.1183	0.1271	0.1121
P_{G2}	0.2880	0.3055	0.3145	0.2971	0.2977	0.2171	0.2979
P_{G3}	0.5792	0.5972	0.5826	0.6123	0.6140	0.6256	0.5692
P_{G4}	0.9875	0.9808	0.9860	0.9514	0.9466	1.0018	0.9775
P_{G5}	0.5255	0.5142	0.5264	0.5080	0.5103	0.5278	0.5491
P_{G6}	0.3564	0.3541	0.3450	0.3780	0.3471	0.3346	0.2640
Cost	605.9817	607.77	607.79	607.6872	607.6896	606.6184	**605.7128**
Emis.	0.22019	0.2198	0.2201	0.2179	0.2176	0.2231	0.2198

Table 2. Best results out of 20 trials for emission

	BB-MOPSO	NSGAII	FCPSO	MOEA/D	SMS-EMOA	MBFO	MCLBFO
P_{G1}	0.4103	0.4101	0.4063	0.4034	0.4002	0.5000	0.4071
P_{G2}	0.4661	0.4631	0.4586	0.4589	0.4521	0.4552	0.4621
P_{G3}	0.5432	0.5434	0.5510	0.5420	0.5394	0.5508	0.5421
P_{G4}	0.3883	0.3895	0.4084	0.3973	0.4101	0.3838	0.3891
P_{G5}	0.5447	0.5438	0.5432	0.5381	0.5413	0.4754	0.5401
P_{G6}	0.5168	0.5151	0.4974	0.4008	0.4909	0.4688	0.5112
Cost	646.21	646.06	643.58	645.16	643.58	653.67	646.16
Emis.	**0.194179**	**0.194179**	0.194212	0.194198	0.194219	0.194955	**0.194179**

MCLBFO could obtain the smallest fuel cost in comparison to all six compared algorithms in terms of emission objective, and there is no big difference between studied methods in terms of fuel cost. FCPSO obtain the comparatively biggest fuel cost, and NSGAII, SMS-EMOA as well as MOEA/D generate more fuel cost, while the remaining methods BB-MOPSO and MBFO get smaller values. Besides, Table 1 also verifies the effectiveness of the MCLBFO for being got the smallest emission, while MBFO is difficult in solving the complex objectives.

Through the experiment results on the standard IEEE 30-bus 6-generator test system, the effectiveness of the novel algorithm has been investigated. All in all, the performance of the novel technique MCLBFO has confirmed its superiority compared to reported traditional multi-objective optimization algorithms in prior literature.

5 Conclusion and Future Work

In this paper, a multi-objective algorithm extended from bacterial foraging optimization is presented to handle the Environmental/Economic Dispatch problem. Through the experiments on the IEEE 30-node, 6-generator system, the proposed algorithm (MCLBFO) has proved to be promising in dealing with multi-objective optimization problems. Studies on solving EED optimization problem are helpful for choosing and modifying bacterial based algorithms for the optimization of multi-objective optimization problems. Even so, BFO, with its own limitations, has a high computation complexity which cannot be ignored. In future, more work will be done to overcome this computational problem, and more powerful and complicated multi-objective bacterial foraging optimization variants will be studied.

Acknowledgments. This work is partially supported by The National Natural Science Foundation of China (Grants Nos. 71571120, 71271140, 71461027, 71471158) and the Natural Science Foundation of Guangdong Province (Grant Nos. 1614050000376, 2014A030310314).

References

1. Lu, Y., Zhoun, J., Qin, H., Wang, Y., Zhang, Y.: Chaotic differential evolution methods for dynamic economic dispatch with valve-point effects. Eng. Appl. Artif. Intell. **24**, 378–387 (2011)
2. Neyestani, M., Farsangi, M.M., Nezamabadi-Pour, H.: A modified particle swarm optimization for economic dispatch with non-smooth cost functions. Eng. Appl. Artif. Intell. **23**, 1121–1126 (2010)
3. Wang, L., Singh, C.: Reserve-constrained multiarea environmental/economic dispatch based on particle swarm optimization with local search. Eng. Appl. Artif. Intell. **22**, 298–307 (2009)
4. Talaq, J.H., El-Hawary, F., El-Hawary, M.E.: A summary of environmental/economic dispatch algorithms. IEEE Trans. Power Syst. **9**, 1508–1516 (1994)
5. Cai, J.J., Ma, X.Q., Li, Q., Li, L.X., Peng, H.P.: A multi-objective chaotic particle swarm optimization for environmental/economic dispatch. Energ. Convers. Manag. **50**, 1318–1325 (2009)
6. Wang, L., Singh, C.: Balancing risk and cost in fuzzy economic dispatch including wind power penetration based on particle swarm optimization. Electr. Power Syst. Res. **78**, 1361–1368 (2008)
7. Ramesh, S., Kannan, S., Baskar, S.: An improved generalized differential evolution algorithm for multi-objective reactive power dispatch. Eng. Optim. **44**, 391–405 (2012)
8. Wu, L.H., Wang, Y.N., Yuan, X.F., Zhou, S.W.: Environmental/economic power dispatch problem using multi-objective differential evolution algorithm. Electr. Power Syst. Res. **80**, 1171–1181 (2010)
9. Jadhav, H.T., Roy, R.: Gbest guided artificial bee colony algorithm for environmental/economic dispatch considering wind power. Expert Syst. Appl. **40**, 6385–6399 (2013)
10. Ghasemi, A.: A fuzzified multi objective interactive honey bee mating optimization for environmental/economic power dispatch with valve point effect. Int. J. Electr. Power **49**, 308–321 (2013)

11. Passino, K.M.: Biomimicry of bacterial foraging for distributed optimization and control. IEEE Control Syst. Mag. **22**, 52–67 (2002)
12. Guzmán, M.A., Delgado, A., De Carvalho, J.: A novel multiobjective optimization algorithm based on bacterial chemotaxis. Eng. Appl. Artif. Intell. **23**, 292–301 (2010)
13. Lu, Z.G., Feng, T., Liu, Z.Z.: A multiobjective optimization algorithm based on discrete bacterial colony chemotaxis. Math. Prob. Eng. (2014)
14. Niu, B., Wang, H., Wang, J., Tan, L.: Multi-objective bacterial foraging optimization. Neurocomputing **116**, 336–345 (2013)
15. Zhao, Q.S., Hu, Y.L., Tian, Y.: An improved bacterial colony chemotaxis multi-objective optimisation algorithm. Int. J. Comput. Sci. Math. **4**, 392–401 (2013)
16. Pandi, V.R., Mohapatra, A., Panigrahi, B.K., Krishnanand, K.R.: A hybrid multi-objective improved bacteria foraging algorithm for economic load dispatch considering emission. Int. J. Comput. Sci. Eng. **11**, 114–123 (2015)
17. Pandi, V.R., Panigrahi, B.K., Hong, W.C., Sharma, R.: A multiobjective bacterial foraging algorithm to solve the environmental economic dispatch problem. Energ. Source Part B **9**, 236–247 (2014)
18. Panigrahi, B.K., Pandi, V.R., Das, S., Das, S.: Multiobjective fuzzy dominance based bacterial foraging algorithm to solve economic emission dispatch problem. Energy **35**, 4761–4770 (2010)
19. Beume, N., Naujoks, B., Emmerich, M.: SMS-EMOA: multiobjective selection based on dominated hypervolume. Eur. J. Oper. Res. **181**, 1653–1669 (2007)
20. Deb, K., Pratap, A., Agarwal, S., Meyarivan, T.: A fast and elitist multiobjective genetic algorithm: NSGA-II. IEEE Trans. Evol. Comput. **6**, 182–197 (2002)
21. Qingfu, Z., Hui, L.: MOEA/D: a multiobjective evolutionary algorithm based on decomposition. IEEE Trans. Evol. Comput. **11**, 712–731 (2007)
22. Liang, J.J., Qin, A.K., Suganthan, P.N., Baskar, S.: Comprehensive learning particle swarm optimizer for global optimization of multimodal functions. IEEE Trans. Evol. Comput. **10**, 281–295 (2006)
23. Abido, A.A.: A new multiobjective evolutionary algorithm for environmental/economic power dispatch. In: Power Engineering Society Summer Meeting, vol. 1262, pp. 1263–1268 (2001)
24. Agrawal, S., Panigrahi, B.K., Tiwari, M.K.: Multiobjective particle swarm algorithm with fuzzy clustering for electrical power dispatch. IEEE Trans. Evol. Comput. **12**, 529–541 (2008)
25. Zhang, Y., Gong, D.W., Ding, Z.H.: A bare-bones multi-objective particle swarm optimization algorithm for environmental/economic dispatch. Inform. Sci. **192**, 213–227 (2012)

Modified Brain Storm Optimization Algorithms Based on Topology Structures

Li Li[1], F.F. Zhang[1], Xianghua Chu[1(✉)], and Ben Niu[1,2(✉)]

[1] College of Management, Shenzhen University, Shenzhen 518060, China
x.chu@hotmail.com, Drniuben@gmail.com
[2] Department of Mechanism Engineering, Hong Kong Polytechnic University,
Kowloon, Hong Kong

Abstract. An algorithm performs better often due to its communication mechanisms. Different types of topology structures denote various information exchange mechanisms. This paper incorporates topology structure concept into brain storm optimization (BSO) algorithm. Three types of topology structures, which are full connected, ring connected and star connected, are introduced. And three novel modified optimization algorithms based on topology structures are proposed (BSO-FC, BSO-RI, BSO-ST). Unimodal and multimodal criteria functions are employed to verify the effectiveness of the raised algorithms. In addition, both the original BSO algorithm and bacterial foraging optimization (BFO) algorithm are selected as contrastive algorithms to expose the optimization capacity of the proposed algorithms. Experimental results show that all of the modified algorithms have better performance than the original BSO algorithm, especially the BSO-ST algorithm.

Keywords: Brain storm optimization · Topology structures · Population-based optimization · Mutation operator · Gaussian mutation

1 Introduction

Optimization problem is an important branch of modern management. In many years, people are trying their best to find better ways to solve these problems. In recent years, optimization algorithms based on the population have been extensively investigated. In contrast to algorithms based on single-point, for example, hill-climbing algorithm, population-based optimization algorithms do its jobs by communicating and competing with each other. Nowadays, population-based optimization algorithms are generally classified as swarm intelligence algorithms.

There are many swarm intelligence algorithms, such as particle swarm optimization (PSO) [1], bacterial foraging optimization (BFO) [2], artificial bee colony optimization (ABC) [3], ant colony optimization (ACO) [4], etc. But all of them are just inspired by simple animals or insects, such as ants, bees, birds, etc.

As a new type of swarm intelligence, brain storm optimization (BSO) was first proposed by Shi in [5, 6]. BSO is motivated by the most intelligent organisms, the human being. After that, many scholars have conducted research on this algorithm because of its excellent performance.

© Springer International Publishing Switzerland 2016
Y. Tan et al. (Eds.): ICSI 2016, Part II, LNCS 9713, pp. 408–415, 2016.
DOI: 10.1007/978-3-319-41009-8_44

Some people do research on the parameters of this algorithm. According to the dynamic range in the iteration, reference [7] proposed a dynamic step-size strategy, which was dynamically changed in each iteration. Reference [8] made investigation on the parameters in the BSO to see how they affected the performance of BSO. In order to reduce the computation time, reference [9] implemented only several iterations of the k-means clustering method in each iteration of BSO, instead of achieving a convergent clustering by using k-means clustering method. Reference [10] adopted the concept of preda-tor-prey into BSO to make full use of the global knowledge.

In addition, the BSO algorithm is also applied to solve the practice problems. Reference [11] found the optimal solution of economic dispatch problem considering wind power based on the BSO algorithm. Reference [10] applied the modified BSO to optimally design a DC brushless motor to maximize its efficiency.

In the BSO algorithm described by Shi [5], it contains two interesting characters. One character is the clustering process that all the ideas are divided into several groups by k-mean clustering method. The other character is the individual generation process that produces new idea by mixing a Gaussian random noise in. As we all know, BSO algorithm has been studied by many researchers, but it is just in its infancy now, and lots of studies are needed.

In this present paper, we concentrate on the character of the individual generation process to come up with three modified BSO algorithms based on the topology structures. Experiments show that the raised algorithms can improve the capability of the original BSO, obviously.

The rest part of this paper is organized as follows. Section 2 demonstrates the BSO algorithm. The modified BSO algorithms are presented in Sect. 3. Benchmark functions are used to verify the proposed algorithms in Sect. 4. Finally, Sect. 5 summarizes the concluding remarks.

2 Brain Storm Optimization Algorithm

As a creative problem-solving method, brain storm process was systematized by Ostorn in 1939. In a brainstorming process, the idea generation often follows four principles. The detailed description is listed in Table 1. When facing the complicated problem which a single people cannot solve, human being get together to brain storm, especially people with different background. This is the central idea of this method and this process can improve the probability of solving the difficult problem.

Table 1. Osborn's original rules for idea generation in a brainstorming process

Rules	Content
Rule1	Suspend Judgment
Rule2	Anything Goes
Rule3	Cross-fertilize
Rule4	Go for Quantity

Inspired by the process of human being idea formation, Shi came up with a new algorithm called brain storm optimization (BSO) [5]. The pseudo code of the BSO algorithm is shown in Fig. 1.

In original BSO algorithm, new ideas are formed by mixing Gaussian noise in. The main formulas are enumerated as follows, where X_{i+1} denotes the new individual, X_i denotes the selected idea, $N(\mu, \sigma)$ is Gaussian function, $logsig()$ is a logarithmic sigmoid transfer function, and k is a coefficient, $max_{iteration}$ is the maximum number of

```
Algorithm BSO
Begin
   Initialize variables and generate N ideas;
   While
      Cluster;
      Select cluster center;
      If(random(0,1)<p_replace)
         Selected a cluster randomly and replace
         the  cluster  center  with  a  randomly
         generated idea;
      End
   FOR(each ideas)
      If(random(0,1)<p_one)
         Randomly select a cluster with a
         probability p;
         If(random(0,1)<p_one_center)
            Add  random  values  to  the  selected
            cluster  center  to  generate  a  new
            idea;
         Else
            Add random values to a random idea of
            the selected cluster to generate a
            new idea;
         End If
      Else
         Random select two clusters
         If(random(0,1))<p_two_center
            Combine  the  two  selected  cluster
            center and add with random values to
            generate a new idea;
         Else
            Combine two random ideas from the two
            selected clusters and add with random
            values to generate a new idea;
         End If
      End If
   End For
   End While
End
```

Fig. 1. Pseudo-code of the BSO algorithm

iterations, $current_{iteration}$ is the current iteration number, and $rand()$ is a random value within (0,1).

$$X_{i+1} = X_i + \varepsilon \times N(\mu, \sigma) \tag{1}$$

$$\varepsilon = logsig\left(\frac{0.5 \times max_{iteration} - current_{iteration}}{k}\right) \div rand() \tag{2}$$

3 The Modified Brain Storm Optimization Algorithms

The excellent performance of the swarm intelligence optimization often dues to the information communication mechanism, like the bees cooperate and compete with each other. The introduction of information exchange mechanisms is aimed to prevent population from local optimum. Each topology structure contains a special information communication strategy. Reference [12] introduced the topology structures in detail. Commutating with each other, population can get more information from others and share information to others, which can improve their behavior in the process of searching optimization solution. In this paper, we incorporates the concept of topology structure into individual generation process instead of the Gaussian function.

Three topology structures, including full connected topology, ring connected topology, star connected topology, are used in this paper. The elementary diagram of the topology structures are presented in Fig. 2 [13].

According to the topology structures, the proposed algorithms based on topology structures are named BSO-FC, BSO-RI, and BSO-ST, respectively.

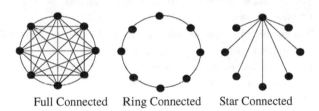

Full Connected   Ring Connected   Star Connected

Fig. 2. Three types of topology structures

4 Experiments and Discussions

4.1 Benchmark Functions

Six benchmark functions are applied to test the virtue of the BSO-FC, BSO-RI and BSO-ST. Among them, Sphere, SumPowers, Schwefel221 are unimodal functions, while Apline, Rastrigin, Griewank are multimodal functions. These benchmark functions are employed to verify the capability of the proposed algorithms. The benchmark functions are introduced in Table 2.

Table 2. Search ranges of the benchmark functions

Attribute	Name	Function	Range
Unimodal	Sphere	$f(x) = \sum_{i=1}^{n} X_i^2$	$[-100,100]$
	SumPowers	$f(x) = \sum_{i=1}^{n} |X_i|^{i+1}$	$[-10,10]$
	Schwefel221	$f(x) = max|X_i|$	$[-10000,10000]$
Multimodal	Apline	$f(x) = |X_i \times sin(X_i + 0.1 \times X_i|$	$[-10,10]$
	Rastrigin	$f(x) = \sum_{i=1}^{n} (X_i^2 - 10cos(2\pi X_i) + 10)^2$	$[-5.12,5.12]$
	Griewank	$f(x) = \frac{1}{4000}\sum_{i=1}^{n} X_i^2 - \prod_{i=1}^{n} cos(\frac{X_i}{\sqrt{i}}) + 1$	$[-600,600]$

4.2 Parameter Settings

In the experiments, each test is run for 1000 iterations and 20 replications are conducted for each test to collect the statistics. All the functions mentioned above are tested with 20 dimensions. We set the population size is 20. The notations used are summarized in Table 3. The parameter settings primarily for algorithms are shown in Table 4.

Table 3. Parameters for the algorithms

Parameters	Description
Ped	The probability of demise
Fed	The interval iterations to perform demise operation
C	Chemotactic step
Ns	The number of executions when a bacteria looking for a good food source
p_replace	Probability to determine whether a dimension is disrupted or not
p_one	Probability for select one individual, not two, to generate new individual
p_one_center	Probability of using one cluster center
k	The slope for changing *logsig()* function
p_two_center	Probability of using two cluster centers

Table 4. Algorithm configurations

Alogithm	Parameter settings		
BFO	Ped = 0.2 Fed = 5	C = 0.05 Ns = 8	
BSO			
BSO-FC	p_replace = 0.2	p_one = 0.8	
BSO-RI	p_one_center = 0.4	k = 20	
BSO-ST	p_two_center = 0.5		

4.3   Comparison on Solutions

The mean and standard deviation for the different functions given by BFO, BSO, BSO-FC, BSO-RI and BSO-ST are shown in Tables 5 and 6. The best values given by the five algorithms are signed as bold. As listed in Tables 5 and 6, BSO-ST indicates better performance than the other four algorithms on Ackley, Schwefel221, Griewank and Rastrigin functions, while the BSO-FC reveals better capability on Sphere and SumPowers functions slightly.

For the three unimodal functions, Table 5 shows that BSO-ST is the best algorithm for Schwefel221 and the second best algorithm for Sphere and SumPower. For the three mulmodal functions, Tables 6 tells that BSO-ST performs the best on Ackley, Rastrigin, and Griewank. Owing to the commutation mechanism, which is open minded, BSO-ST reveals excellent capability on multimodal functions. It is similar to the brainstorming process that any thought of your mind should not be ignored.

The convergence progress of the mean fitness values are shown in Fig. 3. It is obvious that for most functions, the BSO-ST converges the fastest at the beginning and finds the optimum solution soon.

Table 5. Test data based on unimodal criteria functions

Algorithm		Sphere	SumPowers	Schwefel221
BFO	Mean	1.44368e+000	1.68039e−002	5.45828e−001
	Std.	3.98885e−001	1.20481e−002	5.52406e−002
BSO	Mean	9.48757e−005	9.73018e−005	7.11372e−002
	Std.	1.0325e−004	1.7128e−004	1.84648e−002
BSO-FC	Mean	**3.11238e−008**	**4.01170e−008**	2.66992e−003
	Std.	1.39190e−007	1.79409e−007	9.57804e−003
BSO-RI	Mean	3.56433e−003	3.14258e−003	3.98185e−002
	Std.	7.64096e−003	8.54150e−003	4.00449e−002
BSO-ST	Mean	1.74583e−007	1.74377e−007	**9.53348e−005**
	Std.	**1.37360e−008**	**1.4302e−008**	**4.1704e−006**

Table 6. Test data based on multimodal criteria functions

Algorithm		Ackley	Rastrigin	Griewank
BFO	Mean	2.38442e+000	7.62865e+001	7.61453e−002
	Std.	1.90104e−001	8.92492e+000	2.07259e−002
BSO	Mean	4.87586e−002	5.00343e+000	2.50661e−004
	Std.	1.61261e−001	1.76937e+001	3.6427e−004
BSO-FC	Mean	1.12938e−003	2.83987e+001	2.26755e−006
	Std.	5.05073e−003	2.71628e+001	9.50192e−006
BSO-RI	Mean	5.55449e−002	1.68623e+001	1.94659e−004
	Std.	8.33758e−002	3.70785e+001	4.3508e−004
BSO-ST	Mean	**3.74755e−004**	**2.94473e−005**	**1.37943e−008**
	Std.	**1.2816e−005**	**3.0775e−006**	**1.6663e−009**

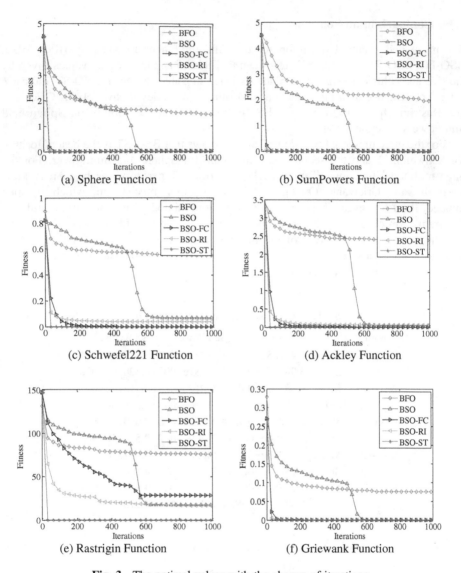

Fig. 3. The optimal values with the change of iterations

5   Conclusions

This paper presents three novel modified brain storm optimization algorithms based on different types of topology structures. Topology structures are used to generate new individuals instead of the Gaussian Function. Because of the good commutation mechanism of the topology structure, we expect that the proposed algorithms can achieve better performance.

The capability of the raised algorithms are tested by criteria functions and compared with BFO and BSO. These functions contain three unimodal functions and three multimodal functions. Experiment results have indicated that the solutions obtained by BSO-FC, BSO-RI and BSO-ST are better than these produced by the original BSO and BFO, obviously, especially the BSO-ST.

Acknowledgments. This work is partially supported by The National Natural Science Foundation of China (Grants nos. 71571120, 71271140, 71461027, 71471158, 71501132) and the Natural Science Foundation of Guangdong Province (Grant nos. 1614050000376).

References

1. Kennedy, J., Eberhart, R., Shi, Y.: Swarm Intelligence. Morgan Kaufmann Publisher, Burlington (2001)
2. Passion, K.M.: Bacterial foraging optimization. Int. J. Swarm Intell. Res. **1**(1), 1–16 (2010)
3. Karaboga, D.: An idea based on honey bee swarm for numerical optimization. Technical report-TR06, Erciyes University, Engineering Faculty, Computer Engineering Department (2005)
4. Dorigo, M., Stützle, T.: Ant Colony Optimization. The MIT Press, Cambridge (2004)
5. Shi, Y.: Brain storm optimization algorithm. In: Tan, Y., Shi, Y., Chai, Y., Wang, G. (eds.) ICSI 2011, Part I. LNCS, vol. 6728, pp. 303–309. Springer, Heidelberg (2011)
6. Shi, Y.: An optimization algorithm based on brainstorming process. Int. J. Swarm Intell. Res. **2**, 35–62 (2011)
7. Zhou, D., Shi, Y., Cheng, S.: Brain storm optimization algorithm with modified step-size and individual generation. In: Tan, Y., Shi, Y., Ji, Z. (eds.) ICSI 2012, Part I. LNCS, vol. 7331, pp. 243–252. Springer, Heidelberg (2012)
8. Zhan, Z., Chen, W., Lin, Y., Gong, Y., Li, Y., Zhang, J.: Parameter investigation in brain storm optimization. In: 2013 IEEE Symposium on Swarm Intelligence, Singapore (2013)
9. Zhan, Z., Zhang, J., Shi, Y., Liu, H.: A modified brain storm optimization. In: Proceedings of Congress on Evolutionary Computation, pp. 1–8. Brisbane, Australia (2012)
10. Duan, H., Li, S., Shi, Y.: Predator-prey based brain storm optimization for DC brushless motor. IEEE Trans. Magn. **49**, 5336–5340 (2013)
11. Jadhav, H.T., Sharma, U., Patel, J., Roy, R.: Brain storm optimization algorithm based economic dispatch considering wind power. In: 2012 IEEE International Conference on Power and Energy, Kota Kinabalu Sabah, Malaysia (2012)
12. McNabb, A., Gardner, M., Seppi, K.: An exploration of topologies and communicational in large particle swarms. In: Proceedings of the IEEE Congress on Evolutionary Computation, pp. 712–719 (2009)
13. Niu, B., Liu, J., Bi, Y., Tan, L.J.: Improved bacterial foraging optimization algorithm with information communication mechanism. In: Computational Intelligence and Security (CIS), pp. 47–51 (2014)

Brain Storm Optimization for Portfolio Optimization

Ben Niu[1,2(✉)], Jia Liu[1], Jing Liu[1], and Chen Yang[1]

[1] College of Management, Shenzhen University, Shenzhen, China
Drniuben@gmail.com, yangc@szu.edu.cn
[2] Department of Mechanical Engineering, Hong Kong Polytechnic University,
Hong Kong, China

Abstract. In this paper, Brain storm optimization (BSO) algorithm is employed to solve portfolio optimization (PO) problem with transaction fee and no short sales. In addition, simplified BSO (SBSO) and BSO in objective space (BSO-OS) are also utilized in the same model. The potential portfolio proportion is regarded as ideas that individual generated. In the experimental study, three cases with different risk aversion factors are considered. Simulation results demonstrate that both original BSO and modified BSOs obviously outperform PSO and BFO in PO problem. In particular, BSO-OS, which not only saves the computation time but also finds the optimal value, shows an extraordinary performance in all set of test data.

Keywords: Brain storm optimization · Simplified BSO · BSO in objective space · Portfolio optimization

1 Introduction

Portfolio optimization (PO) is a process to decide how to allocate wealth and what assets to invest can gain the maximum returns under the acceptable risk. Since Markowitz proposed the famous Mean-Variance model [1] for PO problem, scholars have put forward more sophisticated PO models, which consider various constraints of real-world investments like liquidity risk, market risk [2] and transaction fee [3]. It has become considerably difficult to solve the NP-hard and nonlinear PO problems using mathematical programing methods. Therefore researchers resort to heuristic methods for obtaining optimal asset allocation, such as particle swarm optimization (PSO) [4], bacteria forging optimization (BFO) [5] and etc.

In this paper, we solve PO problem based on brain storm optimization (BSO) algorithm proposed by Shi in 2011 [6]. In finance, BSO has been successfully used in stock index forecasting [7] and v-SVR problems [8]. As a promising algorithm, BSO is expected to have a high tendency to optimize the PO problem. Specifically, we utilize two modified BSO algorithms, i.e., simplified BSO (SBSO) proposed by Li [9] and BSO in objective space (BSO-OS) proposed by Shi [10] to solve the PO problem at the same time.

The rest of this paper is organized as follows. In Sect. 2, the framework of basic BSO, SBSO and BSO-OS are introduced. Then Sect. 3 formulates the mathematical

© Springer International Publishing Switzerland 2016
Y. Tan et al. (Eds.): ICSI 2016, Part II, LNCS 9713, pp. 416–423, 2016.
DOI: 10.1007/978-3-319-41009-8_45

models of the PO problem. Section 4 reports numerical experiments for evaluating the algorithm. Finally, conclusions are drawn in the last section.

2 Brain Storm Optimization

2.1 The Basic Brain Storm Optimization Algorithm

Derived from human being's behavior of brainstorming, Shi proposed BSO algorithm [6] and gained success. The detailed procedure of BSO is presented as follow:

Step 1: Initialize the population according to

$$P = range\_l + (range\_r - range\_l) * rand(n, d) \tag{1}$$

Where n is the population size, $range\_l$ and $range\_r$ are lower and upper limits, respectively. So n ideas with d dimension are randomly generated. Step 2: Calculate the fitness values of n ideas. Step 3: Cluster n ideas into m clusters by k-means clustering method. Then choose the best idea in each cluster as a cluster center. Step 4: Utilize a newly generated idea to replace a randomly selected cluster center with a small probability P_{5a} to explore more potential solutions. Step 5: If rand $(0, 1)$ is smaller than P_{6b}, randomly choose one cluster. Otherwise, two randomly chosen clusters are utilized to obtain X_{old}. Based on one cluster, a cluster center or a random solution is selected according to P_{6biii}.

$$X_{old} = \begin{cases} X_c(k) \\ X_i(k) \end{cases} \text{if} \ \ \text{rand}() < p_{6biii} \tag{2}$$

On the basis of choosing two clusters, combine two cluster centers or two random solutions from selected clusters according to P_{6c}.

$$X_{old} = \begin{cases} w * X_{c1}(k_1) + (1 - w) * X_{c2}(k_2) \\ w * X_{i1}(k_1) + (1 - w) * X_{i2}(k_2) \end{cases} \text{if} \ \ \text{rand}() < p_{6c} \tag{3}$$

Step 6: After selecting one idea as X_{old}, X_{old} is updated according to

$$X_{new} = X_{old} + \xi * N(0, 1). \tag{4}$$

Where $N(0, 1)$ is the Gaussian random value with mean 0 and variance 1. ξ is an alterable factor which can be expressed as:

$$\xi = \log sig(\frac{0.5T - t}{k}) * rand(0, 1). \tag{5}$$

Where $logsig$ () is a logarithmic sigmoid transfer function, T and t are respectively the maximum and current iteration number, and k is for changing the slope of $logsig$ ()

function. Step 7: Compare the newly generated idea with the previous one and keep a better idea.

2.2   Modified Brain Storm Optimization Algorithms

In this paper, we choose SBSO [9] and BSO-OS [10] which can simplify the process and reduce the computation burden respectively to solve PO problem.

1. **The simplified brain storm optimization algorithm.** In SBSO, simple grouping method (SGM) [11] is utilized instead of k-means clustering method. X_{old} is generated in only one way, i.e. combining the two randomly selected ideas from two clusters respectively. Since it's not necessary to select a cluster center any more, then disruption can be omitted. In order to enhance the potential diversity of the ideas, crossover operation is added before keeping the best ideas. X_{old} crossovers with the original idea with the same idea index to generate two more ideas.
2. **Brain Storm Optimization Algorithm in Objective Space.** Instead of applying an algorithm to cluster solutions, BSO-OS ranks all solutions from best to worst. The top $perc_e$ percent of solutions will be regarded as elitists (X_e) while the other ones as normal (X_n). Then disrupt a randomly chosen individual instead of disrupting a cluster center. The way of selecting is a little different from original BSO algorithm. First, decide whether X_{old} will be generated based on "elitists" or "normal", then determine whether based on one or two individuals. If rand (0, 1) is smaller than p_e,

$$X_{old} = \begin{cases} X_e & \text{if } \mathrm{rand}(0,1) < p_{one} \\ s * X_{e1} + (1-s) * X_{e2} \end{cases} \tag{6}$$

Otherwise,

$$X_{old} = \begin{cases} X_n & \text{if } \mathrm{rand}(0,1) < p_{one} \\ s * X_{n1} + (1-s) * X_{n2} \end{cases} \tag{7}$$

Where s is a random value between 0 and 1. Other operations are the same as the original BSO.

3   Brain Storm Optimization for Portfolio Selection Problems

3.1   Portfolio Selection Model

In this section, we introduce a model considering the transaction fee and no short sales in PO presented in [3]. The model can be described as follow:

$$\min F(X) = \min[\lambda\, g(X) - (1 - \lambda)f(X)]$$

$$s.t. \begin{cases} \sum_{i=1}^{n} X_i = 1 \\ \\ X_i > 0 \end{cases} \tag{8}$$

Investors' return function $f(X)$ and risks function $g(X)$ can be formulated as below.

$$f(X) = \sum_{i=1}^{d} r_i X_i - \sum_{i=1}^{d} \left[\mu * k_i^b * (X_i - X_i^0) + (1 - \mu) * k_i^s * (X_i^0 - X_i) \right]$$

$$and \ \mu = \begin{cases} 1 \ X_i \geq X_i^o \\ 0 \ X_i < X_i^0 \end{cases} \tag{9}$$

$$g(X) = \sum_{i=1}^{d} \sum_{j=1}^{d} X_i X_j \sigma_{ij} \tag{10}$$

Suppose that an investor is going to invest in d assets. λ denotes the risk aversion factor. The larger the value of λ is, the smaller risk the investor can bear. Let X_i be the allocation proportion for investing d assets, X_i^0 be the original proportion, r_i be the expected return of asset i, σ_{ij} be the covariance of r_i and r_j. k_i^b and k_i^s are the transaction fee for buying and selling the ith asset respectively.

3.2 Encoding

We encode the potential portfolio proportion as ideas that individual generated in BSO. Each idea (X_i) represents the holdings of asset i. d dimension denotes d kinds of assets. The idea of the individual with the minimum fitness value is the best selection of PO. The fitness function is calculated according to Eq. (8). To ensure the total proportion equals 1 and every proportion is larger than 0, we use a penalty function method.

4 Experiments and Discussions

In order to verify the performance of BSO algorithms, PSO and BFO were selected for comparison. All the algorithms presented in this paper were coded in MATLAB language and run on an Intel Core i5 processor with 2.2 GHz CPU speed and 4 GB RAM machine, running Windows 8.1.

4.1 Definition of Experiments

In the experiments, we supposed that there are ten assets for an investor to invest and the original proportion in every asset X_i^0 equals 0.1. What's more, the sum of the initialized ratio of each asset equals 1. We utilized three risk aversion factors to identify three different kinds of investors. The related parameters are shown as follow:

$$\lambda = (0.15,\ 0.5,\ 0.85)$$

$$\begin{aligned}
\sigma = [&0.0013, 0.0021, 0.0001, 0.0035, 0.0003, 0.0011, 0.0014, 0.0066, 0.0060, 0.0046;\\
&0.0021, 0.0006, 0.0069, 0.0043, 0.0001, 0.0002, 0.0052, 0.0015, 0.0011, 0.0004;\\
&0.0001, 0.0069, 0.0001, 0.0003, 0.0016, 0.0000, 0.0044, 0.0004, 0.0001, 0.0006;\\
&0.0035, 0.0043, 0.0003, 0.0021, 0.0008, 0.0003, 0.0021, 0.0002, 0.0027, 0.0034;\\
&0.0003, 0.0001, 0.0016, 0.0008, 0.0024, 0.0011, 0.0006, 0.0053, 0.0014, 0.0006;\\
&0.0011, 0.0002, 0.0000, 0.0003, 0.0011, 0.0033, 0.0042, 0.0004, 0.0073, 0.0021;\\
&0.0014, 0.0052, 0.0044, 0.0021, 0.0006, 0.0042, 0.0048, 0.0038, 0.0002, 0.0005;\\
&0.0066, 0.0015, 0.0004, 0.0002, 0.0053, 0.0004, 0.0038, 0.0006, 0.0002, 0.0079;\\
&0.0060, 0.0011, 0.0001, 0.0027, 0.0014, 0.0073, 0.0002, 0.0002, 0.0067, 0.0007;\\
&0.0046, 0.0004, 0.0006, 0.0034, 0.0006, 0.0021, 0.0005, 0.0079, 0.0007, 0.0006]
\end{aligned}$$

$$r = [0.2606, 0.2775, 0.2453, 0.1670, 0.2186, 0.1847, 0.1468, 0.0929, 0.1385, 0.1782]$$

$$k_i^b = 0.00065,\ k_i^s = 0.00075$$

Table 1. Parameter settings for BSO algorithms

Method	n	Max iteration	m	P_{5a}	P_{6b}	P_{6biii}	P_{6c}	k	μ	σ	$perc_e$	P_e	p_{one}
BSO	50	1000	5	0.2	0.8	0.4	0.5	20	0	1	–	–	–
SBSO	50	1000	5	–	–	–	–	20	0	1			–
BSO-OS	50	1000	–	–	–	–	–	–	–	–	0.1	0.2	0.8

Table 2. Experimental results of $\lambda = 0.15$

	BSO	SBSO	BSO-OS	PSO	BFO
Max	−2.3071E-01	−2.2461E-01	**−2.2749E-01**	−1.8295E-01	−1.9222E-01
Min	−2.3403E-01	−2.3134E-01	**−2.3464E-01**	−2.0253E-01	−2.0450E-01
Mean	−2.3193E-01	−2.2873E-01	**−2.3273E-01**	−1.9113E-01	−1.9849E-01
Std.	**8.6667E-04**	1.5737E-03	1.7979E-03	4.4838E-03	3.6070E-03
X_1	1.7324E-02	8.0396E-02	1.1210E-03	3.2461E-01	2.9901E-01
X_2	9.6680E-01	8.6306E-01	9.9753E-01	1.2827E-01	2.9442E-01
X_3	1.5875E-02	3.3952E-02	9.7345E-04	3.1324E-01	1.6488E-01
X_4	9.1851E-12	2.8321E-04	3.3939E-09	1.3169E-04	9.2811E-02
X_5	5.8900E-06	1.6003E-02	4.6187E-07	9.1242E-03	1.0123E-02
X_6	6.2965E-12	1.0607E-03	3.7075E-04	1.1170E-01	9.2811E-02
X_7	6.0183E-12	3.0254E-04	4.4330E-06	4.0701E-03	5.1034E-03
X_8	3.0304E-11	4.0807E-12	7.3202E-09	5.3145E-03	6.4883E-03
X_9	7.9940E-13	8.0404E-05	3.8985E-09	2.0727E-07	2.0218E-02
X_{10}	1.3293E-10	4.8626E-03	1.8742E-06	1.0353E-01	1.4132E-02
Return	2.7548E-01	2.7237E-01	2.7616E-01	2.3854E-01	2.4095E-01
Risk	8.4343E-04	1.1710E-03	6.1533E-04	1.5228E-03	2.0504E-03

The parameter settings for BSO, SBSO and BSO-OS are listed in Table 1 below. In PSO, we set $c_1 = c_2 = 2$, and inertia weight $w = 1$. In BFO, the number of chemotactic (N_c) is set to 1000. The number of reproduction (N_{re}) is set to 5. The number of elimination-dispersal (N_{ed}) is set to 2. The number of swimming (N_s) is set to 4. The swimming length (C) is set to 0.2. The elimination-dispersal frequency is set as: $P_{ed} = 0.25$.

Table 3. Experimental results of $\lambda = 0.5$

	BSO	SBSO	BSO-OS	PSO	BFO
Max	**-1.3439E-01**	-1.3229E-01	-1.3436E-01	-1.0659E-01	-1.1298E-01
Min	-1.3751E-01	-1.3615E-01	**-1.3779E-01**	-1.1585E-01	-1.1819E-01
Mean	-1.3592E-01	-1.3385E-01	**-1.3667E-01**	-1.1071E-01	-1.1511E-01
Std.	8.8040E-04	**8.9421E-04**	1.2086E-03	2.5195E-03	1.6680E-03
X_1	1.5132E-02	1.2691E-01	8.4300E-04	1.6493E-01	2.0789E-01
X_2	9.7993E-01	8.5281E-01	9.9813E-01	3.0596E-01	2.7156E-01
X_3	1.9903E-03	1.4019E-02	3.7151E-04	1.9720E-01	8.8183E-02
X_4	9.2030E-12	8.3944E-06	7.1421E-09	7.0368E-04	6.1102E-03
X_5	3.7770E-04	4.8759E-03	6.5620E-04	6.1495E-02	3.0386E-01
X_6	2.5692E-03	9.8826E-04	2.3710E-09	6.6689E-02	1.3877E-03
X_7	1.7836E-11	8.2311E-05	1.4422E-09	5.1546E-02	4.7631E-03
X_8	1.0635E-11	3.5587E-11	3.4492E-09	3.0066E-04	1.3494E-02
X_9	1.5915E-11	3.6857E-05	1.0800E-08	6.0116E-02	1.4570E-02
X_{10}	2.8627E-06	2.6518E-04	3.0042E-07	9.1059E-02	8.8183E-02
Return	2.7569E-01	2.7339E-01	2.7618E-01	2.3377E-01	2.3787E-01
Risk	6.6685E-04	1.0808E-03	6.0654E-04	2.0761E-03	1.4967E-03

Table 4. Experimental results of $\lambda = 0.85$

	BSO	SBSO	BSO-OS	PSO	BFO
Max	-3.9595E-02	-3.8714E-02	**-3.9831E-02**	-2.9707E-02	-3.2222E-02
Min	-4.0803E-02	-4.0228E-02	**-4.0918E-02**	-3.3531E-02	-3.4523E-02
Mean	-4.0286E-02	-3.9489E-02	**-4.0544E-02**	-3.1819E-02	-3.3300E-02
Std.	3.2813E-04	**4.0686E-04**	3.8219E-04	9.4587E-04	7.1641E-04
X_1	1.4121E-02	5.2886E-02	3.2067E-04	1.9163E-01	2.6804E-01
X_2	9.8098E-01	9.1464E-01	9.9923E-01	1.6941E-01	1.8312E-01
X_3	7.0302E-07	1.2040E-03	6.6234E-09	2.2582E-01	1.9015E-01
X_4	4.5797E-12	9.3142E-04	1.8064E-09	9.1337E-03	4.0147E-03
X_5	1.7917E-03	1.6001E-02	5.4601E-05	1.8331E-01	2.2467E-01
X_6	3.1089E-03	1.6808E-03	7.2944E-06	1.1903E-01	8.5130E-02
X_7	6.4148E-12	6.8299E-08	3.7259E-09	3.4295E-02	1.3211E-02
X_8	1.1567E-11	1.6756E-03	1.1547E-09	2.7961E-10	1.1617E-02
X_9	3.8534E-11	7.2210E-03	5.3446E-09	2.3423E-05	4.3654E-03
X_{10}	6.9045E-08	3.7642E-03	3.8347E-04	6.7362E-02	1.5683E-02
Return	2.7563E-01	2.7254E-01	2.7619E-01	2.3242E-01	2.3858E-01
Risk	6.3757E-04	7.6755E-04	6.0075E-04	1.5664E-03	1.4874E-03

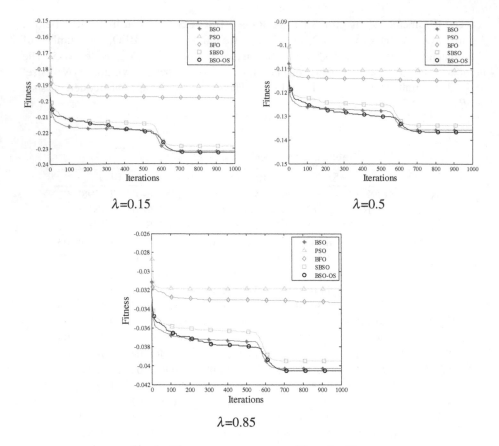

Fig. 1. The convergence curve of five algorithms

The experimental results obtained by five algorithms of $\lambda = 0.15$, $\lambda = 0.5$ and $\lambda = 0.85$ of 20 runs are given in Tables 2, 3 and 4 respectively. The best results are showed in bold. Figure 1 presents the convergence curves of the five algorithms of three different risk aversion factors.

4.2   Experimental Results

Analyzing the tables and figures, it is clear that the performance of three BSO algorithms is obviously better than that PSO and BFO. From the data of the minimum value, it can be seen that BSO-OS can always find the best solution. From the data of the standard deviations in Table 2, we can perceive that the results generated by the basic BSO are the steadiest. However, when $\lambda = 0.5$ and $\lambda = 0.85$, the standard deviations of SBSO are the smallest.

With the increase of the risk aversion factor λ, the fitness values grow up. The results illustrate that high returns involve increase risk. At the same time, asset 2 possesses bigger proportion when λ becomes larger.

5   Conclusion

In this paper, we employed original BSO, SBSO and BSO-OS algorithms to solve PO problem with different risk aversion factors. The obtained results indicate that BSO-OS algorithm is obviously a better choice to solve the difficult PO problem. Further work may focus on the study of obtaining more useful conditions for satisfying demands from the real market to propose new PO model. Based on the results of comparison, BSO algorithms outperformed PSO and BFO for this PO model. As BSO is still in its inception, further study can be conducted to improve it more. Moreover, we expect that BSO will be utilized to solve more complex problems in the future.

Acknowledgments. This work is partially supported by The National Natural Science Foundation of China (Grants Nos. 71571120, 71271140, 71471158, 71461027) and the Natural Science Foundation of Guangdong Province (Grant nos. 1614050000376).

References

1. Markowitz, H.: Portfolio selection. J. Finance **7**(1), 77–91 (1952)
2. Niu, B., Xue, B., Li, L., Chai, Y.: Symbiotic multi-swarm PSO for portfolio optimization. In: Huang, D.-S., Jo, K.-H., Lee, H.-H., Kang, H.-J., Bevilacqua, V. (eds.) ICIC 2009. LNCS, vol. 5755, pp. 776–784. Springer, Heidelberg (2009)
3. Li, L., Xue, B., Tan, L., Niu, B.: Improved particle swarm optimizers with application on constrained portfolio selection. In: Huang, D.-S., Zhao, Z., Bevilacqua, V., Figueroa, J.C. (eds.) ICIC 2010. LNCS, vol. 6215, pp. 579–586. Springer, Heidelberg (2010)
4. Yin, X., Ni, Q., Zhai, Y.: A novel particle swarm optimization for portfolio optimization based on random population topology strategies. In: Tan, Y., Shi, Y., Buarque, F., Gelbukh, A., Das, S., Engelbrecht, A. (eds.) ICSI-CCI 2015. LNCS, vol. 9140, pp. 164–175. Springer, Heidelberg (2015)
5. Niu, B., Bi, Y., Xie, T.: Structure-redesign-based bacterial foraging optimization for portfolio selection. In: Han, K., Gromiha, M., Huang, D.-S. (eds.) ICIC 2014. LNCS, vol. 8590, pp. 424–430. Springer, Heidelberg (2014)
6. Shi, Y.: Brain storm optimization algorithm. In: Tan, Y., Shi, Y., Chai, Y., Wang, G. (eds.) ICSI 2011, Part I. LNCS, vol. 6728, pp. 303–309. Springer, Heidelberg (2011)
7. Sun, Y.: A hybrid approach by integrating brain storm optimization algorithm with grey neural network for stock index forecasting. Abstr. Appl. Anal. **2014**, 1–10 (2014)
8. Shen, L.: Research and Application of V-SVR Based on Brain Storm Optimization Algorithm. Master's thesis, Lanzhou University (2014)
9. Li, J., Duan, H.: Simplified brain storm optimization approach to control parameter optimization in F/A-18 automatic carrier landing system. Aerosp. Sci. Technol. **42**, 187–195 (2015)
10. Shi, Y.: Brain storm optimization algorithm in objective space. In: Proceedings of 2015 IEEE Congress on Evolutionary Computation, (CEC 2015), pp. 1227–1234. IEEE, Sendai, Japan (2015)
11. Zhan, Z., Zhang, J., Shi, Y., Liu, H.: A modified brain storm optimization. In: Proceedings of the 2012 IEEE Congress on Evolutionary Computation (CEC), pp. 1–8 (2012)

Comprehensive Learning PSO for Solving Environment Heterogeneous Fixed Fleet VRP with Time Windows

X.B. Gan[1(✉)], L.J. Liu[1], J.S. Chen[2], and Ben Niu[1]

[1] College of Management, Shenzhen University, Shenzhen 518060, China
ganxb2001@163.com
[2] School of Computer Engineering, Shenzhen Polytechnic,
Shenzhen 518055, China

Abstract. In this paper, the environment heterogeneous fixed fleet vehicle routing problem with time windows (EHFFVRP-TW) has been proposed. The model added the carbon emission factor and time windows based on the heterogeneous fixed fleet vehicle routing problem (HFFVRP) where we consider the cost benefit of carbon trading by selling or purchasing the carbon emission rights. Comprehensive Learning particle swarm optimization (CLPSO) is presented to solve the model, and the performance of CLPSO is estimated by comparing with PSO and GA. We adopt binary encoding way and set 2N dimension of particle to correspond the customer. The first N dimensional coding represents the vehicle number which visited the customer. And the last N dimensional coding correspond to the path order of vehicle. In the experiment, it's demonstrated CLPSO performs best compared with other two algorithm. CLPSO has improved the shortage of premature convergence in PSO and showed the advantages of getting lower cost and better routing.

Keywords: Vehicle routing problem · Comprehensive learning particle swarm optimization · Carbon dioxide

1 Introduction

Climate change has become one part of the international environmental challenges. Greenhouse gases emitted through human activity and in particular Carbon dioxide emissions directly impact on the environment and climate change. Transportation is an important sector to contribute the greenhouse gas emissions and vehicle routing problem is a significant issue in transportation. So, we discuss one variant of vehicle routing problem from a perspective of environment to reduce carbon emission over the paper.

In the past decades, researchers had developed many researches to decline the carbon emission in vehicle routing problem. Kara presented a new function to evaluate the cost of energy minimizing vehicle routing problem based on the load and distance [1]. Xiao et al. had proposed a mathematical optimization model, where fuel con-

© Springer International Publishing Switzerland 2016
Y. Tan et al. (Eds.): ICSI 2016, Part II, LNCS 9713, pp. 424–432, 2016.
DOI: 10.1007/978-3-319-41009-8_46

sumption rate as the load influence factor on the classical capacitated vehicle routing problem had been considered [2]. Tolga Bektas proposed Pollution Routing Problem and construct a comprehensive model, which consider many factors, including load, speed, distance, and so on [3]. Lorena Pradenas adopted the formula Tolga Bektas et al. proposed to estimate the energy consumption to reduce greenhouse gases emission in the vehicle routing problem with backhauls and time windows [4]. In the paper, we adopted the formula of Tolga Bektas proposed to calculate the energy consumption.

Carbon trading scheme also called cap and trade mechanism, which defines carbon as a tradable resource and restricted emission allowances. If companies generating emissions over the allocated allowances, the company would purchase emission allowance in carbon trading market or receive a big amount penalty. This mechanism incentive companies to take proper measures to avoid fines by over carbon-emerging or receive financial reward from selling surplus allowance [5]. Nowadays, many countries have constructed carbon trading scheme. But few paper added it into vehicle routing problem. Kwon et al. considered the carbon trading mechanism to the heterogeneous fixed fleet vehicle routing problem model to minimize the total cost [6]. We added this mechanism to our research, and proposed an environment heterogeneous fixed fleet vehicle routing problem with time windows (EHFFVRP-TW) model.

CLPSO was first proposed by Liang et al., which adopt a novel learning strategy to improve the phenomenon of premature convergence in PSO [7]. In original PSO, the particles updated positon based on the global best and personal best [8]. It's more likely to trap into local optimum. But in CLPSO, global best position doesn't used and each particle take other particle' personal best position as exemplars. Instead of learning from the global best and personal best for all dimensions, each dimension of particle learns from different exemplars' corresponding dimension, and which exemplars to learn is randomly chosen. This strategy expanded the search space and more likely to discover the global optimum of all population, improved the defect of premature convergence in PSO. In former research, scholar has studied the application of CLPSO in real field. However, there are few paper using CLPSO to solve the vehicle routing problem. It has been demonstrated its potential to solve distribution problem when applied in electrical power and energy systems [9, 10]. In the paper, we adopt comprehensive learning particle swarm optimization (CLPSO) [7] to solve the model.

The paper is structured as follows: We presented the proposed model description in Sect. 2. An introduction of the comprehensive learning PSO was presented in Sect. 3. Section 4 provided the algorithm apply the model and result. The summary of the research and further study is developed in Sect. 5.

2 The Model Description

In our model, we add the carbon factor to Heterogeneous Fixed Fleet Vehicle Routing Problem. Bektas and Laporte has proposed a comprehensive formula to evaluate the vehicle energy emission [3], considering many influence factor, including weight, load, speed, distance and other road condition. The energy consumption estimated as follows:

$$P_{ij} \approx \alpha_{ij}(w + f_{ij})\, d_{ij} + \beta\, v_{ij}^2\, d_{ij}, \tag{1}$$

where P_{ij} is the energy consumed on arc (i,j), w is the empty vehicle weight and f_{ij} is the vehicle load on arc (i,j). In the equation, $\alpha_{ij} = a + gsin\theta_{ij} + gC_rcos\theta_{ij}$ is a specific constant, where α is acceleration, g is acceleration of gravity, C_r is the coefficient of rolling resistance and θ_{ij} is the road slope (suppose $\theta_{ij} = 0$ in this paper). $\beta = 0.5C_dA\rho$ is a vehicle specific constant, where C_d is drag coefficient, A is the area of vehicle anterior surface and ρ is the air density.

In this paper, we referred this formula to calculate the energy consumption and calculate the fuel cost. Besides, we add the carbon trading cost into the model. If the carbon emissions exceed the maximum limit, the additional carbon emissions would be purchased form the carbon trading market. Otherwise, carbon benefit cost will be incurred by selling the carbon allowances. The CO2 emissions is calculated using the emission factor of 74,100 kg CO2/TJ, the Calorific value of fuel(diesel) is 35.4 MJ/l, and the unit cost of carbon emission is given at \$25/ton [6]. What's more, we also consider the fixed cost and penalty cost of time windows.

In the model, we hypothesized that vehicles are started and terminated in same depot. The demand of the customers must be satisfied and each customer only be serviced once. Besides, vehicles can't be overweight and all the cargo can be mixed together. We construct a logistic network G = (N, A), N = $\{0, 1, 2, \ldots, n\}$ is the set of nodes, 0 represent the depot, and other nodes denote n customers. A = $\{(i,j), i, j \in N, i \neq j\}$ is the arcs set. Another set K = $\{1, 2, \ldots, K'\}$ represent the type of vehicles. The type k vehicle with Q_k capacity and m_k number. $M_k = \{1, 2, \ldots, m_k\}, k = 1, 2, \ldots, K'$ is the set of vehicle of type k. The total number of vehicles is $K(K = \sum_{k \in K} m_k)$. The vehicle speed is v. The demand of each customer i is q_i, service time is S_i. The time window is $[ETi, LTi]$. The distance of arc (i,j) is d_{ij}. x_{kmij} is denotes as a binary variable. If m_{th} vehicle of type k travels on arc (i,j), x_{kmij} is equals to 1. Otherwise, $x_{kmij} = 0$. We took the waiting cost pe and late cost pl into account if the vehicle arrived to customer earlier or later than the time windows. The fixed cost of each k type vehicle is c_k. In the delivery, it generate fuel cost and carbon trading cost. We assume the fuel cost is c_f per unit, and the net benefit of carbon trading cost is c_e per unit.

In our former research, we present a heterogeneous fixed fleet vehicle routing problem and adopted a series of structure-redesign-based bacterial foraging optimizations to solve [11]. However, in HFFVRP model, it just consider the operation cost and time windows penalty cost. In the base of previous study, we constructed EHFFVRP-TW model, add the carbon factor and fixed cost. The total cost of the model is consist of four part, including fuel consumed cost, carbon trading cost, vehicles' fixed cost and time window cost. The objective function is listed as follows:

$$\text{Min} \sum_{k\in K} \sum_{m\in M_k} \sum_{(i,j)\in A} c_f(\alpha_{ij}d_{ij}\text{w}x_{ikmij} + \alpha_{ij}d_{ij}\text{w}f_{ij} + d_{ij}\beta v_{ij}^2)$$

$$+ c_e\{\sum_{k\in K} \sum_{m\in M_k} \sum_{(i,j)\in A} (\alpha_{ij}d_{ij}\text{w}x_{ikmij} + \alpha_{ij}d_{ij}\text{w}f_{ij} + d_{ij}\beta v_{ij}^2) - \text{AE}\}$$

$$+ \sum_{k\in K} \sum_{m\in M_k} \sum_{j\in N|\{0\}} c_k x_{km0j} \qquad (2)$$

$$+ \sum_{j\in N|\{0\}} \max\{pe(ET_i - T_i), 0, pl(T_i - LT_i)\}$$

Subject to

$$t_j = \sum_{(i,j)\in A} (t_i + d_{ij}/v_{ij} + s_i), \qquad (3)$$

$$\sum_{k\in K} \sum_{m\in K_m} \sum_{i\in N} x_{kmij} = 1 \qquad j = 1.........n, \qquad (4)$$

$$\sum_{k\in K} \sum_{m\in M_k} \sum_{j\in N} x_{kmij} = 1 \qquad i = 1.........n, \qquad (5)$$

$$\sum_{m\in M_k} \sum_{j\in N|\{0\}} x_{kmij} \le m_k \qquad k = 1, 2, ...K', \qquad (6)$$

$$\sum_{j\in N} f_{ji} - \sum_{j\in N} f_{ij} = q_i \qquad i = 1, 2, ...n, \qquad (7)$$

$$\sum_{k\in K} \sum_{m\in M_k} q_i x_{kmij} \le f_{ij} \le \sum_{k\in K} \sum_{m\in K_m} (Q_k - q_i) x_{kmij} \quad (i,j) \in A, \qquad (8)$$

$$x_{kmij} = \begin{cases} 1 \text{ The mth vehicle of type k travels from i to j} \\ 0 \text{ otherwise} \end{cases} \qquad (9)$$

$$f_{ij} \ge 0 \ (i,j) \in A. \qquad (10)$$

The Eq. (3) is used to count the time of vehicle reach to each customer. Equation (4) and Eq. (5) ensure each customer only be serviced once time. Equation (6) is used to guarantee the number of vehicle participated in the delivery is not exceeding the total vehicle quantity in each vehicle type. Equation (7) indicates the flow increasing with the amount of requirement of each serviced customer. Equation (8) is used to limit the vehicle total load, to make the vehicle carries less than its capacity. Equation (9) defines a variable, if the m_{th} vehicle of kth vehicle type travel from i to j, $x_{kmij} = 1$, else, $x_{kmij} = 0$. Equation (10) is the vehicle load is greater than or equal to 0.

3   CLPSO

PSO has been attracted many scholars to research since Kennedy and Eberhart proposed [8]. PSO imitate the behavior of swarm. In the original research, the positon X_i^d and velocity V_i^d updated as follows:

$$V_i^d \leftarrow V_i^d + c_1 * rand1_i^d *(pbest_i^d - X_i^d) + c_2 * rand2_i^d *(gbest_i^d - X_i^d), \qquad (11)$$

$$X_i^d \leftarrow X_i^d + V_i^d, \qquad (12)$$

where $pbest_i^d$ represent the best position discovered previous for the i th particle, $gbest_i^d$ represent best position fond by all particles. c_1 and c_2 are the constants of accelerations, $rand1_i^d$ and $rand2_i^d$ are random number in the scope of [0,1]. In later research, an inertia weight w was added into the original PSO to equilibrate the local and global search abilities [12]. Different from PSO, CLPSO adopt a novel learning strategy [11]. It took other particles' pbest except itself as exemplars for every particle. And particle has the chance to study from these exemplars in the population. Particle update the velocity by learn all other particles' historical best information instead of learning from pbest and gbest. The learning strategy list as follows:

$$V_i^d \leftarrow \text{w} * V_i^d + c_1 * \text{rand}_i^d *(pbest_{f_{i(d)}}^d - X_i^d), \qquad (13)$$

where $f_i = [f_i(1), f_i(2), \ldots, f_i(D)]$ defines the particle learns from which particles' pbests. It introduces a learning probability P_c. In CLPSO, it generates a random number for each dimension of the particle to choose whether learns from exemplars. If the number is smaller than P_c, the dimension will learn from exemplar pbest. Otherwise, it will learn from its own pbest. CLPSO overcome the drawback of premature convergence and has more potential to find the global value compared to PSO.

4   Experiment and Discussion

4.1   Encoding Methods

The EHFFVRP-TW is a NP hard problem. It's hard to settle such complex problem by traditional methods. CLPSO is a heuristic algorithm and have good optimization ability. To solve the problem, corresponding the particle and solution is very important. In this paper, we adopt binary encoding for its decoding easily and operating simple features. And the encoding dimension is 2N. The first N dimensional coding correspond to the vehicle number which visited the customer. And the last N dimensional coding correspond to the path order of vehicle. Such as there are 7 vehicles and 7 customers, the particle dimension should be 14. If one particle position is (2 1 2 2 3 1 3 1.9 0.9 3.2 1.6 2.4 2.8 3.2), we can learned there are 3 vehicles involve the delivery, and the routings are $0 \rightarrow 2 \rightarrow 6 \rightarrow 0$, $0 \rightarrow 4 \rightarrow 1 \rightarrow 3 \rightarrow 0$ and $0 \rightarrow 5 \rightarrow 7 \rightarrow 0$.

4.2   Experiment Assumption

In experiments, we assumed there are 15 customers and 3 type of vehicles. Each type of vehicles' number and capacity are (2, 4ton), (3, 3ton) and (3, 2.5ton) respectively. And we assume the empty weight is almost equal to the vehicle load. We adopt Euclidean distance formula to reckon the distance among each customer and depot. It's estimate approximately: $d_{ij} = \sqrt{(x_i - x_j)^2 + (y_i - y_j)^2}$. The depot coordinate is (0, 50). The wait cost and the delay cost are 30 and 100 yuan per hour. The fuel cost is 11 yuan per liter, and the fixed cost is 400, 300, 250 yuan per vehicle type. The limit of carbon emission AE for this case is set 108,860 units. In the experiments, CLPSO, PSO and GA are used to solve the problem. The depot and customers coordinates (x_i, y_i), the customer demands, corresponding lay time and time windows are listed in Table 1.

Table 1. Customers and depot location coordination, demands, lay time and time windows

Customer	X	Y	Demands (kg)	Lay time (h)	Time windows
1	10	110	1200	1	[8 17]
2	20	28	500	1.5	[8 10.5]
3	60	96	1200	1	[8 17]
4	55	120	1500	4/3	[8 12]
5	90	110	1000	7/6	[8 10]
6	92	66	600	2/3	[12 17]
7	96	100	900	1	[12 17]
8	110	135	800	1.5	[8 10.5]
9	44	160	900	5/6	[8 10]
10	30	150	600	1.5	[8 13]
11	108	32	1400	1	[8 12.5]
12	130	88	800	7/6	[8 17]
13	120	60	1200	3/2	[12 18]
14	30	70	800	1	[12 18]
15	80	70	1600	2/3	[8 17]

To simplify the complexity of the problem, we assume vehicles running at a constant speed of 60 km/h. The internal parameters of the impact of vehicle's engine on energy consumption and greenhouse gas emissions is refer to the Ref [3], shown in Table 2.

Table 2. Energy emission parameter

g	α	C_r	C_d	s	ρ
9.81 m/s^2	0	0.01	0.7	5 m^2	1.204 kg/m^2

4.3   Results and Analysis

The experiments were completed in MATLAB platform. The statistic results are show
in Table 3, which shows CLPSO get the optimum result. The best route calculated by
CLPSO is showed in Table 4. We can get that only 5 number of vehicle joined the
delivery and the vehicle of type 2 didn't participated. Figure 1 gives the mean results of
10 runs by all algorithm over 3000 iterations. It shown CLPSO performs best compared
with other two algorithms. GA gain the faster convergence but soon not update the
value. CLPSO convergence is not performs best but the get the optimum results, which
actually conform to the character that improve the drawback of premature convergence
and more likely to find global best position compared with PSO.

Table 3. The results of all algorithm

	Maximum	Minimum	Mean	Standard deviation
CLPSO	1983.08	**1759.39**	1885.02	112.34
PSO	2072.12	1830.37	1979.78	195.15
GA	2326.76	2260.89	2268.73	20.40

Table 4. The best routing of the problem by CLPSO

Vehicle	Vehicle routing	Total cost
Type 1	$0 \rightarrow 10 \rightarrow 1 \rightarrow 14 \rightarrow 13 \rightarrow 0$	1759.39
	$0 \rightarrow 5 \rightarrow 11 \rightarrow 7 \rightarrow 6 \rightarrow 0$	
Type 3	$0 \rightarrow 9 \rightarrow 4 \rightarrow 0$	
	$0 \rightarrow 12 \rightarrow 15 \rightarrow 0$	
	$0 \rightarrow 8 \rightarrow 2 \rightarrow 3 \rightarrow 0$	

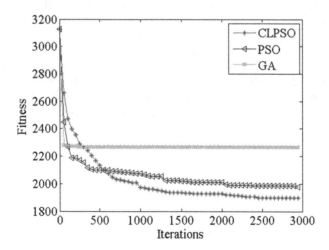

Fig. 1. The mean values of all algorithms run 10 times.

5   Conclusions

In the paper, the environment heterogeneous fixed fleet vehicle routing problem with time windows is developed. We considered the problem on the base of HFFVRP, add the carbon emission and time window constraints, to get the objective of minimize the total cost. To clarify this problem, we constructed a mathematical model, including fuel consumed cost, carbon trading cost, fixed cost and time windows cost. In the carbon trading cost, we get the cost benefit by selling or purchasing the carbon emission rights from carbon trading market. In order to find out the optimal result of the model, we adopted comprehensive learning particle swarm optimization. In the simulation experiments, the results demonstrated the CLPSO performs best compared with PSO and GA and improved the weakness of premature convergence of PSO. In the further research, multi objective model of operation cost and carbon trading cost can be constructed, and other heuristic algorithm can be developed to apply to the problem.

Acknowledgements. This work is partially supported by The National Natural Science Foundation of China (Grants nos. 71571120, 71271140, 71461027, 71471158) and the Natural Science Foundation of Guangdong Province (Grant no. 1614050000376).

References

1. Kara, I., Kara, B.Y., Yetis, M.: Energy minimizing vehicle routing problem. In: Dress, A. W., Xu, Y., Zhu, B. (eds.) COCOA 2007. LNCS, vol. 4616, pp. 62–71. Springer, Heidelberg (2007)
2. Xiao, Y.Y., Zhao, Q.H., Kaku, I., Xu, Y.C.: Development of a fuel consumption optimization model for the capacitated vehicle routing problem. Comput. Oper. Res. **39**(7), 1419–1431 (2012)
3. Bektas, T., Laporte, G.: The pollution-routing problem. Transp. Res. Part B **45**(8), 1232–1250 (2011)
4. Pradenas, L., Oportus, B.: Parada.: Mitigation of greenhouse gas emissions in vehicle routing problems with backhauling. Expert Syst. Appl. **40**(8), 2985–2991 (2013)
5. Zakeri, A., Dehghanian, F., Fahimnia, B., Sarkis, J.: Carbon pricing versus emissions trading: a supply chain planning perspective. Int. J. Prod. Econ. **164**, 197–205 (2015)
6. Kwon, Y.J., Choi, Y.J., Lee, D.H.: Heterogeneous fixed fleet vehicle routing considering carbon emission. Transp. Res. Part D **23**(8), 81–89 (2013)
7. Liang, J.J., Qin, A.K., Baskar, S.: Comprehensive learning particle swarm optimizer for global optimization of multimodal functions. IEEE Trans. Evol. Comput. **10**(3), 281–295 (2006)
8. Eberchart, R., Kennedy, J.: A new optimizer using particle swarm theory. Proc. Int. Symp. Micro Mach. Hum. Sci. **12**(11), 39–43 (1995)
9. El-Zonkoly, A., Saad, M., Khalil, R.: New algorithm based on CLPSO for controlled islanding of distribution systems. Electr. Power Energ. Syst. **45**(1), 391–403 (2013)
10. Mahadevan, K., Kannan, P.S.: Comprehensive learning particle swarm optimization for reactive power dispatch. Appl. Soft Comput. **10**(2), 641–652 (2010)

11. Gan, X., Liu, L., Niu, B., Tan, L.J., Zhang, F.F., Liu, J.: SRBFOs for solving the heterogeneous fixed fleet vehicle routing problem. In: Huang, D.-S., Jo, K.-H., Hussain, A. (eds.) ICIC 2015. LNCS, vol. 9226, pp. 725–732. Springer, Heidelberg (2015)
12. Shi, Y., Eberhart, R.C.: A modified particle swarm optimizer. In: IEEE International Conference on Evolutionary Computation, pp. 69–73 (1998)

Neighborhood Learning Bacterial Foraging Optimization for Solving Multi-objective Problems

Ben Niu[1(✉)], Jing Liu[1], Jingsong Chen[2], and Wenjie Yi[1]

[1] College of Management, Shenzhen University, Shenzhen 518060, China
drniuben@gmail.com
[2] School of Computer Engineering, Shenzhen Polytechnic,
Shenzhen 518055, China

Abstract. Based on the concept of neighborhood learning, this paper proposes a novel heuristic algorithm which is called Neighborhood Learning Multi-objective Bacterial Foraging Optimization (NLMBFO) for solving Multi-objective problems. This novel algorithm has two variants: NLMBFO-R and NLMBFO-S, using ring neighborhood topology and star neighborhood topology respectively. Learning from neighborhood bacteria accelerates the bacteria to approach the true Pareto front and enhances the diversity of optimal solutions. Experiments using several test problems and well-known algorithms test the capability of NLMBFOs. Numerical results illustrate that NLMBFO performs better than other compared algorithms in most cases.

Keywords: Neighborhood learning · Bacterial foraging optimization · Multi-objective problems

1 Introduction

There are a number of multiple objectives optimization problems in real life, such as environment/economic dispatch problem [1], design of shell-and-tube heat exchangers [2] and so on. These optimization problems usually have a set of optimal solutions. The optimal solution is called the Pareto-optimal solution and the set of optimal solutions is called the Pareto front (PF). As intelligent algorithms have been successfully applied in numerous single-objective problems, they are also investigated by researchers to handle multi-objective problems.

One of the representative multi-objective algorithms is Nondominated Sorting Genetic Algorithm (NSGA) proposed by Srinivas and Deb [3] in 1995. After that, Deb et al. [4] presented a modified version of NSGA (NSGA II) which performs better than NSGA. Coello et al. [5] also proposed a multi-objective algorithm (MOPSO) based on Particle Swarm Optimization (PSO) to address multi-objective problems as PSO shows high accuracy of searching.

By simulating *E.coli* bacteria's foraging behavior, Passino presented a novel heuristic algorithm in 2002, namely Bacterial Foraging Optimization (BFO) [6]. Extended from the single objective BFO, we proposed a Multi-objective Bacterial

© Springer International Publishing Switzerland 2016
Y. Tan et al. (Eds.): ICSI 2016, Part II, LNCS 9713, pp. 433–440, 2016.
DOI: 10.1007/978-3-319-41009-8_47

Foraging Optimization (MBFO) in prior paper to handle Multi-objective Problems [7]. Considering that MBFO is not good enough while handling problems with higher dimensional variables, this paper proposes Neighborhood Learning Multi-objective Bacterial Foraging Optimization (NLMBFO) which incorporates neighborhood learning, non-dominated sorting mechanism into the original MBFO to improve MBFO's performance in searching for Pareto-optimal solutions. Two kinds of NLMBFOs are presented according to the difference of neighborhood topologies used.

Several benchmark functions and compared algorithms are used in the experiments of this paper to test the capability of NLMBFOs. And the numerical results demonstrate that NLMBFOs show better performance in terms of coming closer to the true Pareto front and keeping diversity of solutions in most cases.

The rest of the paper is organized as follows: Sect. 2 illustrates the proposed algorithm, Neighborhood Learning Multi-objective Bacterial Foraging Optimization (NLMBFO). Next is the detail of experiment and the simulation results in Sect. 3, followed by conclusion of this paper in Sect. 4.

2 Neighborhood Learning Multi-objective Bacterial Foraging Optimization

Information communication mechanism has been used to improve algorithmic capability such as searching accuracy [8, 9]. The classical bacterial foraging optimization algorithm (MBFO) mainly includes three steps: chemotaxis, reproduction and elimination&dispersal, but it thinks little of information exchange. Thus, this paper integrates neighborhood learning mechanism with MBFO so as to improve global search ability. For neighborhood learning, neighborhood topology structures are supposed to be introduced. According to our previous study [9], the algorithm with ring topology has a faster convergence speed while the algorithm with star topology obtains a better search precision. Thus, ring topology and star topology are adopted for neighborhood learning in this paper. The corresponding novel algorithm is called Neighborhood Learning Multi-objective Bacterial Foraging Optimization with Ring topology (NLMBFO-R) and Neighborhood Learning Multi-objective Bacterial Foraging Optimization with Star topology (NLMBFO-S).

The neighborhood learning mechanism is incorporated into the chemotactic step. When bacteria need to choose the moving direction during the chemotactic step, they could get information form the random direction $Dire_\Delta^i$ and the neighborhood learning $Dire_{neigh}^i$. As we know, each node in Ring topology is connected with its left node and right node. According to this, the bacterium in NLMBFO-R which is based on Ring topology learns from its left neighbor and right neighbor so as to get more information. Similarly, each bacterium in NLMBFO-S gets information from the center of population.

The movement of each bacterium is shown in Table 1 where $\theta^i(j, k, l)$ means the position of the ith bacterium at the jth chemotactic, k^{th} reproductive and l^{th} elimination&dispersal step. Δ indicates a vector in the random direction. c means the learning factor and r means a random number. *ileft/iright* represents the left/right neighborhood

Table 1. The movement of each bacterium

$\theta^i(j+1,k,l) = \theta^i(j,k,l) + c(i)\left(c_1 r_1 Dire^i_\Delta + c_2 r_2 Dire^i_{neigh}\right)$	(1)
$Dire^i_\Delta = \Delta(i) \Big/ \sqrt{\Delta^T(i)\Delta(i)}$	(2)
The neighborhood learning of each bacterium in NLMBFO-R: $Dire^i_{neigh} = c_1 r_1(\theta^{ileft}(j,k,l) - \theta^i(j,k,l))$ $\quad + c_2 r_2(\theta^{iright}(j,k,l) - \theta^i(j,k,l))$	(3)
The neighborhood learning of each bacterium in NLMBFO-S: $Dire^i_{neigh} = c_1 r_1(\theta^{ihub}(j,k,l) - \theta^i(j,k,l))$	(4)

bacterium of the ith bacterium while *ihub* is the hub of the bacteria swarm in star topology. At the same time, two methods which are thought as outstanding strategies in dealing with multi-objective problems are introduced in NLMBFOs, including non-dominated sorting method and crowding distance operation [4].

3 Experimental Results

Generational Distance (*GD*) [10] and Diversity (Δ) [4] are introduced as performance metrics to estimate the algorithmic performance. *GD* is used to evaluate the distance between the Pareto-optimal solutions set found with the true Pareto front while Δ can measure the diversity of solutions.

3.1 Test Problems and Experimental Settings

Five test problems are selected to evaluate the performance of NLMBFO, including SCH1 [11], FON [12], KUR [13], and ZDT1, ZDT2 [14]. Two representative multi-objective algorithms: MOPSO [5] and MBFO [7] are selected to compare with NLMBFO-R and NLMBFO-S. The swarm size in these four algorithms is 100. Detailed parameters setting of MBFO and NLMBFO are presented in Table 2. Each trial is conducted ten times independently.

Table 2. Parameters setting of MBFO and NLMBFOs

Parameter	S	Nc	Ns	Nre	Ned	Ped	C
Values	100	200	4	5	2	0.25	0.1

3.2 Experimental Results and Analysis

Table 3 presents the numerical results obtained by MOPSO, MBFO and our two variants of NLMBFO on five test problems. In Figs. 1, 2, 3, 4 and 5, the Pareto fronts found by these four algorithms are shown as red spots while the true PF are shown as blue lines.

Table 3. Numerical results obtained by competitive algorithms

Test problems	Algorithms	Generational distance GD			Diversity Δ		
		Best	Worst	Mean	Best	Worst	Mean
SCH1	MOPSO	0.0054	0.0060	**0.0052**	0.6305	0.6691	0.6405
	MBFO	0.0077	0.0094	0.0087	0.5321	0.5637	**0.5476**
	NLMBFO-R	0.0061	0.0069	0.0065	0.4988	0.6392	0.5630
	NLMBFO-S	0.0057	0.0066	0.0062	0.5231	0.6604	0.5846
FON	MOPSO	0.1390	0.1460	0.1427	2.2090	3.8179	2.9427
	MBFO	0.1035	0.1105	0.1062	0.5593	0.5943	0.5778
	NLMBFO-R	0.0040	0.0049	0.0044	0.5470	0.5908	**0.5752**
	NLMBFO-S	0.0038	0.0046	**0.0042**	0.5559	0.6469	0.5945
KUR	MOPSO	1.6113	1.8190	1.7584	0.6041	1.1599	0.9472
	MBFO	0.5166	0.8064	0.7082	0.6817	0.9162	0.7622
	NLMBFO-R	0.0479	0.0583	0.0546	0.5901	0.7178	0.6719
	NLMBFO-S	0.0400	0.0444	**0.0425**	0.5930	0.6829	**0.6518**
ZDT1	MOPSO	0.2954	0.3794	0.3123	0.5714	0.6751	0.6145
	MBFO	0.2900	0.3097	0.3052	0.5726	0.6514	0.6213
	NLMBFO-R	0.0284	0.0315	**0.0295**	0.6011	0.6280	**0.6121**
	NLMBFO-S	0.0341	0.0411	0.0376	0.6191	0.7182	0.6671
ZDT2	MOPSO	0.1942	0.2647	0.2216	0.5489	0.7141	0.6730
	MBFO	0.1201	0.2038	0.1651	0.5776	0.6100	**0.5968**
	NLMBFO-R	0.0293	0.0320	**0.0307**	0.6247	0.6709	0.6512
	NLMBFO-S	0.0373	0.0415	0.0399	0.5710	0.7062	0.6337

We can find following conclusions from these tables and graphs: (1) The Pareto front of SCH1 obtained by each competitive algorithm all converges to the true Pareto front, which means the searching capability of each algorithm on SCH1 has no significant difference because of the simplicity of this problem; (2) When it comes to more complex problems such as KUR, FON and ZDT1, both NLMBFO-R and NLMBFO-S show considerable superiority in finding better solutions which are closer to the true Pareto front and have a better diversity; (3) While dealing with ZDT2, NLMBFO-R or NLMBFO-S has the best value of Generational Distance which means the Pareto solutions found by NLMBFOs have a much better accuracy. However, the performance in terms of diversity is not that stable as we expected; (4) The performance of NLMBFO-R and NLMBFO-S has no significant difference which means the category of neighborhood structure doesn't have an obvious influence on the algorithm capability.

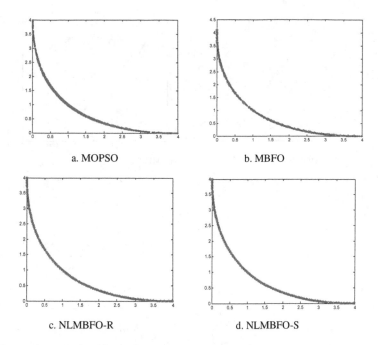

Fig. 1. Pareto front obtained by four algorithms on test problem 1 (SCH1) (Color figure online)

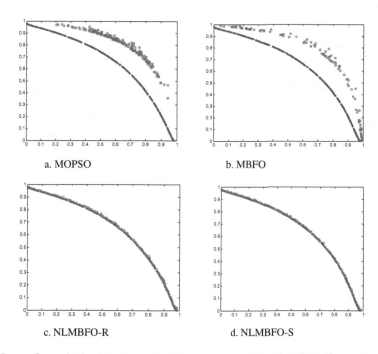

Fig. 2. Pareto front obtained by four algorithms on test problem 2 (FON) (Color figure online)

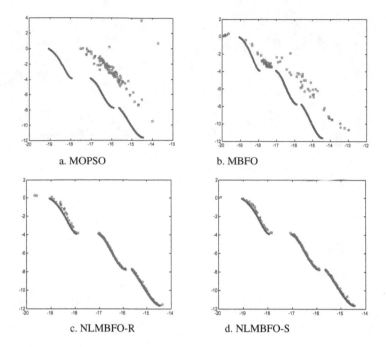

Fig. 3. Pareto front obtained by four algorithms on test problem 3 (KUR) (Color figure online)

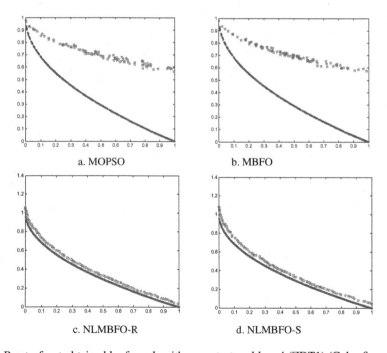

Fig. 4. Pareto front obtained by four algorithms on test problem 4 (ZDT1) (Color figure online)

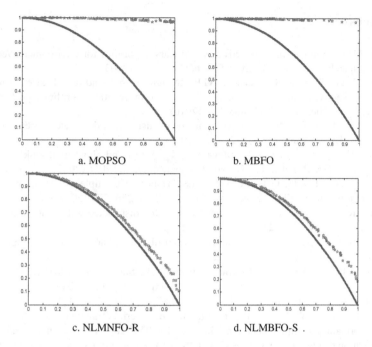

Fig. 5. Pareto front obtained by four algorithms on test problem 5 (ZDT2) (Color figure online)

4   Conclusions

In this paper, two variants of MBFO are proposed based on the neighborhood learning mechanism. According to the neighborhood topology structure used, novel algorithms are called Neighborhood Learning Multi-objective Bacterial Foraging Optimization with Ring topology (NLMBFO-R) and with Star topology (NLMBFO-S). Several well-known test problems are introduced to test the four competitive algorithms: MOPSO, MBFO, NLMBFO-R and NLMBFO-S. Convergence and diversity are esti-mated by Generational Distance (GD) and Diversity (Δ) respectively. Simulation results obtained by numerical experiments illustrate that the overall performance of NLMBFO-R and NLMBFO-S are better than MOPSO and MBFO in most cases. NLMBFOs can easily converge to the true Pareto front regardless of the problem's complexity while NLMBFOs' performance in diversity is not steady. Thus one direction that we would like to further research is the enhancement of diversity maintaining approach which can help Pareto solutions to distribute more uniformly.

Acknowledgements. This work is partially supported by The National Natural Science Foun-dation of China (Grants nos. 71571120, 71271140, 71461027, 71471158) and the Natural Sci-ence Foundation of Guangdong Province (Grant no. 1614050000376).

References

1. Abido, M.A.: A novel multiobjective evolutionary algorithm for environmental/economic power dispatch. Electr. Power Syst. Res. **65**(1), 71–81 (2003)
2. Wong, J.Y.Q., Sharma, S., Rangaiah, G.P.: Design of shell-and-tube heat exchangers for multiple objectives using elitist non-dominated sorting genetic algorithm with termination criteria. Appl. Therm. Eng. **93**, 888–899 (2016)
3. Srinivas, N., Deb, K.: Multiobjective optimization using non-dominated sorting in genetic algorithms. Evol. Comput. **2**(3), 221–248 (1994)
4. Deb, K., Pratap, A., Agarwal, S., Meyarivan, T.: A fast and elitist multi-objective genetic algorithm: NSGA-II. IEEE Trans. Evol. Comput. **6**(2), 182–197 (2002)
5. Coello, C.A.C., Pulido, G.T., Lechuga, M.S.: Handling multiple objectives with particle swarm optimization. IEEE Trans. Evol. Comput. **8**(3), 256–279 (2004)
6. Passino, K.M.: Biomimicry of bacterial foraging for distributed optimization and control. IEEE Control Syst. Mag. **22**(3), 52–67 (2002)
7. Niu, B., Wang, H., Wang, J.W., Tan, L.J.: Multi-objective bacterial foraging optimization. Neurocomputing **116**(20), 336–345 (2013)
8. Rasoulzadeh-akhijahani, A., Mohammadi-ivatloo, B.: Short-term hydrothermal generation scheduling by a modified dynamic neighborhood learning based particle swarm optimization. Int. J. Electr. Power Energy Syst. **67**, 350–367 (2015)
9. Niu, B., Liu, J., Bi, Y., Xie, T., Tan, L.J.: Improved bacterial foraging optimization algorithm with information communication mechanism. In: The 10th International Conference on Computational Intelligence and Security, CIS 2014, Kunming, China, pp. 47–51 (2014)
10. Van Veldhuizen, D.A., Lamont, G.B.: Multiobjective evolutionary algorithm research: a history and analysis. Technical report TR-98-03, Department of Electrical and Computer Engineering, Graduate School of Engineering, Air Force Institute of Technology, Wright-Patterson AFB, Ohio (1998)
11. Schaffer, J.D.: Multiple objective optimization with vector evaluated genetic algorithms. In: Proceedings of 1st International Conference on Genetic Algorithms (ICGA), Pittsburgh, PA, USA, pp. 93–100 (1985)
12. Fonseca, C.M., Flemming, P.J.: Multi-objective optimization and multiple constraint handling with evolutionary algorithms-part II: application example. IEEE Trans. Syst. Man Cybern. **28**(1), 38–47 (1998)
13. Kursawe, F.: A variant of evolution strategies for vector optimization. In: Schwefel, H.-P., Männer, R. (eds.) PPSN I Dortmund. LNCS, vol. 496, pp. 193–197. Springer, Heidelberg (1991)
14. Zitzler, E., Deb, K., Thiele, L.: Comparison of multi-objective evolutionary algorithms: empirical results. Evol. Comput. **8**(2), 173–195 (2000)

Robot Control

Robot Control by Computed Torque Based on Support Vector Regression

Nacereddine Djelal[1(✉)], Isma Boudouane[1], Nadia Saadia[1],
and Amar Ramdane-Cherif[2]

[1] Laboratory of Robotics, Parallelism and Embedded Systems,
University of Science and Technology Houari Boumediene, Algiers, Algeria
ndjelal@usthb.dz, iboudouane@gmail.com,
saadia_nadia@hotmail.com
[2] Laboratory LISV, University of Versailles Saint-Quentin en Yvelines,
Versailles, France
rca@prism.uvsq.fr

Abstract. In this work we propose to control robot using computed torque compensator based on robust estimation method called Support Vector Regression (SVR). This method represent the innovative side of this work; as it is powerful to modeling and identifying the nonlinear systems such as the disturbances that can appear during robot tracking a desired trajectory. The computed torque technique is also used from to pre-compensate the dynamics behavior of the nominal system. In order to demonstrate and show the robustness of the proposed control law, we have tested the system (Puma 560 robot) with a series of tests in simulation environment. The obtained results allow us to validate the proposed control law. As a result, the SVR reacts quickly to reject the errors originating from the disturbances.

Keywords: SVR · Computed torque · Robot · Robust control law · Trajectory tracking

1 Introduction

Robotics systems have a wide area of industry application. This is due to the fact that they are accurate and robust in industry process. However, there are some tasks that provide nonlinearity behavior in which the robot interacts with its environment. In this paper we propose to solve these constraints by the combination of the computed torque and SVR. The computed torque technique is used as a compensator of the dynamics behavior of the system through an appropriate regulation of its proportional and derivative gains. This technique is enough to drive the robot with accuracy in the absence of the disturbances as a condition. However, in the presence of disturbances, we propose to use the SVR in order to model the errors owing by the disturbances.

In this paper we propose a robust control law in order to control a robot of six Degrees Of Freedom (DOF), the PUMA560 performing a trajectory tracking task.

© Springer International Publishing Switzerland 2016
Y. Tan et al. (Eds.): ICSI 2016, Part II, LNCS 9713, pp. 443–450, 2016.
DOI: 10.1007/978-3-319-41009-8_48

The use of this kind of controller provides a good performance in terms of robustness and stability in modeling errors and external disturbances [1, 2]. However, there are constraints of the dynamic and the nonlinearity of the system.

The outline of this paper basically is an implementation of intelligent control law in six DOF robot to track a trajectory. This paper is organized as: first, we present a description of the computed torque control law, and then we give a description about the SVR for the identification and control; after that, we present the proposed control structure.

Finally, a discussion about the obtained results is given as well as and the drawn conclusions.

2   Computed Torque Control Law

The fundamental concept of computed torque control law is the feedback linearization of nonlinear systems. For manipulator robot, the nonlinear dynamic model is given by [3, 4]:

$$\tau = M(q)\ddot{q} + C(q, \dot{q})\dot{q} + F(\dot{q}) + G(q) \tag{1}$$

Where:

- q is the vector of generalized joint coordinates describing the pose of the manipulator,
- $\dot{q}$ is the vector of joint velocities,
- $\ddot{q}$ is the vector of joint accelerations,
- τ is the vector of generalized forces associated with the generalized coordinates q,
- M(q) is the symmetric joint-space inertia matrix, or manipulator inertia tensor,
- $C(q, \dot{q})$ describes coriolis and centripetal effects, the coriolis torques are proportional to $\dot{q}_i, \dot{q}_j$ while centripetal torques are proportional to $\dot{q}_i^2$
- $F(\dot{q})$ describes viscous and Coulomb friction and is not generally considered part of the rigidbody dynamics
- G(q) is the gravity loading

In practical applications it is difficult to know exactly the values of M, C, F and G. Equation 1 can be rewritten:

$$\tau = \hat{M}(q)\tau' + \hat{C}(q, \dot{q})\dot{q} + \hat{F}(\dot{q}) + \hat{G}(q) \tag{2}$$

Where:

If the manipulator is programmed to follow a desired trajectory defined a priori, we can compute τ' based on proportional-derivative (PD) controller [3, 4]:

$$\tau' = K_p(q_d - q) + K_d(\dot{q}_d - \dot{q}) + \ddot{q}_d \tag{3}$$

q_d, $\dot{q}_d$ and $\ddot{q}_d$ are the desired trajectory's position, velocity and acceleration, respectively. K_p and K_d are the proportional and derivative gains respectively. By combining Eqs. 1 and 3 we obtain the expression of the error equation of this system:

$$\ddot{e} + K_d\dot{e} + K_p e = \hat{M}^{-1}(q)[(M(q) - \hat{M}(q))\ddot{q} +$$
$$C(q,\dot{q}) + F(\dot{q}) + G(q) - \hat{C}(q,\dot{q}) - \hat{F}(\dot{q}) - \hat{G}(q)] \qquad (4)$$

Where:

$$\begin{cases} e = (q_d - q) \\ \dot{e} = (\dot{q}_d - \dot{q}) \\ \ddot{e} = (\ddot{q}_d - \ddot{q}) \end{cases} \qquad (5)$$

Craig in the reference [4] shows that the error can be written in the form:

$$\ddot{e} + K_d\dot{e} + K_p e = \zeta \qquad (6)$$

Craig [4] shows that all signals $(\zeta, q, \dot{q}, \ddot{q}, f)$ will be bounded if:

- $\hat{M}^{-1}(q)$ Exist,
- $\|e\|_2 = \|\dot{e}\|_2 = 0$ at $t = t_0$
- $\|I - M^{-1}(q)\hat{M}^{-1}(q)\|_2 < 0.5 - \varepsilon$

The printing area is 122 mm × 193 mm. The text should be justified to occupy the full line width, so that the right margin is not ragged, with words hyphenated as appropriate. Please fill pages so that the length of the text is no less than 180 mm, if possible.

3 Support Vector Regression

A regression method aims to estimate a relationship between the system input and output from the samples or training data. It is desirable that the relationship should be determined so that the system output matches the real value as closely as possible [5]. Once such a relation is accurately estimated, it is used for predicting the system output with the input values.

Let us consider a set of training samples of $(x_i, y_i)_{i=1}^n, i = 1, \ldots, n$, where x_i and y_i are the input and output collected data, and n is the dimension of training data. The general function of SVR estimation takes the form as [6, 7, 8]:

$$f(x) = (w \cdot \varphi(x)) + b \qquad (7)$$

where:

w is a weighting matrix, b is a bias term, φ denotes a nonlinear transformation from n-dimensional space to a higher dimensional feature space, and the dot represents the inner vector product [9].

Equation (7) can be solved by minimizing the regression risk as

$$R_{reg}(f) = \frac{1}{2}\|w\|^2 + C\sum_{i=1}^n \Gamma(f(x_i) - y_i) \qquad (8)$$

Subject to

$$|y_i - w \cdot \varphi(x_i) - b| \le \varepsilon + \xi_i \quad i = 1, 2, \ldots, n \quad \xi_i, \xi_i^* \ge 0 \qquad (9)$$

Where:

$\Gamma(\cdot)$ is a cost function,

ε is the permissible error,

C is a constant which determines the tradeoff between minimizing training errors and minimizing the model-complexity term $\|w\|^2$.

Every vector outside ε-tube is captured in slack variables ξ_i, ξ_i^*, which are introduced to accommodate unpredictable errors on the input training set.

Using a kernel function, the required decision function can be expressed [6]

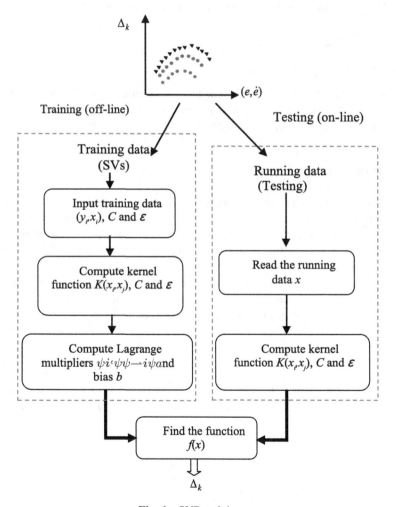

Fig. 1. SVR training.

$$f(x) = \sum_{i=1}^{n} (\alpha_i - \alpha_i^*) \cdot K(x_i, x) + b \tag{10}$$

We propose to use the radial-basis function (RBF) in this work as a kernel [6]

$$K(x_i, x_j) = \exp\left\{ -\frac{|x_i, x_j|^2}{\sigma^2} \right\}. \tag{11}$$

Where the σ is called as kernel parameter.

The schematic presentation of the identification based on the SVR and its procedure of training are illustrated in Fig. 1.

4   Control Structure Implementation and Tests

The validation of the proposed control structure is performed using simulation tests; where we have tested through a robot of six degree of freedom (PUMA 560). The proposed system must perform of trajectory tracking task; for this we propose the control law presented in [10] given by:

$$\ddot{e}_k + K_d \dot{e}_k + K_p e_k = \zeta - \Delta_k. \tag{12}$$

Where: Δ_k is learning term.

Figure 2 illustrates the proposed control scheme.

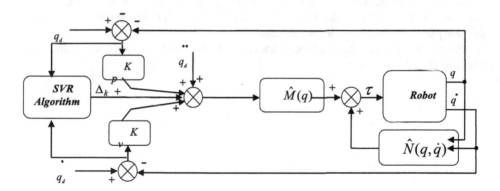

Fig. 2. Structure of the proposed control scheme based on SVR

5   Results and Discussion

In this section we present the results provided by the proposed control structure, so that our simulations are viable.

The dimensions of the simulated mechanical structure (the PUMA 560) are consistent with the real system [11].

Figure 3 shows the simulation results in the case of trajectory tracking in order to test the movement of the robot system according to its six degrees of freedom.

Figure 3 illustrates the six components (e(1)–e(6)) of the error values which are minimized by the proposed control law. The obtained plots show that monitoring is well performed. We observe the error variation, as it was initially between [−0.035, 0.048], and decreases after 200 iterations to [−0.003, 0.002] and hence it can be accepted as an error.

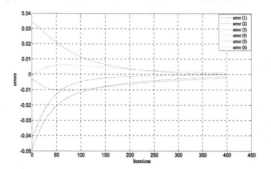

Fig. 3. Convergence of the control law

Figure 4 show the variations of the joint articulations of the robot which monitors the target trajectory. We have noticed that the great variations in the range of the iterations are [0, 200], also the Thata (1), Thata (2) and Thata (4) are stable, otherwise the last joints variants are with small values.

Finally we present the trajectory tracking of the robot end effector during the tracking (Fig. 5).

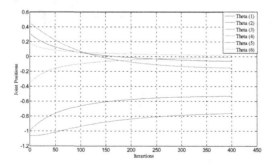

Fig. 4. Variations of the joint articulations

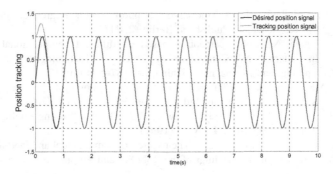

Fig. 5. Position tracking

In this figure we can say that the desired position signal follows the tracking signal. This demonstrates the robustness of the proposed control law.

6 Conclusion

In this work we have implemented an intelligent control law in a robot with six degrees of freedom to perform a trajectory tracking task.

The simulation results showed that the implemented command structure (trajectory tracking) has been validated.

The use of the SVR in the identification process and in the control, allows a good performance in the case of the complexity and the nonlinearity of the systems; in which the SVR control law was able to generate the nominal trajectories with high level of accuracy.

References

1. Saadia, N., Amirat, Y., Pontnaut, J., Ramdane-Cherif, A.: Neural adaptive force control for compliant robots. In: Mira, J., Prieto, A.G. (eds.) IWANN 2001. LNCS, vol. 2085, pp. 436–443. Springer, Heidelberg (2001)
2. Touati, Y., Amirat, Y., Saadia, N.: Artificial neural network-based hybrid force/position control of an assembly task. In: 3rd International IEEE Conference Intelligent Systems, pp. 594–599 (2006)
3. Touati, Y., Amirat, Y., Saadia, N., Ramdane-Cherif, A.: A neural network-based approach for an assembly cell control. Appl. Soft Comput. **8**, 1335–1343 (2008)
4. Craig, J.J.: Adaptive control of mechanical manipulators. UMI Dissertation Information Service, Arbor (1986)
5. Bi, D., Li, Y.F., Tso, S.K., Wang, G.L.: Friction modeling and compensation for haptic display based on support vector machine. IEEE Trans. Ind. Electron. **51**, 491–500 (2004)
6. Cherkassky, V., Miller, F.: Learning from Data: Concepts, Theory, and Methods. Wiley, Hoboken (1998)

7. Smola, A.J., Schölkopf, B.: A Tutorial on Support Vector Regression, Statistics and Computing. Kluwer Academic Publishers, Berlin (2004)
8. Djelal, N., Saadia, N.: Robot perception based on different computational intelligence techniques. In: International Conference of Emerging Technologies for Information Systems, Computing, and Management, China, pp. 587–594 (2013)
9. Ahmed, G.A.K., Lee, D.C.: MPPT control of wind generation systems based on estimated wind speed using SVR. IEEE Trans. Ind. Electron. **55**, 1489–1490 (2008)
10. Dawson, D.M., Qu, Z., Dorsey, J.F., Lewis, F.L.: On the learning control of a robot manipulator. J. Intell. Rob. Syst. **4**, 43–53 (1991)
11. Armstrong, B., Khatib, O., Burdick, J.: The explicit dynamic model and inertial parameters of the PUMA 566 arm. In: Proceedings of the Robotics and Automation, pp. 510–518. IEEE (1986)

Control Nonholonomic Mobile Robot with Hybrid Sliding Mode/Neuro Fuzzy Controller

Mohamed Nabil Houam[1](✉), Nadia Saadia[1], Amar Ramdane-Cherif[2], and Nacereddine Djelal[1]

[1] Laboratory of Robotics, Parallelism and Embedded Systems,
University of Science and Technology Houari Boumediene, Algiers, Algeria
hmnabil@gmail.com, saadia_nadia@hotmail.com,
ndjelal@usthb.dz
[2] Laboratory LISV, University of Versailles Saint-Quentin en Yvelines,
Paris, France
rca@prism.uvsq.fr

Abstract. Many works has been done in the mobile robots research domain, resulting different methods to enhance the performance of the mobile robot. This article will adopt a hybrid approach to improve the performance of a path tracking controller by designing an algorithm that uses two methods: the first one is Sliding Mode [SM], which will have one of its parameters controlled by the second one based on neurofuzzy [NF].

Keywords: Sliding mode · Neural network · Fuzzy logic · Mobile robot

1 Introduction

Various research laboratories have studied the mobile robot control domain, trying to optimize their performance [1] by seeking more effective new control methods [2, 3]. Various methods based on artificial intelligence [4, 5], genetic algorithms, mathematic or heuristics were used [6–8]. But while those methods were being experimented, a new hybrid approach was developed [9] in order to take the best of each methods. This is done by smartly combining them into one controller.

During this work, we tried to optimize the performance of a path tracking controller for a non-holonomic mobile robot. The main controller is based on a robust sliding mode controller [10] which we will try to optimize by dynamically change one of its variables instead of using fixed value. To achieve this we will use a neuro-fuzzy mode controller [11], this to ensures that the variable will have the optimum value.

2 System Presentation

Considering that our system is a non-holonomic mobile robot, its kinematic model should be under the form (Fig. 1):

© Springer International Publishing Switzerland 2016
Y. Tan et al. (Eds.): ICSI 2016, Part II, LNCS 9713, pp. 451–458, 2016.
DOI: 10.1007/978-3-319-41009-8_49

$$\begin{cases} \dot{x} = v_1 * \cos(\theta) \\ \dot{y} = v_1 * \sin(\theta) \\ \dot{\theta} = v_1 * \dfrac{\tan(\varphi)}{lw} \\ \dot{\varphi} = v_2 \end{cases} \tag{1}$$

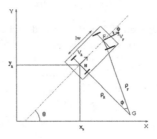

Fig. 1. Mobile robot projection in a 2D space

In order to simplify, the chained form of the system has been used. It is written as follows [12, 14]:

$$\begin{cases} \dot{z}_1 = u_1 \\ \dot{z}_2 = u_1 * z_3 \\ \dot{z}_3 = u_1 * z_4 \\ \quad\vdots \\ \dot{z}_{n-1} = u_1 * z_n \\ \dot{z}_n = u_2 \end{cases} \tag{2}$$

Having a kinematic model of 4th order, our chained form system is equivalent to:

$$\begin{cases} \dot{z}_1 = u_1 \\ \dot{z}_2 = u_1 * z_3 \\ \dot{z}_3 = u_1 * z_4 \\ \dot{z}_4 = u_2 \end{cases} \tag{3}$$

With

$$\begin{cases} z_1 = x \\ z_2 = y \\ z_3 = \tan(\theta) \\ z_4 = \tan(\varphi)/(l * \cos(\theta)^3) \end{cases} \tag{4}$$

And

$$\begin{cases} v_1 = u_1 * \cos(\theta) \\ v_2 = -\dfrac{3}{l} * \dfrac{\sin(\theta)}{\cos^2(\theta)} * \sin^2(\varphi) * u_1 + l * \cos^3(\theta) * \cos^2(\varphi) * u_2 \end{cases} \tag{5}$$

3 Sliding Mode/Neurofuzzy Controller

Based on the sliding mode method [2, 13], a sliding mode controller parameter has been optimized by inserting a neuro-fuzzy controller which will compute dynamically its value. So the main controller (Sliding Mode, SM) will regulate the path tracking, and the second controller (Neuro-fuzzy, NF) will optimize the chosen parameter.

As follow, the used control method (SM controller) [10] and the hybrid controller are presented.

3.1 Sliding Mode Controller

- Path tracking by sliding mode:

  For the path tracking using sliding mode, the following error was modeled:

$$\begin{cases} \dot{\tilde{z}}_1 = 0 \\ \dot{\tilde{z}}_2 = \tilde{z}_3 * u_{r1} \\ \dot{\tilde{z}}_3 = \tilde{z}_4 * u_{r1} \\ \dot{\tilde{z}}_4 = \tilde{u}_2 \end{cases} \tag{6}$$

  We consider the path to follow as:

$$y = f(x) \tag{7}$$

  And the constant generalized velocity as:

$$u1c = K \tag{8}$$

- The first step is to compute the desired states of the system, and their error states.
  - Desired system states:

$$\begin{cases} x_r = x(t) \\ y_r = f(x_r) \\ \theta_r = \arctan \dfrac{\partial f(x_r)}{\partial x_r} \\ \varphi_r = \arctan(l * \dfrac{\partial^2 f(x_r)}{\partial x_r^2} * \cos^3(\theta)) \end{cases} \tag{9}$$

– System Errors:

$$
\begin{cases}
\tilde{z}_1 = x_r - x = 0 \\
\tilde{z}_2 = y_r - y \\
\tilde{z}_3 = \tan(\theta_r) - \tan(\theta) \\
\tilde{z}_4 = \dfrac{\tan(\theta_r)}{l * \cos^3(\theta_r)} - \dfrac{\tan(\theta)}{l * \cos^3(\theta)}
\end{cases}
\tag{10}
$$

– Ideal command variable:

$$
\begin{cases}
u_{1r} = u_{1c} \\
u_{2r} = \dfrac{1}{l * \cos^2(\varphi_r) * \cos^3(\theta_r)} \dot{\varphi}_r + \dfrac{3 * \tan(\varphi_r) * \sin(\varphi_r)}{l * \cos^4(\varphi_r)} \dot{\varphi}_r
\end{cases}
\tag{11}
$$

$$
\dot{\varphi}_r = \cos^2(\varphi_r) * \left(-3 * l * \frac{\partial^2 f(x_r)}{\partial x_r^2} * \cos^2(\dot{\theta}_r) * \sin(\theta_r) * \dot{\theta} + l * \frac{\partial^3 f(x_r)}{\partial x_r^3} * u_c * \cos^3 \dot{\theta}_r \right)
$$

$$
\dot{\theta}_r = \frac{\partial f(x_r)}{\partial x_r} * \cos(\theta_r)
\tag{12}
$$

- The second step is to compute the control variables of the model error.
  – Sliding surface choice:
    Let $\Lambda_1 = -2$ and $\Lambda_2 = -3$ be the poles of our system in sliding mode. Hence should be: $C = \left[6/u_1^2 \; 5/u_1 \; 1 \right]$
    Therefore the slide function of the system is:

$$
S_2 = \frac{6 * z_2}{u_1^2} + \frac{5 * z_3}{u_1} + z_4
\tag{13}
$$

  – Design the control law:
    According to the equation, the control input of the error model is:

$$
u_2 = \frac{-6 * z_3}{u_1} - 5 * z_4 - \varepsilon_2 * sign(S_2)
\tag{14}
$$

- In the third step, the physical control inputs is calculated.
  – Computing the generalized control inputs:

$$
\begin{cases}
u_1 = u_{1c} \\
u_2 = u_{2r} - \tilde{u}_{2r}
\end{cases}
\tag{15}
$$

  – Converting the generalized control variables to physical control variables:

$$
\begin{cases}
v_1 = u_{1c} * \cos(\theta) \\
v_2 = \dfrac{-3 * \sin^2(\varphi) * \sin(\theta) * u_1}{l * \cos^2(\theta)} + l * \cos^3(\theta) * \cos^2(\varphi) * u_2
\end{cases}
\tag{16}
$$

3.2   Hybrid Controller

The choice of the parameter ε_2 in Eq. (14) is essential to ensure the convergence of the sliding mode control algorithm. First, this parameter value was determined empirically. Then, in order to improve the performance of the closed loop system (optimizing its response and adaptability to the task parameters) a neuro-fuzzy controller was implemented to do an online regulation of this parameter during the movement.

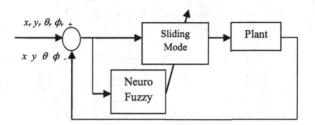

Fig. 2. Structure of hybrid controller

The design of neuro-fuzzy controller starts with the selection of the rules. For this, we started first by developing a fuzzy controller (Fig. 3) to test the performance of the rules. After many tests, we choose the optimal ones and then we implemented them in our neuro-fuzzy network (Fig. 2).

Our simplified controller will have the following general form:

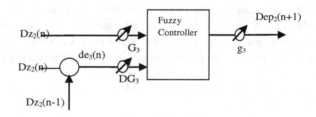

Fig. 3. Proposed fuzzy regulator to perform optimization of ε_2

The proposed fuzzy controller has as inputs variables "Dz_2" and its variation "de_3". And as output it has the correction "$Dep2$" which is applied to the parameter ε_2.

$$\varepsilon_2(n+1) \; = \; \varepsilon_2(n) + Dep_2(n+1) \tag{17}$$

By analyzing the value of the error "Dz_2" and its variation "de_3", the fuzzy controller computes the applied correction "Dep_2" to the parameter ε_2. The value of this correction can be positive or negative, thereby increasing or decreasing the controlled parameter ε_2.

To adopt a multi-layer feedforward neural network representation for the fuzzy controller, a Sugeno adaptive neuro-fuzzy inference system (ANFIS) network spread over five layers was used (Fig. 4). The fuzzification is done by the first layer, the inference by layer 2, 3 and 4. The fifth layer contains the parameter to do the defuzzification.

For the design of this controller a learning method based on the Kalman filter was used.

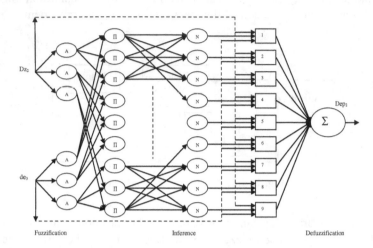

Fig. 4. Neuro-fuzzy network used in the hybrid controller

4   Experimental Results

To evaluate the performance of the neuro-fuzzy/sliding mode hybrid controller, a simulation by software was used. A sinusoidal path tracking task was done to simulate the robot's behavior in closed loop. In the first simulation, the speed of movement of the robot was set to 1.1 m/s. It was clearly noticed that the robot (dotted line) was able to follow the sinusoidal reference path (+ sign) (Fig. 5a).

For the second simulation, the same speed value was kept (1.1 m/s), but an initial position of (3 3) was introduced.

As it can be seen, the robot is able to join the path at the first sinusoidal and to follow it correctly (Fig. 5b).

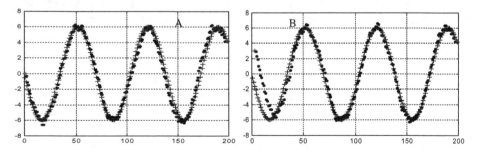

Fig. 5. Sinusoidal follow, (a) V = 1.1 m/s, (b) V = 1.1 m/s (xi = 3, yi = 3)

For the third simulation (Fig. 6), we changed the initial coordinates (xi = 3, yi = 6). The robot was able to join the path and follow it, but with small tracking error value compared to the precedent simulation (Fig. 5b).

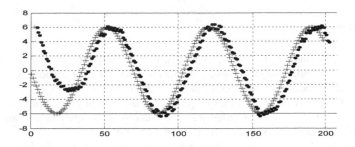

Fig. 6. Sinusoidal follow V = 1.1 m/s (xi = 3, yi = 6)

4.1 Comparative Studies

To compare the performance of the hybrid and the sliding mode implemented command structures, Fig. 7 shows the actual path of the robot for each of these structures (dot sign for hybrid, small triangle sign for SM) compared to the planned path (+ sign) (Speed V = 1.1 m/s and initial position (3 3)).

It was noted that each of the controllers allows to join and to follow the sinusoidal path (Fig. 7a), but in our case the hybrid one has a better follow on the sinusoidal edge (as seen in the 2nd sinusoid) (Fig. 7b).

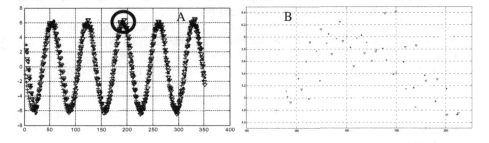

Fig. 7. (a) Comparative between hybrid and SM, Sinusoidal follow, (b) Zoom on the edge

5 Conclusion

In this work, a hybrid control structure was presented based on the regulation of one variable of a sliding mode controller using a Neuro-fuzzy controller.

Various simulations with different initial coordinates have been carried on to test the performance of the hybrid controller. In addition, comparative simulation of performance obtained with both (the hybrid controller and the used sliding mode controller) were also done.

The results of those simulations have shown that the implemented hybrid algorithm improved the following of the reference. This by reducing the value of the tracking error.

References

1. Delmas, P., Bouton, N., Debain, C., Chapuis, R.: Environment characterization and path optimization to ensure the integrity of a mobile robot. In: IEEE International Conference on Robotics and Biomimetics ROBIO, Guilin, pp. 92–97 (2009)
2. Guo, H., Tian, Y., Yao, Y., Liu, C.: Control method and motion path simulation of a mobile robot. In: International Conference on Computer Application and System Modeling ICCASM, Taiyuan, vol. 10, pp. 638–642 (2010)
3. Nganga-Kouya, D., Okou, F.A.: Adaptive backstepping control of a wheeled mobile robot. In: 17th Mediterranean Conference on Control and Automation, Greece, pp. 85–91 (2009)
4. Fierro, R., Lewis, F.L.: Control of a nonholonomic mobile robot using neural networks. IEEE Trans. Neural Netw. **9**(4), 589–600 (1998)
5. Dierks, T., Jagannathan, S.: Control of nonholonomic mobile robot formations using neural networks. In: IEEE 22nd International Symposium on Intelligent Control, ISIC 2007, pp. 132–137 (2007)
6. Liu, N.: Intelligent path following method for nonholonomic robot using fuzzy control. In: Second International Conference on Intelligent Networks and Intelligent Systems, Tianjian, China, pp. 282–285 (2009)
7. Majd, A.A., Yazdanpanah, M.J., Khosrowshahi, G.K.: Tracking control of a mobile robot using a genetically tuned mixed H2/H∞ adaptive technique. In: IEEE Conference on Robotics, Automation and Mechatronics, Singapore, vol. 1, pp. 175–180 (2004)
8. Liang, K., Jian-jun, W.: The solution for mobile robot path planning based on partial differential equations method. In: International Conference on Electrical and Control Engineering, Wuhan, China, pp. 5742–5745 (2010)
9. Zhuang, X., Meng, Q., Yang, S.: Mobile robot control in dynamic environments based on hybrid intelligent system. In: IEEE International Conference on Systems, Man and Cybernetics, Hammamet, Tunisia, vol. 3, p. 6 (2002)
10. Lu, J., Sakhavat, S., Xie, M., Laugier, C.: Sliding mode control for nonholonomic mobile robot. In: Proceedings of the International Conference on Control, Automation, Robotics and Vision, Singapore (SG), pp. 465–470 (2000)
11. Kumar Singh, M., Parhi, D.R., Pothal, J.K.: ANFIS approach for navigation of mobile robots. In: International Conference on Advances in Recent Technologies in Communication and Computing, Kottayam, Kerala, India, pp. 727–731 (2009)
12. Hamerlain, F.: Stabilisation par modes glissants d'ordre superieur de systemes non holonomes: cas de robots mobiles a roues. In: Magister These in Robotic and Process Control Option, USTHB, Algeria (2008)
13. Berni, A.: Contribution à la commande hybride force/position d'un bras manipulateur redondant. In: Magister These in Robotic and Process Control Option, USTHB, Algeria (2004)
14. Nabil, H., Nadia, S.: Neurofuzzy/sliding mode controller for nonholonomic mobile robot. In: Eighth International Conference on Fuzzy Systems and Knowledge Discovery FSKD, Shanghai, China (2011)

Swarm Robotics

Formation Splitting and Merging

Krishna Raghuwaiya[✉], Jito Vanualailai, and Bibhya Sharma

University of the South Pacific, Suva, Fiji
{raghuwaiya_k,vanualailai,sharma_b}@usp.ac.fj

Abstract. This paper presents an approach to swarm split and rejoin maneuvers of a system of multi-robots formations. A post split formation is split into low-degree sub-swarms when the swarm encounters an obstacle. The sub-swarms reestablish links with other sub-swarms and converge into its pre-split formation after avoiding collisions with the obstacles. The leader-follower control strategy is used for maintaining formation shape in the sub-swarms. A set of artificial potential field functions is proposed for avoiding inter-robot, inter-formation and obstacle collisions and attraction to their designated targets. The Direct Method of Lyapunov is then used to establish stability of the given system. The effectiveness of the proposed nonlinear acceleration control laws is demonstrated through a computer simulation.

Keywords: Formation · Lyapunov · Nonholonomic mobile robots · Low-degree

1 Introduction

Recent years have seen considerable attention focussed on the problem of coordinated motion and cooperative control of multi-robot systems. This interest arises from complex systems such as ecology, social sciences, evolutionary biological sciences, control theory in science and engineering systems [1,2]. These systems have individual characteristics in the way they interact with other agents and also reveal certain group phenomena which can give useful ideas for developing control theory for robotic systems. Often observed, biological systems such as groups of ants, fish, birds and bacteria reveal some amazing cooperative behaviors in their motion [1], such as reaching a target or moving in a formation to name a few. Researchers use computer simulations and animations to generate generic simulated flocking creatures called *boids* to study problems in ecology in the context of animal aggregation and social cohesion in animal groups [3]. The pioneering work on this area was done by Reynolds [4].

Now consider a cohesive swarm system in some initial configuration. When the swarm encounters an obstacle, splitting may be useful to maneuver the swarm around the obstacle. In many instances, it is more desirable for some agents to move to one side of the obstacle while the others move to the other side and later merge. The swarm system splits into clustered subswarms, moves around the obstacle and later re-establishes into its pre-split formation. In this paper, we

© Springer International Publishing Switzerland 2016
Y. Tan et al. (Eds.): ICSI 2016, Part II, LNCS 9713, pp. 461–469, 2016.
DOI: 10.1007/978-3-319-41009-8_50

consider creating two or more low-degree post-split sub-formations from some pre-split formation. This is achieved by breaking the link between different post-split sub-formations. However, the links between agents in the post-split sub-formations are still maintained. As such the agents in the same sub-formations maintain their formation structure while avoiding an obstacle. A leader-following based formation scheme is used to maintain the geometric formation shape in each sub-formation. The post-split sub-formations later re-establish links with other sub-swarms and converge into its pre-split formation once collision avoidance with the obstacles is achieved. The control strategy formulates a low degree formation which allows for slight distortions in the sub-formations. These distortions would normally appear if the group encounters an obstacle. Based on artificial potential fields, the Direct Method of Lyapunov is then used to derive continuous acceleration-based controllers which render our system stable.

The remainder of this paper is structured as follows: in Sect. 2, the robot model is defined; in Sect. 3, the artificial potential field functions are defined under the influence of kinodynamic constraints; in Sect. 4, the Lyapunov function is constructed, while in Sect. 5, stability analysis of the robotic system is carried out; in Sect. 6, we demonstrate the effectiveness of the proposed controllers via a computer simulation; and finally, Sect. 7 concludes the paper.

2   Vehicle Model

We will consider h, $h \in \mathbb{N}$ formations with n, $n \in \mathbb{N}$, car-like mobile robots, where the ith agent in the hth formation is denoted by $\mathcal{A}_{hi}$. We denote the hth formation as $\mathcal{A}_h$. Without loss of generalization, we let $\mathcal{A}_{h1}$ represent the leader in the hth formation structure while the others in $\mathcal{A}_h$ take the role of followers. Let (x_{hi}, y_{hi}) represents the Cartesian coordinates and gives the reference point of $\mathcal{A}_{hi}$, which is the midpoint of the centers of the front and rear axles. Moreover, θ_{hi} gives the orientation of $\mathcal{A}_{hi}$ with respect to the z_1-axis of the $z_1 z_2$-plane while ϕ_{hi} gives the steering angle with respect to its longitudinal axis of $\mathcal{A}_{hi}$. L_{hi} represents the distance between the centers of the front and rear axles and l_{hi} is the length of each axle of $\mathcal{A}_{hi}$.

Next, to ensure that each robot safely steers past an obstacle we construct circular regions that protect the robot. Given the *clearance parameters* $\epsilon_1 > 0$ and $\epsilon_2 > 0$, we enclose the each vehicle by a protective circular region centered at (x_{hi}, y_{hi}) with radius $r_{hi} = \frac{1}{2}\sqrt{(L_{hi} + 2\epsilon_1)^2 + (l_{hi} + 2\epsilon_2)^2}$ for $h = 1, \ldots, M$, and $i = 1, \ldots, n$.

These generate the *nonholonomic constraints* on the system. The kinodynamic model of the system, adopted from [6] is

$$\left.\begin{array}{l} \dot{x}_{hi} = v_{hi}\cos\theta_{hi} - \frac{L_{hi}}{2}\omega_{hi}\sin\theta_{hi}, \quad \dot{y}_{hi} = v_{hi}\sin\theta_{hi} + \frac{L_{hi}}{2}\omega_{hi}\cos\theta_{hi}, \\[2mm] \dot{\theta}_{hi} = \frac{v_{hi}}{L_{hi}}\tan\phi_{hi} := \omega_{hi}, \quad \dot{v}_{hi} := \sigma_{hi1}, \quad \dot{\omega}_{hi} := \sigma_{hi2}, \end{array}\right\} \quad (1)$$

for $h = 1, \ldots, M$, and $i = 2, \ldots, n$. Here, v_{hi} and ω_{hi} are, respectively, the instantaneous translational and rotational velocities of $\mathcal{A}_{hi}$, while σ_{hi1} and σ_{hi2}

are the instantaneous translational and rotational accelerations of $\mathcal{A}_{hi}$. Without any loss of generality, we assume that $\phi_{hi} = \theta_{hi}$.

2.1   Leader-Follower Based Formation Scheme

To desire a substantial degree of rigidness in our formation of $\mathcal{A}_h$, we assign a Cartesian coordinate system $(X_h - Y_h)$ fixed on the leaders body of the $\mathcal{A}_h$, as shown in Fig. 1 adopted from [5], based on the concept of an instantaneous co-rotating frame of reference.

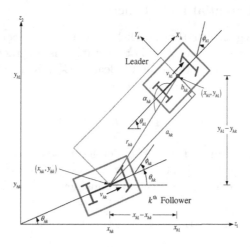

Fig. 1. The proposed scheme utilizing a rotation of axes with axis fixed at the leader.

We consider the position of the kth follower in $\mathcal{A}_h$ by considering the relative distances of $\mathcal{A}_{hk}$ in $\mathcal{A}_h$, from its leader, $\mathcal{A}_{h1}$ along the given X_h and Y_h directions. Thus, we have:

$$A_{hk} = -(x_{h1} - x_{hk})\cos\theta_{h1} - (y_{h1} - y_{hk})\sin\theta_{h1},$$
$$B_{hk} = (x_{h1} - x_{hk})\sin\theta_{h1} - (y_{h1} - y_{hk})\cos\theta_{h1},$$

$$(2)$$

for $h = 1, \ldots, M$, and $k = 2, \ldots, n$ and A_{hk} and B_{hk} are the relative positions with respect to the X_h-Y_h coordinate system of the kth followers in $\mathcal{A}_h$.

2.2   Split/Rejoin Maneuvers

One of the leaders of the sub-formations takes the role of the supreme leader of the entire flock of sub-formations. The rest of the leaders then have a dual role. They become followers to the supreme leader and also take the responsibility to lead its group. To establish an attractive force between the leaders the follower robots are required to follow the supreme leader via a concept known as mobile ghost targets (a_h, b_h) positioned relative to the supreme leader's position.

The supreme leader will move towards its defined target with center (p_{111}, p_{112}), while the ghost targets move relative to the leader's position and the follower robots move towards their designated ghost targets.

3   Artificial Potential Field Function

This section formulates collision free trajectories of the robot system under kinodynamic constraints in a given workspace.

3.1   Attractive Potential Field Functions

Attraction to Target. For the establishment and advancement of the group of M formations having n mobile robots, a target is assigned to the supreme leader. For the attraction of the supreme leader, $\mathcal{A}_{11}$ to its designated target, we consider an attractive potential function

$$V_{11}(\mathbf{x}) = \frac{1}{2} \left[(x_{11} - p_{111})^2 + (y_{11} - p_{112})^2 + v_{11}^2 + \omega_{11}^2 \right]. \tag{3}$$

The leader, $\mathcal{A}_{11}$ will move towards its defined target with center (p_{111}, p_{112}), while the other leader robots, $\mathcal{A}_{h1}$ for $h = 2, \ldots, M$ follow the ghost target relative to the supreme leader. For this we consider

$$V_{h1}(\mathbf{x}) = \frac{1}{2} \left[(x_{h1} - x_{11} + a_h)^2 + (y_{h1} - y_{11} + b_h)^2 + v_{h1}^2 + \omega_{h1}^2 \right]. \tag{4}$$

The leaders, $\mathcal{A}_{h1}$ for $h = 2, \ldots, M$ will be maintaining a desired relative position to supreme leader.

For $\mathcal{A}_{hi}$ for $i = 2, \ldots, n$ to maintain its desired relative position with respect to the leader, $\mathcal{A}_{h1}$, we utilize the following potential function for $h = 1, \ldots, M$ and $i = 2, \ldots, n$

$$V_{hi}(\mathbf{x}) = \frac{1}{2} \left[(A_{hi} - a_{hi})^2 + (B_{hi} - b_{hi})^2 + + v_{hi}^2 + \omega_{hi}^2 \right]. \tag{5}$$

Auxiliary Function. To guarantee the convergence of the supreme leader to its designated target, we define

$$G_{11}(\mathbf{x}) = \frac{1}{2} \left[(x_{11} - p_{111})^2 + (y_{11} - p_{112})^2 + (\theta_{11} - p_{113})^2 \right], \tag{6}$$

where p_{13} is the prescribed final orientation of the supreme leader robot. To guarantee the convergence of the leader mobile robots, $\mathcal{A}_{h1}$ to their designated ghost targets, we design an auxiliary function for $h = 2, \ldots, M$ as

$$G_{h1}(\mathbf{x}) = \frac{1}{2} \left[[(x_{h1} - x_{11} + a_h)^2 + (y_{h1} - y_{11} + b_h)^2 + (\theta_{h1} - p_{h13})^2 \right], \tag{7}$$

where p_{h3} is the prescribed final orientation of the leader robot, $\mathcal{A}_{h1}$ and

$$G_{hi}(\mathbf{x}) = \frac{1}{2}\left[(A_{hi} - a_{hi})^2 + (B_{hi} - b_{hi})^2 + (\theta_{hi} - p_{hi3})^2\right], \tag{8}$$

for $i = 2, \ldots, n$ and $h = 1, \ldots, M$. These auxiliary functions are then multiplied to the repulsive potential field functions to be designed in the following subsections.

3.2   Repulsive Potential Field Functions

We desire the leader, $\mathcal{A}_{h1}$ and its followers, $\mathcal{A}_{hk}$ avoid all fixed and moving obstacles intersecting their paths.

Fixed Obstacles. Let us fix $q \in \mathbb{N}$ solid obstacles within the boundaries of the workspace. We assume that the lth obstacle is a circular disk with center (o_{l1}, o_{l2}) and radius ro_l. For $\mathcal{A}_{hi}$ to avoid the lth obstacle, we consider

$$FO_{hil}(\mathbf{x}) = \frac{1}{2}\left[(x_{hi} - o_{l1})^2 + (y_{hi} - o_{l2})^2 - (ro_l + r_{hi})^2\right], \tag{9}$$

as an avoidance function, where $h = 1, \ldots, M$, $i = 1, \ldots, n$, and $l = 1, \ldots, q$.

Moving Obstacles. To generate feasible trajectories, we consider moving obstacles of which the system has prior knowledge. Here, each mobile robot, has to be treated as a moving obstacle for all other mobile robots in the workspace.

Minimum Inter-Robot Distance. We desire to maintain a minimum inter robot separation distance between the robots. This prevents $\mathcal{A}_{hi}$ from getting very close to (or colliding with) $\mathcal{A}_{hj}$ [6], especially during the re-establishment of the prescribed formation when the system is distorted. We can consider the following obstacle avoidance function

$$MO_{hij}(\mathbf{x}) = \frac{1}{2}\left[(x_{hi} - x_{hj})^2 + (y_{hi} - y_{hj})^2 - (r_{hi} + r_{hj})^2\right], \tag{10}$$

for $h = 1, \ldots, M$, and $i, j = 1, \ldots, n$, with $j \neq i$.

Inter Formation Avoidance. We also desire for each formation structure in the system to avoid any other formation structure in the workspace. For ith body of $\mathcal{A}_h$ to evade the uth body of $\mathcal{A}_m$, we adopt

$$DO_{himu}(\mathbf{x}) = \frac{1}{2}\left[(x_{hi} - x_{mu})^2 + (y_{hi} - y_{mu})^2 - (r_{hi} + r_{mu})^2\right], \tag{11}$$

for $i, u = 1, \ldots, n$ and $h, m = 1, \ldots, M$ with $m \neq h$.

Dynamic Constraints. Practically, the steering angles of the mobile robots are limited due to mechanical singularities while the translational speed is restricted due to safety reasons. Subsequently, we have $|v_{hi}| < v_{\max}$, where $v_{\max}$ is the *maximal achievable speed* of the $\mathcal{A}_{hi}$ and $|\omega_{hi}| < \frac{v_{\max}}{|\rho_{\min}|}$, where $\rho_{\min} := \frac{L_{hi}}{\tan(\phi_{\max})}$. This condition arises due to the boundness of the steering angle ϕ_{hi}. That is, $|\phi_{hi}| \leq \phi_{\max} < \pi/2$, where $\phi_{\max}$ is the *maximal steering angle*. Hence, we consider the following avoidance functions:

$$U_{hi1}(\mathbf{x}) = \frac{1}{2} \left(v_{\max} - v_{hi} \right) \left(v_{\max} + v_{hi} \right), \tag{12}$$

$$U_{hi2}(\mathbf{x}) = \frac{1}{2} \left(\frac{v_{\max}}{|\rho_{\min}|} - \omega_{hi} \right) \left(\frac{v_{\max}}{|\rho_{\min}|} + \omega_{hi} \right), \tag{13}$$

for $h = 1, \ldots, M$ and $i = 1, \ldots, n$.

4   Acceleration Controllers

The nonlinear acceleration control laws for system (1), will be designed using LbCS as proposed in [6].

4.1   Lyapunov Function

We now construct the total potentials, that is, a Lyapunov function for system (1).

$$L_{(1)}(\mathbf{x}) = \sum_{h=1}^{M} \sum_{i=1}^{n} \left[V_{hi}(\mathbf{x}) + G_{hi}(\mathbf{x}) Z(\mathbf{x}) \right], \tag{14}$$

where

$$Z(\mathbf{x}) = \sum_{l=1}^{q} \frac{\alpha_{hil}}{FO_{hil}(\mathbf{x})} + \sum_{s=1}^{2} \frac{\beta_{his}}{U_{his}(\mathbf{x})} + \sum_{\substack{j=1 \\ j \neq i}}^{n} \frac{\eta_{hij}}{MO_{hij}(\mathbf{x})} + \sum_{u=1}^{n} \sum_{\substack{m=1 \\ m \neq h}}^{M} \frac{\gamma_{himu}}{DO_{himu}(\mathbf{x})}.$$

Utilizing the attractive and repulsive potential field functions, continuous time-invariant acceleration control laws, $\sigma_{hi1}, \sigma_{hi2}$, can be generated, that intrinsically guarantees stability, in the sense of Lyapunov, of system (1) as well.

5   Stability Analysis

Theorem 1. *Let* $(p_{h11}, p_{h12}) = (a_h, b_h)$ *for* $h = 2, \ldots, M$ *be the position of the target of the leader in* $\mathcal{A}_h$, *and* p_{hi3} *for* $i = 1, \ldots, n$, *be the desired final orientations of the robots in each* $\mathcal{A}_h$. *Given* a_{hi} *and* b_{hi} *in* $\mathcal{A}_h$, *let* p_{hi1} *and* p_{hi2} *satisfy*

$$a_{hi} = -(p_{h11} - p_{hi1}) \cos \theta_{h1} - (p_{h12} - p_{hi2}) \sin \theta_{h1},$$
$$b_{hi} = (p_{h11} - p_{hi1}) \sin \theta_{h1} - (p_{h12} - p_{hi2}) \cos \theta_{h1},$$

for $i = 2, \ldots, n$ *and* $h = 1, \ldots, M$.

Given $\mathbf{x}_{hi}^* := (p_{hi1}, p_{hi2}, p_{hi3}, 0, 0) \in \mathbb{R}^5$, *if* $\mathbf{x}_e := (\mathbf{x}_{11}^*, \mathbf{x}_{12}^*, \ldots, \mathbf{x}_{hn}^*) \in \mathbb{R}^{5 \times n \times M}$ *is an equilibrium point for (1), then* $\mathbf{x}_e \in D(L_{(1)}(\mathbf{x}))$ *is a stable equilibrium point of system (1).*

Proof. One can easily verify the following, for $i = 1, \ldots, n$ and $h = 1, \ldots, M$:

1. $L_{(1)}(\mathbf{x})$ is defined, continuous and positive over the domain $D(L_{(1)}(\mathbf{x})) = \{\mathbf{x} \in \mathbb{R}^{5 \times M \times n} : FO_{hil}(\mathbf{x}) > 0, l = 1, \ldots, q; DO_{himu}(\mathbf{x}) > 0, m = 1, \ldots, M, u = 1, \ldots, n, m \neq h; MO_{hij}(\mathbf{x}) > 0, j = 1, \ldots, n, j \neq i; U_{his}(\mathbf{x}) > 0, s = 1, 2\}$;
2. $L_{(1)}(\mathbf{x}^*) = 0$;
3. $L_{(1)}(\mathbf{x}) > 0 \; \forall \mathbf{x} \in D(L_{(1)}(\mathbf{x}))/\mathbf{x}_e$.

Next, consider the time derivative of the candidate Lyapunov function along a particular trajectory of system (1), we obtain the following semi-negative definite function

$$\dot{L}_{(1)}(\mathbf{x}) = - \sum_{i=1}^{n} \left(\delta_{hi1} v_{hi}^2 + \delta_{hi2} \omega_{hi}^2 \right) \leq 0,$$

for $h = 1, \ldots, M$ and $i = 1, \ldots, n$, where $\delta_{hi1} > 0$, and $\delta_{hi2} > 0$ are constants commonly known as convergence parameters.

Thus, $\dot{L}_{(1)}(\mathbf{x}) \leq 0 \; \forall \mathbf{x} \in D(L_{(1)}(\mathbf{x}))$ and $\dot{L}_{(1)}(\mathbf{x}_e) = 0$. Finally, it can be easily verified that $L_{(1)}(\mathbf{x}) \in C^1 \left(D(L_{(1)}(\mathbf{x})) \right)$, which makes up the fifth and final criterion of a Lyapunov function. □

Remark 1. This result is in no contradiction with Brockett's Theorem [7] as we have not proven asymptotic stability.

6 Simulation Results

In this section, we illustrate the effectiveness of the proposed continuous time-invariant controllers within the framework of the Lyapunov-based control scheme by simulating a virtual scenario.

We consider the motion of a pair of 4 cars in a split/rejoin formation in a two dimensional space with static obstacles in its path. Each follower robot in each formation structure is assigned a unique position relative to its leader as seen in Fig. 2. This is achieved by assigning appropriate values to (a_{hk}, b_{hk}) to obtain a geometric formation structure. While a leaders $\mathcal{A}_{11}$ and $\mathcal{A}_{21}$ move towards its intended target, the followers, $\mathcal{A}_{hk}$ for $h = 1, 2$ and $k = 2, \ldots, 4$ in each sub-formation is observed to maintain a low-degree formation. The sub swarms split from their initial formation when the swarm encounters an obstacle. The sub-formations reestablish links with other sub-swarms and converge into its pre-split formation after avoiding collisions with the obstacles. The leader-follower control strategy is used for maintaining formation shape in the sub-swarms. Assuming that the appropriate units have been accounted for, Table 1 provides the corresponding initial and final configurations of the two robots and other necessary parameters required to simulate the scenario.

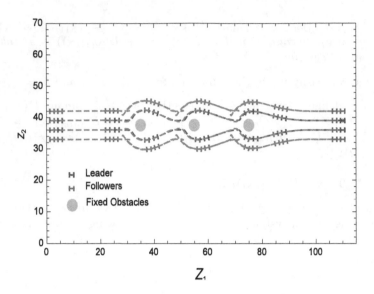

Fig. 2. The proposed scheme showing the split/rejoin formation of the robots. (Color figure online)

Table 1. Numerical values of initial and final states, constraints and parameters.

Initial configuration	
Rectangular positions of leaders	$(x_{11}, y_{11}) = (5, 39), (x_{21}, y_{21}) = (5, 36)$
Desired relative distances of followers	$a_{12} = a_{22} = b_{14} = b_{24} = 0, b_{12} = b_{13} = -3$
	$a_{13} = a_{14} = a_{23} = a_{24} = b_{22} = b_{23} = 3$
Translational velocity	$v_{hi} = 0.5$ for $i = 1, \ldots, 4, h = 1, 2$
Rotational velocities	$\omega_{hi} = 0$, for $i = 1, \ldots, 4, h = 1, 2$
Angular positions	$\theta_{hi} = 0$, for $i = 1, \ldots, 4, h = 1, 2$
Constraints and parameters	
Dimensions of robots	$L_{hi} = 1.6, l_{hi} = 1.2$ for $i = 1, \ldots, 4, h = 1, 2$
$\mathcal{A}_1$ leader's target:	$(p_{111}, p_{112}) = (110, 39), rt_{11} = 0.5$,
Fixed Obstacles	$(o_{11}, o_{12}) = (35, 37.5), (o_{21}, o_{22}) = (55, 37.5)$,
	$(o_{31}, o_{32}) = (75, 37.5)$
Max. translational velocity	$v_{\max} = 5$
Max. steering angle	$\phi_{\max} = \pi/2$
Clearance parameters	$\epsilon_1 = 0.1, \epsilon_2 = 0.05$
Control and convergence parameters	
Collision avoidance	$\alpha_{hil} = 0.1$, for $i = 1, \ldots, 4, h = 1, 2, l = 1, \ldots, 3$
	$\eta_{hij} = 0.01$ for $h = 1, 2, i, j = 1, \ldots, 4, j \neq i$,
	$\gamma_{himu} = 0.01$, for $h, m = 1, 2, i, j = 1, \ldots, 4, h \neq m$
Dynamics constraints	$\beta_{his} = 1$, for $i = 1, \ldots, 4, h, s = 1, 2$,
Convergence	$\delta_{h11} = 12000, \delta_{h12} = 12000$, for $h = 1, 2$
	$\delta_{hij} = 50$, for $i = 1, \ldots, 4, h, j = 1, 2$

7   Conclusion

In this paper we have proposed a leader follower scheme for the coordination of multi robot systems in an environment with obstacles. A split/rejoin maneuver is observed between sub-swarms. After avoiding collisions with the obstacles, the sub-swarms reestablishes its links with the other sub-swarms and converge into its pre-split formation. The leader-follower control strategy is used for maintaining formation shape in the sub-swarms. An advantage of the proposed scheme is that we can have multiple formations structures having different geometric shapes. The approach also considers inter-robot and inter-formation collision avoidance. The effectiveness of the proposed control laws were demonstrated via a computer simulation.

References

1. Qu, Z.: Cooperative Control of Dynamical Systems: Applications to Autonomous Vehicles. Springer, London (2009)
2. Chen, Z., Chiu, T., Zhang, J.: Swarm splitting and multiple targets seeking in multi-agent dynamic systems. In: 49th IEEE Conference on Decision and Control, Atlanta, GA, USA, pp. 4577–4582 (2010)
3. Moshtagh, N., Michael, N., Jadbabaie, A.: Vision-based distribution control laws for motion coordination of nonholonomic robots. IEEE Trans. Rob. **25**(4), 851–860 (2009)
4. Reynolds, C.W.: Flocks, herds and schools: a distributed behavioral model in computer graphics. In: 14th Annual Conference on Computer Graphics and Interactive Techniques, New York, USA, pp. 25–34 (1987)
5. Raghuwaiya, K., Sharma, B., Vanualailai, J.: Cooperative control of multi robot systems with a low-degree formation. In: Sulaiman, H.A., Othman, M.A., Othman, M.F.I., Rahim, Y.A., Pee, N.C. (eds.) Advanced Computer and Communication Engineering Technology. LNEE, vol. 362, pp. 233–249. Springer, Switzerland (2015)
6. Sharma, B.: New Directions in the Applications of the Lyapunov-based Control Scheme to the Findpath Problem, Ph.D. thesis, Fiji Islands (2008)
7. Brockett, R.W.: Asymtptotic stability and feedback stabilization. In: Brockett, R.W., Millman, R.S., Sussmann, H. (eds.) Differential Geometry Control Theory. Springer (1983)

A Grouping Method for Multiple Targets Search Using Swarm Robots

Qirong Tang$^{(\boxtimes)}$, Fangchao Yu, and Lu Ding

Laboratory of Robotics and Multibody System, School of Mechanical Engineering,
Tongji University, Shanghai 201804, People's Republic of China
qirong.tang@outlook.com

Abstract. This paper presents an integrated method based on a modified PSO algorithm and a grouping strategy for swarm robots to search multiple targets simultaneously. The number of robot groups is determined autonomously according to the actual searched environment and the amount of potential targets. A simulation platform is designed to demonstrate the searching process and to verify the method. Comparisons are performed to show the superior of the studied method. Results show that the proposed method has good adaptability and high success rate in the searching multiple targets.

Keywords: Swarm robots · Multiple targets search · Modified PSO · Grouping strategy

1 Introduction

Compared to a single robot, swarm robots have advantages, e.g., bigger range of movement and wider information collection in multiple targets searching. Specifically, for searching multiple targets simultaneously, swarm robots can improve the efficiency greatly. Therefore, it has many potential applications such as searching survivors after disaster, looking for the source of leakage, using micro robots to enter human body to search for pathogens.

Many search methods have been developed for swarm robots to accomplish multiple targets search tasks. A group explosion strategy (GES) for searching multiple targets was proposed in [1]. It shows good performance in search problem with regular fitness values. However, the strategy still remains several shortcomings such as poor scalability. In Derr and Manic's work, the target of each robot is determined by the signal strength [2]. Those robots who have the same target are set into a common group. But there is no coordination among different groups. A method is proposed in [3], which not only considers the travel cost and target weight, but also predicts the target/robot rate, and as well as potential robot redundancy with respect to the detected targets. Furthermore, a dynamic weight adjustment is also applied to improve the search performance. However, it is difficult to estimate the value of target utility accurately. A strategy inspired by division of labor in social wasp is presented in [4]. The distance

© Springer International Publishing Switzerland 2016
Y. Tan et al. (Eds.): ICSI 2016, Part II, LNCS 9713, pp. 470–478, 2016.
DOI: 10.1007/978-3-319-41009-8_51

variable is applied in [5] to response threshold model to solve task allocation in swarm robotics. However, this strategy is easily trapped for local optimum.

In this paper we propose a method for solving the task of searching for multiple targets by swarm robots based on a modified PSO and a grouping strategy. The rest of the paper is organized as follows. Section 2 describes the modified particle swarm optimization method. The grouping search strategy is given in Sect. 3. Simulations of swarm robots motions with the proposed method are given in Sect. 4, while Sect. 5 concludes the paper.

2   Modified Particle Swarm Optimization

Particle swarm optimization (PSO) was first prsented in [6] and used often as a general optimization tool. In this paper, a modified algorithm based on PSO was used to update the velocities and positions of particles. It is assumed that each particle moves in an n-dimensional search space, and the objective function f has n independent variables. The vectors $\mathbf{x}_i = (x_{i1}, x_{i2},..., x_{in})$ and $\mathbf{v}_i = (v_{i1}, v_{i2},..., v_{in})$ represent the current position and velocity of particle i, respectively. That $\mathbf{p}_i = (p_{i1}, p_{i2},..., p_{in})$ is the individual historical best position of particle i, while $\mathbf{g} = (g_1, g_2,..., g_n)$ is the global historical best position among those particles at the moment t. The velocity and position of each particle will be updated according to the following equations

$$v_{ij}(t+1) = \chi(v_{ij}(t) + c_1 r_1(p_{ij}(t) - x_{ij}(t)) + c_2 r_2(g_j(t) - x_{ij}(t))) , \quad (1)$$

$$\chi = \frac{2}{\left|2 - \varphi - \sqrt{\varphi^2 - 4\varphi}\right|} , \quad (2)$$

$$x_{ij}(t+1) = x_{ij}(t) + v_{ij}(t+1) , \quad (3)$$

where i represents the index of each particle, j the j-th dimension, t the current iteration step, $v_{ij}(t)$ and $x_{ij}(t)$ represent the velocity the and position, respectively, $p_{ij}(t)$ is the component of p_i in j-th dimension at t-th iteration, and $g_j(t)$ is the component of g in j-th dimension, c_1, c_2 are referred to as scaling factors, while r_1, r_2 are two independent random parameters between 0 and 1. Here $\varphi = c_1 + c_2$, and $\varphi > 4$. The convergence rate of PSO with limiting factor χ is relatively faster and the error is smaller [7].

3   Grouping Method

3.1   Overall Principle

This section proposes a grouping method, which divides all the particles(robots) into a number of groups. A certain number of particles are generated in a 2-dimensional search space randomly. Each particle's fitness value is calculated according to the objective function. In this study, the fitness value of a particle is assumed to be the objective function value according to its position. In the first stage, all particles move randomly. The positions of particles are updated due to their velocities which are updated with random values in a certain range. Then the fitness values of all the particles are calculated. Meanwhile, each particle records its individual historical best position and best fitness value in iterations. In this stage, the velocity and position of each particle are updated by

$$v_{ij}(t+1) = V_{max}(-1+2r) \ , \tag{4}$$

$$x_{ij}(t+1) = x_{ij}(t) + v_{ij}(t+1) \ , \tag{5}$$

where V_{max} is the maximal velocity of a particle, r is a random value between 0 and 1. Other symbols hold the same meanings with those in Sect. 2. This stage ends when the specified number of iterations are reached.

After the first stage, particles begin to grouping. Firstly, some search points are found according to the individual historical best positions of the particles. The method of how to determine the search points will be introduced in detail in Sect. 3.2. Secondly, if some of the particles are closer to one search point than other search points, these particles construct a group and search the area near the search point. The grouping situation is shown in Fig. 1.

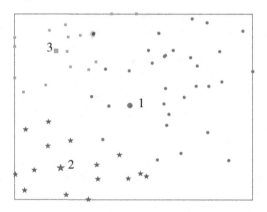

Fig. 1. Particles grouping situation

After this, the particle swarm goes to the second stage. In this stage, the particles which belong to the same group will iterate using the modified PSO algorithm separately. In each group, the global best position and fitness value will be first updated with the position and fitness value of the corresponding search point. Whereas the search point will be replaced during iteration afterwards by the global best position in each group. All particles are re-grouped according to this principle. Then the particle swarm goes into the next iteration. In Fig. 1, the search points are marked by a bigger circle, a bigger square or a bigger pentagram. The particle have same shape to their corresponding search points belong to the same group.

3.2 Grouping Strategy

The key to divide particles into different groups is to find the so called search points. A strategy is proposed to determine which positions can be selected to be the search points. In this paper, targets are distributed at those positions with the minimum fitness values. Suppose the number of recorded individual historical best positions is n. Those individual historical best positions whose fitness values are larger than the predefined threshold of $f_{threshold}$ are first removed. Then the remaining individual historical best positions are sequenced in accordance with their fitness values from small to large. The first m_s positions are verified by a certain condition. The positions which satisfy the condition are regarded as search points.

A parameter l_d which represents the distribution distance had been used as the judgement condition. The distance between any two search points must satisfy the condition

$$d_{s_i, s_j} > l_d, i \neq j, 1 \leq i, j \leq m_s \ , \tag{6}$$

where i and j represent the index of search points, s_i and s_j represent any two search points, l_d is the distribution distance. However, the parameter l_d is difficult to be set. An inappropriate parameter will lead to a rapid decline to the search success rate.

An improved grouping strategy is proposed following to determine the search points without the parameter l_d. The $f_{threshold}$ is still used to prevent some positions with larger fitness values from becoming search points, meanwhile, the parameter l_d is omitted. The strategy is described as follows by which search points are generated much more autonomously.

A contour map of the objective function with five minimum points is shown in Fig. 2. The asterisks (*) represent the positions of the targets. The deeper of the blue color, the smaller of the objective function value, while the red contour

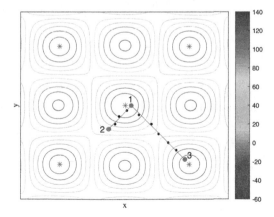

Fig. 2. Determing of search points (Color figure online)

means the larger value of the objective function. There are five minimums which can be the targets on the map. The area around a target is called as a valley. There is only one target in each valley. So it is expected that only one search point is generated in each valley. Suppose that points 1, 2, 3 (with vertical coordinates z_1, z_2, z_3) are the individual historical best positions after the first iteration stage. According to the proposed method, point 1 will be the first search point (since it has the smallest fitness value), then point 2 and 3 will be determined not only by their fitness values but also by the relationships to point 1. The relationship is judged by placing N assistant points (with coordinates (x_1, y_1), (x_2, y_2),..., (x_N, y_N)) on the line which links points 1 and 2 to see whether they are in different valleys. It is verified if their coordinates satisfy

$$\exists i \in N \ make \ f(x_i, y_i) > z_1 \ \& \ f(x_i, y_i) > z_2 \ . \tag{7}$$

If none of the N points satisfy the condition, then, point 2 won't be chosen as a search point. However point 3 is considered to be in a different valley satisfies (7) if they are chosen properly. Therefore, point 3 is an appropriate search point. The determining of search points is shown in Fig. 2.

3.3   Algorithm Flow

The work process of the modified PSO algorithm is that a group of particles cooperate with each other to find a better position. It is very similar to a group of robots

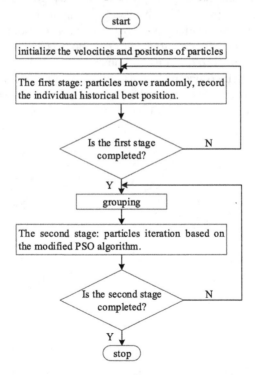

Fig. 3. Flow chart of the proposed algorithm

searching for a target. So the modified PSO algorithm is possible to be used guiding robots to search targets. Indeed, all the parameters of the modified PSO need to be mapped with corresponding ones to the actual physical system [8]. However, the core content of this paper is to simulate the process of swarm robots searching for multiple targets, and it focuses on the multiple targets search method. The physical properties such as mass and volume of the robot are ignored in this study. The modified PSO and the grouping strategy are integrated to form a grouping search algorithm. Its flow chart is shown in Fig. 3.

4   Simulation Result and Discussion

4.1   Test Functions

The proposed method is implemented within MATLAB. It uses two test functions, for the 5 targets and the 9 targets, respectively. The 5 targets function is governed by

$$f_1 = \frac{x^2 + y^2}{10} - 100 \cos \frac{1}{2} \cos \frac{7y}{10\sqrt{2}} + 40 \ . \tag{8}$$

There are 5 targets in this function, distributed at coordinates $(0,0)$, $(6.23, 6.30)$, $(6.23, -6.30)$, $(-6.23, 6.30)$, $(-6.23, -6.30)$. The fitness value in coordinate $(0,0)$ is -60, while the others are -52.09. The 3-D fitness field and 2-D fitness field of the 5 targets function are shown in Fig. 4.

The 9 targets function is governed by

$$f_2 = -\left| \frac{x^2 + y^2}{10} - 100 \cos \frac{1}{2} \cos \frac{7y}{10\sqrt{2}} - 10 \right| \ . \tag{9}$$

There are 9 targets in this function, distributed at coordinates $(0,0)$, $(6.23, -6.30)$, $(-6.23, 6.30)$, $(6.23, 6.30)$, $(-6.23, -6.30)$, $(0, -6.40)$, $(6.33, 0)$, $(-6.33, 0)$, $(0, 6.40)$. The fitness value in coordinate $(0,0)$ is -110, the following

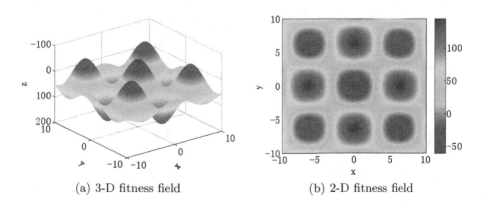

(a) 3-D fitness field   (b) 2-D fitness field

Fig. 4. The fitness field map of the 5 targets function

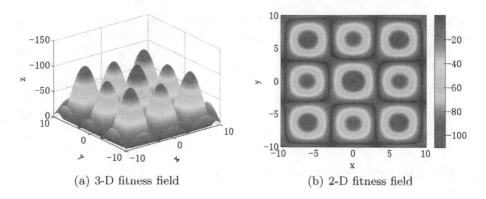

(a) 3-D fitness field   (b) 2-D fitness field

Fig. 5. The fitness field map of the 9 targets function

4 fitness values are -102.09, and the others are -94.06, -93.98, -93.98, -94.06. The 3-D fitness field and 2-D fitness field of the 9 targets function are shown in Fig. 5.

4.2   Simulation Result

It assumes that a search experiment is successful if all of the targets are found. The success rate of experiments in searching multiple targets with different methods is studied. Results are obtained with 100 particles in the swarm and 50 iterations in every experiment. The values of c_1 and c_2 are set as 2.1 and 2.1 respectively to ensure the convergence. Due to the different fitness values, different $f_{threshold}$ are used in the experiments. For 5 targets and 9 targets, experiments using the old method consider l_d are performed for each situation. The results are shown in Table 1.

Table 1. The results of searching for multiple targets using the old method

l_d	$f_{threshold}$		Success rate	
	5 targets	9 targets	5 targets	9 targets
5	-40	-80	98 %	90 %
5	-20	-60	100 %	98 %
5	0	-30	100 %	100 %
3	-40	-80	98 %	92 %
3	-20	-60	100 %	98 %
3	0	-30	100 %	96 %
1	-40	-80	94 %	44 %
1	-20	-60	92 %	8 %
1	0	-30	66 %	0 %

Table 2. The results of searching for multiple targets using the proposed method

number of targets	$f_{threshold}$	Success rate
5	−40	100 %
5	−20	100 %
5	0	100 %
9	−80	94 %
9	−60	100 %
9	−30	100 %

Table 1 shows that the success rate is related to the distribution distance l_d and the threshold $f_{threshold}$ in the old method. The success rate is acceptable when the l_d changes in a reasonable range. It will decline rapidly when the l_d continues to decrease. However, the value of l_d is difficult to be determined in the old method. An inappropriate value of l_d will cause a poor success rate. For 5 targets and 9 targets, 50 runs are performed respectively using the proposed method. The success rate is also studied with the same parameters c_1, c_2, and $f_{threshold}$, see Table 2.

It is clear that this method achieves a good success rate. When searching for 5 targets, all the targets are found in 50 runs under different values of $f_{threshold}$. When searching for 9 targets, the method also achieves a good success rate. The success rate reaches 100 % when the value of $f_{threshold}$ is set as −60 and −30, while it reaches 94 % when the value of $f_{threshold}$ is set as −80. The proposed method omits the parameter l_d, and uses a new strategy to generate the search points. Compared to the old method, this method has better adaptability and scalability, as well as a good success rate.

5   Conclusion

A new method especially a novel grouping strategy is proposed in this paper for swarm robots to search multiple targets simultaneously. It is constructed by a modified PSO algorithm and a grouping strategy which contains an adaptive threshold $f_{threshold}$ what is used to determine the search points. Particles are grouped according to the nearest potential search point. Some issues such as particles moving to a non-target position since the search point generated in an inappropriate position, the poor search capability of particles in a group led by the excessive number of search points, are both considered effectively. Simulation results show that this method has good adaptability and high success rate in searching multiple targets compared to a previously used old method which relies mainly on the distribution distance l_d. As for future work, we plan to apply this method on more complicated search situations by using real physical swarm robots.

Acknowledgements. This research is supported by the Fundamental Research Funds for the Central Universities (No. 2014KJ032, 20153683), by Shanghai Pujiang Program (No. 15PJ1408400) and the Key Basic Research Project of 'Shanghai Science and Technology Innovation Plan' (No. 15JC1403300). Meanwhile, this work is also partially supported by the State Key Laboratory of Robotics and Systems (Harbin Institute of Technology) (No. SKLRS-2015-ZD-03), and the National Science Foundation of China (No. 51579053). All these supports are highly appreciated.

References

1. Zheng, Z., Tan, Y.: Group explosion strategy for searching multiple targets using swarm robotic. In: 2013 IEEE Congress on Evolutionary Computation, Cancun, Mexico, pp. 821–828 (2013)
2. Derr, K., Manic, M.: Multi-robot, multi-target particle swarm optimization search in noisy wireless environments. In: 2nd Conference on Human System Interactions, Catania, Italy, pp. 81–86 (2009)
3. Meng, A.Y., Gan, J., Desai, S.: A bio-inspired swarm robot coordination algorithm for multiple target searching. In: The International Society for Optical Engineering, Orlando, USA,vol. 6964, pp. 696406-1–696406-9 (2008)
4. Song, Y., Zhang, G., Zeng, J., Xue, S.: Research on task allocation in multi-target localization in swarm robotics based on respose threshold model. J. Taiyuan Univ. Sci. Technol. **33**, 262–268 (2012)
5. Shi, Y., Eberhart, R.: A modified particle swarm optimizer. In: 1998 IEEE International Conference on Evolutionary Computation Proceedings, New York, USA, pp. 69–73 (1998)
6. Kennedy, J., Eberhart, R.: Particle swarm optimization. IEEE Int. Conf. Neural Netw. **4**, 1942–1948 (1995)
7. Clerc, M., Kennedy, J.: The particle swarm: explosion, stability, and convergence in a multidimensional complex space. IEEE Trans. Evol. Comput. **6**, 58–73 (2002)
8. Tang, Q., Eberhard, P.: Cooperative motion of swarm mobile robots based on particle swarm optimization and multibody system dynamics. Mech. Based Des. Struct. Mach. **39**, 179–193 (2011)

A Comparative Study of Biology-Inspired Algorithms Applied to Swarm Robots Target Searching

Qirong Tang[1], Lei Zhang[2], Wei Luo[3], Lu Ding[1], Fangchao Yu[1], and Jian Zhang[4(✉)]

[1] Laboratroy of Robotics and Multibody System, School of Mechanical Engineering, Tongji University, No.4800, Cao An Rd., Shanghai 201804, China
qirong.tang@outlook.com
[2] Department of Mechanical Engineering of Tsinghua University, Graduate School at Shenzhen, Shenzhen 518055, China
[3] Department of Mechanical Engineering, University of Stuttgart, Pfaffenwaldring 9, 70569 Stuttgart, Germany
[4] Institute of Advanced Manufacturing, School of Mechanical Engineering, Tongji University, No.4800, Cao An Rd., Shanghai 201804, China
jianzh@tongji.edu.cn

Abstract. In this study, the mechanisms of some creatures' behaviors collaborated in swarm are applied to the coordination of swarm robots, especially for them to search target. Three typical biology-inspired algorithms, i.e., Particle Swarm Optimization, Ant Colony Optimization and Genetic Algorithms, are thus compared, systematically. Corresponding tasks and mathematical models are set up. Based on the experimental work within MATLAB, the performances of the concerned algorithms on the difficulty of task mapping, adaptability for various terrains, as well as convergence and stability are elaborately analyzed and verified, which is helpful for designing real physical swarm robotic systems.

Keywords: Swarm intelligence · Swarm robots · Obstacle avoidance · Particle swarm optimization · Ant colony · Genetic algorithms

1 Introduction

A single robot has restrictions and limitations in acquiring and processing information, thus it is not suitable to deal with complicated works and face to the fast changing environments. However, this situation motivated many researchers to start the study of multi-robot system at the 1980s. Gradually, two specialized subjects, namely swarm robotics and swarm intelligence, were born. Under collaboration, a swarm of robots with inferior abilities for each can fulfill some complicated tasks that a single but superior and expensive robot cannot. As for now, many robots such as e-Puck [1], s-bot [2] are designed and created for the

© Springer International Publishing Switzerland 2016
Y. Tan et al. (Eds.): ICSI 2016, Part II, LNCS 9713, pp. 479–490, 2016.
DOI: 10.1007/978-3-319-41009-8_52

research of swarm robots. A swarm of simple robots can be designed and manufactured more easily and cost less. Moreover, swarm robots also outperform on
robustness, scalability, flexibility [3]. Therefore, swarm robots are used widely.

Swarm robots demonstrate novelties on collaboration among massive robots,
which are often inspired from biologically collective behaviors. In term of
mechanisms of these collective behaviors, many swarm intelligent algorithms
have been constructed, such as, Particle Swarm Optimization/PSO [4], Ant
Colony Optimization/ACO [5], Genetic Algorithms/GAs [6], and Bacterial Forging Algorithm [7]. Having neither centralized control nor complete environment
information, swarm intelligence technology brings a new way of solving complicated problems by taking the advantage of its scalability. Up to now, swarm robotics research has accumulated some benchmark tasks such as aggregation, area coverage, mapping, searching, navigating and locating leakages [8–10].

Different swarm intelligent algorithms have different features. However, to
the best of our knowledge, the research of a systematically comparison of these
algorithms applied to swarm robots have rarely been studied. So this study aims
to discover the differences of PSO, ACO and GAs when they are applied to
swarm robots in searching one or more targets.

Section 2 presents the task mapping models and mathematical models of the
studied algorithms as well as some improvements. In Sect. 3, it makes a quantity
of simulation in various environments containing obstacles and analyzes the performance of the three concerned algorithms. Conclusions are given in Sect. 4.

2   Models

2.1   Particle Swarm Optimization

The task environment of this work is a two-dimensional space with obstacles.
The velocity of the particles in PSO are updated by

$$V_i^{k+1} = \omega V_i^k + c_1 r_{i,1}^k (X_{i,self}^{best,k} - X_i^k) + c_2 r_{i,2}^k (X_{i,local}^{best,k} - X_i^k) , \tag{1}$$

where X_i^k is the position of particle i at kth iteration, V_i^k is the velocity of
particle i at kth iteration. One saves $X_{i,self}^{best,k}$ as the individual best position, and
$X_{i,local}^{best,k}$ is the best position among particles. Here ω is an inertia weight, c_1, c_2
are referred to as cognitive scaling and social scaling factors, $r_{i,1}^k$, $r_{i,2}^k$ are two
independent random parameters. The position of particle i is given by

$$X_i^{k+1} = X_i^k + V_i^{k+1} \Delta t , \tag{2}$$

where Δt is a time step that usually is 1 and thus is omitted often.

Parameter ω represents the inertia of the particles, and be calculated by

$$\omega = \frac{(\omega_2 - \omega_1)}{e} \times j + \omega_1 . \tag{3}$$

Here e is the maximum steps, j is the current step, $\omega_1 = 0.4$, $\omega_2 = 0.9$.

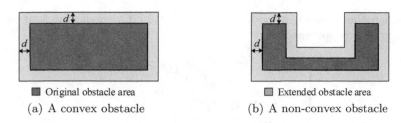

Original obstacle area ☐ Extended obstacle area

(a) A convex obstacle (b) A non-convex obstacle

Fig. 1. Type of obstacles

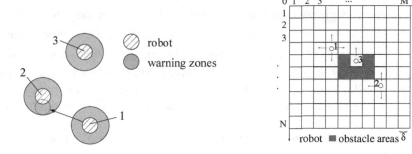

Fig. 2. Choice of rotation direction   **Fig. 3.** The searched environment

The shape of obstacles in real environments is irregular. We shape obstacles with regular boundaries, and categorize them into convex and non-convex, as shown in Fig. 1. Considering braking and signals delay, the extended obstacle areas are set up [11]. When one robot detects obstacles in front, it responds to avoid obstacles by slowing down and rotating itself. This study takes 10° for each rotation.

When updating positions, these robots with physical volume should pay attention so as to keep a safe distance to their peers. In Fig. 2, a mutual collision between robot 1 and robot 2 may happen since robot 1 is entering the warning zone of robot 2. Similar to static obstacles in practical environment, robot 1 takes actions to avoid collision with robot 2. After several rational tries (maximum 18 times), if the robot still can't find a suitable collision free path, it will stay there instead of updating for the current step. In this study, one robot is represented by one particle.

2.2 Ant Colony Optimization

Here, the robot is assumed as a circular with radius R and searches in a two-dimensional space which is divided into an $M \times N(M = N = 10)$ grid, see Fig. 3. All grids have the unit length δ, and consist of barrier grids and free grids. Barrier grids can be used to complement the irregular boundary of the environment area as well, besides of indicating obstacles. One obstacle occupies one or more grids. If some part of the obstacle is not enough to fill a grid, it will be considered as a complete grid. According to [12], if a robot stands at node i, its next node j is governed by

$$j = \begin{cases} argmax\{\tau_{i,s}^{\alpha}(t)\eta_{i,s}^{\beta}(t)\}, s \in A, q \leq q_0 \\ P_{i,s} = \dfrac{\tau_{i,s}^{\alpha}(t)\eta_{i,s}^{\beta}(t)}{\displaystyle\sum_{s \in A} \tau_{i,s}^{\alpha}(t)\eta_{i,s}^{\beta}(t)}, else \end{cases}, \tag{4}$$

where $q \sim (0,1)$ is an random variable, q_0 is a constant ranges from 0 to 1, s is one element of A which contains all candidate nodes for node i, $\tau_{i,s}^{\alpha}(t)$, $\eta_{i,s}^{\beta}(t)$ represent the pheromone concentration and the heuristic information, respectively, from node i to node s in which $\tau_{i,s}^{\alpha}(t)$ and $\eta_{i,s}^{\beta}(t)$ are corresponding coefficients. Here $argmax\{\tau_{i,s}^{\alpha}(t)\eta_{i,s}^{\beta}(t)\}$ means the corresponding node s reach the maximum of $\tau_{i,s}^{\alpha}(t)\eta_{i,s}^{\beta}(t)$, $P_{i,s}$ is the probability that node s is selected as the next position for node i. When the robot is trapped into a valley (see node 3 in Fig. 3), it sets the node 3 as a barrier grid, and goes back to the previous node.

The heuristic information is calculated according to a heuristic algorithm summarized by

$$\eta_{i,s} = D/d_{s,G} , \tag{5}$$

where D is the weight coefficient and $d_{s,G}$ indicates the distance between the node s and target node G. After one iteration, the pheromone concentration needs to be updated by

$$\tau_{i,j}(t+1) = (1 - \varphi)\tau_{i,j}(t) + \Delta\tau_{i,j} , \tag{6}$$

$$\Delta\tau_{i,j} = \sum_{k=1}^{m} \Delta\tau_{i,j}^{k} , \tag{7}$$

where $\tau_{i,j}$ represents the pheromone concentration on the path from node i to node j before updating, φ refers to the pheromone evaporation factor, and m is the robot number, $\Delta\tau_{i,j}$ is the pheromone that robot k leaves on the path from node i to j, $\Delta\tau_{i,j}$ is the sum of pheromone of all robots at current iteration which is a constant Q in this study.

Each robot selects the next node from the four directly adjacent grids. Therefore, the path is composed by a series of perpendicular segments. An optimized path will be obtained when the redundant nodes are deleted, see Fig. 4.

2.3   Genetic Algorithms

Genetic algorithms are applied to solve robot path planning in this section via using the grids map model as same as for ACO. One chromosome represents one path and the genetic information on which is the coordinate values of path nodes. All these path nodes form the complete path.

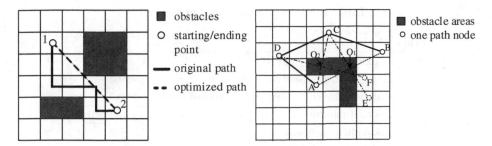

Fig. 4. Remove redundant nodes   **Fig. 5.** The geometric method

A geometric obstacle avoiding method is proposed in [13] by inserting some nodes along the direction perpendicular to the connection line of two nodes. The inserted nodes should be on the same side. This study proposes a new method for judging if the later node is on the same side with the formers. In Fig. 5, since the path from node A to node B collides to one obstacle, node C is inserted firstly, which is in the free area and on the line perpendicular to the connection between node A and node B. After this, node D rather than node F, is inserted since the sign of $\overrightarrow{AD} \cdot \overrightarrow{DC}$ is same to that of $\overrightarrow{AC} \cdot \overrightarrow{CB}$, and the sign of $\overrightarrow{AF} \cdot \overrightarrow{FB}$ is in opposite to $\overrightarrow{AC} \cdot \overrightarrow{CB}$. Finally, a free path $A \to D \to C \to B$ is generated.

To prevent the optimal path from destroying in GA operation, the selection operator endowed with elitist strategy is considered. This study uses the crossover of finding an intersection point(node) of two parent paths, then exchange mutually all of the rest path nodes after this node. And one-point mutation is adopted, which randomly choose one node on the original path, then replace it with another node which is near to the original one and locates in a free grid. The insert mode of genetic operator is chosen as that those outstanding individuals are added into parent population to generate a new one.

3   Simulation and Comparison

Parameter settings of the three algorithms are listed in Table 1. The simulation environment is set as 10m×10m. All of these parameters for each algorithm are carefully selected to fit the current condition and make each algorithm currently optimized.

3.1   Comparison of Environmental Adaptability

The terrains are usually very complex in practical environment, thus the environment adaptability of robot has significant influence on the accomplishment of tasks. This study sets up three kinds of environments to test the concerned algorithms.

(1) Comparison in the environment with convex obstacles

Table 1. Parameters list for experiments

Types	Parameters	Values
PSO	maximal velocity (m/s)	0.1
	safety distance between robots (m)	0.3
	safety distance between robot and obstacle (m)	0.2
	inertial parameter (ω)	$0.4 - 0.9$
	individual weight (c_1)	1.0
	neighbor weight (c_2)	1.0
ACO	pheromone increment factor	0.001
	pheromone evaporation factor	0.1
	probability selection parameter (q_0)	0.8
	pheromone concentration weight (α)	1
	heuristic information weight (β)	1
GAs	selection probability	0.9
	cross probability	0.7
	mutation probability	0.3

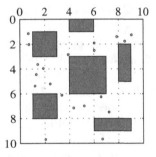

(a) The beginning of the run

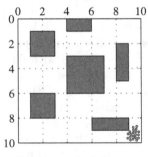

(b) The end of the run

Fig. 6. The motion process of all robots

The first environment contains only convex obstacles, see Figs. 6, 7, 8, and 9. The target is purposely placed at location $(9.5, 9.5)$. The initial position and velocity of PSO are randomly generated, while for GA and ACO the initial points are set at the location $(0.5, 0.5)$, which are far away to the target. Figure 6 shows a running progress with 20 particles based on PSO. At the very beginning, the particles are randomly distributed. With the progress of iteration, all particles gather near to the target and keep a certain distance from each. Figure 8 shows the trajectory of 3 particles (robots), and obviously indicates that all robots have successfully avoided the static obstacles based on PSO. Figure 7 shows the result based on ACO, in which "$\bigcirc$" and "$\triangledown$" represent the start point and the target, respectively. It is obvious that the path in Fig. 7(b) is more smooth than that in Fig. 7(a). GA has also gained a collision free path, see Fig. 9.

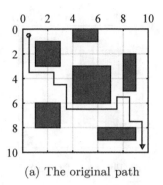 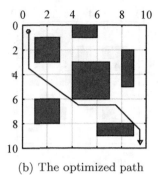

(a) The original path   (b) The optimized path

Fig. 7. Path planning result based on ACO

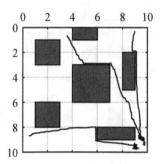 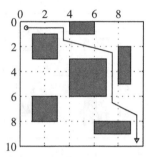

Fig. 8. Trajectories of robots based on PSO   **Fig. 9.** Path planning based on GA

(2) Comparison in mixed-obstacle environment

In this kind of environment, there are convex and non-convex obstacles which increase the difficulty of target searching. In Fig. 10, most of particles based on PSO have reached the target after certain iterations, but a few of particles fail and fall into corners, like the top-right corner in Fig. 10(b). It indicates that PSO has limitation to handle this situation. It is even more restricted in an environment with U-shaped obstacles. Ant colony optimization also performs well in the second environment, see Fig. 11. Compared with the path with redundancy nodes, the optimized one is better on the path length and smoothness, which shows the necessity of removing redundant nodes. Figure 12 displays the path planning result of GA in mixed-obstacle environment. It indicates that the geometric method for obstacle avoidance has a strong generality and adaptability to the environment with convex and non-convex obstacles.

(3) Comparison in the environment with U-shaped obstacles

Compared to the previous two types of obstacles, the U-shaped obstacle brings a lot of difficulties for target searching. To verify the search ability of algorithms as possible as one can, in this paper the initial point is set inside the U-shaped area, at $(4.5, 4.5)$, and the target point is set at outside of the

486   Q. Tang et al.

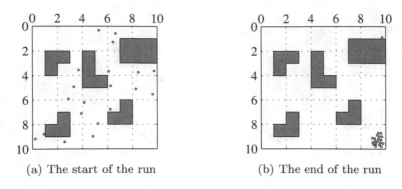

(a) The start of the run   (b) The end of the run

Fig. 10. The motion process of all robots based on PSO

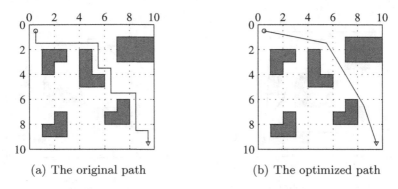

(a) The original path   (b) The optimized path

Fig. 11. Path planning result based on ACO

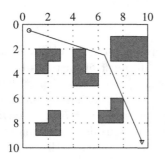

Fig. 12. Path planning of GA in mixed-obstacle environment

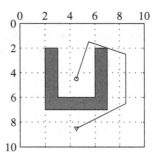

Fig. 13. Path planning of GA in U-shaped obstacle environment

U-shaped, at $(4.5, 8.5)$. Other conditions are same to the previous cases. As shown in Figs. 13 and 14, GA and ACO are also feasible to the big U-shaped obstacle. In Fig. 15, most of particles have arrived the target point, but there are still a certain particles stuck in the U-shaped obstacle.

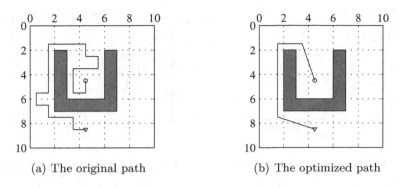

(a) The original path   (b) The optimized path

Fig. 14. Path planning result based on ACO

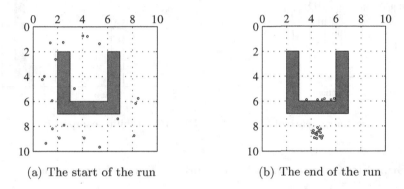

(a) The start of the run   (b) The end of the run

Fig. 15. The motion process of all robots based on PSO

3.2 Convergence Comparison of the Studied Three Algorithms

The convergence speed is one of the evaluation criteria of algorithms and affects to the computational efficiency. As the iteration runs, the trend of the optimal solution can reflect the convergence feature. In Fig. 16, the optimal objective function value approach to 0 after 100 iterations, which means that the target is found. Overall, the convergence trend of PSO is distinct and the converge speed is fast. However, in some complex environments such as the U-shape obstacle environment, some individuals are easily fall into local optimum. ACO has a feature of convergence gradually, shown in Fig. 17. In the first 20 iterations, the path length varies greatly, fluctuates wildly and its convergence trend is not obvious. After all, ACO still converges to the optimal solution at a relatively fast rate. Genetic algorithm hardly converge to an optimum solution, see in Fig. 18, the shortest path is irregularly changing without a tendency of convergence. The reason is that many operations of GA have large randomness, such as the crossover or mutation.

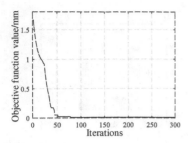

Fig. 16. The convergence of PSO

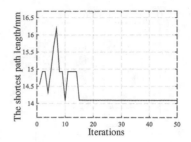

Fig. 17. The convergence of ACO

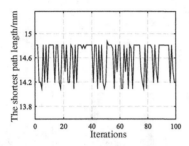

Fig. 18. The convergence of GA

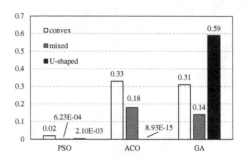

Fig. 19. Stability comparison

3.3   Stability Comparison of the Studied Three Algorithms

Stability is another important evaluation criterion of the algorithms. Each of three algorithms is evaluated independently with 100 runs in the three types of environments. As one can see from Fig. 19, PSO has a good stability in three environments, especially for the mixed-obstacle environment, which has only a standard deviation of 6.23E-04. It shows that swarm robots based on PSO can find stably the target from the perspective of swarm performance. The stability of ACO in the environment with U-shaped obstacles is much better than that in other two environments. The reason is that there are more obstacles in the latter two environments and more various paths are produced after optimization. Therefore, the shortest path is varying heavily. Genetic algorithm demonstrates the instability in three environments on account of its large standard deviation that is related to its big randomness.

4   Conclusion

In this study, PSO, ACO and GAs are applied to the swarm robots target searching and path planning. Through a quantity of simulation and analysis, the performances of the three algorithms are compared on adaptability, convergence and stability.

The results show that the convergence and stability of PSO are good in the environment with convex obstacles. However, when it comes to non-convex obstacles and large U-shaped obstacles, the basic PSO has difficulty in solving this kind of problem in which particles are easily trapped at obstacles. Ant colony algorithm has a good adaptability to the environment change in this study, but it mainly aims at discrete space and its calculation time is longer than others. Compared with the other two, GAs model is more complex and the quality of initial population has a great influence on the final solution. The geometric method for obstacle avoidance which is used in this study has advantages of generality and good adaptability. The future work will concentrate on including more properties of practical robots into the studied algorithms.

Acknowledgements. This research is supported by the Fundamental Research Funds for the Central Universities (No. 2014KJ032, 20153683), by Shanghai Pujiang Program (No. 15PJ1408400) and the Key Basic Research Project of 'Shanghai Science and Technology Innovation Plan' (No. 15JC1403300). Meanwhile, this work is also partially supported by the State Key Laboratory of Robotics and Systems (Harbin Institute of Technology) (No. SKLRS-2015-ZD-03), and the National Science Foundation of China (No. 51579053). All these supports are highly appreciated.

References

1. Cianci, C.M., Raemy, X., Pugh, J., Martinoli, A.: Communication in a swarm of miniature robots: the e-Puck as an educational tool for swarm robotics. In: Şahin, E., Spears, W.M., Winfield, A.F.T. (eds.) SAB 2006 Ws 2007. LNCS, vol. 4433, pp. 103–115. Springer, Heidelberg (2007)
2. Mondada, F., Pettinaro, G.C., Guignard, A., Kwee, I.W., Floreano, D., Deneubourg, J.L.: Swarm-bot: a new distributed robotic concept. Auton. Robots **17**, 193–221 (2004)
3. Şahin, E.: Swarm robotics: from sources of inspiration to domains of application. In: Şahin, E., Spears, W.M. (eds.) Swarm Robotics 2004. LNCS, vol. 3342, pp. 10–20. Springer, Heidelberg (2005)
4. Kennedy, J., Eberhart, R.: Particle swarm optimization. In: International Conference on Neural Networks, Perth, Australia, vol. 4, pp. 1942–1948 (1995)
5. Dorigo, M., Maniezzo, V., Colorni, A.: The ant system: optimization by a colony of cooperative agents. Phys. Rev. Lett. **75**, 2686–2689 (1995)
6. Zhou, L.F., Hong, B.R.: A knowledge based genetic algorithm for path planning of a mobile robot. Acta Electronica Sin. **5**, 4350–4355 (2006)
7. Passino, K.M.: Biomimicry of bacterial foraging for distributed optimization and control. IEEE Control Syst. **22**, 52–67 (2002)
8. Soysal, O., Şahin, E.: A macroscopic model for self-organized aggregation in swarm robotic systems. In: Şahin, E., Spears, W.M., Winfield, A.F.T. (eds.) SAB 2006 Ws 2007. LNCS, vol. 4433, pp. 27–42. Springer, Heidelberg (2007)
9. Payton, D., Estkowski, R., Howard, M.: Pheromone robotics and the logic of virtual pheromones. In: Şahin, E., Spears, W.M. (eds.) Swarm Robotics 2004. LNCS, vol. 3342, pp. 45–57. Springer, Heidelberg (2005)

10. Penders, J., Alboul, L., Witkowski, U., Naghsh, A., Saez-Pons, J., Herbrechtsmeier, S., El-Habbal, M.: A robot swarm assisting a human fire-fighter. Adv. Rob. **25**, 93–117 (2011)
11. Tang, Q., Eberhard, P.: A PSO-based algorithm designed for a swarm of mobile robots. Struct. Multi. Optim. **44**, 483–498 (2011)
12. Luo, D., Wu, S.: Ant colony optimization with potential field heuristic for robot path planning (in Chinese). Syst. Eng. Electron. **32**, 1277–1280 (2010)
13. Li, Q., Feng, J., Liu, Y., Zhou, Z., Yin, Y.: Application of adaptive genetic algorithm to optimum path planning of mobile robots (in Chinese). J. Univ. Sci. Technol. Beijing **30**, 316–323 (2008)

Thrust Optimal Allocation for Broad Types of Underwater Vehicles

Hai Huang[⊠], Guo-cheng Zhang, Yi Yang, Jin-yu Xu, Ji-yong Li,
and Lei Wan

Key Laboratory of Science and Technology for Autonomous Underwater
Vehicle, Harbin Engineering University, 145 Nantong Street, Harbin, China
haihus@163.com

Abstract. Effective thrust allocation for underwater vehicle is very important to realize complex control task. In order to generate optimal thrust allocation according to control command, a thrust optimal allocation scheme has been designed for broad type of underwater vehicles. Corresponding to horizontal and vertical thruster configuration, a force allocation model has been established to realize optimal allocation. Infinity-norm optimization has been combined with 2-norm optimization to construct a bi-criteria primal-dual neural network optimal allocation scheme. In the experiment of open frame remote operated vehicle, bi-criteria primal-dual optimization outperformed 2-norm optimization in the fault tolerance control and accurate path following. In the thrust optimal allocation simulations for underwater vehicle manipulator system, 4 vertical thrust have been optimal allocated to realize diving and attitude maintenance successfully during manipulation process. Thus the thrust optimal allocation has been verified.

Keywords: Thrust allocation · Optimal allocation · Bi-criteria optimization

1 Introduction

As a cost-effective solution for performing complex tasks in the underwater environment, underwater vehicles are attracting significant interest. However, for the vehicles with redundant thrusters, such as open frame vehicles and working class vehicles, thrust optimal allocation problem is particularly important to realize control command, improve fault tolerable ability and reduce energy consumption [1].

Different from land vehicles, underwater vehicle usually have to generate control force through the distribution of actuators [2]. For underactuated underwater vehicles, motion control is realized through compound effects of thrusters and wings [3], while for full actuated ones, motion control is realized through thrusters from the horizontal and vertical planes respectively [4]. A full actuated 6-degree-of-freedom (DOF) underwater vehicle is required to have at least 6 thrusters in order to generate 3 linearly independent translations and 3 linearly independent rotations. In general, there are more than 6 thrusters in any 6-dof underwater vehicle. In other words, the problem of optimal distribution of propulsion forces is necessary to obtain additional power and fault accommodation [5]. Podder and Sarkar exploited the framework of excess number

© Springer International Publishing Switzerland 2016
Y. Tan et al. (Eds.): ICSI 2016, Part II, LNCS 9713, pp. 491–502, 2016.
DOI: 10.1007/978-3-319-41009-8_53

of thrusters to accommodate thruster faults during operation. A redundancy resolution scheme is presented based on weighted least-norm solutions to accommodate thruster faults [6]. Daqi et al. used neural network to accommodate fault and perform an appropriate control reallocation on the basis of redundant thrusters [7].

For vehicles with redundant thrusters, numerous solutions can be obtained in correspondence with one certain control command. Recent research has been carried out to obtain optimal solutions. Based on the self organizing maps [8], the fault accommodation subsystem of reference [9] used weighted pseudo inverse to find 2-norm optimal solution of the control allocation problem. Edin and Geoff has established thruster configuration for open frame underwater vehicle, in order to realize normalization control allocation [10]. However, pseudo inverse solution may generate an unequal propulsion forces distribution and cause a loss of maneuverability because it does not necessarily minimize the magnitudes of the individual thrusts [11, 12]. Serdar et al. demonstrated that minimizing the infinity-norm of the thrust manifold ensures low individual thruster forces [12]. A recurrent neural network has been proposed to realize infinity-norm optimization, however it may encounter a discontinuity problem because of the non-uniqueness of such a solution and the separation of two successive solution-sets [13].

In order to realize thrust optimal allocation for additional control power and fault accommodation, this paper combine infinity-norm optimization with 2-norm optimization in primal-dual neural network to realize thrust allocation for fault tolerance control and accurate path following of open frame remote operated vehicle and realize successfully attitude maintenance during manipulation process of autonomous manipulation underwater vehicle. The rest of this paper is organized as follows. Section 2 will introduce the model of thrust control allocation. Section 3 will issue the bi-criteria thrust optimization algorithm with neural network. Section 4 will make analysis on open frame remote operated vehicle and working class underwater vehicle. We will draw conclusions in Sect. 5.

2 Model of Thrust Control Allocation

For open frame or autonomous manipulation underwater vehicles, there are two common configurations of the horizontal thrusters e.g. X-shaped and Cross-shaped configuration (Fig. 1) [10]. As their position and orientation being illustrated in Tables 1 and 2 respectively, the former configuration can provide greater thrust force at any direction while more direct thrust allocation can be obtained from the latter.

Therefore, vector of forces and moments, exerted by HT, can be expressed as:

$$\tau_i^{HT} = \begin{bmatrix} {}^iF_{\mathrm{h}} \\ {}^iM_{\mathrm{h}} \end{bmatrix} = \begin{bmatrix} {}^iF_{\mathrm{h}}e_{\mathrm{h}i} \\ {}^iF_{\mathrm{h}}(r_{\mathrm{h}i} \times e_{\mathrm{h}i}) \end{bmatrix} = \begin{bmatrix} e_{\mathrm{h}ix} & e_{\mathrm{h}iy} & (r_{\mathrm{h}i} \times e_{\mathrm{h}i})_x & (r_{\mathrm{h}i} \times e_{\mathrm{h}i})_y \end{bmatrix}^T {}^iF_{\mathrm{h}},$$

(1)

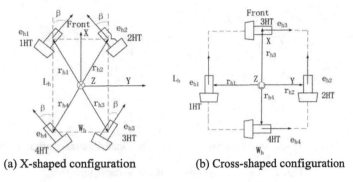

(a) X-shaped configuration   (b) Cross-shaped configuration

Fig. 1. Two configurations of the horizontal thrusters

Table 1. The position for thruster configurations

	1HT	2HT	3HT	4HT
X-shaped configuration	$\left[\frac{L_h}{2} \quad -\frac{W_h}{2}\right]^T$	$\left[\frac{L_h}{2} \quad \frac{W_h}{2}\right]^T$	$\left[-\frac{L_h}{2} \quad \frac{W_h}{2}\right]^T$	$\left[-\frac{L_h}{2} \quad -\frac{W_h}{2}\right]$
Cross-shaped configuration	$\left[0 \quad -r_{h1}\right]^T$	$\left[0 \quad r_{h2}\right]^T$	$\left[r_{h3} \quad 0\right]^T$	$\left[-r_{h4} \quad 0\right]^T$

Table 2. The orientation for thruster configurations

	e_{h1}	e_{h2}	e_{h3}	e_{h4}
X-shaped configuration	$\left[\cos\beta \quad \sin\beta\right]^T$	$\left[\cos\beta \quad -\sin\beta\right]^T$	$\left[\cos\beta \quad \sin\beta\right]^T$	$\left[\cos\beta \quad -\sin\beta\right]^T$
Cross-shaped configuration	$[1 \quad 0]^T$	$[1 \quad 0]^T$	$[1 \quad 0]^T$	$[1 \quad 0]^T$

where ^iF_h is the force exerted by thruster i on underwater vehicle, $^iM_v = {}^iF_v(r_{vi} \times e_{vi})$ is the moment deduced from ^iF_v. The direction of ^iF_v is decided by $e_{vi} = \left[e_{vix} \quad e_{viy}\right]$. We obtain:

$$\tau^{HT} = \sum_{i=1}^{4} \tau_i^{HT} = \sum_{i=1}^{4}\begin{bmatrix} {}^iF_h \\ {}^iM_h \end{bmatrix}$$

$$= \underbrace{\begin{bmatrix} e_{1hx} & e_{2hx} & \cdots & e_{4hx} \\ e_{1hy} & e_{2hy} & \cdots & e_{4hy} \\ (r_{h1} \times e_{h1})_x & (r_{h2} \times e_{h2})_x & \cdots & (r_{h4} \times e_{h4})_x \\ (r_{h1} \times e_{h1})_y & (r_{h2} \times e_{h2})_y & \cdots & (r_{h4} \times e_{h4})_y \end{bmatrix}}_{R_h} \underbrace{\begin{bmatrix} {}^1F_h \\ {}^2F_h \\ {}^3F_h \\ {}^4F_h \end{bmatrix}}_{F_h} = R_h F_h, \quad (2)$$

where $R_h \in \mathbb{R}^{4 \times 4}$ is thrusters configuration matrix.

Similarly for the configurations of vertical thrusters, the configurations are usually cross-shaped and quadrilateral configuration (Fig. 2). Their position and orientation are illustrated in Tables 3 and 4.

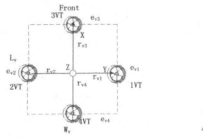
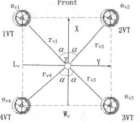

<table>
<tr><td>(a) Cross-shaped configuration</td><td>(b) Quadrilateral configuration</td></tr>
</table>

Fig. 2. Two configurations of the vertical thrusters

Table 3. The position for thruster configurations

	1VT	2VT	3VT	4VT
Cross-shaped configuration	$\left[0 \quad \frac{W_v}{2}\right]^T$	$\left[0 \quad -\frac{W_v}{2}\right]^T$	$\left[\frac{L_v}{2} \quad 0\right]^T$	$\left[-\frac{L_v}{2} \quad 0\right]^T$
Quadrilateral configuration	$\left[\frac{L_v}{2} \quad -\frac{W_v}{2}\right]^T$	$\left[\frac{L_v}{2} \quad \frac{W_v}{2}\right]^T$	$\left[-\frac{L_v}{2} \quad \frac{W_v}{2}\right]^T$	$\left[-\frac{L_v}{2} \quad -\frac{W_v}{2}\right]^T$

Table 4. The orientation for thruster configurations

	e_{v1}	e_{v2}	e_{v3}	e_{v4}
Cross-shaped configuration	$\begin{bmatrix}0 & 1\end{bmatrix}$	$\begin{bmatrix}0 & 1\end{bmatrix}$	$\begin{bmatrix}0 & 1\end{bmatrix}$	$\begin{bmatrix}0 & 1\end{bmatrix}$
Quadrilateral configuration	$\begin{bmatrix}\cos\alpha & -\sin\alpha\end{bmatrix}^T$	$\begin{bmatrix}\cos\alpha & \sin\alpha\end{bmatrix}^T$	$\begin{bmatrix}-\cos\alpha & \sin\alpha\end{bmatrix}^T$	$\begin{bmatrix}-\cos\alpha & -\sin\alpha\end{bmatrix}^T$

Therefore, vector of forces and moments, exerted by VT, can be expressed as:

$$\tau_i^{VT} = \begin{bmatrix} {}^iF_v \\ {}^iM_v \end{bmatrix} = \begin{bmatrix} {}^iF_v e_{vi} \\ {}^iF_v(r_{vi} \times e_{vi}) \end{bmatrix} = \begin{bmatrix} e_{vix} & e_{viy} & (r_{vi} \times e_{vi})_x & (r_{vi} \times e_{vi})_y \end{bmatrix}^T {}^iF_v \quad (3)$$

where ${}^iF_v$ is the force exerted by thruster i on underwater vehicle, ${}^iM_v = {}^iF_v(r_{vi} \times e_{vi})$ is the moment deduced from ${}^iF_v$. The direction of ${}^iF_v$ is decided by $e_{vi} = \begin{bmatrix} e_{vix} & e_{viy} \end{bmatrix}$. We obtain:

$$\tau^{VT} = \sum_{i=1}^{4} \tau_i^{VT} = \sum_{i=1}^{4} \begin{bmatrix} {}^iF_v \\ {}^iM_v \end{bmatrix}$$

$$= \underbrace{\begin{bmatrix} e_{1vx} & e_{2vx} & \cdots & e_{4vx} \\ e_{1vy} & e_{2vy} & \cdots & e_{4vy} \\ (r_{v1} \times e_{v1})_x & (r_{v2} \times e_{v2})_x & \cdots & (r_{v4} \times e_{v4})_x \\ (r_{v1} \times e_{v1})_y & (r_{v2} \times e_{v2})_y & \cdots & (r_{v4} \times e_{v4})_y \end{bmatrix}}_{R_v} \underbrace{\begin{bmatrix} {}^1F_v \\ {}^2F_v \\ {}^3F_v \\ {}^4F_v \end{bmatrix}}_{F_v} = R_v F_v, \quad (4)$$

where $R_v \in \mathbb{R}^{4 \times 4}$ is thrusters configuration matrix.

3   Bi-criteria Thrust Optimal Allocation Algorithm

If we set $F_u = \begin{bmatrix} {}^1F_u & {}^2F_u & {}^3F_u & {}^4F_u \end{bmatrix}^T$ as the upper saturation vector and $F_l = \begin{bmatrix} {}^1F_l & {}^2F_l & {}^3F_l & {}^4F_l \end{bmatrix}^T$ as the lower saturation vector respectively. Thus the 2-norm of thrusters force manifold is $\|F\|_2 = \frac{F^T F}{2}$, while the infinity norm of the thrusters force manifold is defined as:

$$\|\bar{F}\|_\infty = W\max\left\{ |{}^1F| \quad |{}^2F| \quad |{}^3F| \quad |{}^4F| \right\} = l. \tag{5}$$

Therefore, thrust allocation problem can be addressed as the following constrained optimization problem:

$$\text{Minimize } a\|F\|_2 + \frac{1}{2}(1-a)\|F\|_\infty,$$

$$\text{subject to } \tau^{HT} = RF, {}^iF_l W_i \le {}^iF \le {}^iF_u W_i, \tag{6}$$

where $a \in (0, 1]$ is weighting factor. This problem can also be written in matrix form:

$$\text{Minimize } \frac{S^T Q S}{2} \tag{7}$$

$$\text{subject to } B_2 S = d_2, B_1 S \le d_1, S^- \le S \le S^+. \tag{8}$$

where $S = \begin{bmatrix} {}^1F & {}^2F & {}^3F & {}^4F & l \end{bmatrix}^T$, $d_2 = \tau^{HT} \in \mathbb{R}^4$, $Q = \begin{bmatrix} adiag\{1\ 1\ 1\ 1\} & \\ & 1-a \end{bmatrix}$,

$S^- = \begin{bmatrix} F_l \\ 0 \end{bmatrix} \in \mathbb{R}^{4+1}$, $\quad S^+ = \begin{bmatrix} F_u \\ \varpi \end{bmatrix} \in \mathbb{R}^{4+1}$, $\quad I \in \mathbb{R}^{4\times4}$, $\quad I_4 = \begin{bmatrix} 1 & 1 & 1 & 1 \end{bmatrix}^T$,

$B_2 = \begin{bmatrix} R & 0_4 \end{bmatrix} \in \mathbb{R}^{(4+1)\times4}$, τ_u^T and τ_l^T are the upper and lower limits of τ^T respectively,

$B_1 = \begin{bmatrix} I & -1_4 \\ -I & -1_4 \end{bmatrix} \in \mathbb{R}^{(2\times4)\times(4+1)}$, $\varpi \gg 0$ (e.g. 10^{10}) is sufficiently large to replace $+\infty$ numerically $d_1 = 0 \in \mathbb{R}^8$.

(1) When $0 < a < 1$, the objective function (7) is strictly convex due to the positive definite of Q. Provided the feasible region of linear constraints (8) being a closed convex set, the constrained optimal solution to optimal program (6) is unique. In light of the uniqueness property, the continuity of the solution could thus be guaranteed [14].

(2) When $a = 1$, the proposed scheme is reduced to the standard 2-norm scheme. The uniqueness and continuity of such scheme are both guaranteed.

(3) When $a = 0$, the proposed scheme is reduced to the standard infinity-norm scheme. According to [13], the infinity-norm scheme may have non-unique solutions at successive time instants, which may cause discontinuity phenomenon. So we set $0 < a \le 1$ to remedy such discontinuity problem.

If define $u \in \mathbb{R}^4$ corresponding to equality constraint of (8), $v \in \mathbb{R}^8$ corresponding to inequality constraint of (8), the primal-dual decision variable vector g and its bounds $g^{\pm}$ could be constituted as:

$$g = \begin{bmatrix} S \\ u \\ v \end{bmatrix}, g^{+} = \begin{bmatrix} S^{+} \\ 1_4 \varpi \\ 1_8 \varpi \end{bmatrix}, g^{-} = \begin{bmatrix} S^{-} \\ -1_4 \varpi \\ 0_8 \end{bmatrix} \in \mathbb{R}^{12+4+1}. \tag{9}$$

Theorem 1. There exist at least one optimal solution $S^* \in \mathbb{R}^5$, so that the dual problem (7) and (8) can be converted into the linear variation inequalities (LVI) problem:

$$(g - g^*)^T (Hg^* + P) \geq 0, \tag{10}$$

where $H = \begin{bmatrix} Q & -B_2^T & B_1^T \\ B_2 & 0 & 0 \\ B_1 & 0 & 0 \end{bmatrix}, P = \begin{bmatrix} 0 \\ -d_2 \\ d_1 \end{bmatrix}.$

Proof. From [15] the dual problem of (7) and (10) can be derived as:

$$\text{Maximize} \ - \frac{S^T Q S}{2} + d_2^T u - d_1^T v + S^{-T} \chi^{-} - S^{+T} \chi^{+} \tag{11}$$

$$\text{subject to} \ QS - B_2^T u + B_1^T v - \chi^{-} + \chi^{+} = 0. \tag{12}$$

where $\chi \in \mathbb{R}^5$ are the corresponding dual decision variables, $\chi^{-} \geq 0$, $\chi^{+} \geq 0$. Then, a necessary and sufficient condition for optimum value $(S^*, g^*, v^*, v^{\pm *})$ of primal problem (7) and (8) and its dual problem (11)–(12) is the following:

Primal feasibility : $B_2 S^* = d_2, B_1 S^* \leq d_1, S^{-} \leq S^* \leq S^{+},$ (13)

Complementarities :
$v^{*T}(-B_1 S^* + d_1) = 0, \chi^{-*T}(-S^* + S^{-}) = 0, \chi^{+*T}(-S^{+} + S^*) = 0.$ (14)

To correspond with constraint (8) and simplify the above necessary and sufficient formulation, we obtain from (14):

$$\begin{cases} S_i^* = S_i^{+} & \text{if} \quad \chi_i^{+*} > 0, \chi_i^{-*} = 0 \\ S_i^{-} < S_i^* < S_i^{+} & \text{if} \quad \chi_i^{+*} = 0, \chi_i^{-*} = 0 \\ S_i^* = S_i^{-} & \text{if} \quad \chi_i^{+*} = 0, \chi_i^{-*} > 0 \end{cases}.$$

By defining $\chi^* = \chi^{-*} - \chi^{+*}$, dual feasibility constraint (12) becomes:

$$QS^* - B_2^T u^* + B_1^T v^* = \chi^* \begin{cases} \leq 0 & S_i^* = S_i^{+} \\ = 0 & S_i^{-} < S_i^* < S_i^{+} \\ \geq 0 & S_i^* = S_i^{-} \end{cases},$$

which equals the following linear variational inequality:

$$(S - S^*)^T(QS^* - B_2^T u^* + B_1^T v*) \geq 0. \qquad (15)$$

Similarly we obtain: $(v - v^*)^T(-B_1 S^* + d_1) \geq 0$ and $(u - u^*)^T(B_2 S^* - d_2) \geq 0$, where $v* \in \{v | v \geq 0\}$, $u \in \mathbb{R}^4$. Therefore we obtain:

$$\left(\begin{bmatrix} S \\ u \\ v \end{bmatrix} - \begin{bmatrix} S^* \\ u* \\ v* \end{bmatrix} \right)^T \left(\begin{bmatrix} Q & -B_2^T & B_1^T \\ B_2 & 0 & 0 \\ 0 & 0 & 0 \end{bmatrix} \begin{bmatrix} S^* \\ u* \\ v* \end{bmatrix} + \begin{bmatrix} 0 \\ -d_2 \\ d_1 \end{bmatrix} \right) \geq 0. \qquad (16)$$

□

According to [15] the LVI problem (8) is equivalent to the following system of piecewise-linear equations.

$$K_\Omega(g - (Hg + P)) - g = 0. \qquad (17)$$

where $K_\Omega(\bullet) : R^{12+4+1} \to \Omega$ is a piecewise linear projection operator from space R^{12+4+1} onto set Ω with the ith element of $K_\Omega(g)$ defined as:

$$\begin{cases} g_i^- & if & g_i < g_i^- \\ g_i & if & g_i^- \leq g_i \leq g_i^+ \\ g_i^+ & if & g_i > g_i^+ \end{cases} \quad \forall i \in \{1, \ldots, (12+4+1)\}. \qquad (18)$$

Therefore, a primal-dual neural network solver for the thrust force optimal allocation could be developed as the following dynamic equation:

$$\dot{g} = \gamma(I + H^T)(K_\Omega(g - (Hg + P)) - g), \qquad (19)$$

where $\gamma > 0$ is a parameter to scale the network convergence, and should be set as large as possible in the implementation (Fig. 3).

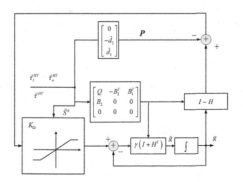

Fig. 3. Block diagram of system solver for (5) and (6)

Equation (16) can be further written as:

$$\dot{g}_i = \gamma \sum_{j=1}^{n} c_{ij} \left(K_{\Omega}^{(j)} \left(\sum_{k=1}^{n} t_{ik} g - P_i \right) - g_i \right), \tag{20}$$

where c_{ij} denotes the ijth entry of scaling matrix $C = I + H^T$, and t_{ik} denotes the ikth entry of matrix $S = I - H$. A more detailed architecture of the neural network based on neuron Eq. (17) is shown in Fig. 4.

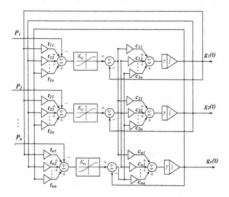

Fig. 4. Network architecture of (20)

4   Simulations and Experimental Results

Fault accommodation experiments and accurate path following have been made based on SY-II open-frame underwater vehicle. The control commands are sent through network communication between surface computer and PC/104 embedded system. SY-II is equipped with a depth gauge, a magnetic compass, 6 thrusters including 2 main thrusters, 2 side thrusters and 2 vertical thrusters. The horizontal thrusters are cross configuration. The hydrodynamic and inertial parameters are illustrated in Tables 3 and 4. Experiments have been made in a $50 \times 30 \times 10$ m tank of Key Laboratory of Science and Technology on Underwater Vehicle, Harbin Engineering University (Fig. 5).

4.1   Fault Accommodation Experiments for Open Frame Remote Operated Vehicle

In the heading control of Fig. 6(a), we have made comparisons on 3 situations: (1) heading control without thrusters fault (no fault); (2) heading control with 2-norm FTC ($a = 0.6$, $\gamma = 10^5$, left main thruster (1HT in Fig. 1(b)) 70 % fault and tail side thruster (4HT in Fig. 1(b)) 100 % fault) (FTC a = 0.6); (3) heading control without FTC, left main thruster (1HT in Fig. 1(b)) 70 % fault and tail side thruster (4HT in

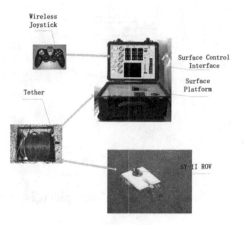

Fig. 5. Composition of SY-II open-frame underwater vehicle

Fig. 1(b)) 100 % fault) (fault without FTC). Cruising experiments have been made with the precondition of left main thruster (1HT in Fig. 1(b)) 40 % fault and tail side thruster (4HT in Fig. 1(b)) 100 % fault. The desired motion trajectory is $(-7,-20)$ $-(-2,-2)-(5,-2)-$ $(5,-22)-(-7,-20)$. From Fig. 6, bi-criteria FTC outperforms 2-norm FTC for fault tolerance control in reducing overshoot during yawing, approaching desired trajectory and cruising against current.

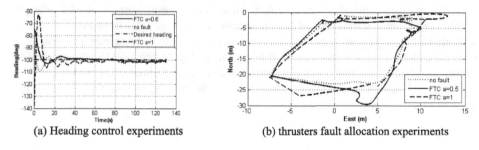

(a) Heading control experiments   (b) thrusters fault allocation experiments

Fig. 6. Experiments of fault tolerant thruster optimal allocation

4.2 Accurate Path Following Experiments for Open Frame Remote Operated Vehicle

During the 3D path-following experiments in the tank polygon path are followed with the disturbances of current speed as $V_{curr} = 0.4$ m/s. This experiment includes two control algorithm e.g. S surface control and model based back stepping control (see ref. [4]). In order to realize accurate path following control in the disturbance, main

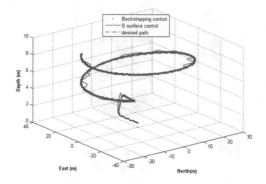

Fig. 7. 3D path-following experiments, spiral path (Color figure online)

thrusters and side thrusters participate heading control at the same time. Bi-criteria optimal algorithm has been applied to optimally allocate the horizontal thrust force (see Fig. 7).

4.3 Simulations for Underwater Vehicle Manipulator System

The designed Underwater Vehicle Manipulator System (UVMS) model for autonomous manipulation is shown in Fig. 8. It is composed of a 6-DOF underwater vehicle and a 4-DOF manipulator. The underwater vehicle is equipped with a main thruster for forward and backward advance, 2 vertical rudders and 2 horizontal wings for heading and diving control during cruise, 2 side thrusters and 4 vertical thrusters for vehicle attitude (the configurations see Fig. 2(a)) and position control during manipulation. The max propulsive forces of thruster is 10 kgf each. In Fig. 9, the UVMS is controlled to complete planned manipulation trajectory. The 4 vertical thrusts are responsible for diving, pitch and roll control simultaneously. The allocation result is close to the control result when a = 0.6.

Fig. 8. UVMS model

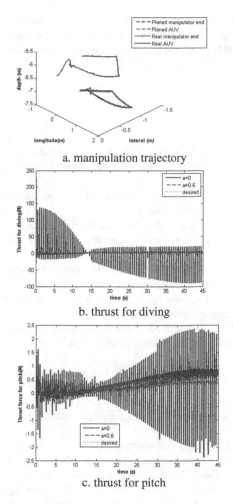

a. manipulation trajectory

b. thrust for diving

c. thrust for pitch

Fig. 9. Simulations for underwater vehicle manipulator system (Color figure online)

5 Conclusions

In order to generate optimal thrust allocation according to control command, this paper has designed a bi-criteria thrust optimal allocation scheme on the basis of horizontal and vertical thruster configuration. Infinity-norm optimization has been combined with 2-norm optimization to construct a bi-criteria primal-dual neural network optimal allocation scheme. When infinity-norm optimal solution is introduced through weighting factor a, the allocation scheme is transformed into bi-criteria optimal scheme. The bi-criteria optimal scheme behaves with better convergence results in compared with infinity-norm or 2-norm scheme in the fault tolerance and path following experiment of open frame remote operated vehicle experiments and UVMS manipulation simulations.

Acknowledgements. This project is supported by National Science Foundation of China under the grants of No. 51209050, No. 5159053, the Doctoral Fund of Ministry of Education for Young Scholar with No. 20122304120003, State Key Laboratory of Robotics and Systems of Harbin Institute of Technology No. SKLRS-2012-ZD-03.

References

1. Corradini, M.L., Monteriù, A., Orlando, G.: An actuator failure tolerant control scheme for an underwater remotely operated vehicle. IEEE Trans. Control Systems Technol. **19**(3), 1036–1046 (2011)
2. Xu, Y.-R., Xiao, K.: Technology development of autonomous ocean vehicle. Acta Automatica Sin. **33**(5), 518–521 (2007)
3. Li, Y., Zhang, L., Wan, L., Liang, X.: Optimization of S-surface controller for autonomous underwater vehicle with immune-genetic algorithm. J. Harbin Inst. Technol. (New Series) **5**(3), 404–410 (2008)
4. Huang, H., Tang, Q., Li, Y., Wan, L., Pang, Y.: Dynamic control and disturbance estimation of 3D path-following for the observation class underwater remotely operated vehicle. Advances in Mechanical Engineering, New Developments in Multibody System Dynamics and Its Applications Special Issue (2013)
5. Sarkar, N., Podder, T.K., Antonelli, G.: Fault-accommodating thruster force allocation of an AUV considering thruster redundancy and saturation. IEEE Trans. Robot. Autom. **18**(2), 223–233 (2002)
6. Podder, T.K., Sarkar, N.: Fault-tolerant control of an autonomous underwater vehicle under thruster redundancy. Robot. Auton. Syst. **34**(1), 39–52 (2001)
7. Daqi, Z., Qian, L., Yongsheng, Y.: An active fault-tolerant control method of unmanned underwater vehicles with continuous and uncertain faults. Int. J. Adv. Robot. Syst. **5**(4), 411–418 (2008)
8. Cuadrado, A., Diaz, I.: Fuzzy inference map for condition monitoring with self-organizing maps. In: International Conference Fuzzy Logic and Technology, Leicester, UK, pp. 55–58
9. Edin, O., Geoff, R.: Fault diagnosis and accommodation for ROVs. In: Sixth IFAC Conference on Manoeuvring and Control Marine Craft, Girona, Spain, pp. 575–588 (2003)
10. Edin, O., Geoff, R.: Thruster fault diagnosis and accommodation for open-frame underwater vehicles. Control Eng. Pract. **12**(2), 1575–1598 (2004)
11. Sarkar, N., Podder, T.K., Antonelli, G.: Fault accommodating thruster force allocation of an AUV considering thruster redundancy and saturation. IEEE Trans. Robot. Autom. **18**(2), 223–231 (2002)
12. Serdar, S., Bradley, J.B., Ron, P.P.: A chattering-free sliding-mode controller for underwater vehicles with fault tolerant infinity-norm thrust allocation. Ocean Eng. **35**, 1647–1659 (2008)
13. Zhang, Y.: A set of nonlinear equations and inequalities arising in robotics and its online solution via a primal neural network. Neuro Comput. **70**, 513–524 (2006)
14. Gravagne, I.A., Walker, I.D.: On the structure of minimum effort solutions with application to kinematic redundancy resolution. IEEE Trans. Robot. Autom. **16**(6), 767–775 (2000)
15. Wai, R.J., Chu, C.C.: Motion control of linear induction motor via petri fuzzy-neural-network. IEEE Trans. Ind. Electron. **54**(1), 281–295 (2007)

Fuzzy Sliding-Mode Formation Control for Multiple Underactuated Autonomous Underwater Vehicles

Hai Huang$^{(\boxtimes)}$, Guo-cheng Zhang, Yue-ming Li, and Ji-yong Li

Key Laboratory of Science and Technology for Autonomous Underwater
Vehicle, Harbin Engineering University, 145 Nantong Street, Harbin, China
haihus@163.com

Abstract. Being a benchmark for autonomous oceanic inspections, long-distance and long-duration surveys, underwater vehicles formation has attracted increasingly attentions on great application potentials. In order to realize formation control for under-actuated AUVs, sliding mode formation control method has been applied on the basis of the AUV's kinematic model. Fuzzy logic system is helpful to maintain formation shape for underactuated AUVs. In the simulations, under-actuated AUV characteristics have been analyzed during the first simulation. In the combination with fuzzy sliding-mode formation control and velocity limits, the controller has successfully realized 3D fold line formation and screw line formation. Therefore the fuzzy sliding-mode formation control strategy for under-actuated AUVs has been verified through the simulations.

Keywords: Formation control · Underwater vehicles · Underactuated

1 Introduction

Multi-robot cooperation is a spontaneous attribute which has drawn significant attentions [1]. Recently since unmanned oceanic vehicles have been systematically researched, cooperative control of multiple Autonomous underwater vehicles (MAUVs) is playing more and more important role in performing underwater tasks, such as ocean sampling, mapping, minesweeping and ocean floor survey [2]. Carrying out MAUVs cooperation missions can increase efficiency and service area, and providing redundancy in case of failure [3]. In compare with one expensive specialized AUV, relatively inexpensive, simple, and small AUVs can solve difficult or complex underwater tasks more quickly. However, Multi-AUVs formation control is interconnected systems with complex dynamics. In compare with wireless communication on the land, acoustic communications has its own limitations such as information propagating with large delays, the speed of sound varies with depth, salinity, the underwater communication channel is very noisy, and transducers for acoustic communications are half duplex.

In the past decade, considerable efforts have been carried out upon the formation control of MAUVs. Various approaches have been proposed, ranging from synchronized path following framework [4], and leader-follower mechanisms [2], to behavioral approach [5]. Since MAUVs formation confronts dynamics uncertainties such as payload

© Springer International Publishing Switzerland 2016
Y. Tan et al. (Eds.): ICSI 2016, Part II, LNCS 9713, pp. 503–510, 2016.
DOI: 10.1007/978-3-319-41009-8_54

variations, hydrodynamics and time-varying oceanic currents, adaptive formation control methods have been suggested to deal with this problem [6]. These approaches allow decentralized implementation and it is natural to deduce control strategies when the robots have multiple competing objects. The algorithms usually required artificial potential trenches to realize the formation scalability [7]. For example, the cross-tracking control was proposed by using potential shaping [8]. Moreover, since most AUVs are underactuated, nonlinear path following control for an underactuated AUV was proposed on path following and formation keeping for underactuated vehicles [9]. However, only low frequency disturbances can be compensated in practical perspective.

On the other hand, sliding mode control is a kind of nonlinear high-speed switched feedback control which is robust to system for uncertainty and external disturbances. Sliding-mode based formation control was investigated for underactuated surface vessels subject to the ocean disturbance [10], where the ocean disturbances are considered in the control design. Particularly, since fuzzy membership can be used to determine the dynamics uncertainties, fuzzy sliding-mode formation controller can be designed as decentralized controllers for individual agents to reach the desired formation in finite time [11]. This paper will make a research on the formation control of underactuated AUV with fuzzy sliding-mode formation controller.

The remainder of the paper is organized as follows: in Sect. 2, the kinematic modeling and control algorithm of the underactuated AUV is presented. In Sect. 3, a fuzzy sliding-mode formation controller is constructed using leader position information. Simulation results are illustrated in Sect. 4. Concluding remarks are given in Sect. 5.

2   Kinematic Modeling and Control of AUV

In order to describe the motion of the ith AUV, vehicle frame $O_v X_v Y_v Z_v$ and navigation frame $O_o X_o Y_o Z_o$ have been used. Thus, the kinematic model of under-actuated AUV is described as:

$$\begin{cases} \dot{x}_i = v_i^H \cos \theta_i \\ \dot{y}_i = v_i^H \sin \theta_i \\ \dot{z}_i = v_i \sin \psi_i \\ \dot{\theta}_i = r_i \\ \dot{\psi}_i = w_i \end{cases} \tag{1}$$

where $q_i = [x_i, y_i, z_i, \theta_i, \psi_i, \phi_i]^T$ represents the pose of the ith AUV v_i is the horizontal velocity of the i-th AUV, $v_i^H = v_i \cos \psi_i$, ψ_i is the pitch angle of the i-th AUV, θ_i is the heading angle of the i-th AUV.

In order to realize autonomous cruising and formation, the S surface controller is applied as the motion controller of AUV [12] (Fig. 1).

$$T_i = \frac{2I}{I + \exp(-K_{cd}\tilde{q}_i - K_{cp}\dot{\tilde{q}}_i)} - I, \tag{2}$$

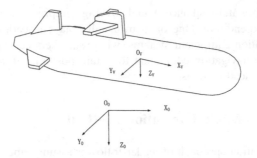

Fig. 1. The navigation frame and body frame of AUV

where, $\mathbf{q}_{id}$ represents the desired AUV position and attitude states, q_i represents the real AUV, $\tilde{q}_i = q_{di} - q_i$ and $\underline{q}_i$ represent the input error and rate error change in normalized form, T_i is the output of the motion controller, which stands for the force in each degree of freedom, K_{cd} and K_{cp} are the proportional and derivative gain matrices, $I = \begin{bmatrix} 1 & 1 & \dots & 1 \end{bmatrix}^T$ is the identity matrix with dimension of the manipulator or AUV DOFs. For the sigmoid function in Fig. 2, the controller commands are loosely considered when the deviation is comparatively large, while strictly treated when the deviation is comparatively small. Thus the kinematic controller indicates the idea of fuzzy control to a certain extent.

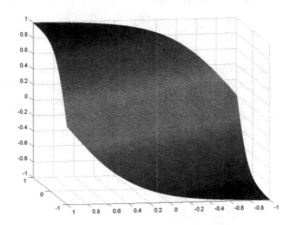

Fig. 2. Sigmoid-function

In order to realize long distance cruising, AUV is usually under-actuated with one thruster for propulsion, whereas, rudders and wings for heading and diving, respectively. Hence formation plan result will provide expected linear velocity, yaw and depth for motion controller of each AUV. Underwater navigation system is responsible for position reckoning through pose, velocity and acceleration perception. Position state is important for AUV motion and formation control. On the realization of multiple AUVs formation, heterogeneous underwater vehicles are usually employed, particularly in case of leader-follower format, to enhance spatial coverage, improve operation accuracy and reduce the entire expenditure. One of them usually plays a leading role with higher performance navigation equipment systems, whereas, others equipped with comparatively lower precise navigation equipment to obtain position information via relative positioning from the leading ones.

3   Fuzzy Sliding-Mode Formation Method

Generally, the formation approach of leader-follower assumes the leader AUV moves along a planned trajectory, while the followers maintain a desired distance and orientation from it. One can replace the virtual reference AUV with a physical AUV acting as the leader. Hence the leader-follower formation approach has been transformed into a dynamic trajectory tracking approach. According to the kinematics model of Eq. (1), the tracking position error between the virtual reference AUV and the leader AUV with respect to the navigation frame is expressed as:

$$
e_i = \begin{bmatrix} e_{i1} \\ e_{i2} \\ e_{i3} \end{bmatrix} = \begin{bmatrix} c\psi_i c\theta_i & c\psi_i s\theta_i s\phi_i - s\theta_i c\phi_i & c\psi_i s\theta_i c\phi_i + s\theta_i s\phi_i \\ s\psi_i c\theta_i & s\psi_i s\theta_i \sin\phi_i + c\theta_i c\phi_i & s\psi_i s\theta_i c\phi_i - c\theta_i s\phi_i \\ -s\psi_i & c\psi_i s\phi_i & c\psi_i c\phi_i \end{bmatrix} \begin{bmatrix} x_r - x_i \\ y_r - y_i \\ z_r - z_i \end{bmatrix}
$$
$$
= \begin{bmatrix} c\psi_i c\theta_i & c\psi_i s\theta_i s\phi_i - s\theta_i c\phi_i & c\psi_i s\theta_i c\phi_i + s\theta_i s\phi_i \\ s\psi_i c\theta_i & s\psi_i s\theta_i \sin\phi_i + c\theta_i c\phi_i & s\psi_i s\theta_i c\phi_i - c\theta_i s\phi_i \\ -s\psi_i & c\psi_i s\phi_i & c\psi_i c\phi_i \end{bmatrix} \begin{bmatrix} e_{ix} \\ e_{iy} \\ e_{iz} \end{bmatrix}
$$
(3)

where e_{ix}, e_{ix} and e_{iz} denote the trajectory-tracking errors between the leader AUV and virtual reference AUV. The coordinates must be integrated as a distance variable so that the number of velocity control inputs is the same as the number of position outputs. On the basis of the Lyapunov stability theory, control inputs concerning speed, yaw and depth, are designed to stabilize the leader AUV through a sliding mode fuzzy tracking controller. Thus we define the sliding surface as:

$$
\begin{cases} s_{i1}(t) = k_{p1} e_{id}(t) + k_{i1} \int e_{id}(t)dt \\ s_{i2}(t) = k_{p2} e_{i\theta}(t) + k_{i2} \int e_{i\theta}(t)dt \\ s_{i3}(t) = k_{p3} e_{iz}(t) + k_{i3} \int e_{iz}(t)dt \end{cases}
$$
(4)

where $e_{id}^2(t) = e_{i1}^2(t) + e_{i2}^2(t)$, $e_{i\theta} = \theta_{ir} - \theta_i$, k_{p1}, k_{p2}, k_{p3}, k_{i1}, k_{i2}, k_{i3} are all positive constants, so that the three sliding surfaces are Hurwitz. If we set $\varsigma_1 = 2k_{p1}e_{i1} \{-c_{i1}\cos\theta_i - c_{i2}\sin\theta_i - c_{i3}(e_{ix}\sin\theta_i + e_{iy}\cos\theta_i)\} + 2k_{p1}e_{i2}\{-c_{i1}\sin\theta_i - c_{i2}\cos\theta_i - c_{i3}(e_{ix}\cos\theta_i + e_{iy}\sin\theta_i)\}$, $\varsigma_2 = -k_{p2}c_{i3}$ and $\varsigma_3 = -k_{p3}c_{i4}$ denotes the terms resulting from the additional uncertainties, the derivatives of three sliding surfaces are expressed as:

$$\begin{cases} \dot{s}_{i1}(t) = k_{p1}\dot{e}_{id}(t) + k_{i1}e_{id}(t) + \varsigma_1 \\ \dot{s}_{i2}(t) = k_{p2}\dot{e}_{i\theta}(t) + k_{i2}e_{i\theta}(t) + \varsigma_2 \\ \dot{s}_{i3}(t) = k_{p3}\dot{e}_{iz}(t) + k_{i3}e_{iz}(t) + \varsigma_3 \end{cases} \tag{5}$$

Since fuzzy systems theory can provide a systematic procedure in order to transform a set of linguistic rules into a nonlinear mapping. The fuzzy logic subsystem for sliding mode fuzzy formation controller is denoted in Fig. 3, which performs a mapping from $X_i \in \Re^2$ to $\Re$. The fuzzy control rules are described as [13]:

$$\text{IF } \bar{s}_i(t) \text{ is } F_{i1}^k \text{ and } \dot{\bar{s}}_i(t) \text{ is } F_{i2}^k, \text{ THEN } u_i(t) \text{ is } G_i^k. \tag{6}$$

where $x_i(t) = \begin{bmatrix} \bar{s}_i(t) & \dot{\bar{s}}_i(t) \end{bmatrix}^T \in X_i \subset \Re^2$, $\bar{s}_i(t) = g_{si}s_i(t)$, $\dot{\bar{s}}_i(t) = \dot{g}_{si}\dot{s}_i(t)$ are the inputs, $\bar{u}_i(t) \in V_i \subset \Re$ are the outputs of the fuzzy logic subsystem respectively. F_{ij}^k, G_i^k are the labels of sets in X_i and V_i respectively. The parameters are chosen so that $\bar{s}_i(t)$ and $\dot{\bar{s}}_i(t) \in [-1, 1]$. On the basis of fuzzy IF-THEN rules in the fuzzy rule base, the fuzzy inference engine performs a mapping from fuzzy sets in $X_i \subset \Re^2$ to fuzzy sets in $V_i \subset \Re$.

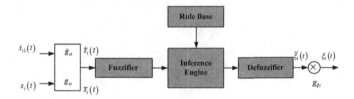

Fig. 3. Fuzzy logic system

Then the outputs of sliding mode fuzzy formation controller are the control inputs i.e. $\xi_1 = v_i$, $\xi_2 = \theta_i$ and $\xi_3 = w_i$:

$$\begin{cases} \xi_1(t) = g_{\xi 1}\bar{\xi}_1(t) = g_{u1}[\bar{s}_1 + \Delta_1 \text{sgn}(\bar{s}_1)] \times \text{sgn}(\bar{b}_1(e_{i1})) \\ \xi_2(t) = g_{\xi 2}\bar{\xi}_2(t) = g_{u2}[\bar{s}_2 + \Delta_2 \text{sgn}(\bar{s}_2)] \times \text{sgn}(\bar{b}_2(e_{i2}, s_1, s_2)) \\ \xi_3(t) = g_{\xi 3}\bar{\xi}_3(t) = g_{u3}[\bar{s}_3 + \Delta_3 \text{sgn}(\bar{s}_3)] \times \text{sgn}(\bar{b}_3(e_{i3})) \end{cases} \tag{7}$$

where $\bar{\xi}_i(t)$ is the fuzzy variable of $\xi_i(t)$, $\Delta_i > 0$, $\bar{b}_1(e_{i1}) = e_{i1}$, $\bar{b}_3(e_{i3}) = e_{i3}$, $\bar{b}_1(e_{i2}, s_1, s_2) = 2s_1k_{p1}e_{i2}d + s_2k_{p2}$.

On the other hand the distance between other AUVs should not be neglected during formation control. Thus the velocity input limit is introduced as:

$$\begin{cases} \xi_1 \leq v_{imax} - (v_{imax} - v_{id})e^{-\left(d_{si}^2/\sigma^2\right)} \\ \xi_2 \leq r_{si} + k\dot{\theta} \\ \xi_3 \leq w_{si} + k\dot{\psi} \end{cases} \tag{8}$$

where v_{imax} is the maximum linear speed of the ith AUV, v_{id} is d_{si} is the distance between the ith AUV and its expected position, σ and k are positive constant.

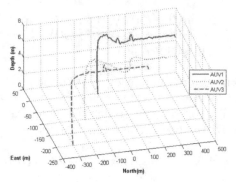

(a) Formation follow the round curve line with depth change

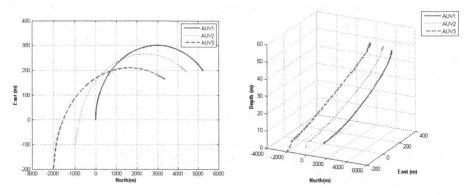

(b) Formation follow the round curve line with depth change

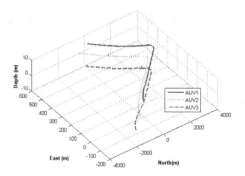

(c) Formation follow the folding line

Fig. 4. Formation control simulations

4 Simulations

Different from other unmanned vehicles, diving and heading control of underactuated AUV must be completed during advancing process. In order to make an analysis, 3 AUVs set out from initial positions respectively, while keeping formation, made a 6 meters dive and 90 degree yaw, then moved along the north direction. Since different AUV with different size, radius of gyration and maneuverability, this process includes changes of relative distance and coupling effects of diving and yawing. In order to reduce these effects, Fuzzy control is employed for the formation control in order to stabilize the fluctuations. In the Fig. 4, these three AUVs were planned to follow round curve and folding line in formation. The followers are planned to maintain the distance between the leader and other AUVs. For the second scenario, vehicles were given linear paths. For the third scenario vehicles were given polygon path to test the performance of regional coverage. Although with different sizes, radii of gyration and manoeuvrabilities, underwactuated AUVs can keep formation while cruising under the strategies proposed in this study.

5 Conclusions

Underwater vehicles formation is of great significance for oceanic inspections, long-distance and long-duration surveys. On the basis of the AUV's kinematic model and basis motion controller, sliding mode formation control method has been applied for multi-AUV formation. In order to maintain the planned formation shape, fuzzy logic system is combined with the sliding mode formation controller on the attitude change during underactuated cruising. In the combination with fuzzy sliding-mode formation control and velocity limits, the controller has successfully realized 3D fold line formation and screw line formation. Therefore the fuzzy sliding-mode formation control strategy for under-actuated AUVs has been verified through simulation.

Acknowledgements. This project is supported by National Science Foundation of China under the grants of No. 51209050, No. 5159053, the Doctoral Fund of Ministry of Education for Young Scholar with No. 20122304120003, State Key Laboratory of Robotics and Systems of Harbin Institute of Technology No. SKLRS-2012-ZD-03.

References

1. Jadbabaie, A., Lin, J., Morse, A.S.: Coordination of groups of mobile autonomous agents using nearest neighbor rules. IEEE Trans. Autom. Control **48**(6), 988–1001 (2003)
2. Cui, R., Ge, S.S., How, B.V.E., Choo, Y.S.: Leader–follower formation control of underactuated autonomous underwater vehicles. Ocean Eng. **37**, 1491–1502 (2010)
3. Kyrkjebo, E., Pettersen, K.: A virtual vehicle approach to output synchronization control. In: Proceeding of 45th IEEE Conference on Decision and Control, San Diego, CA, vol. 1, pp. 6016–6021 (2006)

4. Ihle, I., Arcak, F.M., Fossen, T.I.: Passivity-based designs for synchronized path following. Automatica **43**(9), 1508–1518 (2007)
5. Arrichiello F, Chiaverini S, Fossen, TI.: Formation control of underactuated surface vessels using the null-space-based behavioral control. In: International Conference on Intelligent Robots and Systems, pp. 5942–5947 (2006)
6. Hou, S.P., Cheah, C.C.: Can a simple control scheme work for a formation control of multiple autonomous underwater vehicles. IEEE Trans. Control Syst. Technol. **19**(5), 1090–1101 (2011)
7. Cheah, C.C., Hou, S.P., Slotine, J.J.E.: Region-based shape control for a swarm of robots. Automatica **45**(10), 2406–2411 (2009)
8. Woolsey, C., Techy, L.: Cross-track control of a slender, underactuated AUV using potential shaping. Ocean Eng. **36**(1), 82–91 (2009)
9. Lapierre, L., Soetanto, D.: Nonlinear path-following control of an AUV. Ocean Eng. **34**(11–12), 1734–1744 (2007)
10. Fahimi, F.: Sliding-mode formation control for underactuated surface vessels. IEEE Trans. Rob. **23**(3), 617–622 (2007)
11. Ghasemi, M., Nersesov, S.G.: Sliding mode cooperative control for multirobot systems a finite-time approach. Math. Probl. Eng. **2013**, 1–16 (2013)
12. Li, Y., Zhang, L., Wan, L., Liang, X.: Optimization of s-surface controller for autonomous underwatervehicle with immune-genetic algorithm. J. Harbin Inst. Technol. (New Ser.) **15**(3), 404–410 (2008)
13. Peng, P., Wang, D., Wang, H., Wang, W.: Coordinated formation pattern control of multiple marine surface vehicles with model uncertainty and time-varying ocean currents. Neural Comput. Appl. **25**, 1771–1783 (2014)

Temporarily Distributed Hierarchy in Unmanned Vehicles Swarms

Hong-an Yang$^{(\boxtimes)}$, Luis Carlos Velasco, Ya Zhang, Ting Zhang,
and Jingguo Wang

School of Mechanical Engineering,
Northwestern Polytechnical University, Xi'an, China
yhongan@nwpu.edu.cn

Abstract. There is an increasing number of situations where a group of unmanned vehicles, instead of only one, would optimize efficiency in several military and civilian applications. One of the main limitations with current technology is the fact that the number of unmanned vehicles under control is directly linked with the number of operators controlling each one of them. In this paper the authors propose a system, integrating the social hierarchical behavior of a wolf pack, aiming to obtain an algorithm with hybrid and dynamic features designed to make possible the effective control of a swarm of unmanned vehicles. This system temporarily distributes levels of hierarchy among the members of the drone swarm, which makes possible to control any member of the swarm at any given time. The experimental results in a testbed with three mobile robots prove the efficiency of the proposed system.

Keywords: Unmanned vehicles · Swarm intelligence · Swarm robotics · Distributed hierarchy · Wolf pack algorithm

1 Introduction

An unmanned vehicle (UV) is defined as a mobile robot used to increase the capabilities of human beings. This type of robot usually operates in open environments and on a variety of scenarios performing tasks that are usually done by humans. UVs are controlled by human operators positioned at a remote location using wireless communication links. Operators grant all cognitive processes, which are based on a sensory feedback provided by remote vision systems and remotely sensed data.

Unmanned vehicles are used in both civilian and military operations, these being hostile or difficult for human presence. Many of these tasks can increase response times by assigning the same mission to multiple agents with synchronized collective routines participating under the same purpose. This refers to the concept of "cooperative work", understood in a broad sense as conducting a coordinated action of several participants engaged in a given task. Cooperative work is done by a cooperative system composed of cooperators and a goal.

An alternative to cooperative work is swarm robotics, which is a new approach to the coordination of multi robot systems, mostly composed of large numbers of physically simple robots. It is known that a desired collective behavior emerges from the

© Springer International Publishing Switzerland 2016
Y. Tan et al. (Eds.): ICSI 2016, Part II, LNCS 9713, pp. 511–518, 2016.
DOI: 10.1007/978-3-319-41009-8_55

interactions between robots and the interactions between them with the environment. This concept arose from biological studies of bees, ants and other areas of nature experiencing social behaviors, where swarm intelligence emerges. This paper proposes the utilization of these swarm behavior concepts as an effective method to achieve an active control of an undefined number of UVs.

2   Temporarily Distributed Hierarchy

Recently there have been studies looking for alternatives to robot collective work in tasks of the same purpose. These studies serve as a platform to research new collective operation algorithms in situations where control of multiple UVs is required.

In the past, people discovered the variety of the interesting insect or animal behaviors in the nature. A flock of birds sweeps across the sky. A group of ants forages for food. A school of fish swims, turns, flees together, a pack of wolves hunting in synchrony, etc. [1]. We call this kind of cooperative motion "swarm behavior". Recently biologists, and computer scientists in the field of "artificial life" have studied how to model biological swarms to understand how such "social animals" interact, achieve goals, and evolve. Moreover, engineers are increasingly interested in this kind of swarm behavior since the resulting "swarm intelligence" can be applied in optimization [2], robotics [3, 4], traffic patterns in transportation systems, and military applications [5].

2.1   Swarm Intelligence

A high-level view of a swarm suggests that the N agents in the swarm are cooperating to achieve some purposeful behavior and achieve some goal. This apparent "collective intelligence" seems to emerge from what are often large groups of relatively simple agents. The agents use simple local rules to govern their actions and via the interactions of the entire group, the swarm achieves its objectives. A type of "self-organization" emerges from the collection of actions of the group. Swarm Intelligence is a relatively new branch of Artificial Intelligence that is used to model the collective behavior of social swarms in nature, such as ant colonies, honey bees, and bird flocks.

Although these agents (insects or swarm individuals) are relatively unsophisticated with limited capabilities on their own, they are interacting together with certain behavioral patterns to cooperatively achieve tasks necessary for their survival. The social interactions among swarm individuals can be either direct or indirect. Examples of direct interaction are through visual or audio contact, such as the waggle dance of honey bees. Indirect interaction occurs when one individual changes the environment and the other individuals respond to the new environment, such as the pheromone trails of ants that they deposit on their way to search for food sources. This indirect type of interaction is referred to as stigmergy [6], which essentially means communication through the environment.

2.2   Wolf Pack Social Behavior

Wolves are gregarious animals and have clearly social work division. There is a lead wolf, some elite wolves that act as scouts and some general behavior wolves in a wolf pack. They cooperate well with each other and take their respective responsibility for the survival and thriving of the wolf pack [7].

Firstly, the lead wolf, as a leader under the law of the jungle, is always the smartest and most ferocious one. It is responsible for commanding the wolves and constantly making decisions by evaluating surrounding situations and perceiving information from other wolves. This lead wolf can avoid dangerous situations for the wolf pack and can also command the wolves to smoothly capture a prey as soon as possible.

Secondly, the lead wolf sends some elite wolves to look around for prey and hunt in the probable scope. Those elite wolves are scouts. They walk around and independently make decisions according to the concentration of smell left by prey, where a higher concentration means the prey is closer to the wolves. So they always move towards the direction of getting stronger smell.

Thirdly, once a scout wolf finds the trace of prey, it will howl and report that to the lead wolf. Then the lead wolf will evaluate this situation and make a decision whether to summon the rest of the wolf pack to round up the prey or not. If they are summoned, the other members of the pack will move fast towards the direction of the scout wolf.

Fourthly, after capturing the prey, the prey is not distributed equitably, but in an order from the strongest to the weakest. That is to say that, the stronger the wolf is, the more food it will get. Even though this distribution rule will make some weak wolf die for lack of food, it makes sure that the wolves that have the ability to capture prey get more food, keeping themselves strong and able to capture more prey in the next hunt. This hierarchical behavior avoids that the whole pack starves to death, ensuring its continuance and proliferation [8].

2.3   Temporarily Distributed Hierarchy Paradigm and Its Three Castes

Since the most fundamental characteristic of an unmanned vehicle is the fact that a human operator is the main part of the decision-making loop in any given task, the authors consider this as the base and initial point in the design of the swarm behavior routines oriented to control a group of these robots.

The operator can select and take control of any UV member of the swarm according to the requirements of the active task. The selected UV assumes the highest role among the members of the swarm due to its direct link with the human operator, and it's able to direct orders to the other UVs in the group. The term "Temporarily Distributed Hierarchy" is used to express that this selection of a main coordinator UV is not a permanent decision due to the fact that every member of the swarm is selectable to play this role inside the group, or constantly vary this selection between all the available UVs according to the operator needs.

This Temporarily Distributed Hierarchy paradigm rules the swarm configuration parameters, because it defines three strongly differentiated groups between the members of the UV swarm. This algorithm is based on the social hierarchical behavior of a wolf pack, where these groups are known as "castes", each one of them with different functions. Inside the proposed system these castes are defined as follows:

(1) **Alpha caste:** Only one UV at a time can be selected as an Alpha due to the fact that this drone is the only member of the swarm with a direct link between himself and the operator. This implies that the Alpha drone gets its behavior modifiers directly and exclusively from the operator. At the same time the Alpha drone delivers behavior modifiers to all beta drones in the swarm. See Fig. 1.

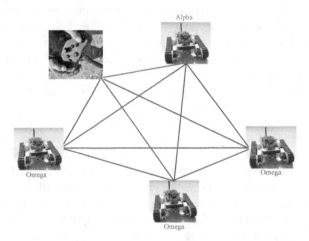

Fig. 1. Alpha Caste: any Omega drone can become the main coordinator of the UV swarm. The green lines indicate the network coverage and all the possible communication links within the swarm. The red line shows the human-alpha drone direct link. (Color figure online)

(2) **Beta caste:** Beta drones get their behavior modifiers from the Alpha drone. This caste can be formed with several UVs from the swarm, and they have a direct communication link with the Alpha drone. Beta drones behavior can be semi-autonomous or tele-operated depending on the task requirements. See Fig. 2.

(3) **Omega caste:** Omega drones wait for behavior modifiers. Any UV that is not being directly controlled by a human operator or the Alpha drone, is considered as member of the Omega caste. The operator can select any Omega drone to assume the Alpha role inside the swarm. Most of the time Omega drones are either on idle state or navigating using semi-autonomous algorithms. See Fig. 3.

3 Algorithms Description

In this section the authors describe the algorithm used in the design of the proposed system in order to make them independent from any programming language and just give a general scope of how they work.

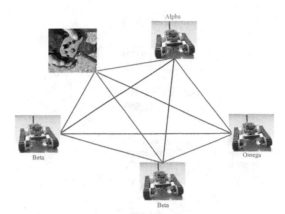

Fig. 2. Beta Caste: Beta drones get their behavior modifiers from the Alpha drone. The green lines indicate the network coverage and all the possible communication links within the swarm. The red line shows the human-alpha drone direct link and the blue lines show the alpha-beta drones link. (Color figure online)

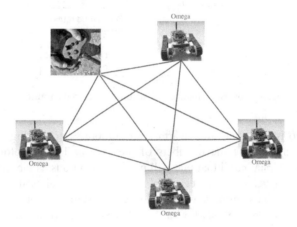

Fig. 3. Omega Caste: Omega drones are on idle state waiting for behavior modifiers. The green lines indicate the network coverage and all the possible communication links within the swarm. (Color figure online)

Firstly, according to the characteristics of the situation, the operator can decide if only one unmanned vehicle will be required or to control more than one drone if the complexity of the situation requires this. If this is not the case, then the operator can proceed to control a single unmanned vehicle with the regular tele-operated control behavior modifiers. Considering that the analyzed situation requires the participation of more than one UV involved in pursuing a common objective, the operator makes use of the proposed swarm behavior routines inspired by a wolf pack [7], like has been previously detailed in the above Sect. 2.2.

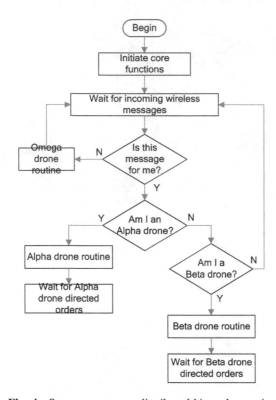

Fig. 4. Swarm temporary distributed hierarchy routine

The next step is to initiate the swarm configuration, establishing the temporarily distributed hierarchy among the members of the swarm. The operator first selects a main drone as the Alpha individual that will perform functions as the main coordinator of the group. Whenever a UV detects that the operator requires this to become the Alpha drone inside the swarm, the proposed system starts a routine configuring the communications topology and starting a direct transference of tele-operated behavior modifiers from the operator to the Alpha UV. If the operator needs to control another drone without closing the direct link with the Alpha individual, this is possible selecting other UVs as Beta drones. In this situation the UVs that identify the Beta drone selection protocol, initiate a routine where they are expecting behavior modifiers coming from the operator through the main coordinator, which is the Alpha drone. See Fig. 4.

We decided to focus only on tele-operated behavior algorithms, due to the fact that the final objective is to control a swarm of unmanned vehicles where a human operator is the main director in the decisions making loop. Nevertheless, limited autonomous behaviors of the non-communicative swarm type are included to show that whenever the situation requires autonomous or semi-autonomous operations, it's just a matter of programming specific routines for the UVs swarm objective requirements. These limited autonomous behaviors included in the demonstration of the proposed system is

Clustering, where all the members of the swarm, like the name suggests, cluster and eventually get all together in one position.

4 Testbed Robot Swarm and Experimental Results

In order to validate the expected results of the proposed algorithms, We designed and built 3 mobile robots to replicate the behavior of a UV Swarm. These 3 robots have the same electronic and mechanical configurations, the same control system and the same algorithms running on each one of them. Their mechanical system is composed by a differential drive configuration installed on 2 caterpillar tracks.

The electronic system of each robot includes a motor driver that controls the speed and direction of rotation of each caterpillar track, the main power source is a 9 V battery pack that feeds the electromechanical and control systems. The control system of each robot is run by an Arduino Micro microcontroller. This microcontroller processes and generates all the digital and analogue signals required by the proposed system. The experimental tests were done using the testbed robot swarm, uploading the same algorithms to each one of the robots and experimenting with the interactions between the members of the swarm.

Figure 5 shows the clustering routine, where all the members of the swarm are separated from each other, after the setup of this initial position the operator sends the

Fig. 5. Swarm clustering routine

Cluster command to all the robots and these start looking for any robots nearby. Eventually locating all the other members of the swarm showing an aggregation behavior which is just a limited example of the swarm intelligence autonomous algorithms that could be programmed on any robot swarm, depending on the requirements of the given objective or situation.

5 Conclusion

There are several situations where controlling only one unmanned vehicle is not enough, and integrating more drones into the same situation following the available protocols would also require the integration of more human operators in the control loop, due to the fact that the number of active drones in a multi-robot system is directly proportional to the number of operators. This directly derives in an exponential increase of the operational costs, since more training for operators would be required.

The proposed system deals with this situation, enabling one single operator to control a group of N unmanned vehicles according to the requirements of the mission. This system uses a swarm intelligence approach, looking to ensure the system's robustness, scalability and low complexity level. The adapted wolf pack behavior successfully contributed with flexibility and scalability to the system, where the proposed castes protocol made possible to control any member of the swarm at any given time. The experimental interaction with the testbed swarm showed satisfactory results and accomplished every objective that was planned for this paper. Therefore, the authors suggest further experimentation with a larger number of robots and different kinds of mobility systems.

Acknowledgments. This research has been supported by the graduate starting seed fund of Northwestern Polytechnical University (Grant No. Z2016083).

References

1. Vicsek, T., Zafeiris, A.: Collective motion. Phys. Rep. **517**, 71–140 (2012)
2. Abidin, Z.Z., Arshad, M.R., Ngah, U.K.: An introduction to swarming robotics: application development trends. Artif. Intell. Rev. **43**, 501–514 (2015)
3. Brambilla, M., Ferrante, E., Birattari, M., Dorigo, M.: Swarm robotics a review from the swarm engineering perspective. Swarm Intell. **7**(1), 1–41 (2013)
4. Barca, J.C., Sekercioglu, Y.A.: Swarm robotics reviewed. Robotica **31**, 345–359 (2013)
5. Bonabeau, E., Dorigo, M., Theraulaz, G.: Swarm Intelligence: From Natural to Artificial Systems. Oxford University Press, Oxford (1999)
6. Holland, O., Melhuish, C.: Stigmergy, self-organization and sorting in collective robotics. Artif. Life **5**(2), 173–202 (1999)
7. Mech, D.L., Boitani, L.: Wolves behavior, ecology and conservation. Am. Fish. Soc. Symp. **1999**(24), 283–284 (2003)
8. Mech, L.D.: Possible use of foresight, understanding, and planning by Wolves Hunting Muskoxen. Arctic **60**(2), 145–149 (2007)

Multi-goal Motion Planning of an Autonomous Robot in Unknown Environments by an Ant Colony Optimization Approach

Chaomin Luo[1(✉)], Hongwei Mo[2], Furao Shen[3], and Wenbing Zhao[4]

[1] Advanced Mobility Lab, Department of Electrical and Computer Engineering,
University of Detroit Mercy, Detroit, MI, USA
luoch@udmercy.edu
[2] Harbin Engineering University, Harbin, China
[3] Nanjing University, Nanjing, China
[4] Cleveland State University, Cleveland, OH, USA

Abstract. An ant colony optimization (ACO) approach is proposed in this paper for real-time concurrent map building and navigation for *multiple goals* purpose. In real world applications such as rescue robots, and service robots, an autonomous mobile robot needs to reach multiple goals with the shortest path that, in this paper, is capable of being implemented by an ACO method with minimized overall distance. Once a global path is planned, a foraging-enabled trail is created to guide the robot to the multiple goals. A histogram-based local navigation algorithm is employed locally for obstacle avoidance along the trail planned by the global path planner. A re-planning-based algorithm aims to generate path while a mobile robot explores through a terrain with map building in *unknown* environments. In this paper, simulation results demonstrate that the real-time concurrent mapping and multi-goal navigation of an autonomous robot is successfully performed under *unknown* environments.

Keywords: Ant colony optimization · Multi-goal motion planning · Autonomous robot · Mapping · Unknown environments

1 Introduction

Motion planning and mapping of autonomous robots is an important issue. Concurrent motion planning and mapping of robotics is to search a suitable collision-free path of an autonomous robot to move from an initial position to a goal designation while the robot builds up a map, in an unknown environment. Research on such point-to-point robot motion planning has been well carried out recently, indeed, which is a sort of single-goal navigation. In real-time world applications, such as rescue robots, service robots, mining rescue robots, and mining searching robots, etc., multi-goal motion planning of autonomous robots is highly desirable. In addition to the service robots, many other robotic applications require multi-goal navigation such as vacuum robots, land mine detectors, lawn mowers, and windows cleaners. The multi-goal path planning aims to search a collision-free path for visiting a sequence of goals with the minimized total route under unknown environments.

Y. Tan et al. (Eds.): ICSI 2016, Part II, LNCS 9713, pp. 519–527, 2016.
DOI: 10.1007/978-3-319-41009-8_56

There have been a large number approaches proposed on autonomous robot motion planning [1–7]. Gu *et al.* [1] proposed an elastic-band-based motion planning model, in which tunability and stability are focused on the robot navigation. Raja *et al.* [2] suggested a gradient function in the conventional potential field method associated with a Genetic Algorithms (GA) based method for robot motion planning. Besides an attractive force, a repulsive force, a tangential force and a gradient force are introduced in the conventional potential field method integrated with a GA path planning model. Davies and Jnifene [3] developed a GA path planner to guide an autonomous robot to reach specified multiple goals with obstacle avoidance. However, a couple of artificial waypoints have to be necessarily predefined to assist in preventing the robot from the deadlock or escaping from the local minima, which causes more timing delay, more length and more cost.

Luo and Yang [4, 5] developed a bio-inspired neural network model that concurrently performs mapping and path planning tasks. Yang and Meng [6] proposed a biologically inspired neural network approach to real-time collision-free motion planning of mobile robots in a non-stationary environment. However, multi-goal navigation studies have not been performed largely for intelligent robot systems. Faigl and Macak [7] developed a self-organizing map method in conjunction with an artificial potential field based navigation function to generate an optimal path of a mobile robot to visit multiple goals, through a Traveling Salesman Problem (TSP) tool. The autonomous robot visits multiple targets just as the traveling salesman problem but with the presence of obstacles, in which mobile robots are navigated with the discrete map representation filled with some exact cell in workspace. Although a multi-goal navigation model of an autonomous robot modeled as a point robot by ACO is implemented by Gopalakrishnan and Ramakrishnan [8], the model lacks of map building component, neither.

In this paper, a real-time concurrent multi-goal motion planning and map building approach of an autonomous robot in *completely unknown environments* is proposed. A local map composed of cells is dynamically built through the histogram-driven local navigator with restricted sensory information such as LIDAR-based data while it is navigated under completely unknown environments with obstacle avoidance to visit multiple requested goals. The sensory information obtained by onboard sensors mounted on the robot is utilized for obstacle avoidance in unknown sceneries.

2 Histogram-Based Local Navigator and Cell-Based Map Building

The navigation system consists of two layers, one is a D*-Lite global path planner with re-planning function and the other is a histogram-based local navigator. Flexibility and efficiency motivate D*-Lite to be adapted to multi-goal path planning. D*-Lite is an incremental heuristic search algorithm extended from A* algorithm by re-utilizing previous search effort in subsequent search iterations for efficient re-planning [9]. A* utilizes a best-first search from a starting point **S**start to the goal **S**goal guided by the heuristic h thus it is able to search an efficiently traversable path between points commonly employed for robotics path planning in known 2D gird-based maps [10].

In this paper, the D*-Lite is employed to generate the global trajectory of an autonomous robot under unknown environments.

The local navigator aims to generate velocity commands for the autonomous mobile robot to move towards a goal. The inclusion of a sequence of markers in the motion planning, which decomposes the global trajectory into a sequence of segments, makes the model especially efficient for the workspace densely populated by obstacles. Ulrich and Borenstein [11] first successfully proposed a Vector Field Histogram (VFH) methodology for navigation. The Virtual Force Field (VFF) approach was initially inspired by potential field method in conjunction with the concept of certainty cells [11, 12]. In this paper, VFH is utilized as our LIDAR-based local navigator. 2D cell-based map filled with cells [12], which are marked as either occupied or free, is built as the mobile robot moves. It is especially beneficial for autonomous robots to perform robust multi-goal navigation in unknown terrains, given the fact that it facilitates the utilization of path planning algorithms to determine the optimal trajectory among waypoints as multiple goals.

3   ACO-Based Multi-goal Visit

In real world applications of multi-goal navigation and mapping, multiple goals as a sequence of waypoints and a relative cost for travelling between each goal to each other are provided. The objective is to search a route through all the waypoints in which all waypoints are visited once, and to find the shortest overall tour. In rescue robot application, for instance, the robot starts at one designated waypoint, visits each other waypoint and then ends at the initial waypoint. Traveling Salesman Problem (TSP) is an optimization problem to minimize the travelling distance in a finite number of cities while the cost of travel between each city is known. The classic TSP is employed to deal with the multi-goal visit problem, in which a sequence of goals is visited so that the total planned length of the route is minimized. The objective of this TSP with regard to multi-goal navigation is to search an ordered set of all the waypoints for the autonomous robot to visit at such that the cost is minimized. A list of waypoints and distances, or cost, between each of them is necessary for TSP. Therefore, goals are called waypoints with GPS coordinates of the goals in latitude and longitude. The multi-goal visit problem has practical applications on some fields such as transportation plan issues where deliveries are required as well as fuel cost and time are to be minimized.

In this paper, ant colony optimization (ACO) is utilized to resolve the TSP for the multi-goal navigation with multiple waypoints [13] programmed in MATLAB. Ants in ACO are agents in the TSP, which traverse from one waypoint to another waypoint navigated by pheromone trails and an a priori available heuristic information. Ant pheromone strength $\tau_{ij}(t)$, a numerical information, defined with each arc (i, j) is updated in the ACO algorithm for TSP, where t is the iteration counter. The agent is initially placed in a waypoint. At each iteration step, a probabilistic action choice rule is applied to an agent, hereafter, a mobile robot, k. The probability of a robot k, currently at waypoint i, which moves to waypoint j at the tth iteration of the algorithm, is obtained as follows.

$$p_{ij}^k(t) = \frac{[\tau_{ij}(t)]^\alpha \times [\vartheta_{ij}]^\beta}{\sum_{l \in \aleph_i^k}[\tau_{il}(t)]^\alpha \times [\vartheta_{il}]^\beta} \text{ if } j \in \aleph_i^k \tag{1}$$

where, $\aleph_i^k$ is the feasible adjacent waypoint of robot k, the set of cities which the robot k has not visited yet. Parameters α and β determine the relative influence of the pheromone trail and the heuristic information. $\vartheta_{ij} = 1/d_{ij}$ is an a *priori* available heuristic value, and d_{ij} is the distance between two waypoints. Parameter α represents importance factor of the pheromone, which matches a classical stochastic greedy algorithm. Parameter β is an importance factor of the heuristics function. If the larger parameter β becomes, the more likely it is that the robot moves to the closest waypoint driven by the heuristic function. If a parameter ρ is defined as the pheromone trail evaporation, $0 < \rho < 1$ to prevent the pheromone trails from accumulating unlimitedly; it is able to allow the ACO algorithm to ignore unreasonably bad decisions previously made.

At each iteration step, $\Delta\tau_{ij}^k(t)$, the amount of pheromone robot k places on the arcs it has visited is dynamically updated by decreasing the pheromone strength on all arcs by a constant factor before enabling each robot to supplement pheromone on the arcs. The pheromone strength τ_{ij} is dynamically updated as follows.

$$\begin{cases} \tau_{ij}(t+1) = (1-\rho) \cdot \tau_{ij}(t) + \Delta\tau_{ij} \\ \Delta\tau_{ij} = \sum_{k=1}^n \Delta\tau_{ij}^k \end{cases}, 0 < \rho < 1 \tag{2}$$

where, the amount of pheromone $\Delta\tau_{ij}^k(t)$, is defined as three modes [13]:

(1). Ant cycle system mode:

$$\Delta\tau_{ij}^k(t) = \begin{cases} \frac{Q}{L^k(t)} & \text{if arc}(i,j)\text{is ued by robot}k \\ 0 & \text{otherwise} \end{cases} \tag{3}$$

(2). Ant quantity system mode:

$$\Delta\tau_{ij}^k(t) = \begin{cases} \frac{Q}{d_{ij}(t)} & \text{if arc } (i,j)\text{is ued by robot } k \\ 0 & \text{otherwise} \end{cases} \tag{4}$$

$$\Delta\tau_{ij}^k(t) = \begin{cases} Q & \text{if arc } (i,j)\text{is ued by robot}k \\ 0 & \text{otherwise} \end{cases} \tag{5}$$

(3). Ant density system mode:
  $L^k(t)$ is the length of the kth robot's tour. $d_{ij}(t)$ is the distance between waypoints i and j. Q is constant representing the total amount of pheromone. The ACO algorithm for TSP to visit multiple waypoints is summarized as Fig. 1 [13].

procedure *ACO algorithm for TSPs*
  Set parameters, initialize pheromone trails
  while (termination condition not met) **do**
    ConstructSolutions
    ApplyLocalSearch
    UpdateTrails
  end
  end *ACO algorithm for TSPs*

Fig. 1. ACO algorithm for TSP with multiple waypoints

After execution of this ACO-based TSP algorithm with the method (1), the minimized total route to connect all the waypoints is obtained. There are fifteen waypoints in the workspace that required a service mobile robot to reach per waypoint. The ACO-based TSP algorithm is applied to this application illustrated in Fig. 2(a). The shortest and mean distances acquired are depicted in Fig. 2(b).

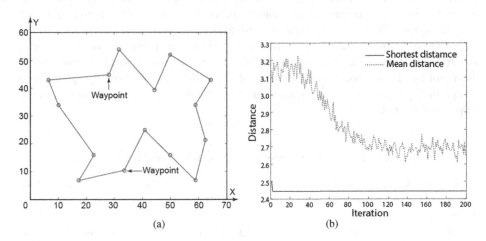

(a)   (b)

Fig. 2. ACO algorithm for TSP applied in a 15-waypoint workspace, (a) TSP route; (b) The shortest and mean distances over iterations

4 Real-Time Concurrent Map Building and Multi-goal Navigator

Given a set of waypoints as multiple goals and possible starting points, a matching *lookup table* for waypoint sequencing is created. The D*-Lite algorithm then provides an initial path marked by markers between pairs of goals, and the VFH navigation algorithm drives the robot alone those markers. A sort of multi-layer software development environment is implemented based on Player/Stage™.

An actual autonomous robot was developed as test-bed for our hybrid system of real-time concurrent multi-goal navigation and map building of an autonomous robot. The robot incorporates six sensors into its compact design as follows: a LIDAR, a DGPS, a digital compass, a camera, and an IMU, each of which is enclosed in a waterproof case and firmly mounted to the robot. A 270° SICK LMS111 LIDAR is configured for the purposes of obstacle detection illustrated.

5   Simulation Studies

In this section, the proposed real-time concurrent multi-goal navigation and mapping of an autonomous robot in unknown environments will be validated on Player/Stage™ simulator.

5.1   Real-Time Concurrent Multi-goal Navigation and Map Building in Unknown Environments Without Obstacles

In order to validate the effectiveness and efficiency of the proposed hybrid system, the model is applied to simulate an autonomous robot under unknown environments in a 15-goal course. The robot is navigated to connect 15 waypoints in a free-space to test the ACO-based multi-goal model with mapping illustrated in Fig. 3. The robot is able to traverse from the initial point to plan the shortest route to visit the 15 waypoint while the robot builds the map with 270° LIDAR. The rout of multi-goal by the robot is shown in Fig. 3(a) whereas the map built is illustrated in Fig. 3(b) at the end of the travel of the robot.

5.2   Navigation in a 15-Goal Course with Obstacle Avoidance

In this simulation, the robot is guided in an unknown environment populated with obstacles with 15 waypoints as multiple goals depicted in Fig. 4(a), which shows that the robot traverses from starting waypoint to connect 15 waypoints obtained by GPS coordinates. The built map while the robot moves in the unknown environment with 270° LIDAR scan is illustrated in Fig. 4(b) while the robot moves on the road from waypoint from waypoint. The black portions are detected obstacles with blue line of the robot path in Fig. 4(b).

In accordance with the planed trajectory, the GPS coordinates of the targets in latitude and longitude in sequence of the goals is dispatched to the D*-Lite based global path planner. As described in the previous sections, from one goal to another goal, the VFH based local navigator is utilized to navigate the robot by following pre-placed markers. Starting from the initial point, the autonomous robot is capable of travelling to the sequence of goals based on the proposed multi-goal navigation system and simulation studies were successfully accomplished.

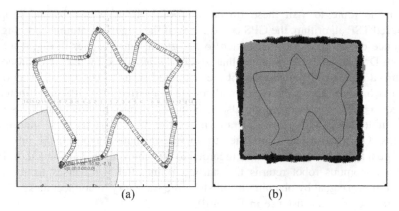

Fig. 3. Simulation result of the robot in unknown workspace. (a) The workspace with obstacles and 15 goals; (b) Built map.

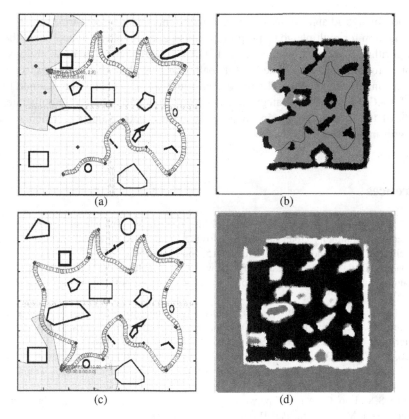

Fig. 4. Multi-goal navigation and mapping by the LIDAR in the ACO-based method, (a) Route planned by the proposed ACO-based TSP model; (b) Built map by the LIDAR; (c) Route planned at the end; (d) Built map by the LIDAR at the end. (Colour figure online)

After the route to visit these fifteen waypoints is generated by the proposed ACO-based TSP algorithm, the GPS coordinates of the targets in latitude and longitude in sequence of the goals are transmitted to the global path planner in Fig. 4. Once a path from D* Lite has been obtained that was early generated by the ACO-based TSP algorithm, a number of points along it are extracted so as to use the VFH algorithm between two goals. These points as markers are converted into GPS coordinates and presented to VFH as consecutive points. VFH then generates motion commands which are transmitted to the drive controllers to move the robot towards these intermediate waypoints. Once the robot approaches to an intermediate waypoint, the next intermediate goal along the desired path is regarded to be achieved and substituted. In Fig. 4 (c), the autonomous robot returns the starting point after it traverses from starting waypoint to every goal by planning a collision-free trajectory to connect these 15 goals with minimized overall distance in Fig. 4(d).

6 Conclusion

A solution by hybrid algorithms was developed in this paper for real-time map building and navigation for multiple goals purpose. In this paper, an alternate approach, the D*-Lite algorithm associated with a local LIDAR-based navigation methodology was developed for multiple goals. The D*-Lite path planning algorithm was used to provide VFH with intermediate goals. The multi-goal route was calculated and planned by the proposed ACO-based TSP strategy. Results from simulation studies demonstrated the benefits of the local navigator in conjunction with a path planner to reach multiple goals with minimized total distance.

References

1. Gu, T., Atwood, L., Dong, C., Dolan, J.M., Lee, J.-W.: Tunable and stable real-time trajectory planning for urban autonomous driving. In: IEEE/RSJ International Conference on Intelligent Robots and Systems (IROS 2015), pp. 250–256 (2015)
2. Raja, R., Dutta, A., Venkatesh, K.S.: New potential field method for rough terrain path planning using genetic algorithm for a 6-wheel rover. Robot. Auton. Syst. **72**, 295–306 (2015)
3. Davies, T., Jnifene, A.: Multiple waypoint path planning for a mobile robot using genetic algorithms. In: IEEE International Conference on Virtual Environments, Human-Computer Interfaces, and Measurement Systems, pp. 21–26, 12–14 July 2006
4. Luo, C., Yang, S.X.: A bioinspired neural network for real-time concurrent map building and complete coverage robot navigation in unknown environments. IEEE Trans. Neural Netw. **19**(7), 1279–1298 (2008)
5. Yang, S.X., Luo, C.: A neural network approach to complete coverage path planning. IEEE Trans. Syst. Man Cybern. Part B **34**(1), 718–725 (2004)
6. Yang, S.X., Meng, M.Q.-H.: Real-time collision-free motion planning of mobile robots using neural dynamics based approaches. IEEE Trans. Neural Netw. **14**(6), 1541–1552 (2003)

7. Faigl, J., Macak, J.: Multi-goal path planning using self-organizing map with navigation functions. In: European Symposium on Artificial Neural Networks, Computational Intelligence and Machine Learning, pp. 41–46, 27–29 April 2011
8. Gopalakrishnan, K., Ramakrishnan, S.: Optimal Path Planning of Mobile Robot with Multiple Targets Using Ant Colony Optimization, pp. 25–30. Smart Systems Engineering, New York (2006)
9. Koenig, S., Likhachev, M., Furcy, D.: Lifelong planning A*. Artif. Intell. J. **155**(1–2), 93–146 (2004)
10. Koenig, S., Likhachev, M.: D*Lite. In: The National Conference on Artificial Intelligence (AAAI) (2002)
11. Ulrich, I., Borenstein, J.: VFH+: Reliable obstacle avoidance for fast mobile robots. In: IEEE International Conference on Robotics and Automation, Leuven, Belgium, pp. 1572–1577, 16–21 May 1998
12. Luo, C., Wu, Y.-T., Krishnan, M., Paulik, M., Jan, G.E., Gao, J.: An effective search and navigation model to an auto-recharging station of driverless vehicles. In: 2014 IEEE Symposium on Computational Intelligence in Vehicles and Transportation Systems, pp. 100–107, Orlando, Florida, USA (2014)
13. Dorigo, M., Stützle, T.: Ant Colony Optimization. The MIT Press, Cambridge (2004)

Robot Indoor Navigation Based on Computer Vision and Machine Learning

Hongwei Mo[1(✉)], Chaomin Luo[2], and Kui Liu[3]

[1] Automation College, Harbin Engineering University, Harbin 150001, China
honwei2004@126.com
[2] Department of Electrical and Computer Engineering,
University of Detroit Mercy, Detroit, MI 48208-2576, USA
luoch@udmercy.edu
[3] Research Institute of Marin Device, Beijing 100000, China
lk399@163.com

Abstract. Autonomous navigation, as a fundamental problem of intelligent mobile robots' research, is the key technology of mobile robot to realize autonomous and intelligent. A method of combing computer vision and machine learning for the problem of robot indoor navigation is proposed in the paper. It realizes robot autonomous navigation through imitating the behavior of experts. Through a camera to perceive environmental information, expert provides some examples of navigation for robot to learn and robot learns a control strategy based on these samples using imitation learning algorithm. When robot is running, the control strategy learned can infer a corresponding control command based on the current perception of environmental information. Therefore, robot is able to mimic the behavior of expert to navigate autonomously.

Keywords: Mobile robot · Indoor navigation · Imitation learning · Visual feature extraction · DA_{GGER} algorithm

1 Introduction

In recent years, many mobile robot platforms emerge. Vision plays an important role in indoor mobile robot navigation and positioning, in recent years vision sensors has become a necessary configuration of the robot. Robot vision navigation and positioning has also become an important feature of the robot [1].

Compared with other on-board sensing technology, since the machine vision can provide more information about the environment that cannot be provided by other sensors, machine vision needs more attention of researchers. The methods of robot indoor navigation based vision can be classified as Fig. 1. In papers [2, 3], the machine vision for mobile robot navigation was reviewed.

The method of this paper belongs to vision-based reactive navigation. The method does not need create complex environment model or robot motion control model, and the robot's autonomous navigation can be realized by imitation learning of robot (learning from the robot operator) [4].

© Springer International Publishing Switzerland 2016
Y. Tan et al. (Eds.): ICSI 2016, Part II, LNCS 9713, pp. 528–534, 2016.
DOI: 10.1007/978-3-319-41009-8_57

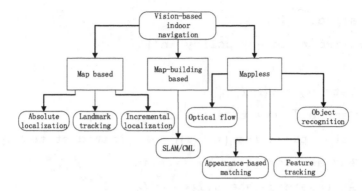

Fig. 1. Classification of indoor navigation technology based on robot vision

We proposed the fusion of machine vision and machine learning, which makes the robot imitate the behavior of the robot operator to achieve autonomous navigation. The method proposed in this paper was testified by experiment in actual indoor environment [5–8].

2 DA_{GGER} Imitation Learning Algorithm

2.1 Imitation Learning

For robot, imitation learning can be simply understood as a task from an expert demonstrate the process of learning a certain strategy, so that the robot can imitate the behavior of experts when performing this task. This method is called Learning from Demonstration [9]. The learner is robot, the teacher is user who operate the robot [10].

2.2 DA_{GGER} Algorithm

DA_{GGER} algorithm is an iterative algorithm that can train a high performance deterministic strategy with its induced state distribution [11]. The main process follows: in first iteration, collect the "state- action" data set D with the strategy of exports, and then record data set D, training a control strategy $\hat{\pi}_2$ which imitates experts. After n times iterations, robot run and collect more "state– action" data set with the strategy $\hat{\pi}_n$, put these collected "state- action" in the data set D, so $\hat{\pi}_{n+1}$ is the best strategy based on all collected data set D. In other word, D_{AGGER} algorithm perform the current strategy and collect data at every iteration, finally, according the accumulation of the data training next strategy [11]. The following is the full process of DA_{GGER} algorithm:

Initialize $D \leftarrow \varnothing$.

Initialize $\hat{\pi}_1$ to any policy in Π .

For $i = 1$ to N do

  Let $\pi_i = \beta_i \pi^* + (1 - \beta_i)\hat{\pi}_i$.

  Sample $T-step$ trajectories using π_i .

  Get dataset $D_i = \{(s, \pi^*(s))\}$ of visited states by π_i .

  And actions given by expert.

  Aggregate datasets: $D \leftarrow D \cup D_i$.

  Train classifier $\hat{\pi}_{i+1}$ on D .

End for

Return best $\hat{\pi}_i$ on validation.

In this paper, the classifier in the DA_{GGER} algorithm is the least squares of error and the classifier. In order to better utilize the expert's strategy, we can use the strategy $\pi_i = \beta_i \pi^* + (1 - \beta_i)\hat{\pi}_i$ instead of strategy $\hat{\pi}_i$ to collect data at the i times iterative, in fact, it is necessary in the initial iteration of a few steps, where π^* represents the expert's strategy, β_i is a scaling factor. Typically, take $\beta_i = p^{i-1}$, the reference to the expert's strategy is to make the reference to the expert's strategy with the increase of the number of iterations and the exponential decay.

2.3 Minimum Error Square Sum Classifier

Suppose a given training set: $T = \{(x_1, y_1), (x_2, y_2), \ldots (x_N, y_N)\}$, $x_i \in R^n$ is input vector, $y_i \in \{-1, 1\}$ is class label of input vector x_i. Minimum error square sum classifier is cost function by minimum:

$$J(w) = \sum_{i=1}^{N} (y_i - x_i^T w)^2 \equiv \sum_{i=1}^{N} e_i^2 \tag{1}$$

Obtain classification hyper plane:

$$w^T x = 0 \tag{2}$$

In Eq. (1), $x_i \in R^{n+1}$ is on the basis of the original $x_i \in R^n$, a dimension is extended, and the value of the dimension is 1, w is parameter vector, $w \in R^{n+1}$, e_i is an error between a class of the expected value and the estimated value of a class. Equation (1) represents the sum of the error that between expected output of all known training feature vectors (in the two category, it is ± 1) and actual output, using this method can reduce the demand for information of probability destiny function [12].

Equation (1) for the minimization of w, the result is:

$$\left(\sum_{i=1}^{N} x_i x_i^T \right) \hat{w} = \sum_{i=1}^{N} x_i y_i \tag{3}$$

Define $X = \begin{bmatrix} x_1^T \\ x_2^T \\ \vdots \\ x_N^T \end{bmatrix} = \begin{bmatrix} x_{11} & x_{12} & \cdots & x_{1(n+1)} \\ x_{21} & x_{22} & \cdots & x_{2(n+1)} \\ \vdots & \vdots & \ddots & \vdots \\ x_{N1} & x_{N2} & \cdots & x_{N(n+1)} \end{bmatrix}$ $y = \begin{bmatrix} y_1 \\ y_2 \\ \vdots \\ y_N \end{bmatrix}$ X is a matrix of

$N \times (n+1)$, its line is a known feature vector for training, y is a vector containing the corresponding expected value, so $\sum_{i=1}^{N} x_i x_i^T = X^T X$, and $\sum_{i=1}^{N} x_i y_i = X^T y$, therefore, Eq. (3) can express:

$$\left(X^T X \right) \hat{w} = X^T y \Rightarrow \hat{w} = \left(X^T X \right)^{-1} X^T y \tag{4}$$

So, the optical weight vector can be obtained by solving the linear equations, and this optimal weight vector is also the only solution to the minimum value of $J(w)$.

In practice, $(n+1) \times (n+1)$ dimensional matrix $X^T X$ usually appears to be a singular case, in that case, the optimal weight vector $\hat{w}$ is obtained by using the following formula:

$$\hat{w} = \left(X^T X + CI \right)^{-1} X^T y \tag{5}$$

I is $(n+1) \times (n+1)$ dimensional identity matrix, C is a very small number of users defined.

3 Design and Implementation of Indoor Navigation System

The first step of learning the DA_{GGER} algorithm is the demonstration of the robot's navigation process. In order to extract the visual feature information in the next step, the video file is separated into a frame of a frame, at the same time, it is required to associate each frame of the image with its corresponding action. In the process of robot navigation action command includes straight, which is represented by integers 2, 3 and 4, respectively, the operator sends the control commands to the robot by the handle on the keyboards. Figure 2 shows the part of the video after the separation of the robot, it can be seen that each frame of the image is associated with the corresponding action.

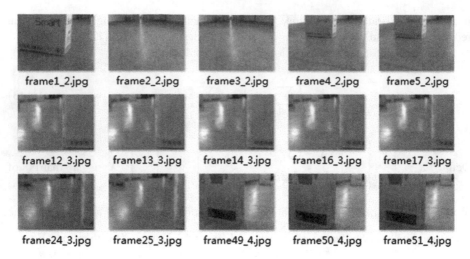

frame1_2.jpg	frame2_2.jpg	frame3_2.jpg	frame4_2.jpg	frame5_2.jpg
frame12_3.jpg	frame13_3.jpg	frame14_3.jpg	frame16_3.jpg	frame17_3.jpg
frame24_3.jpg	frame25_3.jpg	frame49_4.jpg	frame50_4.jpg	frame51_4.jpg

Fig. 2. Separate image of robot running video

4   Experiments

The experimental setup of the case is shown in Fig. 3(a), running path of the robot that avoid obstacles from the starting to the destination by the expert is the red dashed line in the diagram. The parameter of training minimum error square sum C is 0.5, maximum iteration of the DA_{GGER} algorithm N is 10.

Figure 4(a) is a part of the image sequence taken from the autonomous navigation of the robot in the environment of Fig. 3(a), each image give the robot action that inferred by the robot learning strategy π, which marked by the red arrow. It can be seen from the graph that the robot can imitate the behavior of the experts after the imitation learning. In Fig. 3(b), strategy π infer two action including going straight and turn right for the image. From the view of expert, two actions can be executed by robots, straight ahead action can be enforced because of the existence of the robot camera, robots seem to hit obstacles from the image, but there is still a certain distance between them, when the robot go closer to the obstacle. Strategy π is bound to infer the right action, so the robot will not hit obstacles; Turn right can be performed because it is equivalent to the robot in the face of obstacles in advance, but will not hit obstacles.

The experimental setting of the chair environment is shown in Fig. 4(b), running path of the robot that avoid obstacles from the starting to the destination by the expert is the purple dashed line in the diagram.

Figure 4(b) is a part of the image sequence in the process of autonomous navigation of the robot of Fig. 4(b), each image give the robot action that inferred by the robot's learning strategy, that marked by the red arrow. As can be seen from Fig. 4(b), the robot can imitate the behavior of the expert after imitation learning.

(a) Experiment 1   (b) Experiment 2

Fig. 3. Robot navigation experiment

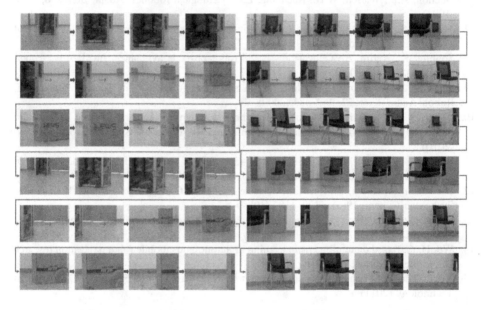

(a) Image sequence 1   (b)Image sequence 2

Fig. 4. Robot auto navigation

5 Conclusions

In this paper, an indoor navigation method based on machine vision and machine learning is presented. And the experiments are carried out on the smart mobile robot platform. Combining with the real-time image of the robot, the standard method for each feature extraction is improved, and the detailed process of improvement is given. The DA_{GGER} algorithm is first applied to the real robot indoor navigation, the three least squares of error square and the classifier are used as the robot navigation system. Experiments show that the proposed method can realize the imitation of the expert's behavior. In the future, we will expand the method to more complex and dynamic environments.

References

1. Gong, Y.L.: Research on the visual navigation and positioning of indoor robot based on embedded. Southwest Jiao Tong University (2012)
2. Desouza, G.N., Kak, A.C.: Vision for mobile robot navigation: a survey. IEEE Trans. Pattern Anal. Mach. Intell. **24**, 237–267 (2002)
3. Bonin, F., Ortiz, A., Oliver, G.: Visual navigation for mobile robots: a survey. J. Intell. Robot. Syst. **53**, 263–296 (2008)
4. Güzel, M.S.: Autonomous vehicle navigation using vision and maples strategies: a survey. Adv. Mech. Eng. **5**, 234–747 (2013)
5. Bernardino, A., Victor, J.S.: Visual behaviors for binocular tracking. Robot. Auton. Syst. 137–146 (1998)
6. Duchon, A.P., Warren, W.H., Kaelbling, L.P.: Ecological robotics. Adapt. Behav. **6**, 1–30 (1994)
7. Temizer, S.: Optical flow based local navigation. M.S. thesis, Massachusetts Institute of Technology, Cambridge, MA, USA (2001)
8. Souhila, K., Karim, A.: Optical flow based robot obstacle avoidance. Int. J. Adv. Robot. Syst. **4**(1), 13–16 (2007)
9. Argall, B.D., Chernova, S., Veloso, M.: A survey of robot learning from demonstration. Robot. Auton. Syst. **57**, 469–483 (2009)
10. Guzel, M.S., Bicker, R.: Optical flow based system design for mobile robots. In: Proceedings of the IEEE International Conference on Robotics, Singapore, pp. 545–550 (2010)
11. Ross, S., Gordon, G.J., Bagnell, J.A.: A reduction of imitation learning and structured prediction to no-regret online learning. In: AISTATS (2011)
12. Borenstein, J., Koren, Y.: Real-time obstacle avoidance for fast mobile robots. IEEE Trans. Syst. Man Cybern. **19**(5), 1179–1187 (1989)
13. Liang, Q.: Research on the autonomous navigation of mobile robot based on reinforcement learning D. Nanjing Agricultural University (2012)
14. Wu, H.Y.: Based on reinforcement learning of autonomous mobile robot navigation research D. Northeast Normal University (2009)
15. Yi, J.: The robot behavior of reinforcement learning based on D. Shenyang University of Technology (2011)

Improved Hormone-Inspired Model for Hierarchical Self-organization in Swarm Robotics

Yuquan Leng[1,2], Xiaoning Han[1,2(✉)], Wei Zhang[1], and Weijia Zhou[1]

[1] State Key Laboratory of Robotics, Shenyang Institute of Automation,
University of Chinese Academy of Sciences, Shenyang 110016, China
{lengyuquan, hanxiaoning, zhangwei, zwj}@sia.cn
[2] University of Chinese Academy of Sciences, Beijing 100000, China

Abstract. More and more robotic systems with lots of robotic individuals severs for human, such as intelligent terminal, intelligent storage, intelligence factory, etc. This is the trend of robotics technology, which will lead robotics system become huger with more individuals. Then, how to organize and manage this huge robotic system will be one important issue. This paper proposes hierarchical self-organizing approach to realize self-management, self-organization. Firstly, hierarchical self-organizing model is put forward and the process of formation is described in detail, which makes the organizing structure of system regularly. Secondly, this paper uses the improved hormone-inspired model (IHM) to establish relation between individuals, which considers topological structure of the organization, supports dynamic reconfiguration and self-organization, and requires no globally certain identifiers for individual robots. Finally, this paper presents the experimental results on swarm robotics system with a large scale of individuals to form a self-organization.

Keywords: Improved hormone-inspired · Hierarchical self-organization · Swarm robotics

1 Introduction

Swarm robotics occurs from artificial swarm intelligence, the biological studies of insects, ants and other fields in nature, so swarm robotics is defined as a new approach to the coordination of multi-robot systems which consist of large numbers of simple physical robots. It is supposed that a desired collective behavior emerges from the interactions between the robots or between the robots and environment [1, 2]. But a fuzzy key word in the concept is the definite number that "large number" means. Besides, the concept of swarm robotics should be extended with the development of robots. We considered that the core of concept should focus on the characteristic of organization structure, not on how many robots which just make it like swarm robotics, we propose that:

> *Swarm robotics is a robotics system with a special organization structure, which has flexibility, unpredictability and mechanism of infinite plus or minus, and consists of any form of robots as long as which could interact with others or environment.*

© Springer International Publishing Switzerland 2016
Y. Tan et al. (Eds.): ICSI 2016, Part II, LNCS 9713, pp. 535–543, 2016.
DOI: 10.1007/978-3-319-41009-8_58

We believe that swarm robotics originates from multi-robot systems, but the flexibility, unpredictability and mechanism of infinite plus or minus make it different. In multi-robot systems, scientists paid more attention to multi-robot cooperation, while in swarm robotics we need to focus on self-organization.

Hierarchical self-organization is wildly researched in sociology, management science, and psychology [3], which not only put focus on how to form one organization also on relation between individuals and how to make the organization more efficient. In hierarchical organization, individuals have tight connection and clearly subordinate relationship which help whole system handle complex task efficiently with suitable planning. In robotic field, hierarchical control has been widely used in multi-robotic system [4], but relation between individuals is built by human in one fixed way not by the self of system. It is not realistic to establish organization by human in swarm system with characteristics that individuals are almost infinite, number of individuals is unpredictably changed, etc.

In the past decade, a series of swarm robotics platforms have been built to imitate the biological motion behaviors using distributed self-organization [5, 6], such as Pheromone Robotics [7], Kilobot [8], I-Swarm [9], E-puck, Swarm-bots [10], TERMES [11], etc. TERMES completely imitates the behavior of termites to establish building using special blocks. Swarm-bots is capable of realizing cooperative transportation, task allocation, and self-reconfiguration. These are all homogeneous and managed by distributed self-organization. Heterogeneous swarm robotics, such as Swarmanoid [12], is shown to imitate between different functional robots and simulate complex human behaviors. It is comprised of numerous autonomous robots of three types: eye-bots, hand-bots, and foot-bots, and has been able to take a book from the shelf successfully. Hierarchical self-organization with large scale individuals is not involved in these systems.

This paper presents the hierarchical self-organization structure (HSS) as a social managing method for robotic swarm system. In this structure, robots are views as Agents that have ability of communication, cooperation, perception, decision, etc. [13]. This structure could manage any number of individuals, from one to infinite and make Agents establish connection only using local information also called "hormone". In this paper, improved hormone-inspired model (IHM) is put forward to elaborate how individuals produce and sense hormone. The hormone is link pheromone between two Agents, which help Agents to build link. Based on the HSS and IHM, some rules are defined to make large scale individuals form one hierarchical self-organization, which are set in every Agents as behavior guide line.

Even using hierarchical structure, one system with large scale individuals could form a great many of organization structure, far more than the number of individuals. In this paper, three performance indexes of self-organization are put forward and defined to evaluate of the merits of the organization, including formation speed, intimacy and longitudinal depth ratio. Using the method mentioned in this paper, one form of organization is proposed for large scale robotic system, and some simulation experiments tests the method.

2 Hierarchical Self-organization Structure

In swarm robotic system, one Agent is defined as a structure with the ability of perception, autonomous decision-making and action which can contain a single robot, sensor, actuator and software unit for solving task problem. Agent technology is an important method to solve the problem of the relation between the tasks and the organization, through which the relations through the whole organization could come into being. Agent technology is helpful to establish relations between upper and lower layers in hierarchical organization structure, which provides interface for date exchange including task decomposition information, Agent capability information, perceived data, execution status, etc.

In the Fig. 1, a hierarchical organization structure is described. Only one top management Agent is included and defined in layer 0 in whole system. This Agent is in the highest layer and represented the whole system. More specifically, this Agent is usually a computer device that can interact with operators, which is convenient for the human to order tasks to swarm robotic system. In generally, this Agent is appointed by human.

The arrows shown in this figure mean that individuals could change their position in organization structure. This characteristic helps organization have ability of dynamic adjustment. In addition, this structure could manage a large number of individuals with fewer layers. Assuming that every Agent could manage q Sub-Agent with n layers in total, then this system could manage the following number of individuals,

$$No_{Ind} = \frac{1 - q^n}{1 - q}, (q > 1, q \in N) \tag{1}$$

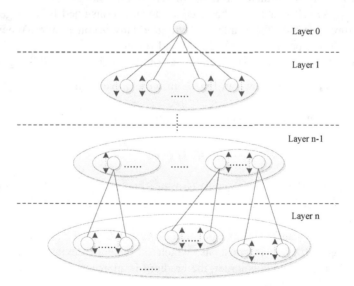

Fig. 1. Hierarchical organization structure

3 Improved Hormone-Inspired Model

The IHM is inspired by three factors: (1) biological discoveries about how animal self-organize by emotional attraction; (2) the existing self-organization models, such as ant colony, flock of birds, etc.; (3) Hierarchical organization method in human society. The basic idea of Improved Hormone-Inspired Model is that a swarm is a network of robots that can dynamically change its links and could complete many tasks arranged by human being. The IHM is proposed based on DHM [14] which is used to distributed control of robotic swarms. IHM is suitable for hierarchical organization and control of swarm robotics by deleting inhibitory factor of DHM and expending activator factor.

Individuals of swarm have been set as Agent, so every Agent has the following information: position, capacity, layer, identification, search radius in initial state, search time and number of Sub-Agent, which could be expressed as

$$Agent : (P, C, L, ID, R_0, T, N_{Sub}).$$

The meaning of every parameters are shown as following:

Position (P): Every Agent in the swarm has a certain geographical position, which is used to calculate the physical distance of two Agents;

Capacity (C): Every Agent has the ability to operate, in this paper, the Agent's ability is assuming a single, and all individuals in the initial state are the same;

Layer (L): The layer of Agent in hierarchical structure and it is null, if Agent is not belong to organization;

Identification (ID): The position of the Agent in the organization;

Search radius in initial state (R_0): The search radius of the Agent at the initial moment is used to calculate the range of Agent's hormones;

Search time (T): It is used to characterize the time consumed by an Agent search link, as time increases, the search area will gradually become larger. When the link is generated, the parameter will be zero and re-time;

Number of Sub-Agent (N_{Sub}): The number of Agents belong to this Agent;

IHM is established as shown in Fig. 2, in which, every Agent produces hormone ring.

Hormone between two agents could be expressed as,

$$H(Agent(ID, ID')) = Ce^{\frac{-(x-x')^2-(y-y')^2}{2\sigma^2}} \tag{2}$$

In Eq. (2), $Agent(ID, ID')$ means the hormone created by $Agent(ID)$ to $Agent(ID')$. In the same way, the hormone created by $Agent(ID')$ to $Agent(ID)$ could be expressed as $H(Agent(ID', ID))$.

In Eq. (2), x, y means coordinate position of $Agent(ID)$, and x', y' means coordinate position of $Agent(ID')$; σ is constants, which is standard deviation of 3D normal distribution. When the value is larger, the gradient is approximately flat, and the value is smaller, the gradient is steeper.

The attraction factor is the sum of hormone one Agent feel to another, and shown as,

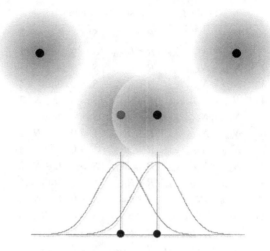

Fig. 2. Improved Hormone-Inspired Model

$$A = H(Agent(ID, ID')) + H(Agent(ID', ID)) \tag{3}$$

Combining to diffusion radius, getting

$$H(Agent(ID, ID')) = \begin{cases} Ce^{\frac{-(x-x')^2-(y-y')^2}{2\sigma^2}} & if(x-x')^2 - (y-y')^2 < R^2_{Agent(ID)} \\ 0 & if(x-x')^2 - (y-y')^2 > R^2_{Agent(ID)} \end{cases} \tag{4}$$

$$H(Agent(ID', ID)) = \begin{cases} Ce^{\frac{-(x-x')^2-(y-y')^2}{2\sigma^2}} & if(x-x')^2 - (y-y')^2 < R^2_{Agent(ID')} \\ 0 & if(x-x')^2 - (y-y')^2 > R^2_{Agent(ID')} \end{cases} \tag{5}$$

For Agent behaviors, IHM is governed by a function $Agent\_B(B|N, R, S, L)$ defined as in Table 1. There are five actions for Agent with the change of state that are following:

(1) Increasing the radius of the perceptual range, and searching again;
(2) Searching in the maximum perceived range;
(3) Establishing connection as Parent-Agent, and set radius of the perceptual range to initial radius;
(4) Establishing connection as Sub-Agent, and set radius of the perceptual range to initial radius;
(5) Keeping status;

In the table, N_{Sub} means number of Sub-Agent, $N_{Sub} < N_{Max-Sub}$ presents this Agent can add other Agents, and $N_{Sub} = N_{Max-Sub}$ indicates this Agent can not to add others; R means radius of the perceptual range; S means the number of Agents which are

Table 1. Action selection table

No.	N_{Sub}	R	S	L	B
1	$N_{Sub} < N_{Max-Sub}$	$R < R_{Max}$	$S = 0$	Null	1
2	$N_{Sub} < N_{Max-Sub}$	$R < R_{Max}$	$S = 0$	$L \in N$	1
3	$N_{Sub} < N_{Max-Sub}$	$R < R_{Max}$	$S > 0$	Null	4
4	$N_{Sub} < N_{Max-Sub}$	$R < R_{Max}$	$S > 0$	$L \in N$	3
5	$N_{Sub} < N_{Max-Sub}$	$R = R_{Max}$	$S = 0$	Null	2
6	$N_{Sub} < N_{Max-Sub}$	$R = R_{Max}$	$S = 0$	$L \in N$	2
7	$N_{Sub} < N_{Max-Sub}$	$R = R_{Max}$	$S > 0$	Null	4
8	$N_{Sub} < N_{Max-Sub}$	$R = R_{Max}$	$S > 0$	$L \in N$	3
9	$N_{Sub} = N_{Max-Sub}$	$R < R_{Max}$	$S = 0$	Null	5
10	$N_{Sub} = N_{Max-Sub}$	$R < R_{Max}$	$S = 0$	$L \in N$	5
11	$N_{Sub} = N_{Max-Sub}$	$R < R_{Max}$	$S > 0$	Null	5
12	$N_{Sub} = N_{Max-Sub}$	$R < R_{Max}$	$S > 0$	$L \in N$	5
13	$N_{Sub} = N_{Max-Sub}$	$R = R_{Max}$	$S = 0$	Null	5
14	$N_{Sub} = N_{Max-Sub}$	$R = R_{Max}$	$S = 0$	$L \in N$	5
15	$N_{Sub} = N_{Max-Sub}$	$R = R_{Max}$	$S > 0$	Null	5
16	$N_{Sub} = N_{Max-Sub}$	$R = R_{Max}$	$S > 0$	$L \in N$	5

perceived; L means Agent layer, if this Agent belongs to organization, $L \in N$, otherwise, L is Null.

4   Experiments

In order to apply the proposed IHM to real robot swarm, it is necessary to realize the method by computer simulation. First of all, a large number of individuals are randomly generated in a box with 100X100 (Length X Width), in which each individual just occupies 1X1 (Length X Width). In this paper, the individuals' density is set as 5 %, so computer can create about 500 individuals in the box at random locations.

In this section, parameters are set as follows: search radius in initial state $R_0 = 3$; maximum search radius $R_{Max} = 9$; maximum search cycle $T_{Max} = 10$; maximum number of Sub-Agent $N_{Max-Sub} = 5$.

In Sect. 2, we describe that the top layer Agent is appointed by human in the initial state. Different locations of Agent have different impacts on self-organization, so we choose three typical locations for testing and comparing, as shown in Table 2, which are corner location, middle location and center location. In the table, green Agent means it does not belong to organization and red Agent means it belongs to organization.

There are 499 Agents is contented in this swarm system, and 488 connections to be built. Complete cycle means the number of total cycles used to before realizing self-organization; deepest layer means the biggest layer in self-organization; Average

Table 2. Self-organization of three typical locations

T	Corner location	Middle location	Center location
1			
10			
20			
30			
40			
50			
	Complete cycle: 45 Deepest layer: 35 Average velocity: 11.0667	Complete cycle: 42 Deepest layer: 28 Average velocity: 11.8571	Complete cycle: 34 Deepest layer: 22 Average velocity: 14.6471

velocity describes velocity of formation. As we can see from the results in Table 2, the closer an agent is to the top Agent location of swarm center, the more favorable this self-organization is.

5 Conclusion and Future Work

With the development of computer technology and industrial technology, an increasing number of robots will enter into human society for servicing human being. These robots, then, will form one system, and connect with each other. It is vital to manage the whole system by using network, composed with relationship units. In this paper, we research the network problem of swarm robotic system, which is called organization.

We present a new concept of swarm robotics, in which emphasis is put on organization structure, rather than quantity. Swarm robotic system which has ability of self-organization is our goal, like a system consisting of vast number of individuals under the condition filled with many bursts and unexpected damages. This paper proposes hierarchical self-organizing approach to realize self-management, self-organization and self-healing. Hierarchical self-organizing model is put forward and process of formation is described in detail, which makes the organizing structure of system regularly. Compared with distributed self-organization, hierarchical self-organization is much easier to deal with complex tasks with sequential logic and is more efficient facing with cooperation tasks.

This paper uses the improved hormone-inspired model (IHM) to establish relation between individuals, which considers topological structure of the organization, supports dynamic reconfiguration and self-organization, and requires no globally certain identifiers for individual robots. From experiment results, we could get three points: (1) Both HSS and IHM are effective to form one swarm system with hierarchical organization; (2) More closely top Agent is located near swarm center, the more favorable the formation of self-organization is.

Intelligent systems such as intelligent terminal, intelligent storage, intelligence factory, artificial war, etc. Fully predict that the future of robotic world is prone to an swarm robotic world. Assuming that, if there is one war, we just need to put on a large number of robots and order the tasks the top robot, then the swarm robotic system will self-organize, manage and complex tasks.

In the further research, we will put our attention on the following questions: (1) How to evaluate one organization to assist the system form one excellent organization by itself; (2) New methods to realize self-organization formation more quickly and efficiently; (3) How to make it easy to adjust organization relations dynamically for special requirements in multi-tasks.

References

1. Mohan, Y., Ponnambalam, S.G.: An extensive review of research in swarm robotics. In: IEEE World Congress on Nature & Biologically Inspired Computing, Coimbatore, 9–11 December, pp. 140–145 (2009)
2. Higgins, F., Tomlinson, A., Martin, K.M.: Survey on security challenges for swarm robotics. In: Conference Record 2009 IEEE International Conference on Autonomic and Autonomous Systems, pp. 307–312 (2009)
3. Pfeifer, R., Lungarella, M., Iida, F.: Self-organization, embodiment, and biologically inspired robotics. Science **318**(5853), 1088–1093 (2007)
4. Kernbach, S.: Structural Self-organization in Multi-Agents and Multi-Robotic Systems. Logos Verlag Berlin GmbH (2008)
5. Viktor, A., Crmen, L.C., Koch, A., Michael, S., Levi, P.: Hierarchical self-organization in swarms of nano-robots. In: Proceedings of the XXV International Conference on Dynamic Days Europe, vol. 29E, pp. 272–273 (2005)
6. Ünsal, C., Bay, J.S.: Spatial self-organization in large populations of mobile robots. In: Proceedings of the 1994 IEEE International Symposium on Intelligent Control, pp. 249–254. IEEE (1994)
7. Purnamadjaja, A.H., Russell, R.A.: Robotic pheromones: using temperature modulation in tin oxide gas sensor to differentiate swarm's behaviours. In: Conference Record 2006 IEEE International Conference on Control, Automation, Robotics and Vision, pp. 1–6 (2006)
8. Rubenstein, M., Ahler, C., Nagpal, R.: Kilobot: A low cost scalable robot system for collective behaviors. In: Conference Record 2012 IEEE International Conference on Robotics and Automation, pp. 3293–3298 (2012)
9. Woern, H., Szymanski, M., Seyfried, J.: The I-SWARM project. In: Conference Record 2006 IEEE International Conference on Robot and Human Interactive Communication, pp. 492–296 (2006)
10. Dorigo, M.: SWARM-BOT: an experiment in swarm robotics. In: Proceedings of the 2005 IEEE Swarm Intelligence Symposium, 2005, pp. 192–200 (2005)
11. Durrant-Whyte, H., Roy, N., Abbeel, P.: TERMES: an autonomous robotic system for three-dimensional collective construction, pp. 257–264. MIT Press, Cambridge (2012)
12. Dorigo, M., Floreano, D., Gambardella, L.M.: Swarmanoid: A novel concept for the study of heterogeneous robotic swarms. IEEE Trans. Rob. Autom. Mag. **20**, 60–71 (2013)
13. Leng, Y., Yu, C., Zhang, W., Zhang, Y., He, X., Zhou, W.: Hierarchical self-organization for task-oriented swarm robotics. In: Tan, Y., Shi, Y., Buarque, F., Gelbukh, A., Das, S., Engelbrecht, A. (eds.) ICSI-CCI 2015. LNCS, vol. 9140, pp. 543–550. Springer, Heidelberg (2015)
14. Shen, W.M., Will, P., Galstyan, A., Chuong, C.M.: Hormone-inspired self-organization and distributed control of robotic swarms. Auton. Robots **17**, 93–105 (2004)

Triangle Formation Based Multiple Targets Search Using a Swarm of Robots

Jie Li[1,2] and Ying Tan[1,2(✉)]

[1] Key Laboratory of Machine Perception and Intelligence,
Ministry of Education, Peking University, Beijing, China
ytan@pku.edu.cn
[2] Department of Machine Intelligence, School of Electronics Engineering
and Computer Science, Peking University, Beijing 100871, China

Abstract. As a distributed system, swarm robotics is well suited for multiple targets search tasks. In this paper, a new approach based on triangle formation and random search is proposed for high efficiency, demonstrating excellent abilities of exploration and exploitation in experiments. In addition, a new random walk strategy of linear ballistic motion, integrated with triangle estimation, is put forward as a comparison algorithm, the performance of which can serve as a benchmark.

Keywords: Swarm robotics · Multiple targets search · Triangle formation · Random search · Exploration and exploitation

1 Introduction

Swarm robotics, inspired from the self-organization phenomena in nature, is a relatively new field, on which people have done lots of various research work [1]. With large number of individuals, swarm robotic system is appropriate for tasks involving area coverage [2], such as searching for multiple targets. When the targets can generate fitness values in certain range and can be collected, it comes to the issue we concern [3]. The multi-target search strategy has a broad prospect of application, such as hunting a submarine [11], searching for victims and wreckage after air crash or shipwreck, monitoring the leak water quality [2], exploring and destroying battlefield targets, and so on.

Behavior-based design methods are commonly used in the task of searching for targets, such as methods based on artificial potential functions [4] or methods adapted from some heuristic algorithms [5,6]. GES [7] and IGES [8] we proposed before borrowed some ideas from the FWA [9], a heuristic algorithm inspired by the firework explosion. Mathematical physics methods are also used to analyze the foraging and migratory behaviors of animals, which is often referred to as "random search" [11,12] or "stochastic optimal foraging theory" [13].

Another thing needed to be introduced here is the formation control for multiple robots or vehicles, and what we used is the behavior-based control [14],

© Springer International Publishing Switzerland 2016
Y. Tan et al. (Eds.): ICSI 2016, Part II, LNCS 9713, pp. 544–552, 2016.
DOI: 10.1007/978-3-319-41009-8_59

where each robot determines its proper position based on a reference point which can be a leader, a neighbor or the unit-center of the whole group.

The rest of the paper is organized as follows. In Sect. 2 , the multiple targets search problem and an idealized model are stated. In Sect. 3, the Triangle Formation Search strategy is described. In Sect. 4, experimental results and discussions are presented. Finally, the work is concluded in Sect. 5.

2 Problem Statement

In the multiple targets search problem, a swarm of robots are delivered into a vast unknown space, where multiple targets are distributed randomly. Robots are expected to search and collect the targets as soon as possible using some collaborative mechanism. In the simplest case, only three kinds of objects are considered: environment space, robots and targets. Obstacles, decoys [10] and inference sources can also be introduced into the problem [3]. Since we focus on the search efficiency in this paper, only the simplest case are studied.

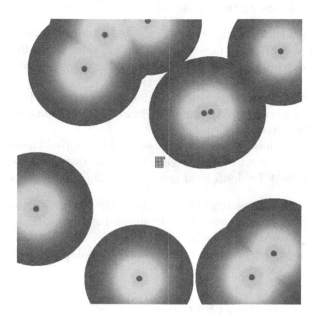

Fig. 1. A screenshot of the problem at the beginning of a simulation. Red rounds stand for the targets. The background color illustrates the fitness value of that position. The robot phalanx is in the center of the figure. (Color figure online)

2.1 An Idealized Model

An idealized model of the problem is shown in Fig. 1. The environment is a 1000*1000 square while the robot is a unit square. Robots can memorize information (positions and fitness values) of 10 iterations. Each target is abstracted to be a round with radius of r_t (10 units), and robots in the round can locate the target directly. Positions and fitness values of targets are generated randomly, and fitness ranges from $F_{max} - 2$ to F_{max} (40 units). Influence scopes of targets are presented as a series of annuli, and each annulus is $0.5r_t$ width which is also the maximum speed limit of robots to ensure the variance of fitness values in two adjacent iterations is small. From the inside out, the fitness value decreases by 1 unit till 0, and greater ones are chosen as the fitness values in overlapped areas. Discrete fitness values are adopted because the hardware design in swarm robotics should be as simple as possible which may lead to low quality sensors and fault sensing results [15]. 10 iterations are required for one robot to collect a target while 10 robots can do that in one iteration. The sphere of local interaction between robots is a round of radius $2r_t$. One problem that has not been considered here is avoiding collisions of robots resulting from route intersection.

3   Triangle Formation Search Strategy

3.1   Characteristics of the Problem

– Compared with each individual robot, the entire search space is vast, so the swarm is supposed to have nice ability of exploration. The influence scopes of all targets cover a large proportion of the entire search space, so the swarm should bear excellent ability of exploitation.
– In order to prevent excessive concentration of resources and give full play to the group exploration ability, the entire swarm should disperse as much as possible in the initial stage.
– The integration of local information is essential to improve the group exploitation ability, so each robot should ensure certain degree of connection with neighboring robots, i.e. form local groups (or niches) with other robots.

In our triangle formation search (TFS) strategy, the swarm is divided into three-robot teams which are arranged in a triangle, including one leader and two other members. According to the conclusions above, the TFS strategy may be a promising approach, for the three-robot teams can balance the exploration and exploitation of the swarm.

3.2   Five Stages of the TFS Strategy

– Initial grouping: Divide the whole swarm into three-robot teams, and robots insufficient to form a team will search alone.
– Initial diffusion: Firstly, the leaders will count the number of neighboring robots and select a sparse direction.
– Search in areas without fitness: The leader will search randomly, and the step lengths are submitted to some type of probability distribution, such as a Lévy or exponential distribution.
– Search in areas with fitness: The leader will estimate the gradient direction according to the information obtained by the team and update its position.

– Target collecting: Robots having found targets will broadcast the information within the team and the other two will move towards the target.

In stages of diffusion and search, members in teams will follow the leader and maintain the formation. Since the strategy involves formation control which increases the complexity of the system, for convenience, we restrict the exchange of information within the team, and the formation cannot be restored once broken.

3.3   Key Issues to be Tackled in the TFS Strategy

Unified Grouping. Taking into account the initial formation of the robot phalanx and the simplicity of implementation, we assign a global ID for each robot in an "S" shape order. Then, the robot whose ID is a multiple of 3 serves as a leader, closely followed by two other members of the team.

Diffusion Control. In order to make full used of the exploration ability of the swarm, an initial diffusion stage is introduced, in which he leaders will monitor the number of neighbors and terminate diffusion if the number has fallen below a certain threshold. We carried out experiments to determine the threshold from 3 to 10, and the optimal value is about 3.

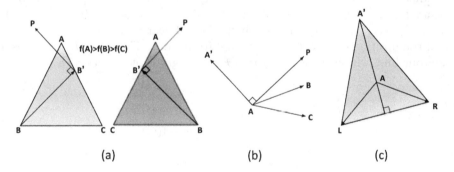

Fig. 2. (a) Calculate the gradient direction when all robots have different fitness values. (b) Robots determine their own roles in role switching process. (b) Robots maintain the triangle formation with the aid of their internal compasses.

Random Search. As is mentioned above, the leaders will search randomly in areas without fitness. And there are already some conclusions drawn from the one-dimensional and bidimensional cases of random search [13].

– When the coverage rate of influence scopes of all targets is high (i.e. dense distribution). "stochastic laws governing *run and tumble* movement patterns come into play and have a clear impact on the search success, with Lévy-like features becoming beneficial [13]."

– When the coverage rate is low (i.e. sparse distribution), ballistic strategies are optimal for non-renewable targets while Lévy flights are optimal strategies for renewable targets.

In our problem model, the coverage rate is high and the targets are non-renewable, so a Lévy-like strategy is suitable and the leader in TFS strategy performs ballistic flights reoriented at exponentially distributed times, and the mean value of the distribution is set to $2 * MapLength$ (i.e. $2 * 1000$).

Gradient Estimation. The leader will integrate the fitness values and positions of the team, and calculate the approximate gradient direction based on the supposition that the fitness value varies almost linearly alone a line in local area. Our basic idea is to construct a vector perpendicular to the local contours, and various cases are presented as follows.

– Case I: all three members share the same fitness value, which means that the team is in the area without fitness, and the leader will search randomly.
– Case II: two robots share the same better fitness value, then the gradient vector equals the center of the two better positions minus the worse one.
– Case III: two robots share the same worse fitness value, then the gradient vector equals the better position minus the center of the two worse ones.
– Case IV: all robots have different fitness values, then the gradient is perpendicular to the local contour line (Fig. 2(a)). In Fig. 2(a), A, B and C stand for the robots in a team, and the fitness values satisfy the inequality $f(A) > f(B) > f(C)$. Based on the local linear variation, $f(B') = f(B)$ and the position of B' can be calculated from Eq. 1. The line BB' serves as a contour line, whose vertical vector $B'P$ is the gradient direction (Eq. 2).

$$BB' = BC + CA \cdot \frac{f(B) - f(C)}{f(A) - f(C)}. \tag{1}$$

$$\begin{cases} B'P \cdot BB' = 0 \\ B'P \cdot B'A > 0 \end{cases}. \tag{2}$$

Role Switching. In order to avoid turning abruptly, to facilitate the maintenance of formation and the estimation of gradient direction, a role switching trick is introduced. At each iteration, the robot with the maximum fitness value serves as the leader while the other two determine their roles (i.e. left wing or right wing) according to their relative positions. As is shown in Fig. 2(b), robot A is the leader and A' is its next position. AP is the right vertical vector of AA'. If $AB \cdot AP > AC \cdot AP$, then robot B serves as the right wing, else the left wing.

Formation Control. Each robot is assumed to be equipped with a compass. The leader will broadcast its next position within the team per iteration, and its members will determine their roles and next positions. As is shown in Fig. 2, given the positions of A and A', the positions of left wing and right wing (i.e. L and R) can be calculated. To maintain the formation, the leader will monitor the distances(D) from itself to its members, and slow down if the distance exceeds a certain threshold (T), otherwise accelerate for high efficiency (Eq. 3, where α and β are factors for deceleration and acceleration). Parameters α and β are set to 0.75 and 1.33 respectively while T is set to $0.8 * Lengh$, where $Length$ $(0.8 * 2r_t)$ is the ideal side length of the triangle.

$$V_{leader} = \begin{cases} V_{leader} \cdot \alpha, D > T \\ V_{leader} \cdot \beta, D \leq T \end{cases}. \tag{3}$$

4   Simulation Results and Discussions

4.1   Algorithms for Comparison

Four searching strategies are chosen as comparison algorithms, which are TFS, IS, RPSO and IGES. All parameters of algorithms are tuned under the same experimental conditions, where the map size is 1000*1000, containing 50 robots and 10 targets. Details and parameters values of TFS strategy are presented in section "Triangle Formation Search Strategy".

The Independent Search (IS) strategy, is a new random walk strategy combining linear ballistic motion with triangle estimation technology. In areas without fitness, robots will move along a straight line until perceiving fitness values. In areas with fitness, robots will estimate the gradient direction using history information and triangle estimation technology introduced in Section "Gradient Estimation". Current position, the best and worst positions in history serve as the vertices. In the TFS strategy, if a robot does not belong to any team, it will search alone according to the IS strategy.

In Robotic Particle Swarm Optimization (RPSO [6]), each robot acts as a particle and determines the gbest individual in a spacial-based topology. And IGES [8] is an improved version of GES [7], and the basic idea for intra-group cooperation is moving the group center towards the center of best positions in the group, or splitting the group when the group size exceeds some threshold value or members within the group share the same fitness value.

4.2   Experimental Setup and Results

The map size is 1000*1000 and 10 targets are generated randomly, covering about 70 % of the map, while other setup information is stated in Section "An Idealized Model". In the experiment, six tests are carried out with 25, 50, 75, 100, 150, 200 robots in turn. In each test, 20 random maps are generated and each strategy is repeated for 20 times, and the results in this section are the average value of these 400 runs.

The criterion for measuring the efficiency of searching strategies, is the number of iterations required for a swarm of robots to collect all 10 targets. Another criterion indicating the computational load is the CPU time used by the swarm in simulation. The experimental results are presented in Table 1 and Fig. 3.

Table 1. Iterations and time costed by each strategy at various population sizes.

Population	RPSO		IGES		IS		TFS	
	Iterations	Time	Iterations	Time	Iterations	Time	Iterations	Time
25	587.77	23.91	294.68	15.39	**275.35**	**9.47**	312.54	14.74
50	417.94	38.90	240.00	20.13	229.74	**7.87**	**211.07**	20.14
75	374.13	64.78	217.62	41.87	208.47	**12.25**	**178.35**	25.06
100	334.41	94.95	205.07	46.09	195.72	**14.52**	**166.21**	38.21
150	295.97	196.71	189.36	81.48	176.76	**20.22**	**147.13**	64.33
200	269.84	347.61	180.45	151.41	167.71	**27.34**	**138.99**	104.85

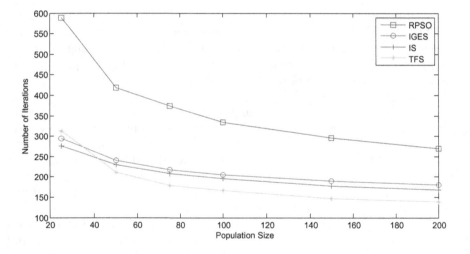

Fig. 3. Iterations costed by each strategy at various population sizes.

4.3   Discussions

As the results show, the efficiency of RPSO is the lowest among four strategies, though we have introduced a random vector to improve its performance by avoid robots' vibrating. We can infer that traditional heuristic algorithms for high dimensional optimization may not apply to swam robotics, which focuses on two or three-dimensional problem scenarios, and the key distinction is the simplicity of gradient estimation in low dimensional cases even with interference. It is worth pointing out that there are few local extrema in the fitness landscape, which plays a critical role to such results.

In the figure, there is a stable difference between IS and IGES. Since IGES has diffusion mechanism and IS works as random walk in areas without fitness, both of them possess excellent exploration ability. So the difference mainly results from the triangle estimation technology, which provides a more accurate direction to approximate the local gradient. Actually, additional experiments show that the IS integrated with independent strategies from IGES instead of triangle estimation technology, shares similar performance with IGES.

When the population size is small, the exploration ability is important for a search strategy, for IS shows the best performance while TFS performs poorly when the number of robots is 25. Under such circumstances, some mechanisms for maintaining connectivity, such as mutual attraction or formation control, may limit the exploration range of the swarm.

As the population size becomes larger (such as 50 or larger), the potential merits of the TFS strategy emerge and it outperforms other three strategies. When the population size is 75, compared with IS and IGES, the search efficiency of TFS increases 14.45% and 18.04% respectively, which means the triangle formation improves the local exploitation ability of the swarm and leads to a more accurate gradient direction than that estimated in IS. In addition, the downtrends of the curves in Fig. 3 demonstrate that both IS and TFS possess excellent scalability like IGES and RPSO. As we can see, three members are enough to construct a proper gradient direction while small team size tends to high exploration performance, so we adopt triangle formation technology.

As to the CPU time costed by each strategy, the IS has the overwhelming superiority, benefiting from its simplicity, while the TFS surpasses the other two strategies. Although the algorithm flow of TFS is kind of complex, its computation load is not heavy, which is a nice property for swarm robotics.

The performance of the IS strategy is qualified as a benchmark for such problem scenario in terms of efficiency and energy conservation.

5   Conclusion

In this paper, a triangle formation search (TFS) strategy and an independent search (IS) strategy were proposed, both of which bear high search efficiency and light computational load compared with RPSO and IGES. Among all four strategies, the TFS is the most efficient while the IS is the most energy-efficient, demonstrating the validity of the triangle estimation technology. In addition, the IFS and IS also show great scalability like RPSO and IGES.

As far as we know, technologies for formation control and rand walk have not been applied to the multiple targets search task in swarm robotics, and the IS strategy is qualified as a benchmark for its simplicity and performance.

Acknowledgements. This work was supported by the Natural Science Foundation of China (NSFC) under grant no. 61375119 and Supported by Beijing Natural Science Foundation (4162029), and partially supported by National Key Basic Research Development Plan (973 Plan) Project of China under grant no. 2015CB352302.

References

1. Tan, Y.: A Survey on Swarm Robotics. In: Li, J. (ed.) Handbook of Research on Design, Control, and Modeling of Swarm Robotics, 1. IGI Global, Hershey (2015)
2. Şahin, E.: Swarm robotics: from sources of inspiration to domains of application. In: Şahin, E., Spears, W.M. (eds.) Swarm Robotics 2004. LNCS, vol. 3342, pp. 10–20. Springer, Heidelberg (2005)
3. Li, J., Tan, Y.: The multi-target search problem with environmental restrictions in swarm robotics. In: 2014 IEEE International Conference on Robotics and Biomimetics (ROBIO), pp. 2685–2690. IEEE, December 2014
4. Gazi, V., Passino, K.M.: Stability analysis of social foraging swarms. IEEE Trans. Syst. Man Cybern. Part B Cybern. **34**(1), 539–557 (2004)
5. Krishnanand, K.N., Ghose, D.: A glowworm swarm optimization based multi-robot system for signal source localization. In: Liu, D., Wang, L., Tan, K.C. (eds.) Design and Control of Intelligent Robotic Systems. SCI, vol. 177, pp. 49–68. Springer, Heidelberg (2009)
6. Couceiro, M.S., Rocha, R.P., Ferreira, N.M.: A novel multi-robot exploration approach based on particle swarm optimization algorithms. In: 2011 IEEE International Symposium on Safety, Security, and Rescue Robotics (SSRR), pp. 327–332. IEEE, November 2011
7. Zheng, Z., Tan, Y.: Group explosion strategy for searching multiple targets using swarm robotic. In: 2013 IEEE Congress on Evolutionary Computation (CEC), pp. 821–828. IEEE, June 2013
8. Zheng, Z., Li, J., Li, J., Tan, Y.: Improved group explosion strategy for searching multiple targets using swarm robotics. In: 2014 IEEE International Conference on Systems, Man and Cybernetics (SMC), pp. 246–251. IEEE, October 2014
9. Tan, Y., Zhu, Y.: Fireworks algorithm for optimization. In: Tan, Y., Shi, Y., Tan, K.C. (eds.) ICSI 2010, Part I. LNCS, vol. 6145, pp. 355–364. Springer, Heidelberg (2010)
10. Zheng, Z., Li, J., Li, J., Tan, Y.: Avoiding decoys in multiple targets searching problems using swarm robotics. In: 2014 IEEE Congress on Evolutionary Computation (CEC), pp. 784–791. IEEE, July 2014
11. Shlesinger, M.F.: Mathematical physics: search research. Nature **443**(7109), 281–282 (2006)
12. Viswanathan, G.M., Buldyrev, S.V., Havlin, S., Da Luz, M.G.E., Raposo, E.P., Stanley, H.E.: Optimizing the success of random searches. Nature **401**(6756), 911–914 (1999)
13. Bartumeus, F., Raposo, E.P., Viswanathan, G.M., da Luz, M.G.: Stochastic optimal foraging theory. In: Lewis, M.A., Maini, P.K., Petrovskii, S.V. (eds.) Dispersal, Individual Movement and Spatial Ecology. LNM, vol. 2071, pp. 3–32. Springer, Berlin, Heidelberg (2013)
14. Balch, T., Arkin, R.C.: Behavior-based formation control for multirobot teams. IEEE Trans. Robot. Autom. **14**(6), 926–939 (1998)
15. Amory, A., Meyer, B., Osterloh, C., Tosik, T., Maehle, E.: Towards fault-tolerant and energy-efficient swarms of underwater robots. In: 2013 IEEE 27th International Parallel and Distributed Processing Symposium Workshops and Ph.D. Forum (IPDPSW), pp. 1550–1553. IEEE, May 2013

A Bio-inspired Autonomous Navigation Controller for Differential Mobile Robots Based on Crowd Dynamics

Alejandro Rodriguez-Angeles[1,2]([✉]), Henk Nijmeijer[2],
and Fransis J.M. van Kuijk[2]

[1] Mechatronics Section, Electrical Engineering Department,
Cinvestav-IPN, Mexico City, Mexico
aangeles@cinvestav.mx, h.nijmeijer@tue.nl, f.j.m.v.kuijk@student.tue.nl
[2] Dynamics and Control Group, Department of Mechanical Engineering,
Eindhoven University of Technology, Eindhoven, The Netherlands
http://www.meca.cinvestav.mx, http://www.tue.nl

Abstract. This article extends ideas from crowd dynamics to a navigation controller for mobile robots. Each mobile robot is considered as an agent, associated to a comfort zone with a certain radius, which creates a repulsive force when this comfort zone is violated by its environment or by another agent, therefore avoiding collisions. Meanwhile, attractive forces drive the agents from their instantaneous position to a goal position. The resulting navigation controller is tested by simulations and experiments. It is found that simulations capture the global dynamic behavior that is shown in experiments, showing robustness of the proposed navigation controller.

Keywords: Crowd dynamics · Behavior · Mobile robot · Obstacle avoidance · Navigation

1 Introduction

When developing autonomous vehicles two major issues have to be considered, first position determination of the vehicle, second obstacle collision avoidance. For obstacle or collision avoidance several methods are proposed, some of them based on potential fields or geometric relations between the vehicle and obstacles. Potential fields are in general based on external sensors, or when the environment is structured and well known, then internal sensors might be used. A potential field is constructed around the obstacle [2], such that the distance to the obstacle determines a repulsive force. Constructing a potential field around an obstacle implies knowing the dimensions of the object and its position on beforehand.

The so called Geometric Obstacle Avoidance Control Method, GOACM, [3], is based on onboard sensors. When one of the sensors detects an obstacle within a safety bound the GOACM is triggered. The GOACM uses the distance and angle related to the sensor to determine a collision free way point, that is commonly located on a specified distance on a perpendicular line to the obstacle.

© Springer International Publishing Switzerland 2016
Y. Tan et al. (Eds.): ICSI 2016, Part II, LNCS 9713, pp. 553–560, 2016.
DOI: 10.1007/978-3-319-41009-8_60

When considering biological navigation systems, we can find schools of fish, herds of quadrupeds, flocks of flying birds, and human crowd motion. A way to model such systems is by considering self-driven particles, see [1]. Research on crowd dynamics helps predicting large crowd behavior at festivals or football matches as described in [4,5]. When information regarding the movement of the crowd is known, it can be used to describe natural phenomena like flocking behavior of birds, herds of animals or bacterial groups.

In this article the goal is to provide autonomy to a differential mobile robot, considered as an agent, in order to move from an initial to a goal position, evading dynamic and static obstacles, such as other agents and environment boundaries. Our controller is based on the crowd dynamics model introduced by Helbing [5], which describes the behavior of multiple agents with a generalized force model. Each agent has its own comfort zone, which when violated by another agent, wall or static obstacle, generates a repulsive force to drive the agent away from the obstacle. When the agent is free of collision, the dominant forces are attractive to a goal position, for that, straight line direction and distance are computed from the actual instantaneous position to the goal position.

In [5] the generalized force model is applied on the dynamics of a mass point, and the goal is to move in a prescribed direction at a desired velocity. Extending the ideas from Helbing [5] to the case of unicycle mobile robot is not immediate, since the last one is generally model at kinematic level and control by translational and angular velocities, while in [5] acceleration of the the mass point is considered. In this article, an extension is proposed resulting in a bio-inspired autonomous navigation controller, such that navigation performance resembles the behavior of an individual in a crowd and constraint environment, when moving to a desired position. The proposed navigation controller is tested by simulations and experiments by using several e-puck mobile robots.

2   Crowd Dynamics Model

The generalized force model as described by Helbing et al. [5] assumes the forces on each agent to consist of both socio-psychological and physical forces. By adding up these forces the total resulting force can be calculated as in (1), where it is considered a particle of mass m_i moving with a velocity vector $\mathbf{v}_i(t)$ and acceleration vector $\dot{\mathbf{v}}_i$, such that v_{x_i}, v_{y_i}, $\dot{v}_{x_i}$ and $\dot{v}_{y_i}$ correspond to the velocity and acceleration vector components in x and y direction respectively, $\mathbf{f}_{ij}$ represent the reaction forces among ith agent and other agents, and $\mathbf{f}_{iW}$ the reactions forces between ith agent and its environment or obstacles.

$$m_i\dot{\mathbf{v}}_i = m_i\frac{v_i^0(t)\mathbf{e}_i^0(t) - \mathbf{v}_i(t)}{\tau_i} + \sum_{j(\neq i)} \mathbf{f}_{ij} + \sum_W \mathbf{f}_{iW} \tag{1}$$

The first term in (1) causes each agent to regulate its velocity $\mathbf{v}_i(t)$ to a desired velocity of magnitude v_i^0, in a desired direction $\mathbf{e}_i^0$, with a characteristic time τ_i. All agents thus move towards their own specified desired direction which is why Helbing et al. [5] describes this term as socio-psychological.

The terms $\mathbf{f}_{ij}$ generate repulsive interaction forces between agents i and j.

$$\mathbf{f}_{ij} = \left[A_i e^{[(r_{ij}-d_{ij})/B_i]} + kg(r_{ij} - d_{ij})\right] \mathbf{n}_{ij} + \kappa g(r_{ij} - d_{ij})\Delta v_{ji}^t \mathbf{t}_{ij} \qquad (2)$$

The tendency for agents to stay away from each other is given by $A_i e^{[(r_{ij}-d_{ij})/B_i]}$, with constants A_i and B_i. The sum of the radii of two agents comfort zones r_i and r_j is r_{ij}. The distance between two agents is $d_{ij} = \|\mathbf{r}_i - \mathbf{r}_j\|$, with $\mathbf{r}_i$ and $\mathbf{r}_j$ their positions. The normalized vector from agent j to agent i is $\mathbf{n}_{ij} = (\mathbf{r}_i - \mathbf{r}_j)/d_{ij}$. The terms $kg(r_{ij}-d_{ij})\mathbf{n}_{ij}$ and $\kappa g(r_{ij}-d_{ij})\Delta v_{ji}^t \mathbf{t}_{ij}$ render normal and tangential forces respectively, that are present when agents get in each others comfort zone. The normal and tangential forces, push and turn the agents away from each other respectively. Function $g(x)$ is zero if $d_{ij} \geq r_{ij}$ this is when the comfort zones aren't violated; and is equal to its argument when $d_{ij} < r_{ij}$. The tangential vector is $\mathbf{t}_{ij} = (-n_{ij,y}, n_{ij,x})$, based on the normal vector $\mathbf{n}_{ij} = [n_{ij,x}\ n_{ij,y}]$. The tangential velocity difference can be expressed as $\Delta v_{ji}^t = (\mathbf{v}_j - \mathbf{v}_i) \cdot \mathbf{t}_{ij}$.

The terms $\mathbf{f}_{iW}$ are interaction forces between agent i and its environment or static obstacles, it depends on the approaching velocity of i-th agent.

$$\mathbf{f}_{iW} = \left[A_i e^{[(r_i-d_{iW})/B_i]} + kg(r_i - d_{iW})\right] \mathbf{n}_{iW} + \kappa g(r_i - d_{iW})(\mathbf{v}_i \cdot \mathbf{t}_{iW})\mathbf{t}_{iW} \quad (3)$$

The distance between agent i and the static obstacle W is d_{iW}, while $\mathbf{n}_{iW}$ is the normalized vector pointing from the obstacle W to agent i; $\mathbf{t}_{iW}$ is the tangential vector. In (2) and (3), k and κ are constant parameters that can be changed to influence the magnitude of the normal and tangential forces respectively.

2.1 Autonomous Navigation Controller Design

Each agent is represented by the kinematic continuous-time model of a unicycle mobile robot (4), with x-position, y-position and orientation angle θ, see Fig. 1. The inputs are the forward velocity and angular velocity (v, ω).

$$\begin{aligned}
\dot{x}_i &= v_i \cos\theta_i \\
\dot{y}_i &= v_i \sin\theta_i \\
\dot{\theta}_i &= \omega_i
\end{aligned} \qquad (4)$$

To extend the generalized force model (1), to the kinematic model (4), the variables v_{x_i} and v_{y_i} (components of vector $\mathbf{v}_i$), are transformed to v_i and ω_i.

Mobile Robot Velocity Inputs from Generalized Force Model. The kinematic model input v_i is the magnitude of the velocity vector $\mathbf{v}_i$, i.e.

$$v_i = \sqrt{(v_{x_i})^2 + (v_{y_i})^2} \qquad (5)$$

Differentiating (4) with respect to time results in acceleration in x and y, and then it can be established that

$$\ddot{y}_i\dot{x}_i - \ddot{x}_i\dot{y}_i = v_i^2\omega_i \qquad (6)$$

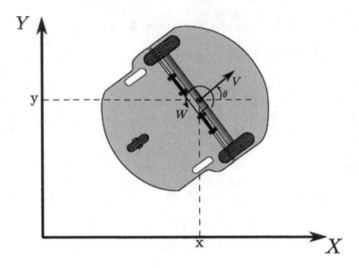

Fig. 1. Differential mobile robot.

From (1) and (4), we can set that $\dot{x}_i = v_{x_i}$, $\dot{y}_i = v_{y_i}$, $\ddot{x}_i = \dot{v}_{x_i}$ and $\ddot{y}_i = \dot{v}_{y_i}$, such that (5) and (6) yield the kinematic control input ω_i (7), where a positive small parameter $\epsilon \approx 0$ is introduced to avoid singularity when $v_i = 0$.

$$\omega_i = \frac{\dot{v}_{y_i} v_{x_i} - \dot{v}_{x_i} v_{y_i}}{\epsilon + \sqrt{(v_{x_i})^2 + (v_{y_i})^2}}, \tag{7}$$

Regulation to a Goal Position. Model (1) focuses on a particle moving along a desired velocity direction $\mathbf{e}_i^0$ with a desired magnitude v_i^0, as long as the i-th agent does not interact with other agents or obstacles. However, our goal is to move from actual (x_i, y_i) to a desired position (x_{d_i}, y_{d_i}). Thus, $\mathbf{e}_i^0$ is proposed as the difference between actual and desired position.

$$\mathbf{e}_i^0 = \begin{bmatrix} x_{d_i} \\ y_{d_i} \end{bmatrix} - \begin{bmatrix} x_i \\ y_i \end{bmatrix} \tag{8}$$

Equation (8) provides the vector and the distance between actual and goal position for i-th agent, thus the agent will move faster as farther from the desired position, and slower as closer to it; being still when getting the desired position. This might result in unsatisfactory performance because of slow convergence, [6]. To render faster convergence, normalization of $\mathbf{e}_i^0$ can be considered. Thus the desired velocity magnitude v_i^0 would be imposed if the agent is free of interaction with other agents or obstacles, and when the desired position is achieved $\mathbf{e}_i^0 = 0$. In this article normalization of $\mathbf{e}_i^0$ is considered for the simulation and experimental results.

3   Simulation and Experimental Results Comparison

To investigate the performance of the proposed controller, (5, 7, 8), six e-puck robots are used, [7]. With a camera system the position and orientation of each robot are determined, in order to compute the control signal for each robot, which is sent by wireless communication. For comparison, a simulator is made in Matlab, that creates a movie showing the movement of each agent. The blue lines and shapes represent the environment which, in this case is represented by multiple walls and a column. The experimental initial positions and orientation angles of the e-pucks, are set as initial conditions in the simulator.

Simulations and experiments are done in two different environments. The first one has a column obstacle in the middle, Fig. 2. The second one has multiple walls with a bottleneck through which the robots pass, Fig. 3. In both simulations and experiments $A_i = 0$ and $B_i = 1$, since these parameters influence the tendency for the agents to stay away from each other and obstacles. Normally the exponential term in (2) and (3) would cause the robots to repulse from obstacles, even though their comfort zones are not violated. By setting $A_i = 0$ this effect is removed.

Column Obstacle Scenery. Comparison between experiment and simulation is shown in Fig. 2. Blue lines are experiment paths while red lines are simulation paths. In this experiment/simulation six e-pucks are initially positioned left of a column obstacle and the robots are trying to move in a straight line to their target position at the right. As soon as an obstacle comes within range of their comfort zone, it is expected that they start to avoid it. The same is expected if the e-pucks are too close with respect to each other. The experiment and simulation parameters are the same for all agents, and are listed at Table 1. Table 2 presents numerical results of the simulation (x_s, y_s) and experiment (x_e, y_e), for initial time $t = 0$ and final time t_f.

Bottleneck Scenery. Figure 3 shows a comparison between experiment and simulation, blue lines are experiment paths and red lines are simulation paths. Notice that in order for the e-pucks to reach their final position at the right, they must pass through the small bottleneck in the middle, causing possible collisions. The parameters for this experiment are the same for all agents and can be seen in Table 3. Meanwhile Table 4 presents the numerical results of the simulation (x_s, y_s) and experiment (x_e, y_e) for initial time $t = 0$ and final time t_f.

Table 1. Experiment and simulation parameters, column obstacle scenery

Parameter	A_i	B_i	k	κ	τ_i	v_i^0 [m/s]	r_i [m]
Value	0	1	2	1	0.5	0.05	0.1

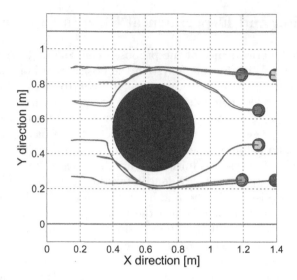

Fig. 2. Comparison between simulations and experiments, column obstacle scenery (Color figure online)

Table 2. Experiment and simulation results, column obstacle scenery

Robot	x_d	$x(0)$	$x_s(t_f)$	$x_e(t_f)$	y_d	$y(0)$	$y_s(t_f)$	$y_e(t_f)$
Green	1.2	0.1492	1.1912	1.1949	0.25	0.2689	0.2491	0.2518
Purple	1.4	0.3046	1.3903	1.4049	0.25	0.2689	0.2493	0.2489
Blue	1.3	0.1463	1.2903	1.2959	0.45	0.2689	0.4505	0.4510
Red	1.3	0.1567	1.2916	1.2961	0.65	0.2689	0.6494	0.6488
Yellow	1.4	0.3174	1.3908	1.4075	0.85	0.2689	0.8507	0.8498
Pink	1.2	0.1470	1.1911	1.1941	0.85	0.2689	0.8508	0.8551

Table 3. Experiment and simulation parameters, bottleneck scenery

Parameter	A_i	B_i	k	κ	τ_i	v_i^0 [m/s]	r_i [m]
Value	0	1	4	2	0.5	0.05	0.1

Table 4. Experiment and simulation results, bottleneck scenery

Robot	x_d	$x(0)$	$x_s(t_f)$	$x_e(t_f)$	y_d	$y(0)$	$y_s(t_f)$	$y_e(t_f)$
Green	1.4	0.2072	1.4002	1.3955	0.8	0.2555	0.7982	0.7936
Purple	1.4	0.1958	1.4008	1.4029	0.6	0.4727	0.6017	0.6082
Blue	1.4	0.1912	1.3987	1.3933	0.4	0.6830	0.3987	0.3786
Red	1.0	0.3761	0.9983	1.0028	0.2	0.7441	0.2001	0.1980
Yellow	1.0	0.3773	1.0014	0.9940	0.8	0.3736	0.8009	0.8015
Pink	1.4	0.1971	1.4015	1.4107	0.2	0.9125	0.2006	0.1912

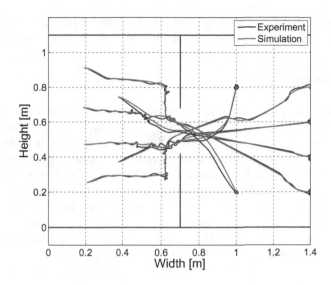

Fig. 3. Comparison between simulations and experiments, bottleneck scenery (Color figure online)

3.1   Results Discussion

Experiment and simulation produce a similar performance, see Figs. 2 and 3. Minimal differences between experiment and simulations, blue and red lines respectively, are noticed. All e-pucks move in a straight line to their target, until an obstacle or other agents get in their comfort zone, when they start turning away and going around, until they are free of collisions. One by one the robots are passing the column obstacle and the bottleneck in the middle until they reach their desired positions avoiding collisions.

Note that although the paths of the e-pucks in the simulation are not perfectly identical to the paths in the experiment, the simulation can roughly predict what the dynamics in the experiments are. The blue lines show that the e-pucks in the experiment have richer dynamics illustrated by the small bumps in their trajectories, Fig. 3. This happens when an e-puck has to make a fast turn, because violation of their comfort zone results in a repulsive force. Since the e-pucks are non-holonomic systems, the only way to react when they are pushed away is by making fast turns and going in the direction of the repulsive force.

Other differences between experiment and simulation are possibly due to the inaccurate model of the e-puck that is considered for the controller design, as well as shortcomings in the setup. The system that detects the position and angle of each e-puck has a poor performance, which could lead to errors. Another flaw in the setup is the dragging-behavior of the e-pucks, which are unbalanced and drag either with its front or back side of their frame over the ground. This results in an extra drag force which isn't accounted for. Another shortcoming is the sampling rate which is limited to 30 Hz due to the frame rate of the camera. This could be a limiting factor when the e-pucks show very fast dynamic behavior.

4   Conclusions

A forced based autonomous navigation controller that steers robots from their actual to a goal position has been designed. The results of applying the proposed navigation controller happens to be a full analogon of what happens in crowd dynamics. Interesting enough- given the simple model, and all further shortcomings, the result features a great similarity of simulation and experiments and is thus very robust. It is shown that interaction forces between agents and environment can be used to react on real time, such that the agent can achieve a goal position by evading other agents and obstacles. Simulations and experiments show that violation of comfort zones leads to very fast turning behavior, instead of moving directly in the direction of the repulsive force. From comparison results, it can be concluded that although the simulations are not completely identical with the experiments, they do capture the global dynamic behavior. In conclusion this type of work is thus highly suitable for understanding and analyzing crowd dynamics, as well as for designing autonomous navigation controllers.

References

1. Vicsek, T., Czirók, A., Ben-Jacob, E., Cohen, I., Shochet, O.: Novel type of phase transition in a system of self-driven particles. Phys. Rev. Lett. **75**(6), 1226–1229 (1995)
2. Kostic, D., Adinandra, S., van de Caarls, J., Wouw, N., Nijmeijer, H.: Collision-free tracking control of unicycle mobile robots. In: Proceedings of the 48th IEEE Conference on Decision and Control and 28th Chinese Control Conference (CDC/CCC), pp. 5667–5672 (2009)
3. Dai, Y., Lee, S.G.: Formation control of mobile robots with obstacle avoidance based on GOACM using onboard sensors. Int. J. Control Autom. Syst. **12**(5), 1077–1089 (2014)
4. Helbing, D., Buzna, L., Johansson, A., Werner, T.: Self-organized pedestrian crowd dynamics: experiments, simulations, and design solutions. Transp. Sci. **39**(1), 1–24 (2005). doi:10.1287/trsc.1040.0108
5. Helbing, D., Farkas, I., Vicsek, T.: Simulating dynamical features of escape panic. Nature **407**, 487–490 (2000)
6. Xuesong, C., Yimin, Y., Shuting, C., Jianping, C.: Modeling and analysis of multi-agent coordination using nearest neighbor rules. In: 2009 International Asia Conference on Informatics in Control, Automation and Robotics (2009)
7. http://www.e-puck.org/

Intelligent Energy
and Communications Systems

Reliability Evaluation of a Zonal Shipboard Power System Based on Minimal Cut Set

Wenzeng Du[✉], GenKe Yang, Jie Bai, Changchun Pan,
and Qingsong Gong

Key Laboratory of System Control and Information Processing,
Department of Automation, Ministry of Education,
Shanghai Jiao Tong University, Shanghai 200240, China
dwz@sjtu.edu.cn

Abstract. A new method of shipboard power system (SPS) reliability evaluation using the minimal cut set approach is developed. The method is dependent on the system network topology, as well as the component reliability parameters. Service interruption rate, indexed to quantify power system reliability, is calculated under normal condition and failure conditions. Under normal condition, the service interruption rate is used to indicate the reliability of each load zone. While under failure conditions where the interruption rate is calculated with conditional probability, the results are able to identify the importance of each element (cable) to specific load power supply.

Keywords: Minimal cut set · Service interruption rate · Reliability evaluation

1 Introduction

With the development of high-energy marine equipment and the insistent power demand for improving the ship maneuverability, researchers from United States first proposed the concept of integrated power system (IPS) [1]. Power systems have then changed their role from the conventional auxiliary status to main power system which set generation, distribution, storage and substation as a whole. Entering the 21 century, with the increasing use of power systems on civilian vessels and military naval ships, the distinct advantages of integrated power system technology have been reflected [2]. Immediately it becomes the mainstream of ship power technology development. The scale of SPS is increasing rapidly together with its complexity.

The evaluation of the overall system reliability is extremely complex as it is hard to include the detailed modelling of both generation and transmission facilities and their auxiliary elements [3]. Various factors such as aging equipment, manual improper operation, battle damage and so on could lead to system failure during operation [4, 5]. A fault of a single component can lead to the failure of power delivery to specific loads or in worst cases a full breakdown of the whole power system [6]. A small oversight could be catastrophic. So it's urging that we put emphasis on the reliability evaluation of SPS and help improved the system design and protection.

Most of the researches on power system reliability analysis use approximation or simplification to degrade the problem into a solvable level. In many cases, static

© Springer International Publishing Switzerland 2016
Y. Tan et al. (Eds.): ICSI 2016, Part II, LNCS 9713, pp. 563–572, 2016.
DOI: 10.1007/978-3-319-41009-8_61

analysis [7–9] can be used to estimate stability margins, identify factors influencing stability, and screen a wide range of system conditions and a large number of scenarios. The risk calculation provided in [10] accounts for both the future uncertainties on the system and the consequences associated with voltage collapse and violation of limits. And it use a cost-based risk index of voltage collapse to price reliability in order to make a trade off between reliability and economics. The application of the fault tree analysis for assessment of power system reliability is proposed in [3, 12], the fault trees are related to disruption of energy delivery from generators to the specific load points. Quantification of the fault tree enables identification of the important elements in the power system. The minimal cut-set method [13] of evaluating load-point reliability indices is proposed but it accounts for only topology of the network.

This work evaluates SPS reliability from the perspective of the overall distribution network topology. The method is specifically applied to a zonal SPS, but it is also applicable to other small-scale distribution system such as microgrid. The purpose of this paper is to develop a convenient method for SPS reliability evaluation, enable identification of the most important elements in the power system and help improve the design and protection work on SPS.

The rest of this paper is organized as follows: In Sect. 2, a description of a zonal SPS model is made. Then a transformed model, bus nodes directed graph, is presented as a research model. In Sect. 3, a reliability calculation method based on minimal cut set is proposed. Section 4 gives the simulation setup and results. Finally, the conclusion of this paper is drawn in Sect. 5.

2   Model Description

The conventional architecture of SPS distribution system is radial, for the capacity of the power system is usually small. While with the development of high-power IPS, the space and weight constraints on shipboard limit the redundancy needed for reconfiguration and restoration purposes. To give a maximum capacity for SPS in a limited space, researchers proposed some innovative distribution architecture such as ring, zonal and net architecture. The object of our study is based on a zonal architecture SPS [14].

2.1   Zonal Shipboard Power System

Referring to the MVDC model [14] put forward by the University of Texas at Austin, we proposed such a zonal SPS scaled-down model as shown in Fig. 1.

The scaled-down zonal architecture SPS model is shown in Fig. 1, which is presented in one-line diagram as a high-level view to favor readability. The generators on the schematic represent four 25 MVA synchronous generators. Each generator produces power at 4.16 kV, 240 Hz and includes a control model. Downstream of each generator are three-phase circuit breakers. Behind each three-phase breaker are AC-DC converters that power the main DC ring bus at 5.5 kV DC. The cables around the main DC bus were modeled as series resistive-inductive sections. Between each AC-DC converter and the main DC bus are disconnect switches. These disconnect switches were included in the model for completeness but they remained in their closed positions.

The main DC bus serves two propulsion loads and five zones. Each of the two propulsion loads was modeled as a three-phase static load served at 3.7 kV AC (5 MVA, 60 Hz). The five zones were each modeled as two parallel DC/DC converters serving a common resistive load (5 MW).

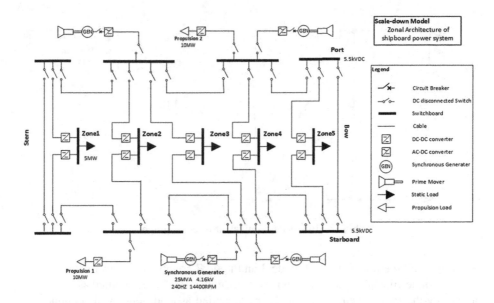

Fig. 1. Scale-down model of zonal SPS

In order to facilitate the following research, two assumptions are presented below:

1. The function of switchboard is to monitor and control the operation of power system. It is regarded that the failure of cables won't spread from one section to another. So we can presume that the fault between each cable is mutual independent.
2. Because of the power capacity redundancy, we can presume that the power supply of the load node is sufficient as long as there is at least one cable connected to generation node.

2.2 Equivalent Directed Graph

When the architecture of the power system is complicated, the general reliability analysis method, fault tree analysis, is hard to figure out the set of the bottom events. Under such circumstances, minimal cut set method is a convenient way to analyze the complex topology structure.

In order to facilitate our research, each switchboard (or bus) is modeled as a node, the cables connecting the switchboard are modeled as edges, then the zonal architecture SPS is abstracted into a bus node equivalent directed graph $G(N,E)$ as shown in Fig. 2, where $N = \{n_i \mid i = 1,2,\ldots,13\}$ is the set of the nodes and E is the set of the edges. Taking into account the current flow in real power system, the edges between $n_1, n_2,\ldots, n_8$ are bidirectional while the edges towards $n_9, n_{10},\ldots, n_{13}$ are unidirectional.

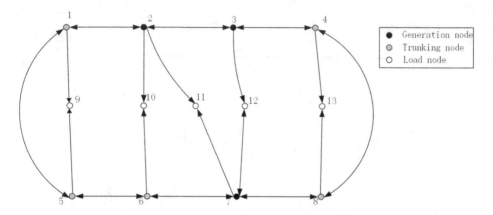

Fig. 2. Bus nodes equivalent directed graph

Among all the nodes, node 2, node 3 and node 7 are abstracted from switchboard directly connecting generators, which means node 2, 3 and 7 are specified as generation nodes. By the same token, node 9 to 13 are specified as load nodes which connecting the five load zones. While other nodes are considered as trunking nodes between generation nodes and load nodes. Therefore, the nodes set N are divided into three subsets, which is the generation node set $N_G = \{n_i \mid i = 2,3,7\}$, the load node set $N_L = \{n_i \mid i = 9,10,11,12,13\}$ and the trunking node set $N_T = \{n_i \mid i = 1,4,5,6,8\}$.

3   Reliability Evaluation Method

From the perspective of qualitative research, system reliability analysis is the evaluation of how often a system is expected to fail under normal condition. In the context of SPS distribution systems, reliability analysis is quantificationally evaluated by the index: service interruption rate [12].

According to assumption 2 above, a service interruption to a specified load is defined as the load being electrically isolated from all generation units. The index service interruption rate is the evaluation of how often a load node is expected to be de-energized (to disconnect from its source), which signifies the expected number of service interruptions that a certain load will experience in a given year.

3.1 Algorithm Design

This paper adopts a minimal cut set algorithm based on adjacent matrix to analyze and assess the reliability of SPS. By analyzing the topology of the bus node diagram we can easily get the adjacent matrix and accessible matrix, then the minimal cut set corresponding to each load node can be computed by matrix operations. Combined with the equipment reliability parameters, the service interruption rate of specific load node can be calculated. The calculation steps are as follows.

Step 1: Input the SPS topology information in the form of adjacent matrix, determine node set N_G, N_L and N_T. Input equipment reliability parameters.

Step 2: For each load node L_i, compute the minimal cut set C_{ij} from G_j to L_i (which $L_i \in N_L, G_j \in N_G$). Then the failure events causing the load service interruption can be calculated by equation:

$$C_i = \bigcap_{j \in G} C_{ij} \tag{1}$$

Step 3: For each load node L_i, Calculate the service interruption rate $P(\bar{L}_i)$ according to the minimal cut sets C_i and equipment reliability parameters, where $\bar{L}_i$ is the event that load node L_i out of service.

Step 4: Remove the edge in the minimal cut set C_i successively, compute the service interruption rate $P(\bar{L}_i|\bar{B}_i)$. Calculate the probability gain $K_{\bar{B}_i}$ by equation:

$$K_{\bar{B}_i} = \frac{P(\bar{L}_i|\bar{B}_i)}{P(\bar{L}_i)} \tag{2}$$

In this paper, probability gain is considered as the evaluation index of system reliability. It can be used to represent the importance of each edge to specific load node.

4 Simulation Setup and Results

4.1 Minimal Cut Sets

Using the method above, we can calculate the minimal cut sets of each load node. For illustration, the reliability diagram of load node 9 is shown in Fig. 3. From Fig. 3 we can see that, the generation node 2, 3 and 7 are aggregated as one node. Node 1, node 5 and node 6 are trunking node connecting generation nodes and load node. The objective of the minimal cut set method here is to enumerate all the edge cut sets that disconnecting generation nodes and load node 9. The diagram shows the typical second order cut (1-2, 6-7) and the typical third order cut (1-5, 1-9, 5-6).

The minimal cut set of each load node to generation lode can be calculated as shown in Table 1. From the results it can be inferred that the failure events of load node 9 is the most. Load node 9 has 3 s order cut and 3 third order cut, which has the maximum number among all the five load nodes. So it can be inferred that load node 9 has more probability to encounter failures, and the work behind has well proved the inference.

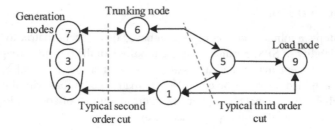

Fig. 3. Reliability diagram of load node 9

Table 1. The minimal cut sets of load node to generation node

Load node	Minimal cut sets
9	{(1-2, 5-6), (1-2, 6-7), (1-9, 5-9), (1-2, 1-5, 5-9), (1-5, 1-9, 5-6), (1-5, 1-9, 6-7)}
10	{(2-10, 6-10), (1-2, 6-7, 2-10), (1-5, 6-7, 2-10), (5-6, 6-7, 2-10)}
11	{(2-11, 7-11)}
12	{(3-12, 7-12)}
13	{(3-4, 7-8), (4-13, 8-13), (3-4, 4-8, 8-13), (4-8, 7-8, 4-13)}

4.2   Component Reliability

In between two main switchboards, there are two auto switches and one main cable (as shown in Fig. 5). In between the main switchboard and distribution switchboard, there are one auto switch, one distribution cable and one DC/DC converter (as shown in Fig. 4).

Fig. 4. Component structure of main bus

Fig. 5. Component structure of distribution bus

The main cable set M = {1-2, 2-3, 3-4, 5-6, 6-7, 7-8, 1-5, 4-8}. Considering the equipment reliability parameters from Table 2, the service interruption rate $P(\bar{B}_m)$ can be calculated by:

$$P(\bar{B}_m) = P_2 + 2 \times P_6 = 0.003, \forall m \in M \tag{3}$$

Distribution cable set D = {1-9, 5-9, 2-10, 6-10, 2-11, 7-11, 3-12, 7-12, 4-13, 8-13}. The service interruption rate $P(\bar{B}_d)$ can be calculated by:

$$P(\bar{B}_d) = P_3 + P_4 + P_6 = 0.005, \forall d \in D \tag{4}$$

The failure rate is defined as the expected number of failures that a given component will occur over the period of one year.

Table 2. Equipment reliability parameters

Equipment name	Symbol	Failure rate (per year)
Generator	P_1	0.1
Main cable	P_2	0.001
Distribution cable	P_3	0.002
Converter	P_4	0.002
Breaker	P_5	0.007
Auto Switch	P_6	0.001
Manual Switch	P_7	0.004

4.3 Service Interruption Rate

For lode node 9, the minimal cut set causing service interruption is {(1-2, 5-6), (1-2, 6-7), (1-9, 5-9), (1-2, 1-5, 5-9), (1-5, 1-9, 5-6), (1-5, 1-9, 6-7)}. The service interruption rate of load node 9 under three typical operation conditions is calculated below:

1. Normal condition (All the power system is intact):

$$P(\bar{L}_9) = P(\bar{B}_{1-2} \cap \bar{B}_{5-6}) + P(\bar{B}_{1-2} \cap \bar{B}_{6-7}) + P(\bar{B}_{1-9} \cap \bar{B}_{5-9}) +$$
$$P(\bar{B}_{1-2} \cap \bar{B}_{1-5} \cap \bar{B}_{5-9}) + P(\bar{B}_{1-5} \cap \bar{B}_{1-9} \cap \bar{B}_{5-6}) + P(\bar{B}_{1-5} \cap \bar{B}_{1-9} \cap \bar{B}_{6-7}) \tag{5}$$
$$= 4.3135 \times 10^{-5}$$

2. Failure condition I (main cable 1-2 failure):

$$P(\bar{L}_9 | \bar{B}_{1-2}) = P(\bar{B}_{5-6}) + P(\bar{B}_{6-7}) + P(\bar{B}_{1-9} \cap \bar{B}_{5-9}) + P(\bar{B}_{1-5} \cap \bar{B}_{5-9}) +$$
$$P(\bar{B}_{1-5} \cap \bar{B}_{1-9} \cap \bar{B}_{5-6}) + P(\bar{B}_{1-5} \cap \bar{B}_{1-9} \cap \bar{B}_{6-7}) = 6.0251 \times 10^{-3} \tag{6}$$

The probability gain under main cable 1-2 failure:

$$K_{\bar{B}_{1-2}} = \frac{P(\bar{L}_9 | \bar{B}_{1-2})}{P(\bar{L}_9)} = 139.1 \tag{7}$$

3. Failure condition II (main cable 5-6 failure):

$$P(\bar{L}_9 | \bar{B}_{5-6}) = P(\bar{B}_{1-2}) + P(\bar{B}_{1-2} \cap \bar{B}_{6-7}) + P(\bar{B}_{1-9} \cap \bar{B}_{5-9}) + P(\bar{B}_{1-2} \cap \bar{B}_{1-5} \cap \bar{B}_{5-9})$$
$$+ P(\bar{B}_{1-5} \cap \bar{B}_{1-9}) + P(\bar{B}_{1-5} \cap \bar{B}_{1-9} \cap \bar{B}_{6-7}) = 3.05 \times 10^{-3} \tag{8}$$

The probability gain under main cable 5-6 failure:

$$K_{\bar{B}_{5-6}} = \frac{P(\bar{L}_9|\bar{B}_{5-6})}{P(\bar{L}_9)} = 69.5 \tag{9}$$

Compare $K_{\bar{B}_{1-2}}$ with $K_{\bar{B}_{5-6}}$, The probability gain under cable 1-2 failure is near twice of the probability gain under cable 5-6 failure. The result suggested that the importance of cable 1-2 for load node 9 is higher than cable 5-6.

The service interruption rate of each load node under normal condition is presented below (Table 3). The interruption rate of load node 9 is the highest, load node 13 takes second place, load node 10, 11 and 12 are almost the same and the lowest. Corresponding to the SPS network, the topological distance between load node 9 and generation nodes is the farthest, load node 10, 11 and 12 are directly connected with generation loads. So the former will encounter more uncertainties and more risks with respect to the latter.

Table 3. The service interruption rate of each load node under normal condition

Load Node	Service interruption rate
9	4.3135e-5
10	2.5135e-5
11	2.5000e-5
12	2.5000e-5
13	3.4090e-5

The interruption rate for every load node under all the failure conditions are presented from Tables 4, 5, 6, 7 and 8 in the descending order, showing the importance of the cable to each load node.

Table 4. The service interruption rate of Load Node 9 under failure conditions

Fault section (cable)	Service interruption rate	Probability gain
1-2	0.006	139.0982
1-9	0.005	115.9152
5-9	0.005	115.9152
5-6	0.003	69.54909
6-7	0.003	69.54909
1-5	8.800e-5	2.040107

Table 5. The service interruption rate of Load Node 10 under failure conditions

Fault section (cable)	Service interruption rate	Probability gain
2-10	0.005	198.9258
6-10	0.005	198.9258
6-7	7.000e-5	2.784961
1-2	4.009e-5	1.594987
1-5	4.009e-5	1.594987
5-6	4.009e-5	1.594987

Table 6. The service interruption rate of Load Node 11 under failure conditions

Fault section (cable)	Service interruption rate	Probability gain
2-11	0.005	200
7-11	0.005	200

Table 7. The service interruption rate of Load Node 12 under failure conditions

Fault section (cable)	Service interruption rate	Probability gain
3-12	0.005	200
7-12	0.005	200

Table 8. The service interruption rate of Load Node 13 under failure conditions

Fault section (cable)	Service interruption rate	Probability gain
4-13	0.005	146.6706
8-13	0.005	146.6706
3-4	0.003	88.00235
7-8	0.003	88.00235
4-8	6.400e-05	1.877383

5 Conclusion

To analyze the reliability of load zones in SPS, the actual model is abstracted into an equivalent bus node directed graph. Considering the network topology, the search for failure events causing service interruption is transformed into the calculation of the minimal cut sets from load node to generation nodes. Referring to the component reliability parameters, service interruption rate of each load node is then calculated. The results demonstrate the reliability of the load node under normal condition.

Further, we considered the interruption rate under various failure conditions which are calculated with conditional probability. The results indicate the importance of each distribution cable to specific load node, enable identification of the most important element in the distribution system, which are significant to the design and protection of SPS.

Acknowledgements. This work is supported by the National Science Foundation of China under Grant 61203178, Grant 61290323 and Grant 61304214.

References

1. Doerry, N.H., Davis, J.C.: Integrated power system for marine applications. J. Naval Eng. J. **106**(3), 77–90 (1994)
2. Gong, Y., Huang, Y., Schulz, N.: Integrated protection system design for shipboard power system. In: Electric Ship Technologies Symposium, 2005, pp. 237–243. IEEE (2005)

3. Volkanovski, A., Čepin, M., Mavko, B.: Application of the fault tree analysis for assessment of power system reliability. J. Reliab. Eng. Syst. Saf. **94**(6), 1116–1127 (2009)
4. Miao, L.H.L.Z.W., Shaoguang, Z.L.J.: Contrasting analysis of shipboard power system topological structures based on reliability model. J. Trans. China Electrotech. Soc. **11**, 010 (2006)
5. Zhiyu, S.C.W.D.Z.: On distribution structure and reliability of shipboard loop area. J. Ship Boat **6**, 011 (2008)
6. Brown, R.E.: Electric Power Distribution Reliability. CRC Press, Boca Raton (2008)
7. Morison, G.K., Gao, B., Kundur, P.: Voltage stability analysis using static and dynamic approaches. J. IEEE Trans. Power Syst. **8**, 1159–1171 (1993)
8. Mashayekh, S., Butler-Purry, K.L.: Assessing the dynamic secure region for an all-electric ship model. In: North American Power Symposium (NAPS), pp. 1–6. IEEE (2012)
9. Ajjarapu, V., Christy, C.: The continuation power flow: a tool for steady state voltage stability analysis. In: Proceedings of IEEE Power Industry Computation Application Conference, pp. 304–311, May 1991
10. Wan, H., McCalley, J.D., Vittal, V.: Risk based voltage security assessment. J. IEEE Trans. Power Syst. **15**(4), 1247–1254 (2000)
11. Huang, J., Ji, X., He, H.: A model for structural vulnerability analysis of shipboard power system based on complex network theory. In: 2012 International Conference on Control Engineering and Communication Technology (ICCECT), pp. 100–104. IEEE (2012)
12. Stevens, B., Dubey, A., Santoso, S.: On improving reliability of shipboard power system. J. IEEE Trans. Power Syst. **30**(4), 1905–1912 (2015)
13. Awosope, C.O.A., Akinbulire, T.O.: A computer program for generating power-system load-point minimal paths. J. IEEE Trans. Reliab. **40**(3), 302–308 (1991)
14. Uriarte, F., Hebner, R., Mazzola, M., et al.: Parallelizing the simulation of shipboard power systems (2015)

Design of DS/FH Hybrid Spread Spectrum System Based on FPGA

Longjun Liu, Hongwei Ding[(✉)], Qianlin Liu, Weifeng Zhang,
and Zhenggang Liu

School of Information, Yunnan University, Kunming 650500, Yunnan, China
{893817028, 767950080, 641846684}@qq.com,
dhw1964@163.com, liuqianlin@sina.com

Abstract. Anti-interference ability is an important index to evaluate the physical fitness of communication system. Compared with general wireless communication systems, the DS/FH Hybrid spread spectrum system has high security and strong ability of anti-interception. This paper presents a design of DS/TH hybrid spread spectrum system based on FPGA. In the design, direct spread spectrum and frequency hopping spread spectrum of the input data is modulated by the spreading code. In the receiver, the pseudo random sequence is used to despread spread spectrum signal after capturing the sync signal. The simulation of the system is implemented by the combination of the hardware circuit description language Verilog HDL and the principle diagram on Quartus II software in the design.

Keywords: DS/FH · Spreading codes · Modulation · Demodulation · FPGA

1 Introduction

With many excellent characteristics, spread spectrum communication system develops rapidly in the fields of communication, navigation, measurement and other fields. Especially, it has strong anti-interference and anti-interception ability, which makes the spread spectrum communication widely used in military and mobile communication. The direct spread system and frequency hopping system are the most widely used in these two fields. In general, frequency hopping system is mainly used in the military communication to counter the deliberate interference, interception and other aspects. In the satellite communication, frequency hopping system is also used for secure communications, and spread spectrum system is mainly a civilian technology.

The combination of direct spread spectrum system and frequency hopping system constitutes a direct spread spectrum/frequency hopping (DS/FH) hybrid spread spectrum system. DS/FH hybrid spread spectrum system increases the function of carrier frequency hopping on the basis of direct sequence spread spectrum. The basic operation principle of the system is direct sequence spread spectrum [1]. The combination of DS and FH makes the system have good resistance to interception capability, communication distance farther than a single spread spectrum system. At the same time, this method can overcome the multipath effect and near-far effect. And it also has

© Springer International Publishing Switzerland 2016
Y. Tan et al. (Eds.): ICSI 2016, Part II, LNCS 9713, pp. 573–580, 2016.
DOI: 10.1007/978-3-319-41009-8_62

less interference to the same frequency equipment, etc. DS/FH hybrid spread spectrum system is recognized the most dynamical anti-interference system in all the world [2].

The method of the combination of microprocessors and digital logic unit is usually used to the investigation of the spread spectrum communication system. Due to the slow processing speed, limited logic resources, which limits the development of communication systems, the traditional methods is not conducive to system simulation [3]. FPGA can overcome all these problems, and it also has the characteristics of reprogrammable, field modifiable design and processing power, etc., which facilitate the research and development of communication systems.

2   The Principle of DS/FH Hybrid Spread Spectrum System

DS/FH hybrid spread spectrum system is a way of communication with the combination the direct sequence spread spectrum (DS) and frequency hopping spread spectrum(FH). In the sender of the system, information data is modulated by direct spread code firstly, and then the signal is modulated by frequency hopping code. And then, we can get the high frequency modulation DS/FH signals, which is sent out after high-pass filtering. At the receiver, the received signal is filtered, and then the modulated signal is demodulated in contrast to the sending end sequence. The low frequency data is captured after low-pass filtering. The despreading requires capturing synchronous signal. The correct solution expansion can not be achieved until the synchronization is achieved [4]. The schematic diagram of DS/FH system is shown in Fig. 1.

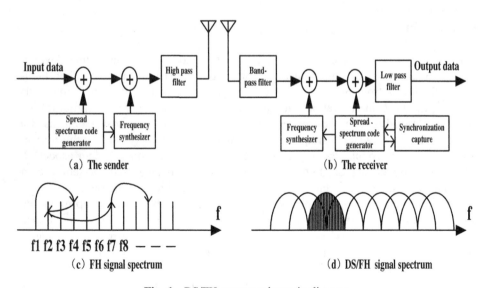

(a) The sender   (b) The receiver

f1 f2 f3 f4 f5 f6 f7 f8 — — —

(c) FH signal spectrum   (d) DS/FH signal spectrum

Fig. 1. DS/FH system schematic diagram

3 The System Overall Scheme

In the sender, firstly, the spread spectrum code is used to direct spread spectrum modulation of information data, and then the information data is modulated by frequency hopping code. After that, the spread spectrum signal mod_out can be got. At the receiver, after achieving synchronization, the modulated signal is demodulated in contrast to the sender sequence. After that, the demodulated signal emod_out can be got.

Comparing emod_out and the sending information data, the correctness and feasibility of the design of the DS/FH hybrid spread spectrum system can be verified.

4 Modular Design of the System

According to the scheme of DS/FH system, DS/FH spread spectrum system can be divided into subsystems of sending and receiving algorithm respectively. In the sending subsystem, it needs the module of spread spectrum code generation, the module of direct s and the module of frequency hopping. In receiving subsystem, it needs the module of local spread spectrum code generation, the module of synchronous capture, the module of frequency hopping despreading and the module of direct pread spectrum despreading.

4.1 Spread Spectrum Sending Subsystem

Spread spectrum sending subsystem is shown in Fig. 2. As is shown in Fig. 2, pn module is the module spread spectrum code generation. The pseudo random sequence is used to modulate of direct spread spectrum and control carrier of frequency hopping. The dsmod module is the spread spectrum modulation module and it completes the modulation of direct sequence spread spectrum. The mod module is frequency hopping modulation module, which achieves the modulation of frequency hopping. At the input, data is input information. At the output, ds_out is output of the direct spread spectrum modulation, mod_out is output of the DS/FH hybrid spread spectrum system. As is shown in Fig. 3, the output ds_out of spread spectrum modulation achieves the spread of the signal spectrum. The modulation output mod_out of DS/FH hybrid spread spectrum System achieves the spread of the frequency spectrum and the frequency hopping of the input information data. That is to say the spread spectrum sending subsystem achieves the designed function of the system.

4.2 Receiving Despreading Subsystem

The top design of DS/FH system is shown in Fig. 4. As is shown in Fig. 4, epn module, tongbu module, emodfh module and emodds module constitute the receiving

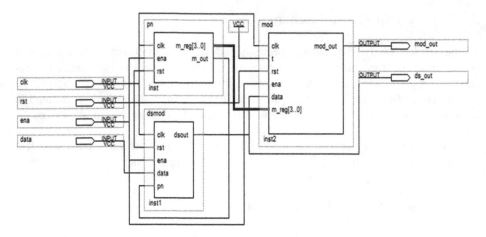

Fig. 2. Spread spectrum sending subsystem

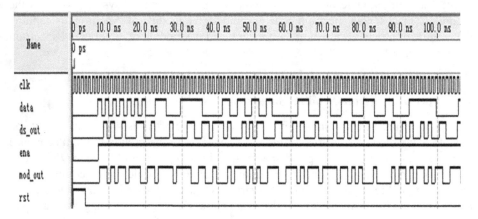

Fig. 3. Simulation of sending subsystem

despreading subsystem. The epn module is the module of local spread spectrum code generation. The pseudo-random sequence is used to demodulate the direct spread spectrum and control the carrier to demodulate the frequency hopping signal. The emodfh module achieves the function of frequency hopping despreading. The emodds module is the function module of direct spread spectrum demodulation. The tongbu module captures the sync signal with using the method of sliding correlation capturing, which controls the synchronization between the receiving end and the sending end strictly [5]. The program of the tongbu module is shown in Fig. 5.

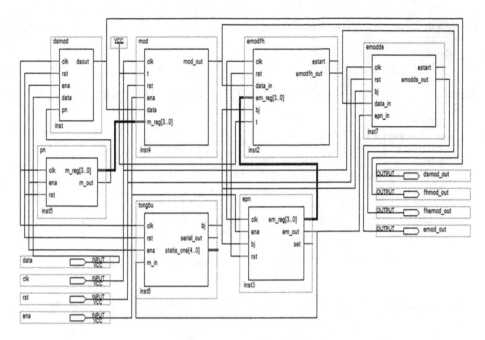

Fig. 4. The top design of DS/FH system

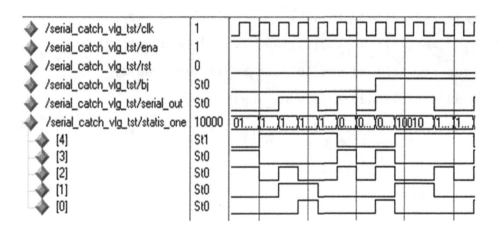

Fig. 5. Simulation of tongbu module

```
always @(posedge clk)
  if(rst)
    begin
      shift_reg[31:0]<=0;

constant_reg[31:0]<=32'b0010_0110_1011_1100_0100_1101_011
1_1000;
      bj<=0;
      add_reg[31:0]<=1;
      serial_in<=0;
    end
  else if(ena==1 && rst==0)
    begin
      serial_in<=m_in;
      shift_reg[31:0]<={shift_reg[30:0],serial_in};
      serial_out<=shift_reg[31];
      for(i=31;i>=0;i=i-1)
        add_reg[i]<=constant_reg[i]^shift_reg[i];

statis_one<=add_reg[31]+add_reg[30]+add_reg[29]+add_reg[2
8]

+add_reg[27]+add_reg[27]+add_reg[25]+add_reg[24]

+add_reg[23]+add_reg[22]+add_reg[21]+add_reg[20]

+add_reg[19]+add_reg[18]+add_reg[17]+add_reg[16]

+add_reg[15]+add_reg[14]+add_reg[13]+add_reg[12]

+add_reg[11]+add_reg[10]+add_reg[9]+add_reg[8]

+add_reg[7]+add_reg[6]+add_reg[5]+add_reg[4]

+add_reg[3]+add_reg[2]+add_reg[1]+add_reg[0];
          if (statis_one<=2'b10)
              begin
                  bj<=1;
              end
    end
```

In Fig. 5, the signal bj changes from 0 to 1, which means that the system achieved synchronization. DS/FH system simulation is shown in Fig. 6. The signal data is information data from the sender. The fhemod_out is the output of frequency hopping. The emod_out is the final output data of the whole system. From the waveform view, compared with the sending information data, emod_out has nearly 4 clock cycles delay. This is mostly due to the sequential circuit's own data transmission and processing, which is inevitable. In the design, the delay is nanosecond that can be received in the general communication system [6]. In addition, compared to the data, emod_out has a

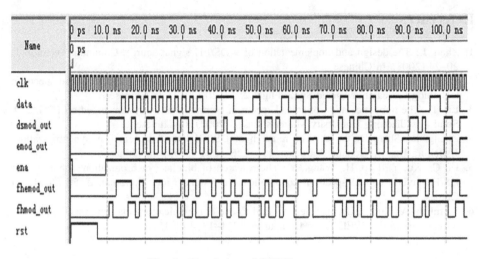

Fig. 6. Simulation of DS/FH system

little error in the initial stage. This is because that the system demodulates mistaken before it reaches synchronization. Beyond this two points, the demodulation output of the system is exactly same with the input information data in waveform. Implementing the designed system on FPGA, the result and simulation result are the same. It means that the design of DS/FH system is correct and feasible.

5   Conclusions

The characteristics and application fields of direct sequence spread spectrum system and frequency hopping spread spectrum system are analyzed in this paper. According to the advantages and disadvantages of the two systems, two systems are combined to form DS/FH hybrid spread spectrum system, and the system is designed on FPGA. In the design, the spread spectrum subsystems of sending and receiving and each function module of each subsystem are designed and simulated by adopting the method of the combination of hardware description language Verilog HDL and schematic diagram in the Quartus II 8.0 software. By simulation and implementation on FPGA of the system, the correctness and feasibility of the design are verified.

Acknowledgments. This thesis was supported by the National Natural Science Foundation of China (61461053,61461054,61072079); the natural science foundation of Yunnan province (2010CD023); the financial support of Yunnan University (NO. XT412004).

References

1. Zhao, J.: The design and implementation of a DS/FH signal source. Commun. Technol. **1**, 20–24 (2008). In Chinese
2. Chang, X., Zhang, X., Liu, Y., Zhan, W.: High jump speed DS/FH hybrid spread spectrum technology research. Manned Space Flight **2**, 14–20 (2012). In Chinese
3. Lu, T.: The implementation of spread spectrum communication system based on FPGA, D. Southwest Jiaotong University Master's thesis, (2003). In Chinese
4. Zhang, B., Shao, D., Li, S.: The research of DS/FH hybrid system fast Synchronization. J. Beijing Univ. Aeronaut. Astronaut. **11**, 1226–1231 (2005). In Chinese
5. Li, C., Xu, Y., Luo, H., Wang., Y.: The research of bipolar synchronization technology of DS/FH hybrid spread spectrum communication system. Mob. Commun. **3**, 50 (2004). In Chinese
6. Saifullah, A., Xu, Y., Lu, C., Chen, Y.: End-to-end communication delay analysis in industrial wireless networks. IEEE Trans. Comput. **5**, 1361–1374 (2015)

The Cost Performance of Hyper-Threading Technology in the Cloud Computing Systems

Xiao Zhang[(✉)], Ani Li, Boyang Zhang, Wenjie Liu, Xiaonan Zhao, and Zhanhuai Li

School of Computer Science, Northwestern Polytechnical University, Xi'an, China
zhangxiao@nwpu.edu.cn

Abstract. Hyper-Threading (or HT, for short) is a technology used in some Intel CPUs. Intel claims that it can use processor resources more efficiently. Many past studies have evaluated the performance of the technology in HPC clusters. In this paper, we discuss the advantages and disadvantages of Hyper-Threading using in the cloud computing systems. We evaluate the performance and energy cost of Intel CPU with Hyper-Threading enabled and disabled on virtualization environment. Our results show that Hyper-Threading technology can get better performance in most cases on a physical machine. The performance of a single core in a virtual machine is slightly lower when HT is enabled. But it doubles the number of available cores.

Keywords: Hyper-threading technology · Cloud computing system · Performance · Energy efficiency

1 Introduction

According to Moore's Law, the performance of CPU doubles every 18 months. The speed of CPU is much faster than memory and IO subsystem. To fill the performance gap between CPU and memory, multi-threading techniques have been used to make the best of the spare time while waiting for data. Simultaneous multi-threading (SMT) exposes more parallelism to the processor by fetching and retiring from multiple instruction streams. Intel's Hyper-Threading technology (Intel HT technology) brings the concept of simultaneous multi-threading (SMT) to the Intel architecture [1]. With Intel's Hyper-Threading technology, one physical processor can run two software threads simultaneously. It was first used on x86 architecture processor in 2002. AMD Bulldozer micro architecture is also a partial SMT implementation. SMT is widely used on high-performance computing (HPC) systems, with at least 63 of the fastest 100 systems in the world containing hardware that support it[1].

Most past researches focus on the performance evaluation of Hyper-Threading. The performance optimization of an application is highly variable

---

[1] http://www.top500.org/.

© Springer International Publishing Switzerland 2016
Y. Tan et al. (Eds.): ICSI 2016, Part II, LNCS 9713, pp. 581–589, 2016.
DOI: 10.1007/978-3-319-41009-8_63

and depends on many factors, such as application characters and the number of threads [2]. In 2013, Intel dropped SMT in favor of out-of-order execution for its Silvermont processor cores, as they found this gave the better performance [3]. Researches evaluated the performance of Hyper-Threading in HPC clusters [2,4,5], and tried to optimize the application to make the most of SMT [6]. There are also some researches focusing on how to schedule virtual machines on the Hyper-Threading platform [7,8].

The Hyper-Threading technology is also widely used in the cloud computing systems. The purpose of this paper is to measure the impact of applying the technology in the cloud computing systems. In order to better understand the performance and energy efficiency of HT, we have designed several comparative experiments. We got some interesting conclusions:

- For individual vCPUs, when HT is enabled, the maximum performance will be reduced.
- Hyper-Threading can get better performance than software virtualization such as QEMU.

The rest of this paper is organized as follows. Section 2 introduces the existing researches on Hyper-Threading. Section 3 describes the experiment environment and presents benchmarks and metrics used in this paper. Sections 4 and 5 present the performance evaluation results on physical machines and virtual machines. The last section ends with conclusions from our works and discusses the future works of this study.

2   Related Works

Intel Hyper-Threading technology first appeared in February 2002 on Xeon server processors and in November 2002 on Pentium 4 desktop CPUs [3]. Recently Intel HT technology is available on the Intel Core processor family and the Intel Xeon processor family [9]. Srivastava et al. pointed out that processor-level threading can be utilized which offers more efficient use of processor resources for greater parallelism and improved performance on today's multi-threading software [1]. Li et al. showed that SMT can provide a performance speedup of nearly 20 % for a wide range of applications with a power overhead of roughly 24 % [10]. Saini et al. found that HT generally improves processor resource utilization efficiency, but does not necessarily translate into overall application performance gained by using four full-scale scientific applications from computational fluid dynamics (CFD) used by NASA scientists [2].

Celebioglu et al. found that the effect of Intel Hyper-Threading (HT) technology on a system's performance varies according to the characteristics of the application running on the system and the configuration of the system [5]. The system-wide benefit of Hyper-Threading technology depends on the application characters and the number of threads. This makes it difficult to predict its benefit. Leo porter et al. presented a methodology for quantifying the performance and the power benefits of simultaneous multi-threading (SMT) for HPC clusters [6]. Oh *et al.* presented a Hyper-Threading aware migration method in the Xen environment according to the workload of VMs [7]. Deng *et al.* pointed out that independent scheduling on SMT platform can be harmful to applications running in VMs [8]. There is not any widely accepted conclusion on whether Hyper-Threading technology should be used in the cloud computing systems or not.

3   Experiment Environment and Benchmarks

3.1   Experiment Environment

Our testbed is shown in Table 1. The server has an Intel processor, 4 GB memory and 2 TB storage. The CPU (Intel core i7-4790) has four Intel Core processors running at 3.6 GHz with base frequency and at 4.0 GHz with max turbo frequency. The operating system is Ubuntu 14.04 LTS and the kernel version is Linux 3.13.0-24-generic. Our virtualization is Kernel-based Virtual Machine (KVM), which is a full virtualization solution for Linux on x86_64 hardware containing virtualization extensions.

Table 1. Configurations of Testbed

Hardware	1 * Intel(R) Core(TM) i7-4790 CPU @ 3.6 GHz(4 core)
	1 * 4 GBytes of RAM (Kingston DDR3 1600 MHz)
	1 * 1 TBytes SATA disk (WDC WD10EZEX-22BN5A0)
Software	Ubuntu 14.04 LTS
	Linux kernel 3.13.0-24-generic
	QEMU emulator version 2.0.0

The power meter used in our experiment is Chroma 66200 Power Meter. It uses an embedded high speed DSP and 16 bits Analog/Digital converter. The voltage range is 150/300/500 Vrms and the current range is 0.01/0.1/0.4/2 Arms. The soft panel has included the required data analysis feature and provided users with easiness to use interface to carry out the standby power measurement according to the international standard. The Soft panel measures and provides test report feature for recording all the key parameters such as Total Harmonics Distortion (THD), waveform crest factor, voltage, current, frequency, active power, apparent power, power factor [11].

3.2   Benchmarks

Here is a brief overview of the benchmarks that we used in this study. UnixBench is the original BYTE UNIX benchmark suite, which was updated and revised by many people over the years [12]. It compares the performance of server under test with a baseline system. UnixBench includes 9 individual applications measure different aspects of the system [12]. File copy test is IO intensive. Dhystone and Whestone are CPU intensive. The rest of applications depend on operating system implements. The benchmark can measure performance of CPU, IO and complex workloads. Because the test results are compared with a baseline system, it is suitable to compare the performance of machine with HT on and HT off.

The benchmark handles multi-core system by running N copies at the same time, where N is the number of cores. So it is a system benchmark, which can measure the performance running a single task and multiple tasks. This feature is suitable to compare system with different numbers of cores. We run the benchmark with HT enabled and disabled. The number of cores is doubled when HT enabled, while the performance of the system does not increase so much. The default maximum copies is 16, we modify the limit so it can run more copies. We also find in some cases, running copies more than the number of cores can get better performance.

From the view of application, the performance is measured by tasks finished in the certain period. For each application in Unixbench, it calculates the numbers of codes executed in a period of time. For example, in the Dhrystone, each thread counts the number of lines running in ten seconds. When the Dhrystone runs over, each thread will report COUNT. With multi-threading, we calculate the performance as:

$$Score = \sum_{i=1}^{n} COUNT_i. \tag{1}$$

where n is the number of threads which run in Dhrystone. As some applications have different parameters, e.g. File copy with different buffer size and max blocks, UnixBench calculates the performance as:

$$Index = Score/Baseline. \tag{2}$$

Which eliminates the differences between different parameters in the same application. In our experiments, the impact of HT is measured by ratio of the differences between HT enabled and HT disabled to performance of HT disabled. With HT enabled and HT disabled, we calculate the performance gain as:

$$P = \frac{Index_{HT} - Index_{ST}}{Index_{ST}}. \tag{3}$$

4   Performance Evaluation on Physical Machines

There are four cores in Intel Core i7-4790. When HT is on, it has eight cores. We modify the limit of UnixBench and it can run up to 64 copies to measure the parallel performance. The final score is the sum of each copy. Figure 1 shows the test results of UnixBench. From the results, we can get the following conclusions in most cases:

- The scores of one to four are similar with HT enabled and disabled.
- When HT is disabled, the score reach maximum with four copies.
- When HT is enabled, the score reach maximum with eight copies.
- Running more copies takes little effect on the final performance.

Which means HT can get better performance in multiple processes environments. It can get 10 % promotion under integer computing benchmark, but it can get maximum 50 % promotion under system call case.

Figure 1(a) shows the graph for Dhrystone. Performance in the case when HT enabled is almost the same with the case when HT disabled for 1, and 2 threads. When there are 4 threads, performance degrades 4.5 % with HT enabled. With HT enabled, the performance promotion is 9.9 %, 9.5 %, 9.5 %, 9.4 % for 8, 16, 32 and 64 threads, respectively.

Figure 1(b) shows the results of Whetstone. It tests the float computing performance. In this plot, the performance keeps increasing when the number of threads increase. The performance of HT enabled and disabled are very similar. The results show that the benchmark can not measure the performance of single core efficiently. The benchmark is CPU-intensive, even a single virtual core can handle several float computing copies. It is suggested by Intel that HT is better disabled for compute-efficient applications, because there is little to promote from HT technology if the processor's execution resources are already well used [9].

Figure 1(g) and (h) present the data of Shell Scripts. In this application, there is a parameter which sets the number of subprocesses. In Fig. 1(g), performance degrades 1.9 %, 3.8 % for 1 and 2 threads, respectively. While in Fig. 1(h), performance gains 32.0 %, 38.9 % for 1 and 2 threads. When the number of threads is 8, 16, 32 and 64, performance gain is almost nearly 35 % both 1 concurrent case and 8 concurrent case. It is suggested by Intel that Intel HT technology adds additional hardware threads to the system [9]. Therefore, to take advantage of Intel HT technology, an application must be able to launch additional threads in order to generate additional parallelism. When the parameter of Shell Scripts is 1 concurrent, application only fork and execute 1 subprocess in one thread. In this case, the application doesn't have additional threads with HT enabled when the number of threads is 1 and 2. When the parameter of Shell Scripts is 8 concurrent, application can fork and execute 8 subprocesses in one thread. In this case, application can benefit from HT technology.

Figure 1(f) shows the graph for Process Creation. The percentage of performance gain is nearly 23 % for 8, 16, 32 and 64 threads. Figure 1(i) shows the graph for System Call Overhead. Performance in the case when HT enabled is almost the same with in the case when HT disabled for 1, 2 and 4 threads. With HT enabled performance gain is almost nearly 53 % for 8, 16, 32 and 64 threads. It gets the most promotion between all the benchmarks.

Figure 2 shows the graph for File copy. When the number of threads is 1, 2, and 4, performance is almost the same with HT enabled and HT disabled. In Fig. 2(a), performance degrades 37.0 %, 21.1 %, 23.9 %, 17.0 % for 8, 16, 32, 64 threads, respectively. In Fig. 2(b), performance degrades 31.4 %, 15.6 %, 16 %,

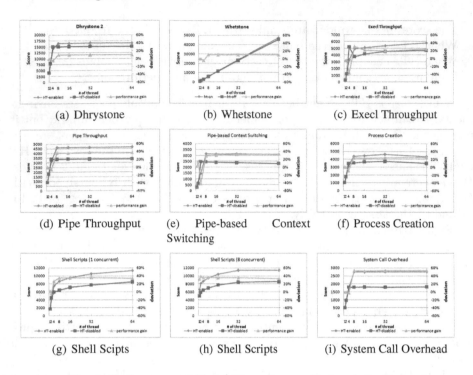

(a) Dhrystone   (b) Whetstone   (c) Execl Throughput

(d) Pipe Throughput   (e)   Pipe-based   Context   (f) Process Creation
Switching

(g) Shell Scipts   (h) Shell Scripts   (i) System Call Overhead

Fig. 1. Performance of Hyper-Threading enabled and disabled

22.6 % for 8, 16, 32, 64 threads. In Fig. 2(c), performance degrades 19.4 %, 20.2 %, 13.1 %, 15.36 % for 8, 16, 32, 64 threads. From the test results, we can find that file copy does not benefit from HT enabled. It is also suggested by Intel that in order to maximize the performance of the CPU(s), sufficient platform resources (e.g., memory, disk I/O, network I/O) must be presented to feed the processor [9]. In this case, performance has reached max when the number of threads is 4. At that time, resources of disk I/O, memory, network I/O are already well used. When HT enabled, a second process which is on the same physical core will force the physical resources to be shared. Because of resource contention, performance degrades when File Copy is running.

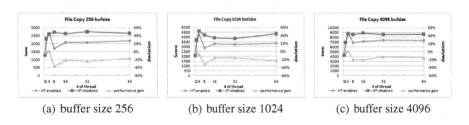

(a) buffer size 256   (b) buffer size 1024   (c) buffer size 4096

Fig. 2. Performance of File copy Hyper-Threading enabled and disabled

5   Performance Evaluation on Virtual Machines

In our test environment, we use QEMU as the virtualization platform. QEMU is a generic and open source machine emulator and virtualizer. When it is used as a virtualizer, QEMU achieves near native performances by executing the guest code directly on the host CPU. QEMU supports virtualization when executing under the Xen hypervisor or using the KVM module in Linux. The vCPU does not map to certain physical CPU. The virtual machines are scheduled according to physical CPU workloads and operating system policy. By using the information from /proc/cpuinfo, we can find the OS distinguishes each core by physical id, core id and apicd. But in a virtual machine, the CPU model name is QEMU Virtual CPU version 1.7.91, and all cores in the virtual machine is counted from 0. Which means we can not find the relation between vCPUs and physical CPUs. When HT is disabled, the number of the cores is four. When HT is enabled, the number of the cores is eight. QEMU can emulate more cores than physical limits, but overcommitting vCPUs can hurt performance.

Marc Fielding tested the performance of virtual machine in Amazon EC2 cloud service [13]. The conclusion is the performance of a vCPU is lower than a physical CPU, but two vCPUs with HT technology get better performance than a physical CPU. In the previous section, we run up to 64 copies on the multi-core system. The results show that in most cases, the benchmark can get maximum score when the number of copies equal to the number of cores. So in this section, we run N copies on each virtual machines, where N is the number of cores of virtual machines.

Figure 3 shows the test results with the different number of CPUs. We compare virtual machines with one, two, four vCPUs with HT enabled and disabled. The difference is less than 1 % when virtual machines use one core. While the degradation comes up to 16 % when virtual machines use four cores. From the results, we can find that physical CPU has the better performance than vCPU. In other words, if cloud computing vendors (such as Amazon) use vCPU to replace CPU, they should provide the service at a reduced price. When HT is enabled, the performance of a vCPU is less than the scenario when HT is disabled.

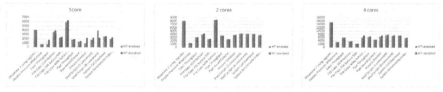

(a) Performance of VM with one core

(b) Performance of VM with two cores

(c) Performance of VM with four cores

Fig. 3. Performance of virtual machines with different number of cores (Color figure online)

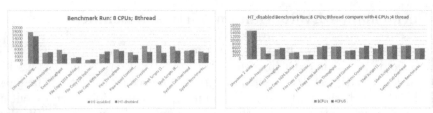

(a) Performance of VM with HT enabled and (b) Performance of VM compared 8 CPUS
disabled   with 4 CPUs when HT disabled

Fig. 4. Performance of virtual machines with eight CPUs (Color figure online)

QEMU can emulate more cores than physical limits, but it may cause heavy
overload. Hyper-Threading technology can double the number of cores. In the
following test, we will evaluate the performance of a virtual machine with eight
cores. When HT is enabled, the number of cores is same with the virtual machine.
When HT is disabled, the number of cores is four. QEMU has to emulate more
cores than physical limits. The Fig. 4 shows the test results. We can find that
when HT is enabled, the virtual machine with eight cores has a better performance (Fig. 4(a)). When HT is disabled, we compare the performance of virtual
machine with eight and four cores. The results are shown in Fig. 4(b), the performance of four vCPUs are better in many cases. This means that schedule technology of Hyper-Threading is better than the default of operating system. And
Hyper-Threading technology can get better performance with more processes.

6   Conclusion

In this paper, we measure the impact of HT in virtualization environments.
We present three sets of experiment to evaluate the performance of physical
machine and virtual machines. Our research shows that the performance of HT
can get better performance in most cases in physical machine. In virtualization
platform, performance of a single physical core is better than a virtual core, but
the performance of all cores is better when HT is enabled.

HT generally improves processor resource utilization, but it does not fit for
all cases. The performance of compute-intensive applications are hard to promote. And the performance of IO-intensive applications are degraded when HT
is enabled because of resources contention. In virtualization environments, HT
technology can provide more cores shared by users. When HT is enabled, the performance of one single core decreases less than 2 %. But the overall performance
of HT enabled is higher.

In the cloud computing systems, HT can provide more cores to users, and it
does not need extra power. Application running inside the virtual machine changes
the characters of workloads, the performance of virtual core is slightly degrade
when HT is enabled. We will measure more virtualization platforms (Xen, VMware
etc.) to evaluate the impact of HT on different virtualization platforms.

Acknowledgments. This work was supported by the National High-tech R&D Program of China (863 Program) under Grant No.2013AA01A215, and the National Natural Science Foundation of China under Grant No.61472323, No.61303037. This work was also supported by the Fundamental Research Funds for the Central Universities under Grant No.3102015JSJ0009.

References

1. Srivastava, N., Awasthi, K., Rizvi, S.Z.: Hyper threading technology in hardware architecture for processor efficiency enhancement. J. S-JPSET **3**, 31–37 (2012)
2. Saini, S., Jin, H., Hood, R., Barker, D., Mehrotra, U., Biswas, R.: The impact of hyper-threading on processor resource utilization in production applications. In: 18th International Conference on High Performance Computing, pp. 1–10. IEEE Press, Bangalore (2011)
3. Hyper-Threading. http://en.wikipedia.org/wiki/Hyper-threading
4. Saini, S., Chang, J., Jin, H.: Performance evaluation of the intel sandy bridge based NASA pleiades using scientific and engineering applications. In: Jarvis, S.A., Wright, S.A., Hammond, S.D. (eds.) PMBS 2013. LNCS, vol. 8551, pp. 25–51. Springer, Heidelberg (2014)
5. Celebioglu, O., Saify, A., Leng, T., Hsieh, J., Mashayekhi, V., Rooholamini, R.: The Performance impact of computational efficiency on HPC Clusters with Hyper-Threading technology. In: 18th IEEE International Symposium on Parallel and Distributed Processing, p. 250. IEEE Press, New Mexico (2004)
6. Porter, L., Laurenzano, M.A., Tiwari, A., Jundt, A., Ward Jr., W.A., Campbell, R., Carrington, L.: Making the most of SMT in HPC: system-and application-level perspectives. J. ACM Trans. Architect. Code Optimization **11**(4), 59 (2015)
7. Oh, C., Ro, W.: Hyper Threading-Aware Virtual Machine Migration. In: 13th IEEE International Conference on Electronics, Information and Communications, pp. 1–2. IEEE Press, Netherlands (2014)
8. Deng, K., Ren, K., Song, J.: Symbiotic Scheduling for Virtual Machines on SMT Processors. In: 2th IEEE International Conference on Cloud and Green Computing, pp. 145–152. IEEE Press, WuHan (2012)
9. Performance insights to intel Hyper-Threading technology. https://software.intel.com/en-us/articles/performance-insights-to-intel-hyper-threading-technology
10. Li, Y., Brooks, D., Hu, Z., Skadron, K., Bose, P.: Understanding the energy efficiency of simultaneous multithreading. In: 4th International Symposium on Low Power Electronics and Design, pp. 44–49. ACM, New York (2004)
11. Model 66200 series Digital Power Meter. http://www.chromaate.com/product/66200_series_Digital_Power_Meter.htm
12. Project of UnixBench. https://code.google.com/p/byte-unixbench/
13. Virtual CPUs with amazon web services. http://www.pythian.com/blog/virtual-cpus-with-amazon-web-services/

Combining Query Ambiguity and Query-URL Strength for Log-Based Query Suggestion

Feiyue Ye and Jing Sun[(✉)]

School of Computer Science, Shanghai University, Shanghai, China
{yefy, rexsunces}@shu.edu.cn

Abstract. The ambiguity of query may have potential impact on the performance of Query Suggestion. For getting better candidates adapting to query's ambiguity, we propose an efficient log-based Query Suggestion method. Firstly we construct a Query-URL graph from logs and calculate the bidirectional transition probabilities between queries and URLs. Then, by taking URL's rank and order into consideration, we make a strength metric of the Query-URL edge. Besides, we conduct random walk with the edge strength and transition probability to measure the closeness among queries. To reflect the influence of query ambiguity, we exploit an entropy-based method to calculate the entropy of each query as a quantitative indicator for ambiguity, making a notion of ambiguity similarity as an available factor in relevance estimation. Finally we incorporate ambiguity similarity with closeness to derive a comprehensive relevance measurement. Experimental results show that our approach can achieve a good effect.

Keywords: Query suggestion · Bipartite graph · Random walk · Query ambiguity · Query-URL strength

1 Introduction

Due to the explosive growth of web information, Internet surfers have more difficulties in finding what they need with search engines. On one hand, information which basically meets the needs of user could be huge, and it is not easy for users to pick the most relevant one. Confronting the shortness [1–3] and ambiguity [4, 5] of queries, search engines will probably get confused when they are trying to understand the exact search intentions. In addition, peoples' ability of abstracting the input query away from their search intention also makes a difference. To solve these problems, almost all search engines apply the technology called Query Suggestion or Query Recommendation to enhance user experience.

So far, as a research focus, Query Suggestion based on logs which contain the prior knowledge of interaction between search engine and users, has been scrutinized much in information retrieval domain. In general, Query-URL graph is considered as an effective tool to do related researches [6–9]. Existing methods mostly concentrate on seeking relevant suggestions simply leveraging the click counts of the Query-URL pairs, making little analysis on query ambiguity or Query-URL edge, while these ignored features can provide some useful information. Researchers have recognized the

© Springer International Publishing Switzerland 2016
Y. Tan et al. (Eds.): ICSI 2016, Part II, LNCS 9713, pp. 590–597, 2016.
DOI: 10.1007/978-3-319-41009-8_64

fact that users who express the same query may not expect the same results and ambiguity of query may have some effects on Query Suggestion [10–14]. Liu et al. [15] noticed that it was not reasonable that Query-URL edge's correlation only depends on the frequency of Query-URL pairs and they used a recursive method to calculate Query-URL edge's strength in suggestion process.

Existing methods analyzing on query ambiguity are mostly in Query Personalization, and they mainly exploited ambiguity merely as a threshold to enhance or reduce the degree of personalization, instead of measuring the relevance among queries in generating suggestions. Liu's method in estimating the strength of Query-URL edge did play a role in generating suggestions of high quality while it could increase the computation burden if the logs were super-scale.

To vanquish these above shortcomings, we propose a novel method for generating query suggestions on a large scale Query-URL bipartite graph, which takes both query ambiguity and strength of Query-URL edge into consideration. Specifically, we first construct a Query-URL bipartite graph from search engine query logs. Then, factors such as transition probability, URL's click-order and rank are comprehensively considered in defining and computing a preliminary relevance among queries. Afterwards, to make the quantitative analysis on query ambiguity, we adopt a method using normalized information entropy based on the distribution of URLs under different queries on the constructed graph. Then, the preliminary relevance and the entropy are integrated together to optimize the measurement of relevance. To demonstrate the effect of our approach, we implement experiments using real search engine logs. Finally we make a conclusion and discuss the future work.

2 Related Work

Nowadays, Query Suggestion technology plays an important role in information retrieval system. Because of the large number of real users' queries and click behaviors, search engine logs have been widely studied by researchers in order to discover relevance among queries. In recent years, many new approaches have been proposed on the basis of query logs, including agglomerative clustering method [16], random walk method using hitting time [8, 9], snippet click model which extracts user interests from URL's snippets [17] and others.

There are several researches on query ambiguity and Query-URL edge's strength. Reference [13] summarized three types of queries: Ambiguous Query, Broad Query and Clear Query, and proposed a machine learning model using search results for query-ambiguity identification. Reference [12] explored the goal behind a search query and made use of the click distribution and other feathers for automatic goal-identification. Reference [18] applied an entropy-biased model to estimate the strength of Query-URL edges. It gave a higher strength to a more specific URL while lower strength to a more general URL and then introduced the inverse query frequency (IQF) concept for measuring the discriminative ability of URLs, which achieved a good result. Reference [10] also utilized the entropy, unlike Ref. [18], it calculated the click entropy of query instead of URL to determine whether to take the strategy of personalization.

In this paper, we implement particular analysis on the Query-URL graph with the purpose of seeking relevant suggestions adapting to origin's ambiguity. In contrast with [15, 18], we use result information such as URL's rank and order to evaluate the strength of Query-URL edge. We also take an entropy method to estimate the ambiguity similarity of two queries in conjunction with relevance computed from the bio-graph, while Ref. [10] uses entropy as a threshold to determine whether some strategies are supposed to be taken.

3   Relevance Measurement Among Queries

We obtain search logs from a famous commercial search engine Sougou. An example record in this logs is '20111230124104, 923f5520919ec5c1e3100e8fb546d17a, NIKE, 1, 1, www.nike.com.cn', which means that an anonymous user ('923...17a') searched 'NIKE' and s/he clicked on the top 1 result 'www.nike.com.cn' at his first click in s/he's current session at a certain time ('2011...4104').

In this paper, the bipartite graph G = ({V1,V2},E) is composed of two types of vertices set: V1 (set of queries), V2 (set of URLs), connected with edges set E (Query-URL edge). We will continue our researches on the graph in the next sections.

3.1   Random Walk on Query-URL Bipartite Graph

In order to calculate the relevance among queries, we make a definition of the transition probability from queries to URLs firstly. Given $i \in V1$, $j \in V2$,

$$p_{ij} = \frac{N_{ij}}{d_i}$$
$$d_i = \sum_{(i,j) \in E} N_{ij} \tag{1}$$

Where N_{ij} means the accumulation when i and j appear in pair. d_i denotes the accumulative times when query i and any connected URL j appear in pair. Likewise, we can calculate the transition probability from URLs to queries by recounting the d_i and N_{ij} in view of URL vertices.

Having these definitions, we can derive a random walk process from formula (1). Given $i, j \in V1$, a 2-step random walk process [5] between two queries can be defined as:

$$RW_{ij} = \sum_{k \in V_2} p_{ik} \times p_{kj} \tag{2}$$

Through formula (2), we can get the arrival probability from one query to another query which can be used to estimate the relevance between two queries.

3.2   Analysis on Query-URL Strength

Experience tells us that after submitting a query, people always prefer the most relevant results, which are usually near the top. That is to say, given one query, if the rank of URL is smaller (rank higher), and the order of URL is smaller (click earlier), the URL may have more importance than those which are clicked later and rank lower. Study [19] showed that for one query, about 85 % of users will only click on the top 10 search results. The features of rank and order for one URL under a specific query may represent the most users' choices. Therefore we define the following formula to give expression to a Query-URL edge's strength:

$$Str_{qu} = \frac{1}{N_{qu}} \times \sum_{s \in U_u} \frac{1}{\alpha \times rank_s + \beta \times order_s} \tag{3}$$

Where U_u is the set of different records whose URL is u in logs. α and β are the weighting coefficients. In our study, we set α and β equal to 0.6 and 0.4 respectively as an empirical consequence, for the reason that we believe that the rank of result returned from search engine is more objective than user's click order. Due to the difficulty of defining a query's rank and order, we simplify the problem by assuming that $Str_{qu} = Str_{uq}$.

3.3   Analysis on Query Ambiguity

We can intuitively discover a phenomenon from graph G that there are some queries which have more centered clicked URLs while some have more diffuse clicked URLs. This phenomenon can be given a reasonable explanation that those who have more centered clicked URLs would represent more search intentions (or conversely, representing fewer search intentions). In other words, different queries have more or less ambiguities. For example, 'apple' may connect to many URLs (we regard them as different search intentions in our work), e.g., iPhone, iPad, Steve Jobs, Fruits, while 'google' is likely to link few URLs, e.g., Google Search, Google Map.

Like some existing methods, we exploit the Shannon Information Entropy [20] for subsequent analysis to describe the query ambiguity. The entropy according to the distribution of a specific query's transition probability is defined as:

$$I(q_i) = -\sum_{j=1}^{n} p_{q_i u_j} \log_2(p_{q_i u_j}) \tag{4}$$

It is necessary to normalize $I(q_i)$ as:

$$I(q_i)_{norm} = \frac{I(q_i)}{\max(I(q_{1:n}))} \tag{5}$$

After this point, we create a new definition for measuring the similarity of entropy. Given q_i, $q_j \in V1$,

$$Sim_{entropy}(q_i, q_j) = \exp(-|(I(q_i)_{norm} - I(q_j)_{norm})|) \qquad (6)$$

If normalized entropies of two queries are approximately equal then the two queries can be thought of similar and vice versa, in this entropy way. Suppose that one query's entropy is extremely large, it indicates that many URLs are clicked for this query or different users have different choices on this query, in other words, this query is ambiguous and represents many search intentions, so we need to enhance the relevance of suggestions who have more search intentions like the original one. Hence as a quantitative evaluation, $Sim_{entropy}$ is provided as one of the influencing factors to measure relevance between queries hereinafter.

3.4   Comprehensive Measurement of Relevance Among Queries

In Sect. 3.1, we have computed the closeness from one query to another using a random walk method. However, that closeness takes no account of the intermediate (the clicked URL) itself apart from the co-occurrence of queries and URLs. Thus we introduce the Query-URL edge's strength Str_{qu} in Sect. 3.2, and with it, we modify formula (2) and derive a preliminary relevance as follow:

$$REL(i,j) = RW'_{ij} = \sum_{k \in V_2} (p_{ik} \times Str_{ik} \times p_{kj} \times Str_{kj}) \qquad (7)$$

As discussed in Sect. 3.3, we then try to modify our preliminary formula by integrating the ambiguity similarity. Given $i, j \in V1$, the improved comprehensive measurement of relevance is defined as follow:

$$CREL(i,j) = Sim_{entropy}(i,j) \times REL(i,j) \qquad (8)$$

So far, we have taken the relevance of random walk method with the consideration of URL's related information and the similarity of query's ambiguity in conjunction to form a comprehensive method (RW-UQ), aiming at measuring two queries' relevance in more aspects. To illustrate the effect of our methodology, we have carried out some experiments in Sect. 4.

4   Experimental Results

We implemented experiments on a sample of one month query logs (more than 40 million records) from a famous commercial search engine. After several preprocessing tasks, we constructed a Query-URL bipartite graph which contained 285280 unique query vertices (V1), 374953 unique URL vertices (V2) and 582768 Query-URL edges (E) between V1 and V2.

For comparison, two algorithms are adopted on the same data set. The baseline method (RW-Base) is a random walk way [21]. Authors in [21] firstly generated candidates with random walk strategy in leveraging click information, and then used

Table 1. Scores of comparative algorithms in MAP, P@5, P@10 and P@20.

Algorithms	MAP	P@5	P@10	P@20
RW-UQ	0.80(+11.1 %)	0.94(+9.3 %)	0.92(+8.2 %)	0.85(+4.9 %)
RW-U	0.78(+8.3 %)	0.92(+7.0 %)	0.90(+5.9 %)	0.83(+2.5 %)
RW-Base	0.72(+0 %)	0.86(+0 %)	0.85(+0 %)	0.81(+0 %)

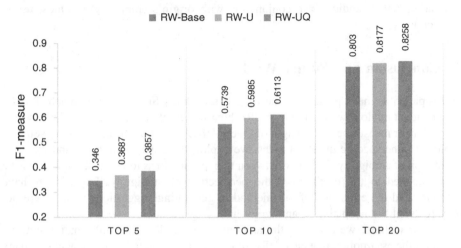

Fig. 1. F1-Measure of three algorithms: RW-Base, RW-U and RW-UQ (Color figure online)

the snippet information as feedback of user intentions to re-ranked candidates. The second method (RW-U) is also a random walk approach like RW-Base while it considers the strength of Query-URL edge in addition as an influencing factors. We take an artificial tagging method using pooling [22] technique to determine whether a suggestion is relevant or not to the origin.

Table 1 summarizes the scores of three algorithms. Taking one with another, our method improves noticeably in all metrics compared to the others. The RW-UQ aside, we can find that RW-U performs much better than RW-Base which has improved about 8 %, 7 % and 6 % in MAP, P@5, and P@10 respectively. Seeing that RW-Base only uses the click information and considers less of the Query-URL pairs, which can lead to the higher rank of some typos, such as 'Gogle (Google)' and 'Baudu (Baidu)', since these wrongly written queries appear frequently and may be recognized correctly by search engine. Given that only fewer users input those misspellings, and people always click the most relevant URLs, expecting to obtain information they need by clicking as fewer URLs as possible, RW-U contains the analysis of URL's rank and order to suggest queries in accord with most users' behavior, which has achieved good performance. On the basic of RW-U, our method shows more effectiveness as a result of the ambiguity analysis.

Figure 1 shows the scores of these algorithms in F1-Measure, it seems that RW-UQ has more effectiveness for the top 10 suggestions. However, as the number of suggestions increases, all methods' disparity decreases. This is probably for the reason that the useful suggestions already have been put near the top, and the total number of relevant queries stays relatively fixed, which constrain the effect if the required number of suggestions approaches the maximum number of origin's relevant queries in logs. In fact, almost all search engines provide no more than ten suggestions for original query in actual need, so it is reasonable for us to focus more on listing relevant suggestions at higher ranks.

Our experimental results show that the proposed suggestion method is capable of generating better candidates in conjunction with origin's ambiguity and most users' click choices.

5 Conclusion and Future Work

In this paper, we have presented a new log-based Query Suggestion approach. We not only conduct random walk on Query-URL bipartite graph, but also exploit URL's rank and order for measuring the strength of Query-URL edge, which stands for most users' choices to one specific query. Besides, we explore the query ambiguity and make an estimation of ambiguity similarity between two queries so as to let those suggestions be more inclined to the origin from the perspective of ambiguity. Experiments have demonstrated the efficiency of our method in generating suggestions which were not only relevant but adaptable in ambiguity with original query.

In our method, we calculate the ambiguity of queries using information entropy based on the assumption that each URL represents a unique topic, while actually many URLs are belonging to the same topic. In addition, when generating suggestions, we pay little attention to the filtration of candidates in our current study which may cause duplication and redundancy. In the future work, we will seek better ways to handle these problems.

References

1. Silverstein, C., et al.: Analysis of a very large web search engine query log. SIGIR Forum **33**, 6–12 (1999)
2. Jansen, B.J., Spink, A., Saracevic, T.: Real life, real users, and real needs: a study and analysis of user queries on the web. Inf. Process. Manage. **36**(99), 207–227 (2000)
3. Wen, J.R., Nie, J.Y., Zhang, H.J.: Clustering user queries of a search engine. In: Proceedings of the 10th International Conference on World Wide Web, pp. 162–168. ACM, China (2001)
4. Krovetz, R., Croft, W.B.: Lexical ambiguity and information retrieval. ACM Trans. Inf. Syst. **10**(2), 115–141 (2002)
5. Baeza-Yates, R., Calderón-Benavides, L., González-Caro, C.N.: The intention behind Web queries. In: Crestani, F., Ferragina, P., Sanderson, M. (eds.) SPIRE 2006. LNCS, vol. 4209, pp. 98–109. Springer, Heidelberg (2006)
6. Song, Y., He, L.W.: Optimal rare query suggestion with implicit user feedback. In: Proceedings of the 19th International Conference on World Wide Web, p. 901. ACM, USA (2010)

7. Craswell, N., Szummer, M.: Random walks on the click graph. In: SIGIR, pp. 239–246 (2007)
8. Mei, Q., Zhou, D., Church, K.: Query suggestion using hitting time. In: ACM Conference on Information & Knowledge Management, pp. 469–478 (2008)
9. Ma, H., Lyu, M.R., King, I.: Diversifying query suggestion results. In: Twenty-fourth AAAI Conference on Artificial Intelligence (2010)
10. Dou, Z., Song, R., Wen, J.R.: A large-scale evaluation and analysis of personalized search strategies. In: Proceedings of the 16th International Conference on World Wide Web, pp. 581–590. ACM, Canada (2007)
11. Teevan, J., Dumais, S.T., Liebling, D.J.: To personalize or not to personalize: modeling queries with variation in user intent. In: Proceedings of the 31st Annual International ACM SIGIR Conference on Research and Development in Information Retrieval, pp. 163–170. ACM, Singapore (2008)
12. Lee, U., Liu, Z., Cho, J.: Automatic identification of user goals in Web search. In: Proceedings of the 14th International Conference on World Wide Web, pp. 391–400. ACM, Japan (2005)
13. Song, R., et al.: Identifying ambiguous queries in web search. In: Proceedings of the 16th International Conference on World Wide Web, pp. 1169–1170. ACM, Canada (2007)
14. Sanderson, M.: Ambiguous queries: test collections need more sense. In: Proceedings of ACM SIGIR, pp. 499–506 (2008)
15. Liu, Z., Sun, M.: Asymmetrical query recommendation method based on bipartite network resource allocation. In: Proceedings of the 17th International Conference on World Wide Web, pp. 1049–1050. ACM, China (2008)
16. Beeferman, D., Berger, A.: Agglomerative clustering of a search engine query log. In: Sixth ACM SIGKDD International Conference on Knowledge Discovery & Data Mining, pp. 407–416 (2000)
17. Liu, Y., et al.: How do users describe their information need: query recommendation based on snippet click model. Expert Syst. Appl. 38(11), 13847–13856 (2011)
18. Deng, H., King, I., Lyu, M.R.: Entropy-biased models for query representation on the click graph. In: Research and Development in Information Retrieval, pp. 339–346. ACM, Boston (2009)
19. Yu, J.H., Liu, Y.Q., Zhang, M., et al.: Research in search engine user behavior based on log analysis. J. Chinese Inf. Process. 21(1), 109–114 (2007)
20. Sloane, N., Wyner, A.: Prediction and entropy of printed english. Bell Syst. Tech. J. 30(1), 50–64 (1951)
21. Luo, C., et al.: Query recommendation based on user intent recognition. J. Chinese Inf. Process. (2014)
22. Jones, S., et al.: Report on the Need for and Provision of an "ideal" Information Retrieval Test Collection. Computer Laboratory, University of Cambridge, UK (1975)

Intelligent Interactive
and Tutoring Systems

Interactive Generator of Commands

Eugene Larkin[1]([✉]), Alexey Ivutin[1], Vladislav Kotov[1], and Alexander Privalov[2]

[1] Tula State University, Tula 300012, Russia
elarkin@mail.ru, alexey.ivutin@gmail.com, vkotov@list.ru
[2] Tula State Pedagogical University, Tula 300026, Russia
privalov.61@mail.ru

Abstract. The analytical model of generator of commands (command generator, CG) to swarm system, based on description of an interactive dialogue between a human operator and computer, is worked out. The model is based on the fundamental theory of semi-Markov process, in particular, on the theory of the ordinary and the 2-parallel semi-Markov processes. In CG under investigation an interactive algorithm is represented as the sequence of 2-parallel processes, which are associated to simple ergodic semi-Markov process, some states of which generate commands to swarm system. Formulae for time characteristics of flow of command to swarm system with use time and stochastic characteristics of elementary actions of human operator and time and stochastic characteristics of operators of computer algorithm are obtained.

Keywords: Human operator · Computer · Command generator · Interactive dialogue · Semi-Markov process · Time characteristics · Stochastic characteristics

1 Introduction

One of features of physical swarm systems is lack of intelligence, that does not allow construct fully autonomous units, which can operate automatically taking into account only his aim task and an environment state [1]. Due to the fact in the near future dominant solution will be remote control of swarm wherein a human operator within the guidelines and control practice will be interact with computer. As a result of dialogue will be formed flow of commands to swarm units.

Both the human operator and the computer operate due their own algorithm every elementary operation of which is executed during an occasional time, and result of execution is occasional too. So, natural and most common approach to modeling of occasional sequence of changing states in time both human and computer is so-called 2-parallel semi-Markov process [2–5].

This approach allows at the preparation of activity both human and computer to logically completed fragments to evaluate time and stochastic characteristics of command generator and further to study behavior of swarm under flow of command with determined characteristics.

© Springer International Publishing Switzerland 2016
Y. Tan et al. (Eds.): ICSI 2016, Part II, LNCS 9713, pp. 601–608, 2016.
DOI: 10.1007/978-3-319-41009-8_65

2　Command Generator

Structure of semi-Markov process, which simulates interactive CG, is shown on Fig. 1. Model consists of $J(\beta)$ simplest 2-parallel semi-Markov processes, every of which includes states $\{a_{j(\beta,1)}, a_{j(\beta,2)}, b_{j(\beta,1)}, b_{j(\beta,2)}\}$, $1(\beta) \leq j(\beta) \leq J(\beta)$ and is described by:

adjacency matrix

$$
{}^{2}\boldsymbol{r}_{j(\beta)} = \begin{bmatrix} 0 & 1 & 0 & 0 \\ 0 & 0 & 0 & 0 \\ 0 & 0 & 1 & 0 \\ 0 & 0 & 0 & 0 \end{bmatrix}, \tag{1}
$$

and semi-Markov matrix

$$
{}^{2}\boldsymbol{h}_{j(\beta)}(t) = \begin{bmatrix} 0 & f_{a[j(\beta)]}(t) & 0 & 0 \\ 0 & 0 & 0 & 0 \\ 0 & 0 & f_{b[j(\beta)]}(t) & 0 \\ 0 & 0 & 0 & 0 \end{bmatrix}, \tag{2}
$$

where $f_{a[j(\beta)]}(t)$ — is density of time of residence the process in state $a_{j(\beta,1)}$ till its switch to state $a_{j(\beta,2)}$; $f_{b[j(\beta)]}(t)$ — is density of time of residence the process in state $b_{j(\beta,1)}$ till its switch to state $b_{j(\beta,2)}$.

After the switching both elementary processes $f_{a[j(\beta)]}(t)$ and $f_{b[j(\beta)]}(t)$ to $a_{j(\beta,2)}$ and $b_{j(\beta,2)}$ correspondingly, outcome the CG from state $\beta_{j(\beta)}$ and switch it to conjugated state $\beta_{n(\beta)}, 1(\beta) \leq j(\beta) \leq J(\beta)$, take place. In principle CG

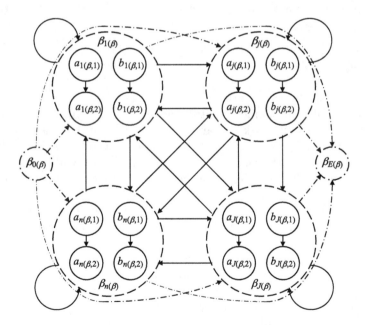

Fig. 1. The model of interactive CG

includes starting $\beta_{0(\beta)}$ and absorbing $\beta_{E(\beta)}$ states (are shown with dash-dot line), but in steady state regime start of the process is not considered and switch the process to state $\beta_{E(\beta)}$ is considered as unlikely event. Thus density of time of outcome from state $\beta_{j(\beta)}$ is determined under the dependence [6,7]

$$f_{j(\beta)}(t) = \frac{d}{dt}\left[F_{a[j(\beta)]}(t) \cdot F_{b[j(\beta)]}(t)\right], \tag{3}$$

In the common model of CG the time density is the shape of the conditional density of time of residence in state $\beta_{j(\beta)}$ with the next switching to state $\beta_{n(\beta)}, 1(\beta) \leq j(\beta) \leq J(\beta)$.

$$f_{j(\beta)}(t) = \sum_{n(\beta)=1(\beta)}^{J(\beta)} h_{j(\beta),n(\beta)}(t); \; h_{j(\beta),n(\beta)}(t) = p_{j(\beta),n(\beta)} \cdot f_{j(\beta),n(\beta)}(t), \tag{4}$$

where $p_{j(\beta),n(\beta)}$ — is the probability of switch from sate $\beta_{j(\beta)}$ to state $\beta_{n(\beta)}$; $f_{j(\beta),n(\beta)}(t)$ — density of time of switch from $\beta_{j(\beta)}$ under condition of switch the process to state $\beta_{n(\beta)}$;

$$\sum_{n(\beta)=1(\beta)}^{J(\beta)} p_{j(\beta),n(\beta)} = 1. \tag{5}$$

Elements $h_{j(\beta),n(\beta)}(t)$ gather on semi-Markov matrix

$$\boldsymbol{h}_\beta(t) = \lfloor h_{j(\beta),n(\beta)}(t) \rfloor, \tag{6}$$

where $1(\beta) \leq j(\beta), \; n(\beta) \leq J(\beta)$.

Semi-Markov process $\boldsymbol{h}_\beta(t)$ is the ergodic one. It contains the states $\beta_{1(S)}$, $\ldots, \beta_{j(S)}, \ldots, \beta_{J(S)}$, which initiate generation of one command to a swarm unit. Without loss of community let us assume that selected states have numbers with the lowest values, i.e. $1(s) = 1(\beta), \ldots, j(s) = j(\beta), \ldots, J(s) = n(\beta) \leq J(\beta)$. The process (6) generates a transaction in one of the next cases:

1. When the direct switch from $\beta_{j(S)}$ to $\beta_{n(S)}$ occurs;
2. When occurs switch from $\beta_{j(S)}$ to $\beta_{j(\beta)}$, $j(\beta) > J(S)$ and subsequent wandering through states $\beta_{j(\beta)}, j(\beta) > j(S)$ to any state $\beta_{j(S)}$.

For evaluation of time intervals it is necessary to replace the walk through states $\beta_{j(\beta)}, j(\beta) > j(S)$ to direct switch with use the procedure of simplification. It is obviously, that in simplified process there are only states, which generate a commands, and every switch generates one command.

For evaluation of time of switch of simplified process from $\beta_{j(S)}$ to $\beta_{n(S)}$ let us split $j(S)$-th state onto two states: starting one $\beta_{j(S,b)}$, marked with index $j(S,e) = j(S)$, and absorbing one $\beta_{j(S,e)}$, marked with index $j(S,e) = J(\beta) + j(S)$. Thus from initial ergodic semi-Markov process non-ergodic process may be formed, in which set of states is the next:

$$B' = \{\beta_{1(S)}, \ldots, \beta_{j(S)}, \ldots, \beta_{J(S)}, \beta_{J(S)+1}, \ldots, \beta_{j(\beta)}, \ldots, \beta_{J(\beta)}, \beta_{J(\beta)+1},$$
$$\ldots, \beta_{J(\beta)+j(S)}, \ldots, \beta_{J(\beta)+J(S)}\}. \tag{7}$$

The semi-Markov matrix (6) is transformed to matrix $h'_\beta(t) = \lfloor h'_{j(\beta),n(\beta)}(t) \rfloor$ as follows:

- columns with numbers from 1 till $J(S)$ should be moved to columns with numbers from $J(\beta)+1$ to columns with number $J(\beta)+J(S)$;
- columns with numbers from 1 till $J(S)$ must be fulfilled with zeros;
- the rows with numbers from $J(\beta)+1$ till $J(\beta)+J(S)$ must be fulfilled with zeros.

Matrix $h'_\beta(t)$ has dimension of $[J(\beta)+J(S)] \times [J(\beta)+J(S)]$.

The weighted density of time of wandering from the state $\beta_{j(S)}$ before switch into $\beta_{n(S)}$ is determined as

$$h''_{j(S),n(S)}(t) = \sum_{k=1}^{\infty} \mathcal{L}^{-1}\left[\ ^r I_{j(S)}\left\{\mathcal{L}\left[h'_\beta(t)\right]\right\}^k\ ^c I_{n(S)}\right], \tag{8}$$

where $^r I_{j(S)}$ is the row vector of size $J(\beta)+J(S)$ in which $j(S)$-th element is equal to one and the remaining elements are zero; $^c I_{n(S)}$ is the column vector of size $J(\beta)+J(S)$ in which $[J(\beta)+J(S)]$-th element is equal to one, and the remaining elements are equal to zero.

Operation (8) should be executed for all $1(S \leqslant j(S), n(S) \leqslant J(S))$. The result is a semi-Markov process

$$h_S(t) = \left[h''_{S(t),n(S)}(t)\right], \tag{9}$$

that includes the states $S = \{s_{1(S)}, \ldots, s_{j(S)}, \ldots, s_{J(S)}\}$ only.

3   The Time Between Commands

Probabilities of residence the process $h'_S(t)$ in the state $s_{j(S)}$ for external with regard to CG observer is defined by the relationship [8,9]:

$$\pi_{j(S)} = \frac{T''_{j(S)}}{\tau_{j(S)}}, \tag{10}$$

where $T''_{j(S)}$ is expectation of time of residence the process $h'_S(t)$ in the state $s_{j(S)}$; $\tau_{j(S)}$ is the time of return the process $h'_S(t)$ into the state $s_{j(S)}$.

The duration of residence the process $h'_S(t)$ in the state $s_{j(S)}$ is determined as

$$T''_{j(S)} = \sum_{n(S)=1(S)}^{J(S)} T''_{j(S),n(S)} p''_{j(S),n(S)}. \tag{11}$$

To determine the duration of return the process $h'_S(t)$ into the state, let us split the state $s_{j(S)}$ into initial and absorbing states. In this case, the set of states takes form $S' = \{s_{1(S)}, \ldots, s_{j(S)}, \ldots, s_{J(S)}, s_{J(S)+1}\}$. Let us assume that in the set S' the state $s_{j(S)}$ is the starting one, and state $s_{J(S)+1}$ is the absorbing one. In this case the semi-Markov matrix (9) should be transformed as follows:

- $j(S)$-th column of the initial matrix should be moved to the column with number $J(S) + 1$ of the transformed matrix;
- $j(S)$-th column of the transformed matrix should be fulfilled with zeros;
- $J(S) + 1$ row of the transformed matrix should be fulfilled with zeros.

The density of time of return to state $s_{j(S)}$ is defined by dependence

$$f_{j(S)}(t) = \sum_{k=1}^{\infty} \mathcal{L}^{-1}\left[\, {}^r\boldsymbol{I}_{j(S)} \left\{ \mathcal{L}\left[\boldsymbol{h}'_S(t)\right]\right\}^k \, {}^c\boldsymbol{I}_{J(S)+1}\right], \tag{12}$$

where ${}^r\boldsymbol{I}_{j(S)}$ is row vector of size $J(S) + 1$, which $j(S)$-th element is equal to one, and the remaining elements are zero; ${}^c\boldsymbol{I}_{J(S)+1}$ is a column vector of size $J(S) + 1$, which $[J(S) + 1]$-th element is equal to one, and the remaining elements are zero.

Expression (12) has the physical meaning of density but not weighted density because the semi-Markov process $\boldsymbol{h}'_S(t)$ has the only one absorbing state $s_{j(S)+1}$, and all wandering trajectories starting in $s_{j(S)}$ sooner or later fall into $s_{j(S)+1}$. Thus,

$$\pi_{j(S)} = \frac{\displaystyle\sum_{n(S)=1(S)}^{J(S)} T''_{j(S),n(S)} P''_{j(S),n(S)}}{\displaystyle\int_0^{\infty} t \sum_{k=1}^{\infty} \mathcal{L}^{-1}\left[\, {}^r\boldsymbol{I}_{j(S)} \left\{ \mathcal{L}\left[\boldsymbol{h}'_S(t)\right]\right\}^k \, {}^c\boldsymbol{I}_{J(S)+1}\right] dt} \tag{13}$$

where numerator defines the duration of residence the process $\boldsymbol{h}'_S(t)$ in the state $s_{j(S)}$ in accordance with (11), and denominator defines expectation of the time of returning the process to the state $s_{j(S)}$, density of such a time was determined in accordance with (12).

Equation (13) defines probabilities of residence a semi-Markov process (9) in states $s_{J(S)}$, $1(S \leqslant j(S) \leqslant J(S))$ for the external (with regard to CG) observer.

If one knows the probability $\pi_{j(S)}$ of staying the instruction generator in the state $s_{j(S)}$, and weighed density of switching from $s_{j(S)}$ to adjacent states, $1(S \leqslant j(S) \leqslant J(S))$, then the density of time between two successive instructions may be defined as

$$g(t) = \sum_{j(S)=1(S)}^{J(S)} \pi_{j(S)} \sum_{n(S)=1(S)}^{J(S)} h''_{j(S),n(S)}(t), \tag{14}$$

where $h''_{j(S),n(S)}(t)$ is weighted density time of switching from state $s_{j(S)}$ to the state $n_{j(S)}$, $\displaystyle\sum_{n(S)=1(S)}^{J(S)} h''_{j(S),n(S)}(t)$ is the density of time of residence the process $\boldsymbol{h}'_S(t)$ in the state $s_{j(S)}$.

So, formula (14) presents average density of time of switching from states $s_{j(S)}$, $1(S \leqslant j(S) \leqslant J(S))$ in steady regime of switches of ergodic semi-Markov process (9).

4   Experiment

Experimental verification of proposed method was carried out on the model of CG shown on Fig. 1, when $J(\beta) = 2$. Simplest 2-parallel semi-Markov processes, are represented by states $\beta_{1(\beta)}$ and $\beta_{2(\beta)}$. Every state is described by the adjacency matrix of form (1) and semi-Markov matrix of form (2). Elements of semi-Markov matrix are the uniform distribution laws (Fig. 2)

$$f_{a[1(\beta)]} = f_{a[1(\beta)]} = f_a(t) = \begin{cases} 1, & \text{when } 0,5 \leq t \leq 1,5; \\ 0, & \text{in all other cases;} \end{cases} \tag{15}$$

$$f_{b[1(\beta)]} = f_{b[1(\beta)]} = f_b(t) = \begin{cases} 2, & \text{when } 0,75 \leq t \leq 1,25; \\ 0, & \text{in all other cases;} \end{cases} \tag{16}$$

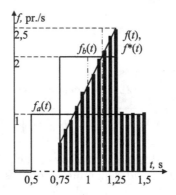

Fig. 2. Densities of components of 2-parallel process and of the process as a whole

Density of time of switch from states $\beta_{1(\beta)}, \beta_{1(\beta)}$ to conjugated state $(\beta_{1(\beta)}, \beta_{1(\beta)})$ is defined in accordance with (3) as (Fig. 2)

$$f(t) = \begin{cases} 4t - 2,5, & \text{when } 0,75 \leq t \leq 1,25; \\ 1, & \text{when } 0,125 \leq t \leq 1,5; \\ 0, & \text{in all other cases.} \end{cases} \tag{17}$$

Expectation and dispersion of $f(t)$ is equal to $T = 1,135417$ s; $D = 0,0337$ s^2.

Dependence (17) was verified experimentally with use the Monte-Carlo method. When carrying out of every computer experiment the randomizer launched twice, and from two time intervals obtained the larger one was selected. The histogram $f^*(t)$ obtained is plotted onto the density chart shown on Fig. 2. Experimental evaluation of expectation of time $T^* = 1,1412$ gives an error 0,5 %. Experimental evaluation of standard deviation $\sqrt{D^*} = 0,1851$ s gives an error 0,8 %.

Time characteristics of CG for the case, when state $\beta_{1(\beta)}$ generates a commands, and state $\beta_{1(\beta)}$ is the auxiliary one, all probabilities of switch the process

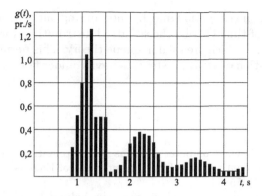

Fig. 3. Histogram of time between commands from CG

from states $\beta_{1(\beta)}, \beta_{1(\beta)}$ to conjugated states $(\beta_{1(\beta)}, \beta_{1(\beta)})$ are equal to 0,5, were determined with use the Monte-Carlo method [10]. When carrying out of every computer experiment a random walk through semi-Markov process under consideration from the state $\beta_{1(\beta)}$ till the state $\beta_{1(\beta)}$ is realized. Time of walk is determined, and statistic of times is stockpiled. Mentioned statistics is shown on Fig. 3.

Expectation and standard deviation of $g(t)$, which is calculated with use of the method described in [7] are the next: $T_g = 2,27$ s, $\sqrt{D_g} = 1,669$ s. Experimental expectation and standard deviation are the next: $T^*g = 2,2635$ s (error 0,28 %), $\sqrt{D^*_g} = 1,685$ s (error 0,9 %). Form of the histogram is quite alike the shifted exponential distribution, that corresponds to theorem by B. Grigelionis [11]. So the experimental results rather precisely correspond to theoretical premise.

5   Conclusion

Thus, an analytical model of generation of control instructions for swarm system is offered. In this model the work of human operator is divided into elementary steps for each of which time response and probability of transition to the next step can be easily measured. So the control system of swarm system can be easily configured for any current conditions. The result can be used in designing interactive algorithms for complex swarm systems.

Further research in this area may be directed to the creation of interactive algorithms, which generate flow of commands with optimal parameters. This implies working out both software for computer and technique of human operator actions procedure. Within the algorithm, the hardware and the software prac-tical evaluation of stochastic and time characteristics of human operator primitive actions should be defined experimentally. Characteristics should be divided onto groups due to degree of training. Also time characteristics of software operation should be evaluated.

Next direction of investigation may be the working out the method of approximation of density of time (14) by the standard distribution, for example by one of those distributions, which are used in queue theory. This permits to use named theory for investigation of swarm systems behavior under flow of commands from point of control.

References

1. Olsson, G., Piani, G.: Computer Systems for Automation and Control (Prentice Hall International Series in Systems and Control Engineering). Prentice Hall (1992). ISBN:0134575814
2. Howard, R.A.: Dynamic Probabilistic Systems, Volume I: Markov Models (Dover Books on Mathematics). Dover Publications (2007). ISBN:0486458709
3. Howard, R.A.: Dynamic Probabilistic Systems, Volume II: Semi-Markov and Decision Processes (Dover Books on Mathematics). Dover Publications (2007). ISBN:0486458725
4. Zaroliagis, C., Pantziou, G., Kontogiannis, S.: Algorithms, Probability, Networks, and Games. Springer, Heidelberg (2015)
5. Limnios, N., Oprisan, G.: Semi-Markov Processes and Reliability. Springer Science & Business Media, New York (2012)
6. Iverson, M.A., Özgüner, F., Follen, G.J.: Run-time statistical estimation of task execution times for heterogeneous distributed computing. In: 1996 Proceedings of 5th IEEE International Symposium on High Performance Distributed Computing, pp. 263–270. IEEE (1996)
7. Ivutin, A., Larkin, E.: Estimation of latency in embedded real-time systems. In: 2014 3rd Mediterranean Conference on Embedded Computing (MECO), pp. 236–239. IEEE (2014)
8. Ivutin, A., Larkin, E., Lutskov, Y.: Evaluation of program controlled objects states. In: 2015 4th Mediterranean Conference on Embedded Computing (MECO), pp. 250–253. IEEE (2015)
9. Ivutin, A., Larkin, E., Kotov, V.: Established routine of swarm monitoring systems functioning. In: Tan, Y., Shi, Y., Buarque, F., Gelbukh, A., Das, S., Engelbrecht, A. (eds.) ICSI-CCI 2015. LNCS, vol. 9141, pp. 415–422. Springer, Heidelberg (2015)
10. Fishman, G.S.: Monte Carlo: Concepts, Algorithms, and Applications. Springer Science & Business Media, New York (1996)
11. Grigelionis, B.: On the convergence of sums of random step processes to a poisson process. Theory Probab. Appl. **8**(2), 177–182 (1963)

A Personalized Intelligent Tutoring System of Primary Mathematics Based on Perl

Bo Song$^{(\boxtimes)}$, Yue Zhuo, and Xiaomei Li

Software College, Shenyang Normal University, No. 253, HuangHe Bei Street,
HuangGu District, Shenyang 110034, China
songbo63@aliyun.com, {602913389,992511723}@qq.com

Abstract. With the popularity of Internet technology, it becomes more difficult for users to retrieve their needed information from so enormous information space. Thus the issue of information overload formed. To solve the problem, recommender system emerged. This paper presents the personalized intelligent tutoring system based on Content-Based Filter Algorithm. The system will be implemented based on Perl with the advantages of developers paying heed to the program logic without caring for data storage, rule of operation and other details. And the use of open source Mojolicious framework can significantly accelerate the development cycle. The function of personalized information recommendation of an intelligent tutoring system will be achieved and personalized information will be recommended to users to improve their learning efficiency.

Keywords: Intelligent tutoring system · Perl · Content-Based Filter · Mojolicious

1 Introduction

The ubiquity and popularity of the Internet have brought a huge number of information to the users. Although it satisfies the demand for information in the Internet era, people can't get their own truly useful part of the information from large amounts of disorderly and unsystematic information. The explosive growth of network information has led to the efficient use of information decreased. This is named information overload problem. To solve the issue, researchers have carried out lots of solutions. Recommender system is one of the great potential. During the recent years, recommender systems have widely used in e-commerce and social network to supply users with personalized information. It is an intelligent and personalized information service system and it could give information to users that they need and are interested in based on a specific recommend approach [1–3]. Content-Based Filter is one of the earliest and successful techniques and it is popular due to its excellent speed and robustness in the field of global Internet. It recommends items according to the similarity of items that users liked in the past. Content-Based Filter is the earliest principal application in information retrieval system, so a lot of information retrieval and information filtering methods can be used in it.

In this paper, we will construct a personalized primary mathematics tutoring system with Content-Based Filter Recommendation Algorithm and will implement the system

Y. Tan et al. (Eds.): ICSI 2016, Part II, LNCS 9713, pp. 609–617, 2016.
DOI: 10.1007/978-3-319-41009-8_66

based on Perl. And the Mojolicious is a Perl real-time web application, which could significantly accelerate the development cycle.

At present, tutoring systems are typically used by large educational institution and focus on supporting instructor in managing and administrating online course. And the majority of these institutions adopt a "one size fits all" approach without considering individual learner's profile. A learner's profile can, for example, consist of his/her learning styles, goals, prior knowledge, abilities, and interests [4]. Recommender system considering users' profiles is used to support the personalized tutoring system and enhance the learning experiences and performance of learners within the course. The proposed system is a type of ITS (Intelligent Tutoring System), which can provide personalized, individual and efficient tutoring for learners. An ITS can maintain an one-to-one relationship with the students and provides personalized and adaptive teaching methods. It can be also to a large extent address the issue of unavailability of skilled teachers [5]. The main key features of the system are:

- It considers learners' profile consisting of learners' learning styles, experience level, prior knowledge and performance to provide advanced personalization.
- It creates a personalized list of recommendations of learning objects to be an individual learner based on the navigation history.
- Information on whether or not a certain learning object was helpful for a particular learner is retrieved through association rule mining among the learning objects visited by him/her instead of asking him/her to provide a rating on the visited learning objects.

2   An Overview of Perl

Perl is sometimes called Practical Extraction and Report Language, but also referred to as Pathologically Eclectic Rubbish Lister now and then. In addition, the abbreviation of the word can be interpreted as a different name because Perl is a backronym rather than acronym. Larry Wall, the founder of Perl, came up with the name first, and then think about how to explain it. In general, Perl represents the program language, but perl is the compiler and interpreter running the program. Perl integrate with many other languages' characteristics (for example, C, Sed, Awk, shell, Scripting and so on). The most important feature is its internal integration of regular expression functions and huge third-party code libraries CPAN. It is very suitable to write "ugly but practice" program in three minutes. What's more, Perl is good at handling problems, which approximately 90 % of the relevant text, and other things related to 10 % for the whole [6].

Perl is a powerful programming language supported by a large and active developer community [7]. At present, thousands of Perl modules and scripts have been developed and archived at the Comprehensive Perl Archive Network (CPAN) (http://www.capn. org/), which is the largest distribution center for Perl module on the Internet. Before using Perl module packages, these modules should be installed at the local computer. This enables the user without having to write a lot of code, just learn how to use these modules can quickly achieve the desired functionality. With an extensive repository of code available for reuse, opportunity to utilize this code clearly abound. Perl is script language, which is compiled each time before running. The Unix knows that it is a perl script and it must be the following header at the topline of every perl script: #!/user/bin/perl where the path to perl has to be correct and the line must not exceed 32 characters. Perl is composed of perl programs that written by C language.

In this paper, we will build a recommender system with Perl language. Perl combines the characteristics of other languages coupled with the support of regular expressions, so that it has very strong word processing capabilities, as well as a good cross-platform portability. Therefore, the key part is completing the content filter using of advantage of Perl.

3 Related Work of Recommender System

Personalized recommendation is an important part in user behavior analysis. As an intelligent personalized information system, recommender systems describe the user's long-term information need by user modeling, based on which it can customize the personalized information with the specific recommendation strategy [8]. In simple term, it is a process of searching the resource which the user might interest in. The technology has developed for decades [9]. It provides users with the most interesting items in the complex network environment based on users' interest and behavior characteristics. This paper talks about a tutoring system based on Content-Based filtering. It generally comprises three steps:

- Item Representation: extract some features (content of the item) to represent this item
- Profile Learning: calculate the user's profile with it's feature data
- Recommendation Generation: recommend to the user a series of the most potential correlated items by comparing user's profile and item's feature

Figure 1 shows the recommender process [8]:

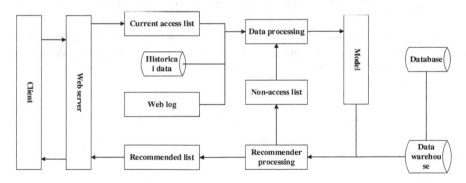

Fig. 1. High level architecture of content-based recommender

As is shown in Fig. 1, firstly we need collect the individual information of users and preprocess the current information resource (historical data, behavior characteristic and so on), extract the features of items. These features can be divided into two parts: structured and unstructured. For the former, we can obtain the users' interested information directly by asking users to fill in the table or questionnaire. For the latter, the system tracks the fetchers' behavior (which is invisible to fetchers), memorize their IP address, inquiry time, and inquiry content, and analyze their interests through Web mining [10]. This approach does not require users to actively participate in the survey, and will not interface their work. So this tutoring system adopts the recessive feature description.

In this system, all the information related to users' browsing time can be obtained, and all the Web pages browsed by users can be found in server cache. Thus we can obtain users' interested pages by analyzing the Weblog. And then the textual documents corresponding with the content of Web Page can be obtained by taking off unrelated structures. The model can be used for structured data, thus we often have to put unstructured data into structured vector. Now we introduce how to structure the items. VSM (Vector Space Model) is adopted to process the problem, which can convert a given document into a vector with multi-dimension. The basic thought is like this: D stands for document, and generally refers to any readable record by computer. T stands for term, usually in form of word or phrase, refers to the fundamental language unit representing the content of the document [11]. A collection of documents can express as $D = \{d_1, d_2, \ldots, d_n\}$, and the terms in these documents refer to $T = \{t_1, t_2, \ldots, t_n\}$, the weight of t_i refer to $(w_1, w_2, \ldots, w_n)$. Every document can be described by a vector, for example, the jth document can be expressed as $d_j = (w_{1j}, w_{2j}, \ldots, w_{nj})$. Here, w_{1j} represents the value of first term t_1 in jth document. And the greater value, the more important. As a result, the key is calculating the w_{kj}. At first, extract the feature words: let $|D|$ be the number of all documents, $|D_t|$ be the number of documents containing the term t. The probability of emergency of t is $P(t)$, which equals to $|D_t|/|D|$. The entropy $H(t)$ of t equal to $-p(t)\log_2 p(t)$, and $H(t)$ is less than a threshold value, term t could be one of the feature word. And then we adopt the method of TF-IDF (term frequency-inverse document frequency). Therefore, the value of kth term in jth document is:

$$TF - IDF(t_k, d_j) = f(t_k, d_j) \times \log(|D|/|D_t| + 0.01) \qquad (1)$$

Here, $f(t_k, d_j)$ represents kth term occurrence frequency in jth document and $|D|$ represents the total number of documents, $|D_t|$ represents the number of documents containing the term t. Eventually, we can express the value of the kth term in the jth document, the formula is:

$$w_{k,j} = \frac{TF - IDF(t_k, d_j)}{\sqrt{\sum_{s=1}^{n} TF - IDF(t_k, d_j)^2}} \qquad (2)$$

The benefit of normalization is the vectors among different documents are normalized to a magnitude, and it facilitates the following operation.

Assume that user has made judgment to some items, then, this step we need to make a model according to users' past preferences. One of the methods is k-NN (k-Nearest Neighbor) [12]. As for a new item, we have to find top-k items that nearest this one. And then we can determine the preference degree of this new item based on user's preference degree of these top-k items. For this method, the key is how to calculate the degree of similarity between two items by the items' properties of the vector. In VSM, the degree of correlation between two documents D_i and D_j, $Sim(D_i, D_j)$ expresses with cosine value of vector, the formula is [13]:

$$Sim(D_i, D_j) = \cos \theta = \frac{\sum_{k=1}^{n} W_{ik} \times W_{jk}}{(\sum_{k=1}^{n} W_{ik}^2)(\sum_{k=1}^{n} W_{jk}^2)} \tag{3}$$

Here, D_i represents the feature vector of the new item, and D_j represents the feature vector of jth neighbor of i. W_{ik} and W_{jk} represent the value of the kth term in D_i and D_j documents respectively. According to the user's interest description, in the category of user is interested in those who have not read Web page recommended to the user and calculate the similarity of corresponding documents, which is the reference value of recommended strength. If the similarity is high, then the first priority to the user.

4   Implementation of the Intelligent Tutoring System

Back in the early days of the web, many people learned Perl because of a wonderful Perl library called CGI. It was simple enough to get started without knowing much about the language and powerful enough to keep you going, learning by doing was much fun. While most of the techniques used are outdated now, the idea behind it is not. Mojolicious is a new endeavor to implement this idea using bleeding edge technologies [13]. This framework's features:

- An amazing real-time web framework, allowing to easily grow single file prototypes into well-structured web applications. Powerful out of the box with RESTful routes, plugins, commands, Perl-ish templates, content negotiation, session management, form validation, testing framework, static file server, CGI/PSGI detection, first class Unicode support and much more for you to discover.
- A powerful web development toolkit, that you can use for all kinds of applications, independently of the web framework. Full stack HTTP and WebSocket client/server implementation with IPv6, TLS, SNI, IDNA, HTTP/SOCKS5 proxy, Comet (long polling), keep-alive, connection pooling, timeout, cookie, multipart and gzip compression support. Built-in non-blocking I/O web server, supporting multiple event loops as well as optional preforking and hot deployment, perfect for building highly scalable web services. JSON and HTML/XML parser with CSS selector support.
- Very clean, portable and object-oriented pure-Perl API with no hidden magic and no requirements besides Perl 5.22.0 (version as old as 5.10.1 can be used too, but may require addition CPAN modules to be installed)
- Fresh code based upon years of experience developing Catalyst, free and open source.

Mojolicious follows the model of a view and a control (MVC) design pattern, which is good at differentiating content processing, presentation, and process control into separate modules. This distinction allows you to modify the code for some aspect of the problem without affecting the solution to other problems. So Mojolicious upgrade the original solution to solve the Web Application module reuse [14]. Below figure is a work flow diagram of Mojolicious shown in Fig. 2.

The Controller receives a request from a user, passes incoming data to the Model and retrieves data from it, which then gets turned into an actual response by the View. In this pattern, View can render the data structure into HTML format, which Controller handled over to the View, and then returned to the client by Controller. Model in Mojolicious can access the database and file system, and is responsible for the related business logic. Controller is a bridge between View and Model, and it can access the underlying HTML protocol transmission details, can carried out to determine the logic of the program and call the relevant Model and Template (View).

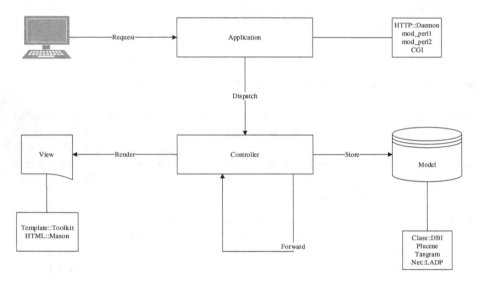

Fig. 2. The architecture of Mojolicious

4.1   A Bird's-Eye View

One of the main differences between Mojolicious and other web frameworks is that it also includes Mojolicious::Lite, a micro web framework optimized for rapid prototyping. You could try it as quickly as possible when you have a cool idea. Full Mojolicious applications on the other hand are much closer to a well organized CPAN distribution to maximize maintainability.

It all starts with an HTTP request like this, sent by browser.

GET/HTTP/1.1
Host: localhost:3000

Once the request has been received by the web server through the event loop, it will be passed on to Mojolicious, where it will be handled in a few simple steps.

1. Check if a static file exists that would meet the requirements.
2. Try to find a route that would meet the requirements.
3. Dispatch the request to this route, usually reaching one or more actions.
4. Process the request, maybe generating a response with the renderer.
5. Return control to the web server, and if no response has been generated yet, wait for a non-blocking operation to do so through the event loop.

With our application the router would have found an action in step 2, and rendered some text in step 4, resulting in an HTTP response being sent back to the browser.

4.2   Model

In Mojolicious we consider web applications simple frontends for existing business logic, that means Mojolicious is by design entirely model layer agnostic and you just use whatever Perl modules you like most.

```
$ mkdir -p lib/TutoringApp/Model
$ touch lib/TutoringApp/Model/Users.pm
$ chmod 644 lib/TutoringApp/Model/Users.pm
```

Our login manager will simply use a plain old Perl module abstracting away all logic related to matching usernames and passwords. The name TutoringApp::Model:: Users is an arbitrary choice, and is simply used to make the separation of concerns more visible.

```
package TutoringApp::Model::Users;
use strict;
use warnings;
use Mojo::Util 'secure_compare';
my $USERS = {
   Jack    => 'ncoivb',
   Mary    => 'bvuyu23',
   Lily    => 'lily7834bg'
};
sub new { bless {}, shift }
sub check {
   my ($self, $user, $pass) = @_;

   # Success
   return 1 if $USERS->{$user} && secure_compare $USERS->{$user}, $pass;

   # Fail
   return undef;
}
1;
```

4.3   Controller

The Controller is a class to implement the certain function. Hybrid routes are a nice intermediate step, but to maximize maintainability it makes sense to split our action code from its routing information.

```
$ mkdir lib/TutoringApp/Controller
$ touch lib/TutoringApp/Controller/Login.pm
$ chmod 644 lib/TutoringApp/Controller/Login.pm
```

All Mojolicious::Controller controllers are plain old Perl classes and get instantiated on demand. In addition, full Mojolicious applications are a little easier to test. Above mentioned it is simulated that the client-side submits request to the Web server-side in this tutoring system [13].

5   Conclusion

In this paper, it proposed the overload information problem, and put forward the recommender system. The Content-Based Filter advanced explored users' unique interest needs with vector space model, recommended the items according to the qualitative value of items information, and automatically adapted to users' feedback information. In this way its comprehensive performance were enhanced. And the space model is used to judge the significance of terms according to its location and occurrence frequency in documents. Besides the algorithm of the intelligent tutoring system of the primary school mathematics is studied, and the concept integration of the development framework of the Mojolicious based on Perl technology is realized.

The content-based filtering recommendation can construct a reliable classifier only when it accumulates adequate evaluation. But it is troublesome to deal with new uses and new items [15]. When the feature of users' interest changed, the recommender system cannot update easily. It must ultimately optimize later.

Acknowledgments. Project supported by the Education Department of Liaoning province science and Technology Research Fund Project (L2013417).

References

1. Gao, J., Song, B.: A personalized e-Learning system based on EGL. Int. J. Mach. Learn. Comput. **4**(5), 411 (2014)
2. Song, B., Gao, J.: The enhancement and application of collaborative filtering in e-Learning system. In: Tan, Y., Shi, Y., Coello, C.A. (eds.) ICSI 2014, Part II. LNCS, vol. 8795, pp. 188–195. Springer, Heidelberg (2014)
3. Song, B., Gao, J.: An e-Learning system based on EGL and Web 2.0. Int. J. Comput. Eng. **7**(2), 162 (2014)
4. Hazra, I., Mohammad, B.Z., Ting, W.C., Kinshuk, S.G.: PLORS: a personalized learning object recommender system. Vietnam J. Comput. Sci. **3**, 3–13 (2016)

5. Hafidi, M., Bensebaa, T.: Architecture for an adaptive and intelligent tutoring system that considers the learner's multiple intelligences. Int. J. Distance Educ. Technol. **13**, 1–12 (2015)
6. Wittmer, K.: EPerl: Perl, C++, and Java. Dr. Dobb's (2005)
7. Lei, R.: Research on some Key Issues of Recommender Systems. East China Normal University (2012)
8. Scbwartz, R.L., Foy, B.D., Pboenix, T.: Learning Perl. O'Reilly, Sebastopol (2011)
9. Xiao, J., He, L.: An expandable recommendation system on IPTV. In: Tan, Y., Shi, Y., Ji, Z. (eds.) ICSI 2012, Part II. LNCS, vol. 7332, pp. 33–40. Springer, Heidelberg (2012)
10. Li, P., Wang, D.S., Chen, K.: A personalization information recommendation system based on VSM. Comput. Eng. Des. **10**, 19–22 (2003)
11. Weihong, H., Yi, C.: An E-commerce recommender system based on content-based filtering. Wuhan Univ. J. Nat. Sci. **5**, 1091–1096 (2006)
12. Oh-Woog, K., Jong-Hyeol, L.: Web page classification based on k-neareat neighbor approach. In: The 5th International Workshop on Information Retrieval with Asian Language (2000)
13. Mojolicious - Perl real-time web framework. http://mojolicious.org/
14. Toll, A.P.: Mojolicious. Ceed Publishing (2012)
15. Huang, X.J., Xia, J., Wu, L.D.: A text filtering system based on vector space model. J. Softw. **14**, 390–402 (2003)

The Construction and Determination of Irreducible Polynomials Over Finite Fields

Yun Song[1] and Zhihui Li[2($\boxtimes$)]

[1] School of Computer Science, Shaanxi Normal University,
Xi'an 710062, China
songyun09@163.com
[2] School of Mathematics and Information Science,
Shaanxi Normal University, Xi'an 710062, China
lizhihui@snnu.edu.cn

Abstract. The approach for constructing the irreducible polynomials of arbitrary degree n over finite fields which is based on the number of the roots over the extension field is presented. At the same time, this paper includes a sample to illustrate the specific construction procedures. Then in terms of the relationship between the order of a polynomial over finite fields and the order of the multiplicative group of the extension field, a method which can determine whether a polynomial over finite fields is irreducible or not is proposed. By applying the Euclidean Algorithm, this judgment can be verified easily.

Keywords: Irreducible polynomial · Finite field · Cyclotomic polynomial · Mersenne prime · Order

1 Introduction

Generating irreducible polynomials and determining their irreducibility is one of the challenging and important problems in the theory of finite fields and its applications, especially computer algebra [1], coding theory [2] and cryptography [3]. For instance, irreducible polynomials are often used as a basic unit in constructing stream ciphers for implementing linear feedback shift register. Since linear feedback shift registers are described by coefficients of polynomials, these polynomials should be treated as the key information. Due to their important role in various applications, recent advances in these areas have awakened an even more interest to the subject of such polynomials [4–9]. Let F_q be the finite field of order $q = p^k$, where p is a prime and k is a positive integer. According to the field structure of F_q, there exists an irreducible polynomial of degree n over finite field F_q [9]. Kyuregyan [10, 11] presented some results on the constructive theory of synthesis of irreducible polynomials over F_{2^s}. Abrahamyan et al. [12, 13] considered recursive constructions of irreducible polynomials over finite fields.

This paper proposes a method for constructing irreducible polynomials over F_q in terms of the number of the roots over the extension field. If $\varphi(q^n - 1) = q^n - q$, the irreducible polynomials of arbitrary degree n over finite fields can be obtained by factoring $Q_{q^n-1}(x)$, because the primitive polynomials over F_q of degree n are all the irreducible polynomials; If $\varphi(q^n - 1) \neq q^n - q$, the irreducible polynomials of arbitrary

Y. Tan et al. (Eds.): ICSI 2016, Part II, LNCS 9713, pp. 618–624, 2016.
DOI: 10.1007/978-3-319-41009-8_67

degree n over finite fields can be obtained by factoring $I(q, n; x)$. Furthermore, The correlation between the order of a polynomial over finite fields and the order of the multiplicative group of the extension field is analyzed, and a sufficient and necessary condition on judging whether a polynomial of arbitrary degree n over finite fields is irreducible or not is presented.

2 The Construction of Irreducible Polynomials Over F_q

In this section, we will construct the irreducible polynomials of arbitrary degree n over finite fields based on the number of the roots over the extension field. Only monic polynomials, i.e., the polynomials whose leading coefficient is equal to 1, are studied in this paper.

Definition 2.1 [14]. Let $f(x) \in F_q[x]$ be a nonzero polynomial. If $f(0) \neq 0$, then the least positive integer e for which $f(x)$ divides $x^e - 1$ is called the order of $f(x)$ and is denoted by $\mathrm{ord}(f(x))$. If $f(0) = 0$, then $f(x) = x^h g(x)$, where $h \in N$ and $g(x) \in F_q[x]$ with $g(0) \neq 0$ are uniquely determined; $\mathrm{ord}(f(x))$ is then defined to be $\mathrm{ord}(g(x))$.

Theorem 2.2 [12]. Let K be a field of characteristic p, n a positive integer not divisible by p, and ζ a primitive nth root of unity over K. Then the polynomial

$$Q_n(x) = \prod_{s=1, \gcd(s,n)=1}^{n} (x - \zeta^s)$$ is called the nth cyclotomic polynomial over K.

By Definition 2.1, the order of any root of $Q_n(x)$ is n.

Theorem 2.3 [10]. Let $(q, n) = 1$. $Q_n(x)$ factors into $\frac{\phi_n(x)}{d}$ distinct monic irreducible polynomials in $F_q[x]$ of the same degree d, where d is the least positive integer such that $q^d \equiv 1 \bmod n$.

Theorem 2.4. Let T_q^n be the number of irreducible polynomials in $F_q[x]$ of degree n, then

$$\frac{\phi(q^n - 1)}{n} \leq T_q^n \leq \left\lceil \frac{q^n - q}{n} \right\rceil.$$

In particular, for the case $\phi(q^n - 1) = q^n - q$, $T_q^n = \frac{q^n - q}{n}$.

Proof. If $f(x)$ is an irreducible polynomial in $F_q[x]$ of degree n, then all roots of $f(x)$ are in $F_{q^n} - F_q$. Thus we have

$$T_q^n \leq \left\lceil \frac{q^n - q}{n} \right\rceil.$$

The number of primitive polynomials in $F_{q^n}[x]$ of degree n is $\frac{\phi(q^n - 1)}{n}$. Since the primitive polynomial over finite fields is irreducible as well, it follows that

$$\frac{\phi(q^n - 1)}{n} \leq T_q^n.$$

Therefore

$$\frac{\phi(q^n - 1)}{n} \leq T_q^n \leq \left\lfloor \frac{q^n - q}{n} \right\rfloor.$$

In particular, if $\phi(q^n - 1) = q^n - q$, $T_q^n = \frac{q^n - q}{n}$.

Corollary 2.5. Let T_2^n be the number of irreducible polynomial in $F_2[x]$ of degree n. If $2^n - 1$ is Mersenne prime, then $T_2^n = \frac{2}{n}(2^{n-1} - 1)$.

Note that all the irreducible polynomials over F_q of degree n are primitive if $\phi(q^n - 1) = q^n - q$ by Theorem 2.4. Besides, the product of all primitive polynomials over F_q of degree n is equal to the cyclotomic polynomial $Q_e(x)$ with $e = q^n - 1$. Therefore, we will present a factorization method for $Q_e(x)$ over F_q.

Lemma 2.6 [14] (Berlekamp Algorithm). Let $f(x)$ be a polynomial over F_q of degree k, $h(x) \in F_q[x]$, and $h(x) = \sum_{l=0}^{m-1} h_l x^l$. Then we have

$$f(x) = \prod_{c \in F_q} \gcd(f(x), h(x) - c)h^q(x) \equiv h(x) \bmod (f(x)).$$

Note that the key to factoring the polynomial of degree k by applying Berlekamp algorithm is to find the polynomial $h(x)$ whose degree is at most $k - 1$. Similarly, the first method of factoring $Q_{q^n-1}(x)$ over F_q is given by the following theorem.

Theorem 2.7. Let $h(x) \in F_q[x]$ and $h(x) = \sum_{l=0}^{m-1} h_l x^l$, where $(m, q) = 1$.

If $h_{lq \bmod m} = h_1 (l = 0, 1, \ldots, m - 1)$, then $h(x)^q = h(x) \bmod Q_m(x)$ and we have $Q_m(x) = \prod_{c \in F_q} \gcd(Q_m(x), h(x) - c)$.

Proof. We first prove that $h(x)^q = h(x) \bmod (x^m - 1)$.

Let $s_l = lq \pmod m$ and $l = 0, 1, \ldots, m - 1$. Since $(m, q) = 1$, then $s_1 \bmod m$ will go through all the elements in $0, 1, \ldots, m - 1$. Thus we have

$$h(x)^q = \sum_{l=0}^{m-1} h_l x^{s_l} \bmod (x^m - 1) = \sum_{l=0}^{m-1} h_{s_l} x^{s_l} = \sum_{l=0}^{m-1} h_l x^l = h(x),$$

$$h(x)^q = h(x) \bmod (x^m - 1).$$

Note that

$$Q_m(x) | x^n - 1,$$

So

$$h(x)^q = h(x) \bmod Q_m(x).$$

The conclusion then follows from Lemma 2.6.

Example 2.8. We construct all irreducible polynomials in $F_2[x]$ of degree 3 in accordance with Theorem 2.7.

Since $2^3 - 1$ is a Mersenne prime, then all the irreducible polynomials over F_2 of degree 3 are primitive, which can be obtained by factoring $Q_7(x)$ over F_2, where $Q_7(x) = x^6 + x^5 + x^4 + x^3 + x^2 + x + 1$. By Theorem 2.7, $l \rightarrow 2l \bmod n$ is a permutation of $\{0, 1, \ldots, n-1\}$.

Let $U_7 = \{0, 1, 2, 3, 4, 5, 6\}$. The permutation of U_7 can be expressed as $\begin{pmatrix} 0123456 \\ 0246135 \end{pmatrix}$, which implies three cycle including $(0), (1, 2, 4)$ and $(3, 6, 5)$. Each of cycles can determine a polynomial satisfying Theorem 2.7, $h_1(x) = 1, h_2(x) = x^4 + x^2 + x, h_3(x) = x^6 + x^5 + x^3$.

By Theorem 2.7 and Euclidean algorithm, we have

$$(Q_7(x), h_2(x)) = x^3 + x + 1,$$

$$(Q_7(x), h_2(x) + 1) = x^3 + x^2 + 1.$$

Hence, $Q_7(x) = (x^3 + x^2 + 1)(x^3 + x + 1)$ and all irreducible polynomials in $F_2[x]$ of degree 3 are $x^3 + x^2 + 1$ and $x^3 + x + 1$.

Remark. If $\phi(q^n - 1) \neq q^n - q$, all irreducible polynomials in $F_q[x]$ can be determined by factoring $I(q, n; x)$ which is the product of all irreducible polynomials in $F_q[x]$ of degree n [14].

3   The Determination of Irreducible Polynomials Over F_q

According to the relationship between the order of a polynomial over finite fields and the order of the multiplicative group of the extension field, and the construction of irreducible polynomials above, we will present a sufficient and necessary condition on judging whether a polynomial of arbitrary degree n over finite fields is irreducible or not.

For irreducible polynomials in $F_q[x]$ of degree n, which satisfy $\phi(q^n - 1) = q^n - q$, we can get the following theorem by the discussion in Sect. 2.

Theorem 3.1. Let $f(x) \in F_q[x]$ be a polynomial over F_q of degree n and $\phi(q^n - 1) = q^n - q$. $f(x)$ is irreducible if and only if $\mathrm{ord}(f(x)) = q^n - 1$.

Lemma 3.2 [12]. For every finite field F_q and every $n \in N$, the product of all irreducible polynomials over F_q whose degree divides n is equal to $x^{q^n} - x$.

Lemma 3.3 [12]. Let c be a positive integer. Then the polynomial $f(x) \in F_q[x]$ with $f(0) \neq 0$ divides $x^c - 1$ if and only if $\mathrm{ord}(f(x))|c$.

Let $f(x)$ be a polynomials in $F_q[x]$ of degree n, and $n = p_1^{t_1} p_2^{t_2} \ldots p_s^{t_s}$ is the prime factor decomposition of n. Let $n_i = n/p_i$.

Theorem 3.4. Let $f(x)$ be a polynomial over F_q of degree n. If $f(x)$ satisfies the following properties:

(1) $\mathrm{ord}(f(x))|q^n - 1$;
(2) For every $c \in F_q, f(c) \neq 0$;
(3) $\gcd(\mathrm{ord}(f(x)), q^{n_i} - 1) = 1, (i = 1, 2, \ldots, s)$

then $f(x)$ is an irreducible polynomial over F_q.

Proof. Since $\mathrm{ord}(f(x))|q^n - 1$, we have $f(x)|x^{q^n - 1} - 1$.

According to Lemma 3.3, $f(x)$ has no repeated factor. Suppose $f(x)$ were reducible over F_q. Then we have a factorization $f(x) = f_1(x)f_2(x)\ldots f_t(x)$, where each $f_j(x)(j = 1, 2, \ldots, t)$ are pairwise relatively prime. Since $f_j(x)|x^{q^n - 1} - 1$, then

$$\deg(f_j(x))|n \quad (j = 1, 2, \ldots, t).$$

we claim that

$$\deg(f_j(x)) \nmid n_i \ (i = 1, 2\ldots, s).$$

Suppose $\deg(f_j(x))|n_k$ for some $1 \leq k \leq s$. Then

$$f_j(x)|x^{q^{n_k}} - x \quad \text{and} \quad f_j(x)|x^{q^{n_k} - 1} - 1.$$

By Lemma 3.2, since

$$\mathrm{ord}(f_j(x))|q^{n_k} - 1 \text{ and } \mathrm{ord}\, f_j(x)|\mathrm{ord}\, f(x),$$

by Lemma 3.3, we have

$$\gcd(\mathrm{ord}(f(x)), q^{n_k} - 1) \neq 1.$$

a contradiction to (3). Therefore, $\deg(f_j(x)) = n$ and $f_j(x) = f(x)$. Hence, $f(x)$ is an irreducible polynomial over F_q.

Theorem 3.5. If $f(x)$ is an irreducible polynomial over F_q, then

(1) $\mathrm{ord}(f(x))|q^n - 1$;
(2) For every $c \in F_q, f(c) \neq 0$;
(3) $\mathrm{ord}(f(x)) \nmid q^{n_i} - 1$.

Proof. Since $f(x)$ is an irreducible polynomial over F_q of degree n, then $f(c) \neq 0$ for every $c \in F_q$, and $\mathrm{ord}(f(x))|q^n - 1$.

Suppose

$$\mathrm{ord}(f(x))|\ q^{n_k} - 1 \quad \text{for some} \quad 1 \leq k \leq s.$$

According to Lemma 3.3,

$$f(x)|x^{q^{n_k}-1} - 1.$$

Then

$$f(x)|x^{q^{n_k}} - x.$$

Hence, $\deg(f(x))|n_k$, a contradiction to $\deg(f(x)) = n$ by Lemma 3.2. Therefore, $\mathrm{ord}(f(x)) \nmid q^{n_i} - 1 \ (i = 1, 2, ..., s)$.

The following results can be implied by above two theorems.

Corollary 3.6. Let $f(x)$ be a polynomial over F_q of degree p, where p is a prime. Then $f(x)$ is irreducible if and only if:

(1) $\mathrm{ord}(f(x))|q^p - 1$;
(2) For every $c \in F_q, f(c) \neq 0$.

Corollary 3.7. Let $f(x)$ be a polynomial over F_q of degree $n = p_1 p_2$, where p_1 and p_2 are prime numbers. Then $f(x)$ is irreducible if and only if:

(1) $\mathrm{ord}(f(x))|q^n - 1$;
(2) For every $c \in F_q, f(c) \neq 0$;
(3) $\mathrm{ord}(f(x)) \nmid q^{p_1} - 1$ and $\mathrm{ord}(f(x)) \nmid q^{p_2} - 1$.

4 Conclusion

Irreducible polynomials over finite fields play an important role in computer algebra, coding theory and cryptography. This paper constructed the irreducible polynomials of arbitrary degree n over finite fields based on the number of the roots over the extension field and determined whether a polynomial over finite fields is irreducible or not in terms of the relationship between the order of a polynomial over finite fields and the order of the multiplicative group of the extension field. Furthermore, a relevant example was analyzed to show the specific construction procedures.

Acknowledgments. This work was supported by the National Natural Science Foundation of China (61373150, 11501343), the Key Technologies R&D Program of Shaanxi Province (2013k0611), and the Fundamental Research Funds for the Central Universities (GK201603087).

References

1. Garefalakis, T., Kapetanakis, G.: A note on the Hansen-Mullen conjecture for self-reciprocal irreducible polynomials. Finite Fields Appl. **35**(4), 61–63 (2015)
2. Magamba, K., Ryan, J.A.: Counting irreducible polynomials of degree r over F_{q^n} and generating Goppa Codes using the lattice of subfields of $F_{q^{nr}}$. J. Discrete Math. **2014**, 1–4 (2014)
3. Ugolini, S.: Sequences of irreducible polynomials over odd prime fields via elliptic curve endomorphisms. J. Number Theor. **152**, 21–37 (2015)
4. Ri, W.H., Myong, G.C., Kim, R., et al.: The number of irreducible polynomials over finite fields of characteristic 2 with given trace and subtrace. Finite Fields Appl. **29**, 118–131 (2014)
5. Kaminski, M., Xing, C.: An upper bound on the complexity of multiplication of polynomials modulo a power of an irreducible polynomial. IEEE Trans. Inf. Theor. **59**(10), 6845–6850 (2013)
6. Fan, H.: A Chinese remainder theorem approach to bit-parallel $GF(2^n)$ polynomial basis multipliers for irreducible trinomials. IEEE Trans. Comput. **65**(2), 1 (2016)
7. Kopparty, S., Kumar, M., Saks, M.: Efficient indexing of necklaces and irreducible polynomials over finite fields. In: Esparza, J., Fraigniaud, P., Husfeldt, T., Koutsoupias, E. (eds.) ICALP 2014. LNCS, vol. 8572, pp. 726–737. Springer, Heidelberg (2014)
8. Nechae, A.A., Popov, V.O.: A generalization of Ore's theorem on irreducible polynomials over a finite field. Discrete Math. Appl. **25**(4), 241–243 (2015)
9. Pollack, P.: Irreducible polynomials with several prescribed coefficients. Finite Fields Appl. **22**(7), 70–78 (2013)
10. Kyuregyan, M.K.: Recurrent methods for constructing irreducible polynomials over F q of odd characteristics. Finite Fields Appl. **12**(3), 357–378 (2006)
11. Kyuregyan, M.K., Kyureghyan, G.M.: Irreducible compositions of polynomials over finite fields. Des. Codes Cryptogr. **61**(3), 301–314 (2011)
12. Abrahamyan, S., Kyureghyan, M.: A recurrent method for constructing irreducible polynomials over finite fields. In: Gerdt, V.P., Koepf, W., Mayr, E.W., Vorozhtsov, E.V. (eds.) CASC 2011. LNCS, vol. 6885, pp. 1–9. Springer, Heidelberg (2011)
13. Abrahamyan, S., Alizadeh, M., Kyureghyan, M.K.: Recursive constructions of irreducible polynomials over finite fields. Finite Fields Appl. **18**(4), 738–745 (2012)
14. Kaliski, B.: Irreducible Polynomial. Springer, New York (2011)

Author Index

Printed in the United States
By Bookmasters

Lecture Notes in Computer Science   10117

Commenced Publication in 1973
Founding and Former Series Editors:
Gerhard Goos, Juris Hartmanis, and Jan van Leeuwen

More information about this series at http://www.springer.com/series/7412